1993 EARTH JOURNAL
ENVIRONMENTAL ALMANAC AND RESOURCE DIRECTORY

1993 EARTH JOURNAL

ENVIRONMENTAL ALMANAC AND RESOURCE DIRECTORY

FROM THE EDITORS OF BUZZWORM MAGAZINE

BUZZWORM BOOKS
BOULDER, COLORADO

1993 EARTH JOURNAL

Editor-in-Chief
Joseph E. Daniel

Editor
Ilana Kotin

Editorial Staff
Elizabeth Darby Junkin
Deborah Houy
Marina Lindsey
Genevieve Auston
Mike Grudowski
Lisa Jones

Design and Production
Steve Harley
Christy Brennand
Jennifer L. Wolcott

Editorial Interns
Michael Eric Bronner
Amy Onderdonk
Eric Patterson
David Kemmerer
Anne Jeffers
Benjamin Allan Hirasawa
Susan Fredrickson
Timothy B. Albinson

Production Intern
Viktor J. Strayer

BUZZWORM BOOKS
Founder & Publisher
Joseph E. Daniel

Co-founder
Peter Stainton

Copyright ©1992 by BUZZWORM BOOKS. All rights reserved. No part of this work may be reproduced or transmitted in any form or by any means, electronic or mechanical, including photocopying and recording, or by any information storage and retrieval system without prior written permission of the publisher.

Earth Journal is published annually in November.

Copies of the *Earth Journal* may be ordered directly by mail. Special pricing available for premium, gift and fundraising bulk orders. Call toll-free (800) 333-8857 for price and shipping information.

Distributed to the trade in the United States by Publishers Group West.

1993 EARTH JOURNAL—ENVIRONMENTAL ALMANAC AND RESOURCE DIRECTORY
ISBN 0-9603722-7-X
ISSN 1059-6488

BUZZWORM BOOKS is an imprint of BUZZWORM Magazine

Earth Journal is a trademark of BUZZWORM, Inc.

BUZZWORM, Inc.
2305 Canyon Blvd., Suite 206
Boulder, CO 80302
(303) 442-1969

 Printed on recycled paper with 10% post-consumer waste

PRINTED IN THE UNITED STATES OF AMERICA
10 9 8 7 6 5 4 3 2 1

"There is a Midrashic story: A man is on a boat. He is not alone, but acts as if he were. One night, he begins to cut a hole under his seat. His neighbors shriek: 'Have you gone mad? Do you want to sink us all?' Calmly, he answers them, 'What I'm doing is none of your business. I paid my way. I'm only cutting a hole under my own seat.' What the man will not accept, what you and I cannot forget, is that all of us are in the same boat."

—*Elie Weisel, winner of the Nobel Peace Prize in 1986*

Table of Contents

Preface xi
Foreword xii

PART 1
EARTH DIARY

Chapter 1
YEAR IN REVIEW 3
Diary 4
Journal 4
- Earth Summit 4
- CFC Replacements 5
- Aids 6
- El Niño 7
- The World's Worst Polluter 9
- Chernobyl 10
- Big Bang 11
- California Earthquakes 13
- Nunavut (Our Land) 14
- Bhopal Disaster 15
- Rare Killer Whale Attack 16
- Early Humans in America 16
- Cholera Epidemic 18
- Chimps Use Herbal Medicine 19
- Parks Shelter Homeless 20
- Rose Parade Protested 21
- Medicinal Plants 22
- Red Gold 23
- Smog Closes Mexico City 25
- Cooling Off 26
- Landfills Stuffed 28
- Elephants and Ivory 29
- Poland's Debt-For-Nature 31
- Drought of the Century 32
- Endangered Species Ax 34
- Huge Forest Reserve Approved in Zaire 34
- Earth Day 35
- $151 Billion for Transit 36
- Food Labeling 37
- Moby Dick, the Sequel 37
- Biosphere 2 Do 38
- Trading in Endangered Species 40
- Voters Go Green in 1992 40

The Worst and Best of 1992 42
Eco-Forecasts for 1993 51
What Happened in 1992 53

PART 2
EARTH PULSE

Chapter II
EARTH ISSUES 59
Life Out of Balance 60
Air Pollution 62
Animal Rights 65
Biodiversity 68
Deforestation 71
Desertification 74
Disease and the Environment 77
Drinking Water 80
Ecofeminism 83
Endangered Species 86
Global Warming 89
Indigenous People 92
Minorities and the Environment 95
Nuclear Weapons, Power and Waste ... 98
Ocean Pollution 101
Ozone Depletion 104
Population 107
Recycling 110
Renewable Energy 113
Sustainable Organic Agriculture 116
Toxic Pollution 119
Transportation 122
Wilderness 125

Chapter III
REGIONAL REPORTS 129
Antarctica 131
Southern and Central Africa 133
East Africa 136
Ivory Coast 138
North Africa 140
Middle East 142
Western Europe 144
Eastern Europe 147
Former Soviet Union 150
South Asia 153
East Asia 156
Southeast Asia 159
South Pacific 162
South America 164
Central America 167
North America 169
The Arctic 172

PART 3
ECOCULTURE

Chapter IV
ARTS & ENTERTAINMENT 177
People 178
    The 10 Most Interesting
        Environmentalists of the Year 178
    Green Stars..178
    What's Green and What's Not.............180
    The Environmental Youth Movement ..181
    Earth Pledge of Allegiance 181
    EcoPhilospher Extraordinaire...............181
    Kid Heroes 182
    Chatwin Revisited 182
    Spirituality and Ecology........................183
    Resources for Ecological Spirituality...183
Publishing 186
    Books 186
    Environmental Magazines 186
    Children's Books....................................197
Media 202
    Environmental Film 202
    Green Radio..202
    Green Television....................................204
    The Winners...206
    Ecolinking 209
    Computer Software................................209
    Are You Computer "Green?" 212
Arts 214
    Painting 214
    Theater 214
    Eco-Photography ,..................................215
    Sculpture 215
Fashion 219
    Fashion Grows Green 219
    Beauty: Is it up in the Air.....................219
    Environmental Armor 219
    Save the Earth, One Button at a Time 220
    Go Organic 221
Music 224
    The Greening of the Pop Music
        World..224
    Songs Like These 224

Chapter V
ECOHOME 231
Home...232
    The Future of Housing..........................232
    Heating and Cooling 232
    Nuke it...232
    Recycling...233
    Household Energy Efficiency...............233
    Energy-Efficient House Plans..............234
Home Care 237
    Household Cleaning..............................237
    Lead in Your China 237
    Pest control Methods and Materials....238
    Renovating Uncovers Lead Paint 238
    Cockroaches: How to Control Them ..239
Food 241
    Food Additives.......................................241
    Shopping for Safe Foods 241
    Vegetable Curry 242
    Drinking Water Safety 242
    Home Water Treatment Devices.........243
    Barbecued Tempeh 244
    Food Resources......................................244
    Where the Buffalo Roam......................245
    Green Wine 245
    A Few Arguments for Minimizing or
        Eliminating Meat from Your Diet...246
Health 247
    Wired for Cancer? 247
    Toxic Medications.................................247
    Natural Remedies..................................247
    The story of Cosmetics 249
    Simple Ways to Reduce Formaldehyde
        in Your Diet 250
Gardening 252
    Gardening at Home..............................252
    Vegetable Growing Tips.......................252
    Creating an Efficient Organic Garden .253
    Climate and Lawn Grasses 253
    Power Lawn Tools.................................254
    Lawn Pesticides.....................................255
    Bugs Adapt, Too....................................255
Cars 256
    Auto-Mania: What it's Doing to Us 256
    Auto-Free Cities 256
    Tire Waste 256
    Bicycles 257
    Public Rail 257
    Batteries 258
    Antifreeze 258
    Ten Worst mpg Cars.............................259
    Ten Best mpg Cars 259
    Oil 259
Urban Ecology 260
    Are Cities Inherently Unhealthy?.........260
    Environmental Stress and Population.261
    Idaho 262

TABLE OF CONTENTS

The Best Place to Live 262
Environmental Research Spending 263

Chapter VI
GREEN BUSINESS 265
The Greening of Business 266
    Small Companies Prevent Environmental Problems 266
    Green Taxes 266
    Earth Share 267
    Green Relations—US and Mexico 269
    Eco-Photography 269
    What Some Businesses Are Doing to Save the Earth 270
    GEMI 270
    Database Waste 271
    CERES Principles 271
    Scrap and Reclaim 273
    Green Careers 276
Green Investing *278*
    Profile: Green Century Funds 278
    Green Funds 278
    The Five Principles of Using Your Principles 280
Green Consumerism 282
    When Products are Tied to Causes 282
    Green Catalogs 282
    Federal Trade Commission Guidelines 286
    Cruelty Free 289
    Junk Mail Kit 289
    Shopping for a Better World 290
    A Shopping Checklist 290
    Cartons of Eggs and Milk 291
    Directory to New Green Products 292
    Green Business Resources 297

Chapter VII
ECOTRAVEL 299
Ecotravel News 300
EcoHotels 301
Types of Ecotravel 302
Ecotourism in Costa Rica 303
Travel Magazines 303
Guide Books 305
Monitoring Ecotourism 306
People-to-People Outreach 308
Organizations Promoting Ecotravel . 308
Resources 309
Directory to EcoTravel Outfitters ... 310

Chapter VIII
ECOCONNECTIONS 321
Volunteer! 322
How to Write Congress 322
Eco-Awards ....
Directory to Environmental Groups 344
Directory to Environmental Education Programs 366

PART 4
ECOVOICE

Chapter IV
EARTH DIGEST 393
Rio and the New Millennium 394
Moving From a Material World 396
Green Guilt and Ecological Overload 398
Snapshots of a World Coming Apart 402
Bush's Polluter Protection Isn't Pro-Business 404
An Urban Environmental Justice Perspective 406
How the People Saved the Earth Summit 408
The Face of Gaia 410
Goodbye, Old Desert Rat 413
Do We Really Need Zoos? 417
The Real "Superwomen" 419
Act Locally 421

Index 425

Preface

It's easy to see that the world is in an environmental crisis. Species extinctions occur faster than at any other time in history, some think as fast as one mammal or bird per day, plus countless plants, insects, amphibians and others. World population grew by 100 million people in the last year. The ozone hole expanded over the North American continent. People all over the world now lack clean air to breathe, food to eat, water to drink, or access to healthcare.

In the face of such obvious environmental degradation, there are efforts to deny the seriousness of the situation. This last year saw a backlash against the environmental movement. Loopholes were created in the Clean Air Act by the Council on Competitiveness, led by then-Vice President Dan Quayle. That same agency helped redefine "wetlands" to allow development of up to 50 percent of the nation's wetlands. The Bush Administration issued a moratorium on new environmental regulations. Yet the very efforts to deny and undermine the seriousness of our environmental situation lend credence to the gravity of it. The reality is hard to grasp and difficult to swallow.

But there is much to be hopeful about. The Earth Summit in Rio de Janeiro made millions of world citizens conscious of the issues, and of the connection between the environment and the economy, between rich nations and poor, between North and South. The Earth Summit was a remarkable achievement, despite the disappointing performance of the US. And, America may again be moving in the right direction. The election of Bill Clinton and Al Gore is great news for the US environment and perhaps the world as well. Al Gore, who wrote the foreword to this book, is also author of *Earth in the Balance*, a best-selling book about the environment. What these two young politicians understand, and the organizers of the Earth Summit understood—and what the Bush Administration seemed to have forgotten—is that the well-being of the environment is crucial to the health of the economy and the nation. There is hope, after all.

And to help you translate that hope into action, the editors of *Buzzworm* magazine present the *1993 Earth Journal*. *Buzzworm* magazine was founded in 1988 to explore complex environmental issues in an unbiased manner. Its goal is to help you feel informed instead of overwhelmed by environmental issues such as the ozone hole, tropical deforestation, extinctions, desertification and air pollution. In the same spirit, we offer this second annual *Earth Journal*. After seeing so much complex information from such disparate sources, we decided it would be most useful to bring it all together in one easy-to-read book. We hope *Earth Journal* becomes a book by which to live.

—*Ilana Kotin*

Foreword

- By Vice President Al Gore -

Human civilization has reached a turning point, a defining moment in history. We stand at an environmental crossroads, and the path we choose to follow, the decisions we make now and in the next few years, will determine much of what happens in the decades that follow—decades that belong to our children. Our response now to the environmental challenges we face will affect future generations as profoundly as the industrial revolution has affected our lives.

For the past 150 years, industrial development has enriched humanity in immeasurable ways. But in the process, it has taken a toll on the environment. Indeed, the crisis is visible around the globe: the disappearance of the Aral Sea, the burning of the rainforests, the disappearance of living species at a rate 1,000 times faster than the natural extinction rate, the garbage crisis, the oil spill in Alaska, the dead dolphins in the Gulf of Mexico, the dead seals in the North Sea and the dead starfish in the White Sea, and the 37,000 children who die each day from preventable disease and malnutrition.

We now recognize these phenomena as symptoms of an underlying crisis: a relatively recent and dramatic change in our relationship to Earth, which has led to a collision between industrial civilization and the ecological system of the planet. In the past, we have thought that what we did—or did not do—to our environment had no effect or could be easily fixed. But the truth is that we are not separate from the Earth. In fact, civilization is a powerful natural force like the winds and the tides, and our voracious consumption of Earth itself threatens to push it out of balance.

To change the current trend of environmental destruction, we must gain enough momentum to overcome the two obstacles that lie in our path. The first is the obstacle of denial that prevents us from recognizing the problem as it truly exists. But even after we break through that barrier and realize the enormity of the challenge before us, we run headlong into the second obstacle, even more formidable than the first. It is called despair. How do you tell people to have hope in the face of these enormous tragedies unfolding around the world? How can we have hope when we see how large these tragedies are and how enormous these challenges have become?

We can have hope because we are human beings and because we are part of the Earth. Archbishop Camara never gave up hope when democracy was snuffed out in Brazil. Nelson Mandela never gave up hope when apartheid threatened to destroy South Africa. The men and women of the formerly communist countries never gave up hope when they saw the opportunity to restore freedom in their lands. We have no right to give up hope. We must reach out to freedom, and we must understand the concept of sustainable freedom. As the Reverend Martin Luther King, Jr., said, "Whenever freedom is denied to anyone, freedom is threatened for everyone."

The United Nations Conference on Environment and Development, also known as the Earth Summit, also gives us cause for hope. It was a turning point for a world trying to confront an unprecedented global ecological crisis. People from around the world came together to address the crisis we face and to make commitments to change the way we view and treat our world.

At the Earth Summit, there was an understanding that we are a community of separate nations and shall remain so, but we face a global environmental crisis that demands an international response. The conference was a success because it laid the groundwork for meaningful changes in policies in every nation to stop the destruction of the Earth. Leaders around the world are now thinking about the same challenge at the same time in a new way.

The Earth Summit was also a very powerful learning process, moving us to a better understanding of how future progress is inextricably linked to environmental protection and sound stewardship of natural resources. No economy can flourish if the natural resource base that supports it has been ruined.

Germany and Japan understand this well. They have already recognized and are poised to take advantage of what they call the largest new business opportunity in the world market—the market for the technologies and processes that encourage economic growth while protecting the environment. The global market for environmental technologies is currently at about $270 billion a year and is growing at 7 percent a year. Meanwhile, we now import 70 percent of our pollution control technologies, and we are net importers of solar and wind energy technologies.

Our own economic figures confirm the potential for the new products and processes that foster growth without harming the environment. One sector of our economy that has shown impressive growth over the last four years is the environmental products and services industries. Environmental businesses grew six times as fast as the overall US economy during this period. And, in 1991, America's 40 largest environmental firms reported payroll increases of nearly 9 percent.

If the US were as energy efficient as Japan, we would save $200 billion every year—money that could be used to create jobs and reinvest in America. What was once called too expensive is now a bargain too good to pass up.

Here in the US, we must seize this opportunity to invest in our future by protecting our environment. If we fail to act, we lose. We lose jobs, we lose opportunities, we lose the ability to remain competitive in the international marketplace and we lose our natural resources.

We must also commit to greater energy efficiency, greater use of renewable energy sources, real preservation and protection of forests, and a meaningful effort not only to clean up the pollution in our environment, but also to stop the pollution at its source. We can and should do better.

We care about the environment and the natural bounty of this biologically rich land, because we care about the future of America and our children. Our environment is literally the common ground of generations past and to come. It does not belong just to us, and, therefore, we have an obligation to protect it.

The solutions we seek will be found in a new faith in the future of life on Earth after our own, a faith in the future that justifies action in the present, a new moral courage to choose higher values in the conduct of human affairs. We will need to steer by the stars, not by the lights of each passing ship, when setting the future course of our planet, and ultimately, human civilization. But we shall overcome this great struggle. And we can begin to overcome this struggle by reading the *1993 Earth Journal*. We can use this resource to inspire, inform and motivate us to make changes.

PART I
EARTH DIARY

YEAR IN REVIEW

"Hope is . . . not the conviction that something will turn out well, but the conviction that something makes sense, regardless of how it turns out," says playwright and statesman Vaclav Havel. Environmental news started making sense this year, from the Earth Summit in Rio de Janeiro (where thousands gathered to discuss environmental action), to the election of an American vice president who is perhaps the most famous environmentalist in the country. Problems remain, but the world is beginning to make sense again.

Diary

The entries in this column document daily environmental events around the world for one year starting in October 1991. They are presented in chronological order to provide a record of the significant environmental happenings on our planet. These news briefs are culled from a variety of sources, including *The New York Times* (NYT), Associated Press (AP), Reuters (R), *The Los Angeles Times* (LAT), *International Herald Tribune* (IHT), *The Wall Street Journal* (WSJ), *The Washington Post* (WP) and United Press International (UPI). The source and date are provided for ease in consulting the original story for more information (—*NYT 10/1* would indicate an article in *The New York Times* on October 1, 1991).

October 1991

1 HAVANA, Cuba—For the first time in 30 years, American and Cuban scientists are combining efforts to **explore and catalog a plentiful diversity** of plant and animal species that are believed to make Cuba the richest island, biologically speaking, in North and South America. —*NYT 10/1*

2 ROWE, MA—The **Yankee Rowe nuclear power plant** was closed by its owners after federal regulators recommended that it be shut down

Journal

From earthquakes to the Earth Summit, *Journal* offers news and analysis on the most important environmental issues of the last year.

EARTH SUMMIT

Rio de Janeiro—"Rio must be seen as the start of a process, not the end," says Maurice Strong, Secretary General of the United Nations Conference on Environment and Development (UNCED), which met

The Earth Summit in Rio de Janeiro drew worldwide attention to the relationship between environment and development.

in Rio de Janeiro in June 1992. The conference, also known as the Earth Summit, was the largest assembly of heads of state meeting about the environment (118, including US President George Bush) in history. More than 9,000 news reporters (the largest media group in history) covered the Rio events, which included UNCED and the Global Forum—a "people's" assembly of thousands from all over the world, representing nongovernmental environmental groups from every corner of the globe.

Major action at the Earth Summit included passage of Agenda 21, a nonbinding 800-page blueprint to protect the environment and encourage sustainable development, adopted by a consensus of the 180 nations represented at UNCED. A nonbinding Declaration of Environment and Development (27 broad principles emphasizing the safeguarding of ecological systems and giving priority to needs of

Q: *What percentage of urban America is comprised of roads, parking lots and alleyways?*

developing countries) was also adopted.

A binding treaty to limit emissions of "greenhouse" gases like carbon dioxide and methane was passed, but the US insisted that no specific timetables be included. Other industrial countries, however, promised to cut carbon-dioxide emissions back to 1990 levels by the year 2000.

A legally binding treaty passed, requiring inventories of plants and wildlife, including endangered species. The US was the only UNCED participant that refused to sign the biodiversity treaty, claiming it could adversely affect US biotechnology corporations.

CFC REPLACEMENTS

The Stratosphere—When introduced in the early 1930s, CFCs (chlorofluorocarbons) were all the rage. Versatile, durable, nontoxic and highly efficient, CFCs have since been incorporated into a wide range of industrial processes and products, from refrigeration and air conditioning to fire prevention. However, after having been identified in the late 1970s as playing a primary role in the depletion of the atmosphere's vital ozone shield, and after being deemed responsible for creating ozone holes observed over Antarctica in 1985 and in 1992 over the east coast of North America, CFCs now float in quite a different light.

The durability of CFCs, once considered an asset, constitutes the heart of the problem. The stable compounds remain intact through the Earth's lower atmosphere, rising into the stratosphere, where

"The only difference between a pigeon and the American farmer today is that a pigeon can still make a deposit on a John Deere."

Jim Hightower, former Texas agriculture commissioner

intense solar radiation ultimately pries them apart. This process frees chlorine where it is most destructive to ozone molecules.

A 1987 international agreement—the Montreal Protocol—determined that CFCs should be phased out of use entirely by January 1, 2000. The deadline was recently moved ahead to December 31, 1995.

immediately for safety reasons. The Nuclear Regulatory Commission said the closing isn't necessarily permanent. —*WSJ 10/2*

4 GOLDEN, CO— **Coors Brewing Co.** will pay a $700,000 fine to settle allegations by the EPA that it contaminated soils and water under the brewery from 1981 to 1984. —*WSJ 10/4*

5 GORE, OK—Federal regulators have ordered the indefinite shutdown of the **Sequoyah Fuels Corp. nuclear fuel plant** in Gore, Oklahoma, where uranium was found to be leaking into the ground and where executives allegedly lied about the case to investigators. —*WP 10/5*

6 WASHINGTON—The amount of protective **ozone in the atmosphere** over Antarctica fell to its lowest level in 13 years of recorded data, NASA said. —*LAT 10/10*

7 WASHINGTON—Federal officials cut by more than half the allowable threshold for **lead exposure in children's blood**, which adds 3.5 million children under age 6 considered to be suffering from lead poisoning. —*WP 10/8*

8 ANCHORAGE, AK— US District Judge H. Russel Holland approved the **$1.125 billion agreement** between Exxon Corporation and the US and Alaskan governments to settle criminal and civil cases

A: *40 percent*
Source: Campaign for New Transportation Priorities

stemming from the *Exxon Valdez* oil spill in Prince William Sound in 1989. —*LAT 10/9*

11 WASHINGTON—Conoco Inc. announced that, for economic reasons, it is abandoning its **plans to develop oil** in Ecuador's tropical rainforest. —*WP 10/12*

11 NEW YORK—Bristol-Myers Squibb Co. reached a $50,000 settlement with 10 states on charges that it **misrepresented the environmental benefits** of certain cosmetic and household products. —*WSJ 10/11*

13 KETTLEMAN CITY, CA—More than 500 environmentalists, led by the Reverend Jesse Jackson, converged on Kettleman City, California, to protest plans to build California's largest commercial hazardous waste incinerator. Jackson called the plans for an incinerator in the town populated mainly by Latino farm workers **"toxic racism."** —*LAT 10/13*

17 BOISE, ID—Tribal police dispatched by the Shoshone-Bannock tribes in Idaho **intercepted a truck** carrying spent nuclear fuel and, for 12 hours, refused to allow it to cross their reservation. —*NYT 10/17*

19 WASHINGTON—The number of **whooping cranes** in the wild has dropped for the first time in more than a decade, from 146 last

AIDS

Geneva—The World Health Organization (WHO) announced in 1992 that more than 1 million people contracted the human immunodeficiency virus (HIV) between April 1990 and February 1992, and that 90 percent of all cases worldwide are contracted through heterosexual contact. In a report published in February 1992, WHO predicted that AIDS will eventually become the main cause of premature death in Western cities. By early 1992, 10 million to 12 million people worldwide had contracted HIV. Experts say that by the year 2000, that number could reach 30 million to 40 million.

African AIDS cases make up 60 percent of all cases worldwide. One scientist predicts that the populations of some African nations, such as

> *"The environment receives far less than the sustained attention the urgent danger signs deserve."*
>
> Senator Tim Wirth, Dem.-Colorado.

Uganda, Zimbabwe and Tanzania, could decline as a result of the epidemic. Roy M. Anderson, of the Imperial College of Science and Technology in London, says that in 15 years Uganda could have 20 percent fewer people than if the epidemic had not happened.

African health officials are beginning to confront social habits that have allowed the spread of AIDS. In some countries, older men, many of whom are infected with AIDS, seek young girls, who are less likely to carry the disease than older women. The health minister of Zimbabwe has called for a "cultural revolution" that would encourage adults to maintain monogamous relationships, teach children and teens safe sexual practices, and increase condom use. Condom sales in Zimbabwe in 1991 were up to 44 million from 500,000 sold in 1986. Prophylactics are now free in Zimbabwe's government health clinics.

In America, approximately 1 million cases have been reported and 1,500 to 2,000 HIV-infected infants

Q: *How many pounds of garbage do New Yorkers throw away each day?*

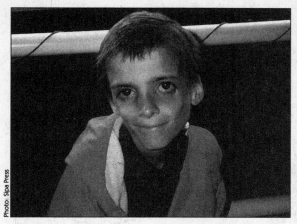

A dying hemophiliac boy who was infected with HIV during a blood transfusion.

are born annually. A congressional study released in April 1992 indicated that the number of teenage and young adult AIDS cases has increased by 62 percent over the past two years. The study concludes that federal efforts to fight AIDS among young Americans are underfunded and inadequate. It suggests that safe sex behavior should be taught in grade school, yet only 300 US schools have health programs that address AIDS.

EL NIÑO

Boulder, CO—El Niño is a name given to an invasion of warm surface water from the western part of the Pacific Ocean into the area off the coasts of Peru and Ecuador. It refers to the Christ child and

> "Americans did not fight and win the wars of the 20th century to make the world safe for green vegetables."
>
> Richard Darman, Director of the Office of Management and Budget

was so named by Peruvian fishermen because it usually begins about Christmastime. The invading warm water displaces nutrient-rich cold water that is normally in the coastal area, bringing with it

A: *About six pounds per person*
Source: Garbage *magazine*

year to 142 this year. —*IHT (WP) 10/19*

23 WASHINGTON—**Thinning of the ozone layer**, once thought to be a seasonal polar phenomenon, is occurring over the US and the rest of the world in spring and summer, an international scientific panel concluded. —*LAT 10/23*

23 NEW YORK—New York's State Court of Appeals upheld the 1989 agreement to close the **Shoreham nuclear power station,** removing one of the few remaining obstacles to the final dismantling of the plant. —*NYT 10/23*

26 MARBLE FALLS, TX— Billy Dale Inman was sentenced to two months in prison and fined $23,100 for **killing an endangered whooping crane.** —*IHT 10/26*

27 ANCHORAGE, AK—The Alaska Department of Environmental Conservation announced new, more **stringent oil-spill prevention** and response standards governing tankers, terminals and the Trans-Alaska Pipeline System. —*WP 10/27*

30 ANCHORAGE, AK—The **Exxon Corporation is suing** the makers of an automatic steering device that the oil company says caused the *Exxon Valdez* oil spill of 1989. —*NYT (AP) 10/31*

31 NEWPORT, RI—The world's oceans are undergoing an **explosion of algae blooms,**

including toxic "red tides," scientists say. These scientists suspect that the phenomenon should be taken as an early warning of a massive ecological breakdown. —*WP 10/31*

November 1991

1 WASHINGTON—Because he failed to inform Congress that the National Park Service believed development could threaten the park, Interior Secretary Manuel Lujan was accused by House Democrats of manipulating the conclusions of a report to Congress that favored development of **a geothermal energy plant** near Yellowstone National Park. —*WP 11/1*

3 LONDON—The leaders of **the Quorn Hunt** in Leicestershire, central England, resigned after animal rights campaigners produced a videotape showing hunters breaking Master of Foxhounds Association rules by digging a fox pup out of its hole so it could be torn apart by a pack of hounds. —*LAT 11/3*

4 WASHINGTON—President Bush's plans to cut the US nuclear arsenal has led the Energy department to **delay for two years** its decision on a new $6 billion nuclear reactor to make weapons material. —*WSJ 11/4*

6 SAN JOSE, CA—The discharge of toxic fumes that sent 36 people to the hospital and required

heavy rains to areas that are usually arid. At the same time, rainy regions in the western part of the Pacific Ocean, such as Indonesia and the Philippines, are plagued by drought. Since the 1982-1983 El Niño episode in the central and eastern equatorial Pacific Ocean—the largest in a century—the phenomenon seems to have become a household word.

El Niño events recur, with mild ones (a small increase in the sea's surface temperature) taking place every two to three years and major ones every eight to 11 years.

The most recent El Niño event occurred in 1991-1992. According to some scientists, the most devastating impact occurred along the eastern coast of the African continent and in southern Africa, with some areas experiencing their worst drought-related food shortages since the 1920s. Brazil was plagued by drought in its northeastern regions and floods in its southern states. The south-central part of the US was inundated as a result of heavy rains.

The eruption of Mt. Pinatubo in the Philippines in 1991 reduced global temperatures by ejecting millions of tons of climate-cooling aerosols into the atmosphere, mitigating El Niño's effects.

—*Michael Glantz, National Center for Atmospheric Research, Boulder, Colorado*

Q: *On an average college campus, how much of the waste stream is made up of paper?*

YEAR IN REVIEW

THE WORLD'S WORST POLLUTER

Toronto—The world's military is the worst polluter on Earth, according to a Canadian study published in June 1992. The report claims that the world's armed forces cause 10 percent to 30 percent of environmental damage worldwide and misuse land and energy. The US Army was charged this year with dumping toxic wastes at Love Canal, and its plans to incinerate mustard gas have stirred controversy in Maryland. Yet, due to US pressure, government officials at the Earth Summit in Rio de Janeiro did not discuss the impact of military pollution on the world's natural resources.

The Canadian study, conducted by researchers at the University of Toronto, says that the Pentagon leads the US in oil consumption. In 1989, it bought 200 billion barrels of oil, enough to keep all US public transit systems operating for 22 years. It takes less than 60 minutes for an F-16 jet to expend two times the amount of fuel the average American motorist burns in one year. The world's military consumes more aluminum, copper, nickel and platinum than all developing nations combined.

The report also asserts that the armed forces abuse land it should be protecting. The US military controls land equal to the size of Virginia. The world's military is responsible for 6 percent to 10 percent of global air

"Japan's out there killing whales and running driftnets . . . and we're twisting ourselves into knots over how many jobs to abolish to save a subspecies of owl."

Republican adviser to President Bush, speaking on condition of anonymity

pollution and two-thirds of all CFC-113, a greenhouse gas. In West Germany, 58 percent of all air contaminants released by aircraft is generated by military planes. The US and former Soviet military are the top producers of hazardous waste. The former Soviet military was forced to seal the floor of Lake Karachay with concrete because it had dumped so much nuclear waste there.

the evacuation of the Aromat Corp. manufacturing plant was blamed on a worker who tried to improperly dispose of **hazardous waste in an oven.** —*LAT 11/6*

6 BURGAN, Kuwait— **The last of 732 wells** torched by retreating Iraqi forces at the end of the Persian Gulf War was extinguished and capped. —*WSJ 11/7*

7 MEXICO CITY—The Mexican government moved to revise its agrarian reform so that for the first time millions of **cooperative farmers** will be able to own, rent or sell their plots. —*WP 11/8*

7 LOS ANGELES—Los Angeles officials unveiled the first working prototype of the LA301, **an electric car.** —*LAT 11/8*

10 LOS ANGELES— With little congressional review, the Department of Defense is reimbursing big contractors for millions of dollars in **environmental cleanup costs.** —*LAT 11/10*

10 LOWER ALLOWAYS CREEK TOWNSHIP, NJ—A fire in a non-nuclear portion of the **Salem II nuclear power plant** in New Jersey caused extensive damage to the main generator, closing the plant indefinitely, federal Nuclear Regulatory Commission officials said. —*NYT 11/11*

12 WASHINGTON— The nuclear power industry has begun a

A: *Half*
Source: The Student Environmental Action Guide, *EarthWorks Press*

three-year, $9 million public relations and advertising campaign in Nevada designed to "neutralize" political opposition to the **high-level nuclear waste repository at Yucca Mountain**, 100 miles northwest of Las Vegas. —*NYT 11/13*

13 NEW YORK—The United Arab Emirates (UAE) will introduce **unleaded gasoline** next year in a bid to cut pollution. This would make the UAE the first Arab state to sell unleaded gas at the pump. —*WSJ 11/13*

14 WASHINGTON— The National Marine Fisheries Service formally designated the Snake River **sockeye salmon** an endangered species. —*WP 11/15*

17 WASHINGTON— The EPA, at the urging of the White House, is **withholding assessments** showing that a large number of the nation's wetlands will lose federal protection under the new guidelines proposed by the Bush Administration. —*LAT (AP) 11/17*

18 WASHINGTON— Federal officials proposed removing the **California gray whale** from the endangered species list. It is the first time that a marine creature once near extinction was found to have recovered sufficiently to be delisted. —*LAT 11/19*

18 GENEVA—Environment ministers or their

CHERNOBYL

Chernobyl, Ukraine—Previously classified Soviet Politburo documents and a Ukrainian government study implicate the former Soviet leadership in covering up the scope of the effects of the explosion at Chernobyl Atomic Energy Station on April 26, 1986, now considered the world's worst nuclear accident.

On December 11, 1991, a Ukrainian parliamentary commission charged that less than 12 hours after reactor No. 4 at Chernobyl exploded, the Politburo, including Mikhail Gorbachev, was aware of massive amounts of radiation spilling out of the burning reactor, but failed to inform the public. Moscow officials have maintained that the Ukrainian leadership kept them in the dark about the magnitude of the disaster. Four days after the disaster, Dr. Anatoly Kasianenko presented a report to the Council of Ministers warning of a dramatic increase of radiation registered in Kiev, 80 miles south of Chernobyl. According to then-chair of the Ukrainian Parliament, Valentina Shevchenko, the Ukrainian Communist Party Politburo ordered schoolchildren and workers to march down the fallout-showered streets of Kiev in a May Day parade designed to show the world that nothing was wrong in the Ukraine.

Secret Politburo documents, published in *Izvestia* in April 1992, confirmed many of the commission's claims and revealed that many more people were exposed to lethal doses of radiation than were reported by the Soviet leadership. The papers

A fire in the still-operating portion of the Chernobyl Nuclear Plant destroyed a generator in October 1991.

Q: *By showering, what percentage of water do you save as compared with taking a bath?*

uncovered a Politburo-approved plan that increased by ten the amount of radiation considered safe, in order to reduce the number of people treated and hospitalized. The Politburo also allowed radiation-contaminated meat and milk to be mixed with non-contaminated food so that it could be sold throughout the country. Some 47,500 tons of meat and 2 million tons of milk with contamination levels higher than previously approved for human consumption were transported from the region between 1986 and 1989.

While the official death toll has remained at 32, Yuri Spizhenko, a Ukrainian health official, estimates that 6,000 to 8,000 people in the Ukraine have died as a result of Chernobyl. Spizhenko claims that as of January 1, 1992, 1.5 million people had undergone follow-up medical tests, and that each year fewer receive a healthy diagnosis.

> *"Nature is not a pretty, manicured place maintained for human beings. It is a dynamic continuum, often a violent one."*
>
> Dave Foreman, founder of Earth First!

BIG BANG

Deep Space—The discovery of "ripples in the fabric of space-time" by an American physicist this year has untangled a 27-year-old hitch in the Big Bang theory of the creation of the universe.

First proposed in 1927 by Georges Lemaitre, a Belgian priest, the Big Bang theory deduces that the universe was born around 15 billion years ago in the explosion of an infinitely dense ball of matter, and since has existed in a state of expansion. All raw material for the universe as it is known today, thought to have originated from this ball, is believed to have been smaller than the point of a pin.

The hitch in the Big Bang theory corresponded to the 1964 observation of a strange atmospheric hiss by Bell Laboratories scientists Arno Penzias and Robert Wilson, while testing a new, highly sensitive microwave detector. After eliminating all nearby sources as possible causes for the hiss, Penzias and

deputies from some 30 countries adopted the United Nations-sponsored **agreement to cut or freeze their emissions** of volatile organic compounds, which stem mainly from motor vehicles. —*IHT (R) 11/19*

20 LOS ANGELES—**Air pollution** in the nation's smoggiest cities dropped 10 percent during the past decade, and 10 million fewer Americans are breathing unhealthy air, EPA chief William K. Reilly announced. —*LAT 11/21*

20 WASHINGTON—The US and 12 Soviet republics have reached agreement on **$1.5 billion in economic aid** to help the Soviet republics feed themselves through the winter, with most of the money coming from private loans guaranteed by the US government. —*LAT 11/21*

23 BUENOS AIRES—Great Britain says it is opening the waters around the Falkland Islands to **oil exploration**. —*NYT 11/24*

24 TOKYO—The Japanese government will start **importing shipments of plutonium** sometime next year in lightly defended sea convoys, each containing enough material to make 100 to 150 atomic bombs. —*NYT 11/25*

26 TOKYO—Japan agreed to comply with a United Nations moratorium on the use of huge **fishing nets** in the

A: *66 percent*

Source: *Protect Our Planet Calendar, Running Press Book Publishers*

northern Pacific Ocean that scientists say are responsible for widespread destruction of marine life. —*NYT 11/27*

26 TOKYO—North Korea announced it will permit international inspection of its **secret nuclear installations** if the US also allows inspections to guarantee the removal of all nuclear weapons from South Korea. —*NYT 11/27*

26 WASHINGTON—Federal District Judge John Garrett Penn issued an injunction that **indefinitely bars** Energy Secretary James D. Watkins from opening the New Mexico Waste Isolation Pilot Plant, the nation's first permanent repository for nuclear wastes. —*NYT 11/28*

26 WASHINGTON—The Senate voted to spend $500 million in funds out of the Pentagon budget to **help the Soviet Union dismantle** its nuclear weapons. —*WSJ 11/26*

27 WASHINGTON—Violators of environmental laws were given **more jail time** in the 1991 fiscal year than in previous years, the EPA reported. —*NYT 11/29*

December 1991

1 RIO DE JANEIRO—Brazilian President Fernando Collor de Mello set aside 19,000 square miles of **virgin rainforest** (an area slightly larger than Switzerland) as a

Drawing by Ziegler; © 1992. The New Yorker Magazine, Inc.

"*And this is our son Danny's room. Danny is being raised by wolves.*"

Wilson concluded that they had found the background radiation, or remnant "glow," of the Big Bang. Their research ultimately gained acceptance by most physicists and cosmologists, and in 1978 won them the Nobel Prize for Physics.

While Penzias's and Wilson's findings affirmed the notion of a uniform explosion, or Big Bang, they initiated a search to explain the discrepancy between the probable result of such an explosion—a smooth and shapeless universe—and the "lumpy" composition of galaxies, clusters of galaxies and empty space in the known universe. Just as a dust particle is needed to initiate the condensation of water vapor into a raindrop, a disturbance in the uniformity of the background radiation must have existed to begin the conglomeration of matter in the universe.

After a quarter of a century of searching for this wrinkle in the "primeval glow," upon which hinged the plausibility of the Big Bang theory, the first

---

"We need more trees and fewer Bushes."

Dr. Helen Caldicott, environmental activist

---

quantitative evidence was announced in April 1992 by George Smoot, a young physicist from the Lawrence Berkeley Laboratory. Smoot and his team of scientists detected minute temperature variations, as small as .00001 of a degree, through ultra-sensitive instruments

Q: *In the past 20 years, what is the estimated number of species that have vanished from the world's tropical forests?*

mounted on NASA's Cosmic Background Explorer satellite. In charting the borders of these zones of temperature variation, the team mapped a series of "ripples," or massive wisps of gas, each greater than 500 million light years long. These constitute the largest and oldest structures ever observed in space, and give a big boost to the Big Bang theory.

CALIFORNIA EARTHQUAKES

Los Angeles—Major earthquakes struck California this year, heightening fears that the long-feared massive tremor is imminent. Scientists believe that the San Andreas fault has become more dangerous as a result of the quakes.

Two earthquakes struck Southern California on June 28, 1992—one in Landers and another near Big Bear City, just east of Los Angeles. The first quake, which was felt from Colorado to Los Angeles, struck Landers, a tiny desert town north of Palm Springs, in the early morning, but the aftershocks rumbled through the region all day. Measuring 7.4 on the Richter scale of ground

> *"Eco-pessimism persists . . . in part because it serves a political program. Some environmentalism is a 'green tree with red roots.' It is the socialist dream—ascetic lives closely regulated by bossy visionaries—dressed up as compassion for the planet."*
>
> George Will, syndicated columnist

motion, the quake was the strongest one to strike California since a 7.7 magnitude trembler struck near Bakersfield in 1952. This quake was almost three times as powerful as the 1989 quake that killed 65 people in San Francisco.

The second earthquake, which hit Big Bear Lake near San Bernardino three hours later, measured 6.5 on the Richter scale. The tremor blocked roads to the resort town and caused landslides that sent clouds of dust rising above the San Bernardino Mountains.

A scientist at Caltech who flew over the region the same day reported that the ground along the San Andreas fault 43 miles north of Landers had moved

homeland for 500 Kaiapo natives. —*NYT 12/1*

2 MOSCOW—The International Chetek Corp., a Moscow-based company, has offered to destroy chemical, biological and nuclear wastes using **underground nuclear explosions.** —*LAT 12/2*

5 NEW YORK—US and Japanese scientists reported that **global warming** is causing a significant increase in accumulated ice in eastern Antarctica, rather than an expected decrease. —*IHT 12/5*

6 WASHINGTON—Cuban President Fidel Castro said his country's **energy crisis** is "the biggest test of the survival of our revolution." Bicycles and horse-drawn carts have replaced automobiles, for which there is not enough fuel, on many Cuban streets. —*IHT 12/6*

8 NEWARK—New Jersey's state environmental regulators hammered out a set of guidelines to cover cleanup of almost 200 **toxic compounds.** —*NYT 12/8*

9 WASHINGTON—The federal court of appeals threw out two of EPA's major **hazardous-waste regulations,** ruling that EPA did not seek public comment when it imposed the regulations 11 years ago. —*WSJ 12/9*

10 MESCALERO, NM—The **Mescalero**

A: *1 million species*
Source: Time *magazine*

Apache tribe has accepted a $100,000 federal grant to study a plan to temporarily store highly radioactive spent reactor fuel on their 460,000-acre reservation.
—*LAT 12/10*

14 WASHINGTON— Energy Secretary James D. Watkins authorized Westinghouse Electric Corporation to resume production of **tritium**, a material used in nuclear warheads, at the Savannah River nuclear weapons site in South Carolina. —*LAT 12/14*

17 WASHINGTON—The Bush Administration announced plans to drastically reduce government **nuclear weapons** production, leaving four production plants and one test site in operation. Energy Secretary James Watkins said recent political events have diminished the need for nuclear weapons. —*NYT 12/17*

19 BEIJING—China announced that its **first nuclear power plant**, a 300,000-kilowatt reactor at Qinshan, will begin trial operations in preparation for going on line in June. —*NYT 12/19*

21 UNITED NATIONS— The United Nations adopted a resolution banning **driftnet fishing** worldwide, starting in 1993. Three countries that opposed the resolution in the past—Japan, South Korea and Taiwan—agreed to obey the ban on the 30-mile-long "curtains of death," which

18 feet. Two weeks later, seismologists analyzing data from the tremors concluded that the two quakes dramatically increased the possibility that the southern portion of the San Andreas fault will produce a powerful earthquake of magnitude 8 or greater. Researchers say that instead of relieving stress, the quakes increased seismic strain on the fault, which runs from the Mexican border to Cape Mendocino, and they fear that the so-called "big one" could occur within the next five years.

NUNAVUT (OUR LAND)

Northwest Territories, Canada—In December 1991, the Canadian government announced an agreement with the Tungavik Federation of Nunavut ("our land") in which native Inuit will gain political domain over 770,000 square miles of northernmost Canada.

The Tungavik Federation of Nunavut, a 15-year-old organization fighting for recognition of Inuit claims to the vast stretches of land, has landed the largest land claim settlement for native people in Canada's history. While Canada's history is a mere blip when compared to the 4,000 years of Inuit activity in the area, the agreement represents a significant step toward two primary goals of the Inuit: self-government, and control over resource development in the region.

In the area designated as Nunavut, which represents one-fifth of Canada, the 17,500 Inuit residents will gain legal title to only 136,000 square miles of land, or about 20 percent of their claim, though their majority in the rest of the region will essentially grant them self-government. They will control mineral rights for an

> *"Everybody knows that if it gets hot, all you do is cool it off."*
>
> Radio personality Rush Limbaugh's advice on global warming, to Ross Perot

even smaller portion of land, but will receive royalties from the Ottawa government on all development of natural resources in the area, which is believed to be rich in oil, natural gas and precious metals. The settlement also includes a sum of over $1 billion, to be paid to the Inuit over a period of 14 years.

An area that was once described by an American

Q: *According to a recent poll, how many Americans strongly oppose more investments in nuclear power?*

serviceman as "nothin' but miles and miles of miles and miles," Nunavut represents a large chunk of Canada's Northwest Territories. Some Inuit hope that the agreement with the Canadian government will facilitate the return to a more traditional lifestyle, and

---

"I'm an environmentalist, always have been."

President George Bush

---

an end to the paternalism and chronic social problems that have accompanied the Inuit's extended contact with modern society.

BHOPAL DISASTER

Bhopal, India—Eight years after a Union Carbide chemical leak killed more than 4,000 people and injured 20,000 in Bhopal, a judicial magistrate in that city ordered seizure of all Indian assets of the Union Carbide Corporation.

In 1991, the Indian Supreme Court ordered Union Carbide to pay $470 million to the Indian government as compensation for Bhopal victims. Special courts have been set up to hear claims, which range from $3,840 to $11,530 for victims' next of kin, and $1,920 to $3,840 for injured survivors. Union Carbide agreed to pay the $470 million compensation.

In April 1992, Union Carbide announced plans to sell its Indian holdings to raise $17 million for a hospital in Bhopal. But on May 1, Magistrate Gulab

In addition to people, thousands of animals were killed in the 1984 Bhopal chemical disaster.

trap and kill whales, seals, dolphins and seabirds, as well as fish. —*LAT 12/21*

25 GAUHATI, India—Wildlife officials said 31 people were killed by **elephants** in 1991 in the Indian state of Assam. In at least six incidents, wild elephants wandered into villages in search of homemade rice beer, then drunkenly trampled people, officials said. —*LAT 12/25*

29 HILTON HEAD ISLAND, SC—Authorities shut down water supplies from the Savannah River after samples showed **tritium** in excess of EPA standards. Contaminated coolant from the Site K Reactor was accidentally released into the river. —*AP 12/29*

31 ROME—An Italian prosecutor said that **Soviet uranium** and plutonium, the raw materials for building a nuclear bomb, are being offered for sale abroad. A Swiss businessman tried to sell an Italian citizen a small amount of uranium, which was seized and later identified as Soviet in origin. —*WP 12/31*

31 UNITED NATIONS—Acute respiratory infections are the most frequent cause of **death in children** under age 5, worldwide, according to the United Nations. —*WP 12/31*

January 1992

1 WASHINGTON—The US Fish and Wildlife Service added the

A: *62 percent*
Source: Greenpeace

Louisiana black bear to the list of threatened species protected by the Endangered Species Act. —*NYT 1/3*

9 NEW YORK—Scientists say they have detected at least **two planets** orbiting a star in the Milky Way galaxy. If confirmed, the planets would be the first known to exist outside our solar system. —*AP 1/9*

9 WASHINGTON—Greenpeace said a powerful explosion was set off to **destroy a chemical dump** at the US McMurdo Station in Antarctica. Greenpeace officials voiced fears that toxic chemicals may have been scattered over a wide area of Antarctica. —*NYT 1/10*

13 MULBERRY, FL—An $8 million Vindicator plant has opened in Florida to **irradiate foods** with cobalt-60 gamma rays. The process protects against spoilage and disease-causing organisms. Opponents of food irradiation are organizing protests and boycotts. —*IHT 1/13*

15 FILLMORE, CA—Two zoo-bred **California condors were released** into the wild as part of a $25 million effort to save the birds from extinction. Natives of the Chumash tribe, who believe the condor is sacred, sang and played drums as the birds were released. —*NYT 1/15*

15 WASHINGTON—Federal Judge Thelton E. Henderson ordered the Sharma ordered seizure of all Union Carbide Indian assets, which have a market net worth of about $71 million.

RARE KILLER WHALE ATTACK

San Francisco—Scientists witnessed a pod of killer whales kill and eat a young California gray whale, an event rarely observed. Researchers and amateur photographers captured the attack on videotape, only the second filming of a whale-on-whale attack and possibly the best record of this behavior. In a rarely-seen display of cooperative hunting, a group of about ten orcas stalked a female gray whale and her offspring, separated the calf from its mother and killed it. Witnesses watched two killer whales overturn the carcass and yank at its pectoral fins. A large

> *"I believe that man is at the top of the pecking order. I think that God gave us dominion over these creatures."*
>
> Secretary of the Interior Manuel Lujan, discussing the Endangered Species Act

bull repeatedly struck the dead calf with its flukes, and other animals tore off large pieces of flesh. When one researcher arrived at the scene, large chunks of blubber floated on the water's surface, while orcas leapt into the air and slapped the water with their fins.

While mature orcas can be as long as 28 feet, California gray whales often reach 46 feet. The orcas gain the upper hand by outnumbering, exhausting and separating their prey from the rest of its pod. These whales belong to a classification of orcas that feed primarily on marine mammals like sea lions, harbor seals and porpoises. A second classification, almost identical in appearance, feeds on fish. Normally the mammal-eating class remains in the waters of southern Alaska and British Columbia, but is occasionally known to swim farther south.

EARLY HUMANS IN AMERICA

Chicago—The discovery of ancient human relics in a New Mexico cave has led archaeologists to theorize that humans arrived in North America earlier than

: *How many licensed, operational nuclear power plants exist in the US?*

serviceman as "nothin' but miles and miles of miles and miles," Nunavut represents a large chunk of Canada's Northwest Territories. Some Inuit hope that the agreement with the Canadian government will facilitate the return to a more traditional lifestyle, and

---

"I'm an environmentalist, always have been."

President George Bush

---

an end to the paternalism and chronic social problems that have accompanied the Inuit's extended contact with modern society.

BHOPAL DISASTER

Bhopal, India—Eight years after a Union Carbide chemical leak killed more than 4,000 people and injured 20,000 in Bhopal, a judicial magistrate in that city ordered seizure of all Indian assets of the Union Carbide Corporation.

In 1991, the Indian Supreme Court ordered Union Carbide to pay $470 million to the Indian government as compensation for Bhopal victims. Special courts have been set up to hear claims, which range from $3,840 to $11,530 for victims' next of kin, and $1,920 to $3,840 for injured survivors. Union Carbide agreed to pay the $470 million compensation.

In April 1992, Union Carbide announced plans to sell its Indian holdings to raise $17 million for a hospital in Bhopal. But on May 1, Magistrate Gulab

Photo: AP/Wide World Photos

In addition to people, thousands of animals were killed in the 1984 Bhopal chemical disaster.

A: *62 percent*
Source: Greenpeace

trap and kill whales, seals, dolphins and seabirds, as well as fish. —*LAT 12/21*

25 GAUHATI, India—Wildlife officials said 31 people were killed by **elephants** in 1991 in the Indian state of Assam. In at least six incidents, wild elephants wandered into villages in search of homemade rice beer, then drunkenly trampled people, officials said. —*LAT 12/25*

29 HILTON HEAD ISLAND, SC—Authorities shut down water supplies from the Savannah River after samples showed **tritium** in excess of EPA standards. Contaminated coolant from the Site K Reactor was accidentally released into the river. —*AP 12/29*

31 ROME—An Italian prosecutor said that **Soviet uranium** and plutonium, the raw materials for building a nuclear bomb, are being offered for sale abroad. A Swiss businessman tried to sell an Italian citizen a small amount of uranium, which was seized and later identified as Soviet in origin. —*WP 12/31*

31 UNITED NATIONS—Acute respiratory infections are the most frequent cause of **death in children** under age 5, worldwide, according to the United Nations. —*WP 12/31*

January 1992

1 WASHINGTON—The US Fish and Wildlife Service added the

EARTH JOURNAL

Louisiana black bear to the list of threatened species protected by the Endangered Species Act. —*NYT 1/3*

9 NEW YORK—Scientists say they have detected at least **two planets** orbiting a star in the Milky Way galaxy. If confirmed, the planets would be the first known to exist outside our solar system. —*AP 1/9*

9 WASHINGTON—Greenpeace said a powerful explosion was set off to **destroy a chemical dump** at the US McMurdo Station in Antarctica. Greenpeace officials voiced fears that toxic chemicals may have been scattered over a wide area of Antarctica. —*NYT 1/10*

13 MULBERRY, FL—An $8 million Vindicator plant has opened in Florida to **irradiate foods** with cobalt-60 gamma rays. The process protects against spoilage and disease-causing organisms. Opponents of food irradiation are organizing protests and boycotts. —*IHT 1/13*

15 FILLMORE, CA—Two zoo-bred **California condors were released** into the wild as part of a $25 million effort to save the birds from extinction. Natives of the Chumash tribe, who believe the condor is sacred, sang and played drums as the birds were released. —*NYT 1/15*

15 WASHINGTON—Federal Judge Thelton E. Henderson ordered the Sharma ordered seizure of all Union Carbide Indian assets, which have a market net worth of about $71 million.

RARE KILLER WHALE ATTACK

San Francisco—Scientists witnessed a pod of killer whales kill and eat a young California gray whale, an event rarely observed. Researchers and amateur photographers captured the attack on videotape, only the second filming of a whale-on-whale attack and possibly the best record of this behavior. In a rarely-seen display of cooperative hunting, a group of about ten orcas stalked a female gray whale and her offspring, separated the calf from its mother and killed it. Witnesses watched two killer whales overturn the carcass and yank at its pectoral fins. A large

> *"I believe that man is at the top of the pecking order. I think that God gave us dominion over these creatures."*
>
> Secretary of the Interior Manuel Lujan, discussing the Endangered Species Act

bull repeatedly struck the dead calf with its flukes, and other animals tore off large pieces of flesh. When one researcher arrived at the scene, large chunks of blubber floated on the water's surface, while orcas leapt into the air and slapped the water with their fins.

While mature orcas can be as long as 28 feet, California gray whales often reach 46 feet. The orcas gain the upper hand by outnumbering, exhausting and separating their prey from the rest of its pod. These whales belong to a classification of orcas that feed primarily on marine mammals like sea lions, harbor seals and porpoises. A second classification, almost identical in appearance, feeds on fish. Normally the mammal-eating class remains in the waters of southern Alaska and British Columbia, but is occasionally known to swim farther south.

EARLY HUMANS IN AMERICA

Chicago—The discovery of ancient human relics in a New Mexico cave has led archaeologists to theorize that humans arrived in North America earlier than

: *How many licensed, operational nuclear power plants exist in the US?*

YEAR IN REVIEW

previously thought. Scientists have long believed that humans entered the New World from Asia about 11,500 years ago. Yet new evidence discovered by archaeologist Richard S. MacNeish suggests that Asians arrived at least 28,000 and possibly 38,000 years ago.

MacNeish reported at a meeting of the American Association for the Advancement of Science in February 1992 that he had discovered human palm- and fingerprints on clay shards found in a cave on the grounds of Fort Bliss in New Mexico. MacNeish reported that the clay pots were imbedded in a layer of earth dated 28,000 years old. He also found several hearths embedded in earth as old as 38,000 years.

> "The Earth isn't a machine, it's a living organism. It's like getting to know a person. When you want to get to know someone, you don't try to control them. Instead, you engage in a dialogue."
>
> Fritjof Capra, physicist

Many of the hearths still contained burnt logs that were too large for animals to have brought into the cave. Charcoal-stained rocks surrounded many of the hearths, and an analysis of the stones revealed that they had been heated hotter than natural fire.

Commerce department to **ban yellowfin tuna imports.** Yellowfin tuna often swim with dolphins, who are caught and killed in fishing nets. —*NYT 1/15*

16 NEW YORK—Officials studying **food safety in New York City** and Chicago found PCBs, mercury and pesticides in samples of fish bought in city markets. —*NYT 1/16*

17 ATLANTA—Scientists working with the Hubble Space Telescope believe they have found a **"supermassive" black hole** with a dense concentration of matter equal to 2.6 billion suns. —*WP 1/17*

20 WASHINGTON—There are **no antelope living wild** on the North American continent, James D. Yoakum says in an upcoming book on ungulates. What Americans have long called antelope are actually pronghorns, which cannot be classified with the African and Asian antelope. —*WP 1/20*

21 NEW YORK—Two researchers say the Northern Hemisphere's **ice sheets will expand** as much as they did at the start of the last Ice Age, if global warming proceeds as forecast. The scientists used mathematical models of the world's climate to predict that the average global temperature will rise 3 to 8 degrees Fahrenheit in the next century. —*NYT 1/21*

A: 113
Source: Union of Concerned Scientists

22 WASHINGTON— The Energy department acknowledged that several million pounds of **radioactive wastes** have been improperly shipped from federal nuclear weapons facilities to waste-treatment plants. —AP 1/22

22 WASHINGTON— President Bush named Andrew H. Card, Jr., as the new **Secretary of Transportation**. Card, a former state legislator in Massachusetts, has no experience in the transportation field. —WP 1/23

23 ATHENS— Thousands of **dolphins** worldwide are being killed by a viral epidemic and by toxic waste, United Nations officials said. —IHT(R) 1/23

24 WASHINGTON— The Bush Administration is preparing a plan ensuring full employment for the estimated **2,000 nuclear scientists** in the former USSR. The plan includes a multinational effort to provide jobs in civilian research institutes. —AP 1/24

31 LOS ANGELES— In the days following President Bush's State of the Union announcement that "By the grace of God, America won the Cold War," three companies have organized a venture to **implement worldwide nuclear disarmament**. Lockheed Corp., Olin Corp. and McDermott International Inc. have formed International Disarmament Corp. —WSJ 1/31

CHOLERA EPIDEMIC

Iquitos, Peru—The first outbreak of cholera in this century struck Peru in January 1991. Since then, the disease has spread to 12 Latin American nations, and health officials predict that most Latin American countries will be infected by 1993. Some 500,000 new cases of cholera were reported in South America in 1991, with 300,000 cases in Peru alone. The World Health Organization and Pan-American Health Organization called the outbreak the worst epidemic of any kind ever to hit Latin America.

If left untreated, a person infected with cholera can die in six hours, due to severe dehydration caused by

Cholera bacillus thrives in Lima, Peru, as sewage pumped directly into the ocean drifts to nearby beaches, contaminating waters.

diarrhea, vomiting and fever. When treated properly with rehydration and antibiotics, only 1 percent to 2 percent of those infected become seriously ill. Of the approximately 500,000 cases reported, about 5,000 have resulted in death.

The disease is spread mainly through food and water contaminated with human waste. Efforts to educate Latin Americans about cholera are complicated by ancient traditions that increase its transmission. In Bolivia, when a person dies of cholera, the body is washed and a vigil held for 24 hours. Because many peasants are unaware how cholera spreads, many take no precautions when handling water and cooking food.

The medical establishment in Latin America hopes that the lessons learned from the cholera outbreak will eventually lead to the elimination of the less dramatic, but more insidious, problems associated with

 How much coal is necessary to keep an electricity-generating plant operating?

dirty water—such as gastroenteritis. The fear of a large-scale cholera epidemic has forced many countries to improve water purification systems and sewage treatment.

CHIMPS USE HERBAL MEDICINE

Chicago—Chimpanzees use medicinal plants to treat illnesses, a practice that could lead to new insights about early human medicine. According to scientists at a meeting of the American Association for the Advancement of Science in February 1992, wild African chimps eat plants containing natural drugs to fight intestinal parasites and cure constipation. More than one-quarter of all drugs used today are made with wild plants, and scientists say that the chimps' behavior could lead researchers to other herbs useful for human medicine.

Dr. Richard Wrangham, a Harvard University scientist, observed that when chimpanzees in Tanzania awoke in the morning, some would occasionally seek out the plant *Aspilia* instead of going to their usual feeding ground. The chimps would choose only young leaves and, after carefully putting them in their mouths, would swallow them without chewing. Wrangham says that *Aspilia* is not a normal part of the chimps' diet, and leaves found in the droppings were almost completely intact. Eloy Rodriguez of the University of California at Irvine analyzed *Aspilia* and presented his findings at the February meeting. Rodriguez discovered that *Aspilia* leaves contain

---

"The importance of asking the proper question is often overlooked."

Robert W. McFarlane, ornithologist

---

chemicals that destroy roundworms, hookworms, fungi and retroviruses. Leaves found in chimpanzee dung no longer contain these compounds.

Another researcher reported seeing chimps find the plant *Vernonia* when they were lethargic, not eating, constipated and emitting dark urine. The animals were careful not to eat the leaves, but would peel the bark back, eat the pith or the center of the stem, and spit out the fibers. Michael A. Huffman of Kyoto University noticed that, after one day, the

February 1992

1 RESERVE, NM—The **Mexican spotted owl**, cousin to the northern spotted owl of the Pacific Northwest, is the focus of a dispute between the federal government, loggers and environmentalists. The US Forest Service has temporarily forbidden logging on thousands of acres of federal land in the Southwest while studies determine whether the estimated 2,000 Mexican spotted owls are endangered. —NYT 2/1

1 WASHINGTON—The EPA, under pressure of a federal lawsuit, agreed to extend federal standards to 16 categories of industry, replacing state controls. The new controls are expected to reduce by millions of tons per year the **toxics dumped** into lakes, rivers and coastlines. —WP 2/1

4 WASHINGTON—US District Judge John Garrett Penn ordered the Energy department to obtain a hazardous waste storage permit from New Mexico before shipping plutonium waste to the **Waste Isolation Pilot Project**, a network of caverns carved into rock salt near Carlsbad. Environmental lawyers called the ruling a setback for Energy Secretary James Watkins. —WP 2/4

7 TOKYO—A Japanese court has ruled that government authorities were not responsible for

Over 500 tons per hour
Source: Protect Our Planet Calendar, Running Press Book Publishers

discharges of mercury that killed more than 1,000 people and crippled others who had eaten contaminated fish. The **mercury discharges** by Chisso Corporation occurred in the 1950s and 1960s. —*IHT 2/8*

8 WASHINGTON—The Bush Administration announced that all CIS republics, except Russia, will eliminate all nuclear-tipped strategic missiles within seven years. Only Russia will have **ballistic missiles** capable of striking the US. —*WP 2/8*

12 TOKYO—Japan launched its first **environmental survey satellite.** The satellite, which will orbit for two years 353 miles above the Earth, will observe weather patterns and measure changes such as diminishing rainforest areas. —*WSJ 2/12*

14 WASHINGTON—The EPA reversed its plan to ban most uses of pesticides containing EBDC (ethylene bisdithiocarbamate). The decision to **reinstate EBDC** use was based on a new method of assessing risk, which measures the amount of the chemicals remaining on produce in the supermarket, rather than on crops in the field. —*WP 2/14*

14 WASHINGTON—A new study estimated that 1,600 people die from **air pollution** in the Los Angeles area each year. The estimates were drawn from a model that evaluated health effects

chimps seemed revived. Huffman says that analysis of the plant revealed that while the leaves and bark contain lethally poisonous compounds, the pith has substances that act against intestinal parasites like roundworm, hookworm, giardia and organisms that cause schistosomiasis.

The findings indicate that humans may have used medicine as early as 5 million or 6 million years ago, when chimpanzee and human evolutionary lines separated.

PARKS SHELTER HOMELESS

Cottage Grove, OR—A group of homeless people in Oregon have a place to stay, thanks to a campsite in a national forest. Several families set up shelters in May 1992 in the first campground on public lands reserved exclusively for the homeless. The site, located in an abandoned quarry pit in the Umpqua National Forest 40 miles southeast of Eugene, will house up to 25 people for an indefinite period of time. The three-acre campground, named Blodgett, is considered unfit for recreational use by the Forest Service because it lacks basic facilities found at other campgrounds.

The Forest Service normally does not allow campers to remain in one spot for more than 14 consecutive days, although homeless people have squatted in national forests for years. One 72-year-old woman was forced to move into the new Oregon site by Forest Service officials who had let her live in a recreational campsite because the $580 Social Security check she gets each month would not cover rent and expenses in town. Other residents of

"It is never easy, it is never easy to stand alone on principle, but sometimes leadership requires that you do. And now is such a time."

President Bush, explaining his positions at the Earth Summit

Blodgett include a family of four whose father was one of 600 local timber workers who have lost logging jobs over the past five years. Most of the people in the campsite face similar financial problems.

Another campground for the homeless is being

: *In Athens, Greece, what is the* **nephos**, *the monster who is responsible for an average of a dozen deaths a day during its peak times?*

YEAR IN REVIEW

"They're such cute little animals I hate to hunt them. But if we don't thin out their numbers they'll simply starve themselves out of existence!"

Cartoon by Ed Fisher. Reprinted by permission.

considered north of the Cottage Grove site in the Willamette National Forest. Forest Service officials believe that if the Oregon campgrounds work, they could be tried in other forests nationwide. When the oil-based economies of Colorado and Wyoming crashed in the early 1980s, campgrounds throughout these states were inundated with homeless people, and a clash between recreational campers and squatters ensued. Some Forest Service officials believe the pilot program is bringing conditions that have existed for years out in the open. They see these official campgrounds as a creative new approach to an old problem.

ROSE PARADE PROTESTED

Pasadena, CA—The year 1992 marked the 500th anniversary of Columbus's voyage to the "New World," and the quincentennial of the journey did not go unnoticed by Native American activists.

At sunrise on New Year's Day, 1992, a few hours before the annual Rose Parade in Los Angeles, a group

A: *"The cloud"* of smog that hangs over the city
Source: Time magazine

of atmospheric ozone and particulate matter encountered in a year by the 12 million residents of the area. —*AP 2/14*

15 LOS ANGELES—The federal government announced that it no longer has enough water to supply farmers in California who are suffering from a **six-year drought.** The Federal Bureau of Reclamation said that, barring more heavy precipitation, one-third less water will be available for farming than in normal years. —*NYT 2/15*

15 UNITED NATIONS— Representatives of 37 island nations voiced concern that **ocean levels are rising** due to global warming and melting polar ice caps. In addition to island inundation, a three-foot rise in ocean levels would render 72 million people homeless in China, 11 million in Bangladesh and 8 million in Egypt, a researcher said. —*NYT 2/17*

17 MOSCOW—The US will supply Russia with 25 specially fitted railroad boxcars to **transport nuclear warheads** to storage sites. The US will also supply $25 million to set up an institute employing former Soviet nuclear-bomb designers in peaceful projects, Russian President Boris Yeltsin said after meeting with US Secretary of State James Baker. —*NYT 2/18*

24 WAUCHULA, FL— Farmers are blaming dead or dying crops on a

EARTH JOURNAL

Du Pont **fungicide called Benlate.** Du Pont has already paid more than $205 million to settle damage claims by growers. A horticulturist at Iowa State University, Dr. David Koranski, said, "We are getting numerous symptoms from this chemical, not just stunting . . . but stem elongation and water-soaked spots." —*NYT 2/24*

24 SOFIA, Bulgaria—A 1,000-megawatt **nuclear reactor** was unplugged from the national energy grid in Bulgaria when a fault was detected in a turbine outside the reactor building. Officials said the turbine problem will force Bulgaria to halve its electric power supply and impose blackouts. —*NYT 2/24*

24 MIAMI—Federal District Court Judge William M. Hoeveler ordered regulators and sugarcane growers to begin an 11-year plan to clean up pollution caused by farm chemicals in the **Florida Everglades.** — *NYT (AP) 2/26*

24 WASHINGTON—The General Accounting Office released a report stating that federal laws and regulations offer inadequate protection to farm workers **exposed to pesticides.** The EPA estimates that hired farm workers suffer up to 300,000 acute illnesses and injuries each year from exposure to pesticides. —*NYT 2/25*

24 TORONTO—Canada announced it will

of protesters representing the Southern California indigenous Gabrielino Nation drew the waiting crowd's attention with traditional costumes, drums and dances. The protest culminated with an all-night vigil by Native American and Hispanic activists.

Though the group expressed outrage, their protests were aimed at the Rose Parade's organizational committee, who had selected a direct descendant of Christopher Columbus himself as a Grand Marshal of the Parade.

Judith Cuauhtemoc, a member of the Cuauhtemoc Dance Troupe that joined the protest, said that "what they [the parade organizers] are celebrating as a symbol of discovery is a symbol of genocide, of destruction, for the natives of the whole continent."

Los Angeles has followed Berkeley, California's move to change the October 12 holiday's name from Columbus Day to Indigenous Peoples Day. Berkeley Mayor Lori Hancock says that the explorer's name has been expunged from the calendar "to celebrate the important place that indigenous people hold in this country."

MEDICINAL PLANTS

Brazil—The promotion of sustainably harvested "rainforest products" represents an effort to strike a balance between development and conservation of remaining rainforests, which are currently lost at a rate of 55,000 square miles per year. Until recently, however, the majority of sustainably harvested

The Pacific yew tree is a source for the cancer-fighting drug, taxol. Scientists hope to synthesize the drug to protect the Pacific yew trees from being over-harvested.

: *How many mammals does the fishing industry kill in driftnets each year?*

rainforest products, such as the Brazil nut, have been luxury items. Now, interest in botanical medicines by major pharmaceutical companies offers indigenous people and others incentive to protect the forests.

Traditional healers throughout human history have used botanical medicines to heal their people. Even today, three-fourths of the global population relies on the medicinal function of plants as a primary form of health care.

Perhaps the most famous botanical agent incorporated into modern Western medicine is rosy periwinkle, used to treat childhood leukemia and Hodgkin's disease. And, recently, researchers in India derived a compound from the native guggal tree that reduces levels of cholesterol and other blood fats without side effects.

Many companies are researching the rainforests to find other medicinally valuable plants. The Costa Rican Research Institute and Merck & Company, the world's largest pharmaceutical company, have joined forces to collect and screen as many plants, microorganisms and insects as possible, to assess their medicinal properties. Merck & Company is financing the collection of samples, and will share with Costa Rica the profits of any drug derived from a Costa Rican sample. The project provides jobs in Costa Rica, as well as money to fund conservation projects.

> "If we let people see that kind of thing, there would never again be any war."
>
> Pentagon official explaining why US military censors refused to release video footage showing Iraqi soldiers being killed by US gunfire

RED GOLD

Russia—In attempts to ride out the wave of chaos swelling in the wake of the disintegration of the Soviet Union, Russians are scrambling for a piece of the energy sector of that nation's splintered economy.

Yields from Russian oil fields, which brought the former Soviet Union its status as the largest producer of oil in the world, have declined significantly in the past few years. Since 1990, daily production has dropped more than 1 million barrels, from 11.7

resume promoting the hunting of **harp seals.** Seals have been blamed for the decreasing levels of cod available to fishers. —WP 2/25

25 LOS ANGELES—President Bush announced a joint plan by the US and Mexico to spend $700 million on **pollution cleanup** and safer drinking water along the shared border. Mexico has pledged $460 million and the US has promised $241 million. —WP (AP) 2/26

26 ROWE, MA—The owners of a 32-year-old nuclear power plant, **Yankee Rowe,** in Rowe, Massachusetts, announced they will permanently close the plant to avoid the high cost of determining whether it is safe to operate. The plant, near the New York and Vermont borders, was the nation's oldest operating power reactor. —NYT 2/27

27 NEW YORK—The Recording Industry Association of America announced that its members (including MCA, Polygram, Sony, Warner and others) will replace the long cardboard or plastic display boxes encasing **compact discs** with more efficient packaging. The new packaging will have few if any disposable components, according to the association's announcement. —NYT 2/28

28 RIO DE JANEIRO—An appeals court has ordered a retrial of Darly

A: *Over 200,000*
Source: Newsweek magazine

Alves da Silva, who was convicted of planning the 1988 murder of **Francisco (Chico) Mendes,** a Brazilian rubber tapper and conservationist. Da Silva's son, Darci, confessed to killing Mendes, and was sentenced to 19 years in prison. —*NYT 2/29*

March 1992

3 ASHEVILLE, NC—The National Climatic Data Center reported that this winter was the warmest in the US in the 97 years they have been keeping records. The season's average temperature for the contiguous states was 36.87 degrees F., and some 93 percent of the Lower 48 states had **warmer-than-normal** temperatures. Nearly half the country was "much warmer" than usual, according to reports. —*WP (AP) 3/4*

4 MOSCOW—A Russian nuclear power plant was shut down by an electrical fire that took approximately 40 minutes to extinguish. Russian energy officials said **no radiation** was released. The fire was apparently due to a short circuit near a reactor at the Balakovskaya power plant. —*WP 3/5*

8 BRASILIA—Brazil's population growth rate has fallen below 2 percent for the first time in 50 years, Brazil's census takers report. Nearly 30 percent of married Brazilian women of childbearing age have been **sterilized,** compared to 17 percent in the US and 7 percent in

million barrels per day to 10.3 million.

The directors of the state-owned oil companies deem the decline a direct reflection of three decades of inefficient centralized planning. Until May 1992, the oil companies were required to sell 80 percent of their yield to the state for a mere 80 cents per barrel, rather than on the open market for the relatively hearty rate

> *"Untenable levels of population will be staved off only by famine, disease or wars. How can anyone think those preferable to the condom, the intrauterine device or the birth control pill?"*
>
> Anna Quindlen, columnist, *The New York Times*

of $18 per barrel. With insufficient funds for necessary parts and repairs coming back to the companies, directors of oil facilities watched their operations, as well as plans for future drilling and exploration, dwindle into the swamps and barren tundra.

Confident, however, in the expertise of their workers, as well as in the extent of oil reserves waiting in untapped sites, the directors of oil companies in Russia are attempting to transform their outfits into private stock companies. Though Russian bureaucrats insist that these "oil generals" are deluding themselves in a "game without rules," and that ownership of oil operations in fact remains securely in the hands of the state, it may be the bureaucrats who are deluded.

In January 1992, a new company calling itself the People's Industrial Investment Euro-Asian Corporation began selling shares to the public from a makeshift office in an empty wing of the Museum of the History of Moscow. Far from convinced of the company's ability to yield returns on their sparse rubles, people waited hours in line to invest.

If oil can be sold at more reasonable prices (and the government in May 1992 authorized a fivefold increase in price for oil sold within Russia), the oil directors believe they can raise enough capital to repair existing rigs, develop tremendous reserves and bring Russian oil production to the forefront of the world market.

 Out of every $10 Americans spend on food, how much of it pays for packaging?

SMOG CLOSES MEXICO CITY

Mexico City—The world's largest city shut down on March 17, 1992, due to incapacitating air pollution. For the first time in its history, Mexico City, home to some 16 million people, closed schools, ordered industries to reduce operations and banned almost half of all cars from the streets. All road work came to a grinding halt; half of the government's vehicles were put out of circulation; and 225 industries cut production by 50 percent to 70 percent. On any given day, one-fifth of all cars are barred from entering the city, but when the mayor of Mexico City imposed emergency measures, another one-fifth of automobiles were barred.

Mexico City is perhaps the most polluted city in the world. The metropolis is surrounded by

A street vendor capitalizes on Mexico City's severe pollution problem.

A: $1
Source: Environmental Defense Fund

the rest of Latin America.
—*NYT 3/8*

12 MEXICO—Mexico's population **grew by 16 million,** or 25 percent, in the 1980s, according to census figures. The current population is estimated at 81 million.
—*WSJ 3/12*

12 TOKYO—Over strong objections by the US delegation, the Convention on International Trade in Endangered Species voted to require that the **American black bear,** a species that is reportedly thriving, be protected.
—*WP 3/13*

18 TOKYO—Japan announced it will continue to ban birth control pills. Japan's Health and Welfare Ministry, citing a growing number of AIDS cases in Japan, hopes to encourage condom use by **banning the pill.**
—*NYT 3/19*

24 MOSCOW—A nuclear power plant near St. Petersburg released "six times the permissible amounts of radioactive gases and ten times the allowable radioactive iodine" into the air when a water channel ruptured, Russia's State Committee for Emergencies said. It was the most **serious nuclear power incident** in the CIS since the 1986 Chernobyl accident, authorities said.
—*LAT 3/25*

25 SAN FRANCISCO—A federal appeals court voted 3 to 0 that a nuclear weapons plant in Idaho must accept

nuclear waste shipments from out of state. John McCreedy, Idaho's deputy attorney general, said Idaho may seek a new injunction on the ground that the shipments would violate federal environmental laws. The disputed shipments come from Colorado's Ft. St. Vrain nuclear power plant, which shut down in 1989.
—*NYT 3/25*

25 WASHINGTON— The Supreme Court ruled that Congress can temporarily modify its environmental laws to allow cutting of old growth forests where endangered **spotted owls** live. The ruling will allow the sale of timber in areas previously designated as off-limits.
—*LAT 3/26*

26 WASHINGTON— Rockwell International has agreed to pay an **$18.5 million criminal fine** to settle charges that it violated federal environmental protection laws at its Rocky Flats nuclear weapons plant in Colorado. The government alleged that Rockwell burned hazardous waste and dumped cancer-causing chemicals into streams at the plant near Denver.
—*WSJ 3/26*

27 ALBANY, NY—New York Governor Mario Cuomo cancelled a contract to buy energy from a proposed hydroelectric complex in Canada's **James Bay** area in northern Quebec. Analysts said New York's

mountains, which cause thermal inversions during winter. This layer of cold air hangs over the city and traps pollution underneath. The fact that only a small number of cars use unleaded gasoline exacerbates this problem. On bad days, the air quality in Mexico City has been compared to that in New York City's Holland Tunnel. In fact, on the morning of March 17, ozone levels reached three times that of the worst ozone day recorded in New York City, in 1990.

The mayor of Mexico City, Manuel Camacho Solis, has taken heat from opposition politicians who accuse him of failing to address the worst air pollution on record and of holding back on strict

> *The heads of state at the Earth Summit are "the ugliest environmental pigs I have ever seen."*
> Spense Havlick, a Boulder, Colorado, city councilman, at the Earth Summit

anti-pollution legislation in order to protect his popularity. Yet Camacho has done more than any former mayor to combat the city's smog problems. He has urged commuters to use public transportation and forced industries to use cleaner fuels and pollution-fighting equipment.

COOLING OFF

The Philippines—When the Philippine volcano Mt. Pinatubo erupted in June 1991, dumping up to 20 million tons of sulfur dioxide and an unknown amount of ash into the stratosphere, climate modelers jumped at the chance to test their computer software. They predicted that in the next few years, global warming due to the emission of greenhouse gases, such as carbon dioxide, would be more than compensated by global cooling, due to reflection of sunlight from tiny droplets of diluted sulfuric acid derived from the sulfur dioxide injected by the volcano at 40,000 to 90,000 feet altitude. Recent measurements by National Oceanic and Atmospheric Administration satellites indicate that by May 1992, the global average temperature had decreased by 1 degree Fahrenheit, effectively balancing the global warming that had occurred in the past 20 years.

: *Which country has 16 percent of the world's population, yet consumes 3 percent of its energy, and emits 3 percent of all CO_2 produced?*

Ash fall from Mt. Pinatubo created this ghostly scene at a roadside stand in the Philippines.

However, since the small acid droplets succumb to gravity, albeit very slowly, they will eventually leave the stratosphere and no longer have a cooling effect. Models predict that the globe will return to its warming ways by the late 1990s.

An environmental side effect of the volcano is that it produces "clouds" in the stratosphere similar to those that form naturally in the cold Antarctic stratosphere, and converts normally benign human-produced CFC-related molecules into ones that, when exposed to sunlight after the long dark polar winter, destroy ozone and form the spring Antarctic "ozone

> "To the average British farmer, organic farming is about as relevant as caviar and a flight on Concorde."
>
> Oliver Walston, *The Observer*

hole." During 1991, the Southern Hemisphere received a "double dose" with the eruption of Mt. Hudson in Chile in August adding to that of Mt. Pinatubo. Evidence from ozone measurements at the South Pole last austral spring suggest that ozone depletion was more severe than usual in regions that volcanic aerosol affected. Although models predict global ozone reductions from the massive eruption of Pinatubo, the effects are not large, and careful data analysis of this "once in a century" event will test

A: *India*
Source: Time magazine

withdrawal is a severe setback for the controversial project. —*NYT 3/28*

April 1992

1 WASHINGTON—President Bush backed down from his original plan to put his **wetland proposal** into effect today. Thousands of angry letters forced Bush to delay his proposal that, to qualify for protection, a wetland must have standing water for 21 consecutive summer days. —*AP 4/1*

3 WASHINGTON—The National Oceanic and Atmospheric Administration announced that a virus that killed more than **17,000 seals** in Europe in 1988 has been found in three species of seals on America's northeast coast. —*WP 5/5*

7 VIENNA—A United Nations inspection team has convinced Iraq to destroy its **Al Atheer nuclear installation**, a complex of more than 100 buildings believed to contain nuclear weapon research materials. —*NYT 4/8*

8 WASHINGTON—Colorado has more sites contaminated by **radioactive materials** than any other state, the EPA reports. California ranks second. A total of more than 45,300 "hot spots" are scattered across the US, according to the EPA. —*LAT 4/9*

8 PARIS—France announced it will

suspend its nuclear weapons testing programs in the South Pacific for the remainder of the year. —LAT 4/9

15 DALLAS—Medical researchers report that lead and other toxic pollutants may play a role in 11 percent of all congenital heart defects. The study said **exposure to toxins** appears to cause heart problems in families with a genetic susceptibility to heart defects. —NYT 4/17

17 ANCHORAGE—A new state and federal report says the *Exxon Valdez* oil spill has caused long-term damage to sea otters, killer whales, harbor seals, seabirds and fish that is **worse than expected.** The report directly contradicts a national advertising campaign by Exxon that contends that Prince William Sound's ecological health has been restored. —NYT 4/18

20 MAPUTO, Mozambique—A Greek ship, the *Katina P*, spilled an estimated **1 million gallons of oil** near Mozambique's coast, causing the worst environmental disaster in Mozambique history. Oil spilled onto beaches and drifted into Maputo Bay, authorities said. —LAT 4/21

21 CHICAGO—The Chicago Board of Trade has received permission from the Commodity Futures Trading Commission to set up a market in government-granted

ozone depletion models and give scientists additional assurance in their predictions of future degradation of the Earth's protective ozone layer.
—*Dave Hofmann, National Oceanic and Atmospheric Administration, Boulder, Colorado*

LANDFILLS STUFFED

US—American landfills are closing at an alarming pace. Currently, fewer than 4,000 of the 14,000 landfills that operated nationwide in 1977 are still open, and that number is expected to drop to 1,800 by the year 2000. Yet Americans churn out 160 million tons of garbage a day, twice as much per capita as other industrial nations. The Environmental Protection Agency (EPA) says that by 2000, the US will produce five pounds of garbage per person per day, more than twice the rate of 30 years ago. With two landfills closing every day in the US, the country is in desperate need of new sites.

To complicate matters further, new environmental regulations have made the cost of building new landfills skyrocket. As of September 1991, the EPA mandates that all landfills have plastic and clay liners, liquid collection and treatment systems, and other expensive equipment. Old dumps that do not meet these requirements by 1993 will be shut down. It now costs at least $10 million to build a landfill that meets the new standards. Waste disposal has moved into the hands of a dozen national corporations that will operate about 1,000 landfills across the country, and towns will have to ship waste to these massive regional dumps. Already, 42 states import and export

> *"Poverty is the ultimate form of pollution in a world out of balance with people's needs and its own future."*
>
> Bella Abzug, Center for Our Common Future, *Network '92*

trash, and many local governments are finding that waste disposal is their largest expense.

Waste corporations are having problems finding communities willing to host a landfill. Huge waste disposal companies like Browning-Ferris Industries and Waste Management are increasingly seeking out poor communities who find the economic benefits of

Q: *According to the Department of Transportation, how many gallons of gasoline are consumed daily by American automobiles?*

YEAR IN REVIEW

"These pristine mountain lakes used to be swarming with frogs. None seem to have adapted to today's toxic environment!"

the $130-billion-a-year industry more compelling than the drawbacks of a dump in the backyard. While many towns have rejected landfill proposals, companies woo public approval with "community partnership" programs that offer gifts like new community centers, new sports facilities for local high schools and scholarship money.

Currently, 80 percent of all garbage ends up in landfills; another 10 percent is burned; and the remaining 10 percent is recycled. As of 1991, 26 states and the District of Columbia have comprehensive recycling laws. While New York City recycles only 6 percent to 8 percent of its trash, Seattle, with the highest recycling rate for large American cities, recycles 34 percent of its solid waste and 77 percent of the waste in its residential areas. Recycling has great potential for alleviating some of the landfill problems, and some figures indicate that recycling, combined with composting, incinerating and salvaging, could reduce the waste stream by 90 percent.

ELEPHANTS AND IVORY

Zambia—On February 14, 1992, Zambia marked its allegiance to its endangered elephant population

rights to pollute the air. Environmentalists hope industries will try to reduce their pollution output so they can sell leftover pollution "rights" for a profit. —*WP 4/22*

21 WASHINGTON—The Supreme Court overturned a lower court ruling and decided that federal agencies **do not have to pay fines** for violations of federal pollution laws. Justices Byron White, Harry Blackmun and John Paul Stevens filed dissenting opinions. —*NYT 4/23*

24 WASHINGTON—The Bush Administration said the US can reduce its annual output of **carbon dioxide** from 325 million to 200 million metric tons by the year 2000, if output is reduced 7 percent to 11 percent from previous projections. —*LAT 4/25*

25 SANTA MONICA, CA—A new study says only one-tenth of the money spent by insurance companies to settle **Superfund environmental claims** is actually spent on cleanup. About 90 percent of the money goes for legal fees and related costs, the report said. —*NYT 4/26*

29 WASHINGTON—The United Nations Population Fund released a report estimating that **world population** will reach 5.8 billion this summer, 6 billion by 1998 and 10 billion by 2050. —*WP 4/30*

 Over 200 million gallons of gas per day
Source: 50 Simple Things You Can Do to Save the Earth, *Earth Works Press*

May 1992

3 OCEAN COUNTY, NJ—A brushfire forced **Oyster Creek Nuclear Generating Station** to close. The fire, fanned by winds, roared through more than 5,000 acres near the plant. —AP 5/4

4 WASHINGTON—A Gallup poll of citizens in 22 countries found that a majority of people polled **favor environmental protection,** even at the risk of slowing economic growth. —WP 5/5

5 MOSCOW—Firefighters struggled to contain fires in an area contaminated by the 1986 Chernobyl nuclear disaster. "**Substances containing radiation** are being carried considerable distances, contaminating clean settlements and fields," the Itar-Tass news agency reported. —WSJ (R) 5/5

6 LOS ANGELES—Environmental experts warned that more than half of the 5,000 buildings burned in the **Los Angeles uprisings** may contain toxic chemicals and asbestos. Authorities said toxic materials, soot and debris poured from the burning buildings into storm drains that empty into the sea. —LAT 5/6

6 WASHINGTON—The EPA admitted that it violated a **hazardous waste law** by approving a permit for an Ohio toxic waste incinerator in 1983. —NYT 5/8

8 NEW YORK—Joseph J. Hazelwood, who was

EARTH JOURNAL

with the stroke of a match. The African nation, which burned nine tons of confiscated ivory, joined Kenya, Dubai and Taiwan in making the statement of non-tolerance to poachers and the rest of the world. The tradition began on July 18, 1989, in Nairobi National Park, when Kenyan President Danial arap Moi set fire to 12 tons of ivory, or about 2,500 elephant tusks.

In an effort to save the species, whose population dwindled from 1.3 million in the 1970s to 600,000 in 1989, the seventh Convention on International Trade in Endangered Species (CITES) in 1989 imposed an international ban on the sale of ivory. With the primary consumers of ivory—the US, Japan, the European Community, Canada, Australia and Switzerland—consenting to the ban, CITES (a 103-nation league) hoped to discourage ivory hunters by cutting off demand. The African elephant received "Appendix 1" status as an endangered species, which prohibits trade in all "elephant products."

Just three years after ratification of the ban, however, the governments of five nations in southern Africa called for a lifting of the ivory ban.

Endangered African elephants, protected under CITES, graze peacefully in Kenya.

Zimbabwe, Botswana, Malawi, Namibia and South Africa, which have large elephant populations, insist that the regulated sale of ivory from the thinning of their herds is integral to the funding of wildlife conservation programs. In South Africa's Kruger National Park, government rangers kill approximately 520 elephants annually so that the herds do not exceed the park's capacity. South

Q: *How much of the Earth's rainforests are destroyed each second?*

Africa's National Parks Board, which earns $2 million each year from the sale of ivory, says that the "culling" provides funding for efforts to combat poaching and eases the burden on rural people who must coexist with the elephants.

At the 1992 CITES Conference in Kyoto, Japan, Zimbabwe, Botswana, Malawi and Namibia proposed

> *"Destroying rainforest for economic gain is like burning a Renaissance painting to cook a meal."*
>
> Edward O. Wilson, quoted in *Time*

downlisting the elephant from Appendix 1 to Appendix 2, which would render the elephant "threatened" rather than "endangered." The change in classification would reopen markets for elephant meat, hides and ivory. South Africa proposed lifting the ban as well.

While acknowledging that the southern nations' elephant populations are in better shape than those in eastern Africa, CITES upheld the ban, stating that African unity is integral to the success of the treaty. As elephants frequently cross national borders, so must the ban protecting them. Also, should the southern nations be exempted from the ban, they could too easily become a clearinghouse for "laundered" elephant products from animals slaughtered in the east and smuggled to southern ports.

Since the gestation period for an elephant is 22 months, it is still too early to tell how successful the ban will be in rejuvenating East Africa's elephant population. Some biologists say it may take up to 40 years to restore the great herds.

POLAND'S DEBT-FOR-NATURE

Warsaw—Poland, like other former Communist-bloc nations, stressed rapid industrialization at the expense of clean air and water, a policy that has left the nation's environment in a shambles. In 1992, a government document on the state of the environment called Poland one of the most polluted countries in the world. The report claims that one-third of Poles live in environmentally hazardous areas, the nation's

captain of the *Exxon Valdez* when it struck a reef and spilled 11 million gallons of crude oil into Prince William Sound, has been hired to teach at the New York Maritime College. Hazelwood will teach students how to stand watch on a ship's bridge. *—WP 5/8*

8 UNITED NATIONS—One billion people breathe unhealthy air, skin cancer is increasing and malnutrition is prevalent, according to a new report by the United Nations Environment Programme. The report concluded that **environmental problems are much worse** than they were 20 years ago. *—LAT 5/8*

9 UNITED NATIONS—Negotiators from 143 countries approved a **global warming treaty** that asks nations to report the amount of carbon dioxide (and other gases) they release into the atmosphere, and to "return . . . to their 1990 levels of these emissions." *—AP 5/10*

11 PASADENA, CA—A California Institute of Technology professor, Dr. Joseph Kirschvink, said his research shows that human brain cells contain tiny crystals of a highly **magnetic mineral** known as magnetite. Scientists conjectured that these crystals could explain why exposure to electromagnetic fields results in increased risk of cancer. *—AP 5/12*

 An area equivalent to the size of a football field
Source: Protect Our Planet Calendar, Running Press Book Publishers

EARTH JOURNAL

14 HILL CITY, SD— Federal agents seized the bones of a **Tyrannosaurus rex** dinosaur from a private institute that was accused of stealing the bones from Sioux tribal lands. The bones are believed to be the most complete dinosaur skeleton ever found. —*LAT (AP) 5/15*

14 QUITO, Ecuador— Months of negotiation ended as President Rodrigo Borja granted legal title of more than 3 million acres to **indigenous tribes** in the Amazon basin. Sixty tribal leaders, many decked in bright feathers and carrying spears, received the land titles at a special ceremony at the presidential palace. —*WP 5/15*

20 WASHINGTON— The House passed a provision that cuts the time required to obtain a **nuclear power plant** construction license from 14 years to six years. The Senate passed identical legislation in February. —*WSJ 5/21*

21 WASHINGTON— China set off its **largest nuclear test** ever, equivalent to 70 times the explosive power of the atomic bomb dropped on Hiroshima. —*NYT 5/22*

22 LOS ANGELES— Southern California Gas Company launched the nation's first commercially operating fuel cell generator. Pure water and heat by-products are recycled to the fuel processor as steam, **increasing energy**

average life expectancy is decreasing and "civilization-related" ailments are increasing dramatically.

In response, Poland has begun development of a multinational "eco-conversion" plan, using the almost $48 billion it owes foreign creditors. This debt-for-nature scheme, called the "environment fund," will convert part of Poland's huge debt to fund environmental protection and cleanup over the next 18 years.

The idea for the swap came in April 1991 when the Paris Club, a group of 17 Western government

> *"[Energy conservation] is simply a euphemism for reducing your standard of living."*
>
> Herb Schmertz, quoted in *Washington Post*

creditors, signed an agreement to forgive at least half the $35 billion owed by Poland. The Paris Club agreed to earmark 10 percent for environmental cleanup and established the environment fund. The program allows Poland to convert part of its debt into local currency and use it exclusively for environmental improvement. So far, only the US has committed itself to the scheme.

While Poland spent 1.1 percent of its gross domestic product (GDP), or $1 billion, on environmental protection in 1991, some scientists believe that a country must commit 1.5 percent to 3 percent of its GDP merely to maintain previous environmental conditions and to prevent future degradation.

DROUGHT OF THE CENTURY

Africa—A severe drought which has been called the worst of the century is causing severe food shortages, and drying lakes, rivers and wells throughout southern and eastern Africa. The Agency for International Development (AID) predicts that close to 30 million Africans face malnutrition and starvation in 1992, in large part due to the drought. Farmers throughout the region face crop losses of 70 percent to 90 percent, authorities say. Almost no rain has fallen since January 1992, "obliterating all hope for successful crops this year," AID reports.

The drought is devastating agricultural land in South Africa, Zimbabwe, Kenya, Zambia, Malawi, Mozambique, Namibia, Angola, Botswana, Lesotho,

  A commercial incinerator operating at a 99.99 percent destruction and removal efficiency still releases how many unburned chemicals each year?

Swaziland, Somalia and Uganda. In Zimbabwe, wild animals have been herded into corrals and slaughtered with assault rifles so the meat can be used to feed hungry children. Colin Saunders, head of the animal culling operation, said the country's wildlife program calls for elimination of at least 5,000 impalas, 2,000 elephants and 1,500 buffalo before November 1992, when the next rainy season should begin. In Somalia, the combination of drought and warfare has left thousands of people dead, many from starvation. The traditional Somalian nomadic lifestyle—which depends on cattle, camels and sheep—has collapsed due to drought and war. And South Africa, which once ranked just behind the US and Argentina as a grain-exporting country, has imported more than 5 million tons of corn and grain this year because of crop failures due to drought.

AID officials said $1.1 billion in aid will probably be needed, including at least 3.4 million metric tons

A malnourished child drinks milk at a refugee camp in Kenya.

efficiency to more than 80 percent. —*LAT 5/22*

25 AIKEN, SC—The Savannah River nuclear weapons plant's K Reactor sprang a leak of **radioactive water** during tests by the Department of Energy. The reactor has been closed since 1988, due to safety problems. —*NYT 5/27*

29 KIGALI, Rwanda—Mrithi, a 24-year-old male silverback gorilla who was featured in the film **"Gorillas in the Mist,"** about the life of gorilla researcher Dian Fossey, was shot with an assault rifle and found dead. —*NYT 5/29*

June 1992

1 DENVER—A federal district judge accepted Rockwell International's guilty plea in criminal charges of **mishandling poisonous waste** (including allowing toxic waste to leak from cardboard boxes stored outdoors) at Rocky Flats nuclear weapons plant near Denver. Rockwell agreed to pay $18.5 million in fines. —*NYT 6/2*

1 WASHINGTON—President Bush said he will increase US international aid for **forest conservation** by more than $100 million per year. "Halting the loss of the Earth's forests is one of the most cost-effective steps we can take to cut carbon dioxide emissions," Bush said. —*AP 6/1*

2 WASHINGTON—The US Supreme Court struck down a Michigan

A: *7,000 pounds annually*
Source: Greenpeace Action

law that had barred **landfill operators** from accepting solid waste generated outside the landfill's county. The court also ruled that Alabama cannot impose a special fee on hazardous waste brought into the state. —*WSJ 6/2*

3 WASHINGTON—After three years of study, the US Forest Service announced it will direct its foresters to substantially reduce the amount of **timberland harvested** by clearcutting. —*NYT 6/4*

5 CAPE CANAVERAL—NASA engineers said they have been forced to redesign space station shields because "so much **junk is littering space**." The added protection will raise the cost of the space station by millions of dollars, officials said. —*WP 6/25*

6 RIO DE JANEIRO—Former Soviet President Mikhail Gorbachev accepted the chairmanship of a new environmental group, the **International Green Cross**. The group announced its official formation during the Earth Summit. —*AP 6/7*

11 SOUTH CAROLINA—South Carolina banned public display of **dolphins and whales** in theme parks, becoming the first state to protect the marine mammals from exploitation for human entertainment. —*AP 6/11*

15 WASHINGTON—The US, Venezuela and

of donor-financed corn, plus flour, legumes, sorghum, sugar and vegetable oil.

ENDANGERED SPECIES AX

Washington—For the second time in the Endangered Species Act's 19-year history, an exemption has been granted that puts development ahead of the protection of an endangered species—in this case, the spotted owl. The Bush Administration's so-called God Squad, headed by Interior Secretary Manuel Lujan, voted 5 to 2 in May 1992 to waive protection requirements and allow logging on 13 federally owned timber tracts (totaling 1,700 acres) in Oregon.

The endangered bald eagle. Less-photogenic but equally endangered species suffer considerable pressures and neglect.

The Bush Administration is also proposing an amendment to the Endangered Species Act that would allow logging on millions of acres in the Pacific Northwest, where an estimated 3,000 pairs of spotted owls remain. Environmentalists claim that the administration's proposal is a direct attack on the 1973 act.

HUGE FOREST RESERVE APPROVED IN ZAIRE

Zaire—The government of Zaire has ordered the creation of the 5,000-square-mile Okapi Wildlife Reserve. The new reserve (roughly the size of Connecticut) is located within the Ituri Forest in northeastern Zaire. Leopards, elephants, okapi,

Q: *Which country is home to over 34,000 abandoned toxic waste sites?*

hundreds of bird species, ten species of antelope and many other wild animals live in the reserve.

John Hart, a Wildlife Conservation International scientist who has worked in the area for 20 years, says the Ituri Forest has a long history as a wildlife sanctuary. "For over 50 years, traditional Mbuti chiefs protected areas of the Ituri Forest for okapi," Hart says. The Mbuti pygmies, whose lives are described in anthropologist Colin Turnbull's famous book, *The Forest People*, hunt and farm.

The okapi, a giraffe-like creature unknown to the outside world until 1902, is the national symbol of Zaire.

EARTH DAY

Los Angeles—From Budapest to Winnipeg to Ecuador's rainforests, Earth Day 1992 (April 22) was celebrated with concerts, parties and demonstrations. Rainforest Action Network supporters marched in the streets of Los Angeles, while rock musicians and anti-nuclear protesters gathered in the Nevada desert for the 100th Monkey Project to Stop Nuclear Testing.

More than 5,000 Amazon Natives marched to Quito, Ecuador, to protest rainforest destruction. In the Netherlands, more than 200 towns participated in a campaign emphasizing public transportation and bicycles.

At Drexel University in Philadelphia, presidential hopeful Bill Clinton made an Earth Day speech calling for higher fuel-efficiency standards for automobiles, tax policies that would reward conservation and punish "energy wasters," national legislation to

> "The market . . . knows all about prices but nothing about values."
>
> Octavio Paz, 1990 Nobelist in Literature

increase recycling through refundable deposits on bottles and cans, and a ban on oil drilling in the Arctic National Wildlife Refuge.

Clinton said: "I have come to learn something that George Bush and his advisers still don't understand: I've come to reject the false choice between economic growth and environmental protection."

A: **Germany**
Source: Newsweek *magazine*

Mexico reached an agreement to ban the setting of nets around schools of tuna that include **dolphins**. Tens of thousands of dolphins are tangled in tuna nets each year. —*NYT 6/16*

16 WASHINGTON— President George Bush and Russian President Boris Yeltsin agreed today to reduce their nations' stocks of long-range **missile warheads** and to eliminate all multiple-warhead missiles based on land. —*NYT 6/17*

17 SHOREHAM, NY— The Nuclear Regulatory Commission has approved plans to dismantle the Long Island, New York, **Shoreham nuclear** power plant. The $5.5 billion plant generated electricity for approximately 30 hours. The Shoreham plant is 60 miles east of Manhattan, and local and state governments felt emergency evacuation plans were unworkable. —*WP 6/18*

19 WASHINGTON— The Supreme Court ruled that the federal government "may not conscript state governments as its agents" and held that a law aimed at making states responsible for **radioactive waste** generated within their borders is unconstitutional. New York Governor Mario Cuomo said the decision will put more pressure on companies that generate nuclear waste to develop disposal sites or to face continued storage problems. —*NYT 6/19*

26 WASHINGTON—The House Appropriations Committee voted to end funding for a so-called streamlined **cattle inspection** method that places more responsibility on meatpackers' employees. Consumer advocates say the inspection method has resulted in impure beef products, endangering public health. —*WSJ 6/26*

26 MOSCOW—*Izvestia* reported that Soviet scientists detonated a **nuclear bomb** next to a Ukrainian coal mine in 1979, then sent thousands of miners back to the mine one day later, without informing them of the explosion. The nuclear bomb was detonated in hopes of clearing the mine of dangerous methane gas. —*AP 6/28*

July 1992

4 OMAHA, NE—An alert was called at the Ft. Calhoun Station nuclear power plant, where thousands of gallons of radioactive **reactor coolant spilled** onto the floor of a containment building. On the same day as the Ft. Calhoun Station accident, the Peach Bottom nuclear power plant in Delta, Pennsylvania, shut down automatically after a transformer exploded at a nearby electrical substation. —*AP 7/5*

8 WASHINGTON—A federal court of appeals ruled that the EPA must remove from the market any **pesticides** that have

$151 BILLION FOR TRANSIT

Washington—President Bush signed a $151 billion transportation bill in December 1991, calling it "the most significant revolution in American transportation history." Representative Bud Shuster (Rep.-Pennsylvania) said, "We will look back on this bill as the beginning of integration of transportation," and explained that the new system will coordinate every form of US transportation—roads, rail, air and pedestrian and bicycle paths. Jessica Matthews, vice president of World Resources Institute, said the bill will "seat dozens of new decision-makers at tables once reserved for highway engineers. Air quality, oil needs, congestion and rational land use will all be determined by the outcome."

Bush emphasized that the bill will fund as many as 600,000 jobs in construction and transportation industry during the coming year. California will receive the highest share of funds, $10.5 billion, with money

"This is not the sort of America I envisioned, Tom—an America in which the middle class has to use public transportation."

Q: *How much crankcase oil pollutes America's waterways each year?*

earmarked for highways, Bay Area Rapid Transit in San Fransisco and Metro Rail systems in Los Angeles.

FOOD LABELING

Washington—On November 6, 1991, the Food and Drug Administration and the Department of Agriculture announced the most sweeping set of food labeling rules in US history. In what is known as the "truth-in-food initiative," producers must now meet specific criteria before passing off their products as "Low Fat," "Fresh," "Light" or even "Lite."

Also relegated to stricter definition is the term "serving size." In the past, producers have been prone to manipulate the serving size used on the label in order to portray nutritional information in the best light, which can be misleading to the consumer. Certain manufacturers of soda pop, for example, have labeled products "low sodium," though the serving size used in reaching this designation represents only half the can.

The 2,000-page reform plan is under public review and subject to revision before going into effect sometime in 1993. In November 1992, Secretary of Agriculture Edward R. Madigan blocked publication of new food ruling regulations. Senator Howard Metzenbaum (Dem.-Ohio), who helped write the labeling law, blamed the meat industry for the impasse. Metzenbaum said that "at the behest of the meat industry," the Bush Administration was trying to block the regulation requiring "full disclosure about the fat content of food."

MOBY DICK, THE SEQUEL

Scotland—The International Whaling Commission (IWC) Conference, held in Scotland in June 1992, resulted in the crumbling of a six-year-old international moratorium on commercial whaling.

During the conference, Norway announced that it will resume hunting minke whales in the summer of 1993. Norway claims the hunt is scientific research to assess the potential negative impact live minke whales may have on fish stocks in Norwegian waters. Both Norway and Iceland agree that whales and seals are eating too much fish and should be culled as part of a management program to protect fish stocks.

Japan claims that its harvest of 300 minke whales

the potential to both cause cancer and leave residues in processed foods. The court directed the EPA to begin action to remove such residues from thousands of products. —*NYT 7/9*

9 ATLANTA—A three-judge panel in Atlanta threw out Labor department **workplace-exposure** limits for 428 toxic chemicals, saying the approach used by the department is flawed. The court decision is expected to lend impetus to labor union efforts to enact new legislation to set limits. —*WSJ 7/9*

10 WASHINGTON—A federal appeals court upheld a permanent injunction and unanimously ruled that the government cannot ship **nuclear waste** to an underground storage facility near Carlsbad, New Mexico. —*AP 7/11*

13 JAPAN—Industrial emissions blowing in from China have caused severe **acid rain** throughout Japan. Scientists found acidity in snow almost equal to that in orange juice, in a study earlier this year. —*WSJ 7/13*

23 DENVER—The Colorado Court of Appeals upheld benefits in the cancer death of an employee at Rocky Flats nuclear weapons plant. A witness for operator Dow Chemical claimed the victim's **radiation exposure** was equal to that of a person who "worked at Dairy Queen." —*AP 7/24*

2.1 million tons annually
Source: 50 Simple Things You Can Do to Save the Earth, *EarthWorks Press*

27 SALINAS, CA—United Farm Workers Union leader Cesar Chavez led about 2,000 **farm workers** and supporters on a march to protest low wages for the workers. "It's time to fight back," Chavez said. —*AP 7/27*

30 NEW YORK—Toxic algae that may be responsible for the deaths of millions of fish are spreading rapidly, scientists said. The **"phantom" algae,** which have not yet been named, apparently use straw-like tubes to suck up flecks of fish tissue. —*NYT 7/30*

August 1992

1 WASHINGTON—James Overbay, deputy chief of the Forest Service, announced that logging in national forests where the **spotted owl** lives will remain at a standstill for at least two years, as the Forest Service revises its plan to protect the bird. —*NYT 8/2*

3 PETROLIA, CA—Scientists say the April 25 California **earthquake** jacked up a 15-mile stretch of California beach by as much as four feet. Biologists estimate it will take three to five years for all tide-pool and intertidal organisms to return to their newly shifted habitats. —*LAT 8/3*

3 WASHINGTON—The Senate approved a nine-month moratorium on **nuclear testing** and a permanent halt to testing in 1996. President Bush

annually is also scientific research, although the meat is sold in Tokyo markets. Japan did not confirm whether it would side with Norway and continue to ignore the moratorium.

Since 1982, the US and several other nations have worked to establish a Revised Management System to regulate kill quotas for various whale species. Minke whales have been protected since 1985, due to a significant population decline resulting from commercial whaling. The whaling commission estimates that there are 86,000 minkes in the Northern Hemisphere and 600,000 in the Southern Hemisphere. But little is known about minke whales—not even where they mate or calve.

In September 1992, Norway, Iceland, Greenland and the Faeroe Islands formed the North Atlantic Marine Mammals Commission. The commission will set new guidelines for hunting small whales and seals in the North Atlantic.

Japan, Norway and Iceland have repeatedly broken International Whaling Commission regulations protecting several endangered whale species. The IWC has

---

"We did it for duty, honor and country."

Ted Kleisner, president of Greenbriar Resort, site of the $14 million nuclear bomb shelter for Congresspersons

---

no enforcement or penalty provisions. However, the Burger King and Long John Silver Restaurant chains have cancelled fish exportation contracts with Norway in protest of the resumed hunt.

BIOSPHERE 2 DO

Oracle, AZ—Despite controversy, Biosphere 2, the 3.15-acre glass dome that is home to eight people and 3,000 species of plants and animals, celebrated its first year of enclosed operation on September 26, 1992. The Biospherians—four men and four women—will continue to raise their own food and to study the sustainability of plant, animal and human life inside their shimmering glass house for a second full year.

In July 1992, a scientific panel headed by Dr. Thomas Lovejoy of the Smithsonian Institution recommended that Biosphere 2 hire a director of science, to ensure scientific credibility for the

---

 In 1989, how many pounds of hazardous waste were burned in more than 1,100 American incinerators, kilns and industrial boilers?

YEAR IN REVIEW

Biosphere 2's synergistic collection of plants, animals and climates enters its second year sealed from Bio 1, the Earth.

Biospherians' research in their artificial ocean, desert, rainforest and other biomes. "This is in condition to run for maybe 100 years, so there is plenty of time to do some really good stuff," Dr. Lovejoy said.

Biosphere 2, privately funded by Space Biosphere Ventures and Texas billionaire Ed Bass, is a $150-million miniature biosphere meant to replicate Biosphere 1 (Earth). Internationally known scientist Carl Sagan compares its scientific relevance to Disney World's, and others criticize a proposed golf course and new hotel, gift shop and cafe for the thousands of tourists who pay $11.95 each to peer into the glass "ark."

Project leaders say they are learning from their past mistakes—including the public relations nightmare that occurred soon after enclosure, when the *Village Voice* reported the "secret" installation of a CO_2 scrubber at Biosphere 2. Project directors deny that they tried to cover up the scrubber's existence. Margret Augustine, the project's chief administrator, admitted they have made "10,000 mistakes," but said the first year of any long-term project (and Biosphere 2 is expected to operate for a century) is bound to be a shakedown period.

"I feel as though I am far out in space," Biospherian Dr. Roy Walford told a reporter. Walford and the other Biospherians are not due back to our Biosphere until September 1993.

has threatened to veto the bill, saying the US needs to continue nuclear testing "to determine the safety" of the weapons. —*NYT 8/4*

5 EVANS CREEK, OR— Almost 9,000 firefighters battled flames in Oregon, California, Washington, Idaho and Nevada, while **fires** were contained in Utah and Colorado. More than 200,000 acres were in flames across the West this week. — *NYT (AP) 8/5*

6 WASHINGTON— States routinely test **milk** for only four out of a possible 82 drugs that are used on milk cows, the General Accounting Office reported. —*WSJ 8/6*

6 OSLO—The former Soviet Union dumped 12 ships' **nuclear reactors** in the Barents Sea, Norway reported. At least some of the reactors (from old submarines and icebreakers) contained radioactive fuel, Norway said. —*WP 8/6*

6 WASHINGTON—In a blow to environmentalists, the Senate agreed to limit the public's ability to block **sales of timber** from national forests. The Senate agreement was not as restrictive, however, as a Bush Administration proposal that would close off all appeals to the US Forest Service except for court actions. —*AP 8/7*

6 WASHINGTON—The Senate defeated a proposal that would have allowed **logging** in

A: **7.6 billion pounds**
Source: Greenpeace

Northwest forests by waiving the Endangered Species Act protection of the spotted owl. —*LAT 8/7*

7 COLORADO SPRINGS, CO—Interior Secretary Manuel Lujan said the Endangered Species Act is too stringent, and warned that the potential for conflict grows as more plant and animal species are listed. He called the spotted owl debate "a big headache." —*AP 8/8*

8 MOSCOW—Sixty people died in Russia and the Ukraine after eating possibly mutant, toxic mushrooms. Scientists were uncertain how the mushrooms, which had long been considered edible, suddenly became poisonous. More than 190 cases of **mushroom poisoning** have been reported in the vicinity of a nuclear power plant 350 miles south of Moscow, authorities said, but they were unsure whether the proximity to the nuclear plant is relevant. —*LAT 8/9*

12 WASHINGTON—The US, Mexico and Canada unveiled the **North American Free Trade Agreement,** a three-nation pact that will merge 360 million consumers into a $6 trillion market spanning the North American continent. Environmentalists warned that unless Mexico is required to improve its enforcement of environmental laws, US and Canadian companies may flee to

TRADING IN ENDANGERED SPECIES

Kyoto, Japan—The Eighth Convention on the International Trade of Endangered Species (CITES) met in Kyoto, Japan, in March 1992. The 112 CITES member-nations voted on the fate of endangered and threatened species worldwide, including the African elephant, the North American black bear, and the Western Atlantic bluefin tuna.

> *"We have to start realizing that* we *have the power, not them."*
>
> Ben Cohen, founder, Ben & Jerry's Ice Cream

Five Southern African nations (Zimbabwe, Botswana, Malawi, South Africa and Namibia) fought to lift the ban on elephant ivory trade, but the ban was upheld for another two years.

Denmark successfully pushed through an initiative to uplist the North American black bear from Appendix 3 (wherein member nations practice self-regulation) to Appendix 2 (wherein limited trade is permitted under specific controls). The US and Canada (where the bears range) refused to support Denmark's proposal, claiming that the number of black bears in each country remains stable, despite worldwide trade in bear gall bladders and paws. (One serving of bear paw soup sells for $800 in Asia, where it is believed to cure ailments.) The Danish proposal was first defeated, then passed.

The Swedish delegation moved to prohibit all trade in Western Atlantic bluefin tuna, which has been overfished in many areas. But the US, Canada, Japan and Mexico fought to preserve their fishing rights, and the Swedish proposal failed.

Forty-five new species were added to the CITES appendices at the conference, and 35 species were uplisted.

VOTERS GO GREEN IN 1992

Little Rock, AR—Borrowing a phrase from cartoon characters Ren and Stimpy, "Happy, happy, joy, joy!" was the cry of environmentalists across the US on election night, November 3, 1992. Both President-elect Bill Clinton and Vice President-elect Al Gore

: *How many pounds of grain are needed to produce one pound of pork in the US?*

YEAR IN REVIEW

Newly-elected President Bill Clinton and Vice President Al Gore wave to supporters on election night.

reiterated their commitment to "the environment" during their victory speeches in Little Rock, Arkansas, on election night. Democrats' victory was also a victory for environmental groups (including the Sierra Club, who enthusiastically endorsed Clinton and Gore) and for the growing percentage of American voters (22 percent, in June) who told pollsters they choose candidates based on environmental stances. "After 12 dark years of environmental ignorance and broken promices in the White House, America is ready for a new era of environmental enlightenment," said Sierra Club President H. Anthony Ruckel.

Mexico to avoid more stringent requirements. —*LAT 8/12*

12 NEW YORK—*The Wall Street Journal* said in an editorial that the 59 percent of Americans who told pollsters they would choose **"environment over growth"** should "stop whining and find a way to enjoy the high-cost environment they said they're so happy to pay for." —*WSJ 8/12*

13 WASHINGTON—The Environmental Protection Agency released new **worker protection regulations** that will affect 4 million Americans. The new rules apply to greenhouse, forest, nursery and farm workers who handle pesticides. Under new regulations, workers must have a place to wash, access to emergency medical care, instruction on using pesticides, and protective equipment such as gloves and goggles. —*WSJ 8/14*

14 WASHINGTON—Researchers at the University of North Carolina believe they can produce a plastic-like material that can **replace most plastics,** and that will cause much less water and air pollution during production. Du Pont and 3M are providing corporate backing. —*WSJ 8/14*

15 FAIRBANKS, AK—CIA Director Robert M. Gates said that many years and billions of dollars will be needed to

A: *6.9 pounds*
Source: Worldwatch *magazine*

The Worst and Best of 1992

IN 1992, THE WORLD'S ENVIRONMENT CHANGED at an alarming pace. Environmental stories—some good, some bad—hit the front pages of newspapers across the country. In an effort to assess the most significant environmental events of the year, *Earth Journal* independently polled 19 environmental and special interest groups, together known as the Green Group. The CEOs of each group meet roughly once a year to discuss issues that impact them. The participating groups believe the Green Group alliance makes them more effective and gives them additional lobbying power.

Sixteen of the 19 groups responded to our questionnaire; (the other three groups pleaded too busy). Their responses are indicative of the issues that are most important to each group: The Environmental Defense Fund, Natural Resources Defense Council and the Sierra Club Legal Defense Fund tended to respond about environmental regulations; wildlife groups, such as Defenders of Wildlife, National Audubon Society and the National Wildlife Federation, felt more strongly about issues that affected wildlife; and the Native American Rights Fund offered a focus clearly missing from the other groups. Due to the timing of the survey, respondents could not include results of the presidential election, since endorsement would threaten the groups' tax-exempt status.

Listed below are the members of the Green Group. For mission statements and contact information see the "Directory to Environmental Groups," page 344. The results of the survey are rated by the average order of significance given to the issue by the respondents. Following the top ten issues are the many other issues the groups felt were significant, listed in the order they were rated.

*Children's Defense Fund
Defenders of Wildlife
Environmental Defense Fund
Friends of the Earth
Izaak Walton League of America
National Audubon Society
National Parks and Conservation Association

---

clean up **radioactive pollution** in the former Soviet Union. He cited Lake Karachy, where radiation levels are so high that one hour's exposure at the shoreline can be fatal.
—*WP 8/17*

16 NEW YORK—The 16-year-old **Trojan nuclear power plant** in Rainier, Oregon, announced plans to close, making it the third US nuclear plant to announce closure this year. After Trojan closes, 109 commercial nuclear power plants will remain in the US, including 2 that are older than 25 years; 12 that are 20 to 25 years old; and 35 that are 15 to 20 years old.
—*NYT 8/16*

17 LOS ANGELES—Stock analysts said cleanup contracts awarded to Fluor Corp. and Jacobs Engineering to do $2.2 billion worth of cleanup on the Fernald, Ohio, **nuclear waste site** offer a return on investment "that is just incredible." Analysts believe the Department of Energy may award over $100 billion in nuclear cleanup contracts over the next decade.
—*LAT 8/17*

18 WASHINGTON—Scientists are testing a **contraceptive vaccine** on 30 white-tailed does at the National Zoo. They hope the new method can be used to control burgeoning deer populations throughout the Northeast.
—*WP 8/18*

 What main ingredient necessary for the treatment of childhood leukemia is found solely in the Madagascar rainforest?

YEAR IN REVIEW

*National Toxics Campaign
National Wildlife Federation
Native American Rights Fund
Natural Resources Defense Council
Planned Parenthood
Population Crisis Committee
Sierra Club
Sierra Club Legal Defense Fund
Union of Concerned Scientists
Wilderness Society
*World Wildlife Fund
Zero Population Growth

*Did not respond to questionnaire.

WORST ENVIRONMENTAL DISASTERS OF 1992

1. **US performance at the Earth Summit**—President Bush failed to step forward and take a leadership role. The US refused to sign the biodiversity treaty and was generally obstructive to the Earth Summit process of cooperation.

2. **"God Squad" decision**—An exemption to the Endangered Species Act allows logging in certain ancient forests in the Northwest that provide habitat for the threatened northern spotted owl.

3. **Citizen action**—The US Supreme Court ruling in *Lujan* v. *Defenders* shuts doors on many environmental lawsuits. There have been efforts to limit citizens' rights to appeal government decisions involving public lands management.

4. **Wetlands regulations**—The Bush Administration released a wetland delineation manual that would allow development of 50 percent of the nation's valuable wetlands, including parts of the Florida Everglades. The effort was spearheaded by Vice President Dan Quayle and his Council on Competitiveness.

5. **Further ozone depletion**—The ozone layer continued to decline. Holes are now developing over parts of the US and the entire South Pole region.

6. **Clean Air Act**—Vice President Dan Quayle and his Council on Competitiveness have created massive

18 ANCHORAGE—The Mt. Spurr **volcano erupted** in Alaska, showering light ash and steam. The 11,100-foot volcano is across Cook Inlet from Anchorage. —*LAT 8/19*

21 LOS ANGELES—After two years of improvement, the Los Angeles area's air quality diminished this year, exceeding federal health standards for ozone levels on 96 days between January 1 and August 19. First-stage **smog alerts** were administered 33 times during that period, compared with 21 times in the same period in 1991. The higher rate was nonetheless an improvement over the 1970s, when first-stage alerts were issued for 121 days in 1977 and 120 days in 1979. —*LAT 8/21*

21 NEW YORK—Scientists reported increasing evidence that **speeding objects from space** crashed into Earth 370 million and 65 million years ago, causing global climate change and mass extinctions. Scientists theorized that the death of the dinosaurs may be blamed on an asteroid crashing into Earth and releasing huge quantities of dust and carbon dioxide, forming a greenhouse effect that wiped out plant and animal life. —*NYT 8/21*

21 WASHINGTON—The US State Department says it supports Japan's plan to **ship weapons-grade**

: *The rosy periwinkle flower*

Source: Protect Our Planet Calendar, Running Press Book Publishers

plutonium to France and Britain, where the plutonium will be reprocessed for nuclear power plants. In 1988, the US and Japan signed a nuclear cooperation agreement establishing guidelines for international transportation of plutonium.
—*WP 8/22*

22 CHACOCENTE, Nicaragua—Nicaraguan soldiers are guarding thousands of **olive ridley sea turtles** now arriving on the nation's beaches to lay eggs. The soldiers' presence deters Nicaraguans from gathering the eggs, which sell for about 25 cents each. In the past, entrepreneurs have smuggled eggs to shampoo factories, which use the turtle eggs as a "protein rich" ingredient.
—*AP 8/22*

22 MOSCOW—Russia's chief prosecutor says in a new book that President Mikhail Gorbachev lost control of the **"nuclear suitcase"** containing codes for Soviet nuclear weapons on the afternoon before the August 1991 attempted coup d'état. The Soviet military could have triggered a reactive strike without the president's approval, Valentin Stepankov says.
—*NYT 8/23*

26 WASHINGTON—The Bush Administration asked a US appeals court to lift a judge's **ban on timber harvesting** in national forests where spotted owls live. The Agriculture department

loopholes for industries to undermine the amendments of the Clean Air Act.

7. **Rise of the Wise Use movement**—Anti-environmentalists are organizing. The so-called Wise Use advocates believe all public lands should be used for profit, with no restrictions on logging, mining, recreation, etc.

8. **Vice President Dan Quayle and the Council on Competitiveness**—An extraordinarily influential group that exists to review regulatory burdens on American business. The group has successfully intervened to weaken important environmental regulations.

9. **World population**—The population of the world has grown by another 100 million people in the last year. Population planning efforts are still inadequate.

10. **Loss of biodiversity**—Another 50,000 invertebrate species are going extinct, and among the larger species, such as birds and mammals, at least one a day is becoming extinct, according to estimates from the world's leading wildlife biologists. The loss represents 1 percent to 2 percent of the Earth's species over the past year.

11. Bush Administration issues moratorium on all environmental regulations.
12. Clarence Thomas appointed to the Supreme Court.
13. Exxon pays less than one-tenth of actual Alaska oil spill cost.
14. The recession.
15. Development by the Bush Administration of the North American Free Trade Agreement is devoid of environmental safeguards.
16. Executive branch, Congress and courts all cut public out of the process of nuclear licensing once a new nuclear power plant is constructed. Congress expedites licensing procedures for nuclear power plants.
17. Bluefin tuna trade ban fails at CITES meeting.
18. Mercury contamination throughout East Coast of US.
19. Chlorine found more concentrated than expected in Northern Hemisphere.
20. US Supreme Court rules that federal agencies do

: *By switching from incandescent to compact fluorescent lighting, how much would Americans save on utility bills in one year?*

not have to pay fines for violations of federal pollution laws.
21. *Dzil nchaa sí ia* (Mt. Graham)—Construction underway of a telescope on Mt. Graham in Arizona—an Apache sacred site and home of the endangered red squirrel.
22. National energy legislation still contains many bad provisions (i.e., no way to deal with high oil consumption after the Persian Gulf War, and aid to nuclear industry).
23. Red Butte, Arizona—The US Forest Service issues uranium mining leases on a Havasupai Tribe sacred site near the Grand Canyon.
24. Scientists report oceans are undergoing explosion of algae blooms, which are likened to toxic red tides. Scientists believe this could be a front-runner to a massive ecological breakdown.
25. Congress' failure to increase auto fuel-efficiency standards.
26. Vatican succeeds in striking contraception from Earth Summit "Agenda 21."
27. World Health Organization estimates 10 million AIDS cases worldwide.
28. Businesses that do not care about and are careless with the only life-sustaining environment we have.
29. Refusal of the House and Senate to pass legislation strengthening federal standards for hazardous waste.
30. Woodruff Butte, Arizona—A private landowner is razing sacred Hopi and Zuni shrines for a gravel mining operation.
31. Nearly every Department of Defense site is severely contaminated with toxic and radioactive waste.
32. Uptick in US fertility and population growth.
33. Canada resumes harp seal hunting.
34. Environmental racism—Distribution of pollution continues to be heaviest in areas with a majority of people of color.
35. Failure of EPA to fund Native American tribes for development of environmental programs necessary to ensure health and integrity of their people.
36. Japan announces it will continue to ban birth control pills.
37. 36-inch fuel pipeline ruptures in South Carolina,

filed the request for harvesting in Oregon, Washington and California. —*AP 8/27*

27 ATLANTA—Federal Centers for Disease Control officials said they will begin a national tracking system to measure possibly dangerous **lead levels in children's blood.** Spokeswoman Suzanne Binder said: "We know that throughout the country, children of all races and ethnicities and income levels are being affected by lead." —*WP 8/28*

September 1992

1 WASHINGTON—After 20 years of negotiations, a treaty that would ban the production, use and stockpiling of **chemical weapons** is ready to be endorsed by the United Nations. The UN General Assembly is expected to pass a resolution endorsing the treaty, and the accord will go into effect in about two years. —*NYT 9/2*

3 WASHINGTON—The Interior department issued rules giving the **National Park Service**—rather than concessionaires—the responsibility for defining which services are needed in each park. —*WSJ 9/3*

3 HARTFORD, CT—The largest fine ever assessed under the **federal Clean Water Act** has been levied on the Dexter Corp., a paper products manufacturer. Dexter was fined $4 million in criminal penalties

A: *Approximately $11 billion*
Source: Time *magazine*

45

and $9 million in civil penalties for discharging hazardous waste into the Connecticut River. —AP 9/4

9 SAN FRANCISCO— President Bush asked federal agencies to scale back requirements for environmental review of **"salvage" logging operations**. The Bush Administration said the new policy would reduce forest fire danger in the West and increase the amount of timber that can be logged on federal lands. —LAT 9/10

9 LOS ANGELES— DDT, the deadly toxin that has been banned in the US for 20 years, is still showing up in Southern California marine life (including the white croaker fish, which is harvested and eaten). California's Palos Verdes Peninsula is home to the nation's largest **ocean graveyard of DDT,** covering almost 20 square miles on the ocean floor and containing an estimated 200 tons of the pesticide. —LAT 9/9

14 BOMBAY— Authorities believe that more than one-third of **Bombay's 100,000 prostitutes** (who often charge customers as little as 50 cents) have been infected with the HIV virus that causes AIDS. India's efforts to educate citizens about AIDS protection are hampered by cultural taboos against public discussion of sex, as well as by widespread poverty, which forces people into prostitution

spilling 400,000 gallons of fuel into nearby waterways.
38. News of coral bleaching throughout the world.
39. Montana National Forest Management Act.
40. EPA proposes exemptions of certain hazardous wastes from further regulation.
41. Renewed consideration by the Chinese government of the Three Gorges Dam on the Yangtze River.
42. The Bush Administration's "Revised" Yellowstone Vision Document.
43. Preparations for a new ozone treaty. Not making good progress, particularly with methyl bromide.
44. Pat Williams of Montana introduces Drinking Water Regulations Exemption Bill in Congress— affects 90 percent of drinking water in US.
45. At least three-fourths of all deaths worldwide are caused by diseases related to the environment, according to a report released by the World Health Organization.
46. Norway and Iceland plan to resume whale hunting.
47. President Bush takes credit for the agreement on Grand Canyon air quality.
48. Muzzling of government whistle blowers.
49. Campaign in West to punch roads through national parks and other public lands.
50. Food and Drug Administration relaxes regulations for genetically altered foods, putting consumers at the disadvantage of not even knowing if foods are genetically engineered and disregarding potential risks of genetically altered organisms.
51. Persian Gulf oil pollution in air, water and land still not cleaned up, and little is being done about it.
52. Greek tanker spills 2.9 million gallons of crude oil into Indian Ocean.
53. China sets off largest nuclear test ever, 70 times the explosive power of the atomic bomb dropped on Hiroshima.
54. House votes to allow hunting in Mojave Desert.
55. Propane explosion in a chicken factory in North Carolina kills employees trapped in a locked building.
56. The current GATT draft would eliminate many environmental protections.
57. Explosion of underground gas lines in Guadalajara kills several and injures hundreds.
58. Bottling up of California Desert Protection Act.

Q: *Due to technical and financial limitations, what percentage of plastics are currently recycled?*

BEST ENVIRONMENTAL SUCCESS STORIES OF 1992

1. **Earth Summit (despite US performance)**—The Earth Summit in Rio de Janeiro successfully focused world attention on global environmental problems. The summit linked the development needs of the South and the future of the global environment.

2. **ANWR removed from energy bill**—The energy bill reintroduced to the Senate in February 1992 did not include provisions for oil and gas drilling in the Arctic National Wildlife Refuge.

3. **Al Gore**—Bill Clinton's selection of Senator Al Gore as his running mate. Gore is author of *Earth in the Balance*, a best-selling book about the environment.

4. **Endangered Species Act success**—Two endangered species in captive breeding programs were reintroduced into the wild: the black-footed ferret and the California condor.

5. **Ban on Clearcutting**—California's Board of Forestry approved a ban on clearcutting in ancient forests and restricted other logging activities. Court ruling created an injunction on logging sales until northern spotted owl recovery plans are in place.

6. **Cleaner air at the Grand Canyon**—The air is cleaner around the Grand Canyon following a September 1991 ruling to reduce sulfur dioxide emissions by 90 percent at the Navajo Generating Station 16 miles north of the canyon.

7. **UN bans driftnet fishing**—The United Nations called for an end to global driftnet fishing by the end of 1992. Driftnet fishing has been responsible for indiscriminate killing of countless sea creatures.

8. **Biodiversity convention agreed to**—Increasing awareness of biodiversity and its importance and signing of treaty at Rio. Some 154 nations sign biodiversity treaty.

9. **Everglades protection**—Reconfiguration of water releases around Everglades in order to restore wildlife.

or into selling their blood to survive. —*WP 9/14*

14 LONDON—Scientists say the amount of **ultraviolet radiation** striking the ground in New Zealand is twice that striking Germany, although the countries lie at comparable latitudes in their respective hemispheres. Researchers theorize that the seasonal ozone depletion over Antarctica drains upper-atmosphere ozone over New Zealand, and that Germany's intense air pollution absorbs ultraviolet radiation before it can strike the ground. —*WP 9/14*

16 WASHINGTON—Military and health care officials told Congress they have found no link between Kuwaiti oil fires set during the **Persian Gulf War** and the skin rashes, liver disorders, elevated blood pressure and other symptoms appearing in more than 300 veterans of the war. The panel said there is "no common syndrome that can be identified." —*WP 9/17*

16 NEW YORK—Astronomers have detected a reddish, glowing object about 120 miles in diameter, located in the Kuiper Belt beyond Neptune and Pluto. Scientists believe it may be a small planet; they've named it **"1992 QB-1."** An astronomer who spotted the object said, "We were delirious with joy." —*NYT 9/16*

18 WASHINGTON—The Department of

A: *Less than 10 percent*
Source: Greenpeace Action

Agriculture approved the use of **radiation** to control harmful bacteria in poultry. The National Broiler Council, which represents chicken producers, said, "Normal cooking kills the same harmful microorganisms." A spokesperson for Perdue Farms commented: "We are not going to use the system until the consumer supports it." —WP (R) 9/19

18 MIDDLETOWN, PA—A huge steam cloud containing trace amounts of radiation was released while operators were making adjustments at the **Three Mile Island** nuclear power plant. A spokesperson said, "We can't really measure it, but we know it's there," and described the radiation as "insignificant." —AP 9/19

19 GOLDEN, CO—A team of independent experts assembled by the US Department of Energy will assess the **Rocky Flats nuclear weapons plant** and determine what must be done before the plant can switch from a weapons facility to a plutonium storage and shipment facility. —AP 9/19

22 GLAND, Switzerland—Peter Jackson of the World Conservation Union says poachers in India are killing **wild tigers** at an alarming rate (20 in one park last year) to supply the Chinese demand for tiger bone wine. The "tiger wine" is believed to cure ulcer, malaria,

10. **Nuclear weapons reductions**—Arms agreements between the US and CIS reduce risk of nuclear war by reducing number of warheads. The Strategic Arms Reduction Treaty is scheduled to be ratified by the end of 1992.

11. Resignation of John Sununu as White House chief of staff.
12. Nations signed global climate treaty at Rio.
13. Approval of increased funding for military base cleanup.
14. International ban on ivory trade is upheld by CITES.
15. Kuwait oil well fires were extinguished faster than originally anticipated.
16. House passes the California Desert Protection Act, the largest land acquisition legislation since 1908.
17. House votes to take away funding from the Council on Competitiveness.
18. New Clean Air Act amendments to massively reduce air pollution, with the potential to save million of dollars.
19. Passage of national energy legislation.
20. The millions of Americans who identify themselves as environmentalists.
21. Kazakhstan and US reached agreement to eliminate 104 long-range Soviet missiles from Kazakhstan. Kazakhstan will become nonnuclear within 10 years.
22. Organizations who make up the Green Group and the millions of Americans who are members.
23. First National Indian Environmental Summit, November 18, 1991.
24. Demise of Animas-La Plata dam project in Colorado.
25. The ineffectiveness of Bush/Quayle and how they couldn't destroy the environment no matter how much they tried.
26. Stepped-up environmental education in US schools.
27. Delay (at least) of Hawaii geothermal project.
28. Businesses that recognize the importance of environmentalism and support both corporate and civic involvement.
29. Waste industry strategy—Tribes reject attempt by waste industry to use Native reservations as dumps for toxic waste to avoid state regulations.

Q: *What percentage of food bought at the supermarket by Americans is thrown away?*

YEAR IN REVIEW

30. Issuance of Integrated Resources Planning Framework order for Hawaii utilities.
31. Yankee Rowe nuclear power plant in Rowe, Massachusetts, closed.
32. Japan agrees to stop driftnetting.
33. Senate acknowledges the need for reforms in the International Monetary Fund with the passage of amendment for the Russian aid bill.
34. Growing involvement of people of color in environmental protection.
35. EPA policy that supports exercise of tribal jurisdiction in managing and implementing programs on Native American lands.
36. Joint statement on overpopulation by the National Academy of Sciences and Royal Society of London.
37. Health and Human Service Secretary Louis Sullivan recommends that all children have blood tested for lead by age one.
38. Mt. Pinatubo cools planet for awhile.
39. Extension of requirements for use of turtle excluder devices to inshore waters and year-round, based on findings that the devices are overwhelmingly effective and do not decrease shrimp catches.
40. Environmental racism is identified and acknowledged.
41. House passes a moratorium on nuclear testing.
42. Los Angeles water districts prepare to sign an agreement that would effectively conserve 500,000 acre-feet of water annually.
43. European move to greener technology.
44. New York Governor Mario Cuomo pulls plug on contract with Hydro-Quebec.
45. Depo-Provera (injectable contraceptive) gets nod from US FDA Scientific Advisory Panel.
46. More women political candidates.
47. National Park Service's "Vail Agenda" and Bruce Vento's National Parks and Landmark Conservation Act.
48. InterAmerican Tropical Tuna Commission's 10-nation accord to protect dolphins from tuna nets.
49. Brazil's population growth rate falls below 2 percent for the first time in 50 years.
50. Fisheries conservation in New England, California, Pacific Northwest and elsewhere.
51. Oil industry moves to cleaner auto fuels, raising

rheumatic pain and burns. "The wild tigers [7,000 worldwide] may indeed be doomed," Jackson says.
—*NYT 9/22*

22 MONTEREY, CA— Three hundred fifty miles of California coast have been designated the **Monterey Bay National Marine Sanctuary**. EPA chief William Reilly said during dedication ceremonies: "*Sanctuary* is a word that indicates reverence, respect and permanence." The new sanctuary is home to 22 endangered or threatened species, including sea otters, seven types of whales and four kinds of sea turtles. —*LAT 9/23*

23 WASHINGTON— The Interior department announced that beginning in October, **oil and gas drilling** can proceed while appeals of the agency's decisions are under review. The new rules affect 270 million acres under the aegis of the Bureau of Land Management.
—*WP 9/24*

25 ANCHORAGE—A university researcher has found out that more than 15,000 pounds of soil contaminated with **radioactive fallout** from a Nevada nuclear test were shipped to Cape Thompson, Alaska, in 1962 and buried. Cancer rates in the region have risen steadily over the past 30 years.
—*LAT 9/25*

25 WASHINGTON—An independent, federally appointed panel of

A: *10 percent*
Source: The Recycler's Handbook, *EarthWorks Press*

scientists announced their findings that **dioxin** is not a large-scale cancer threat unless persons are exposed to unusually high levels in factories or in accidents. The panel said exposures to high levels of dioxin seem to increase the risk of lung cancer and sarcoma. —*NYT 9/26*

27 WASHINGTON— The Bush Administration is pushing the Interior department to open millions of acres of national parks and forests to **strip mining for coal**. The new regulations would eliminate Congress-approved protection against mining in national parks and forests, except to those who hold "a valid existing right" to coal. The phrase has been reinterpreted by the courts and by the Bush Administration. —*NYT 9/28*

29 WASHINGTON— NASA announced that satellite measurements show that the **ozone hole over Antarctica** is now almost three times larger than the total area of the US. It is the largest ozone hole on record. —*AP 9/30*

prices slightly.
52. Continuing "residue management" revolution in American farming.
53. Freedom of Choice Act.
54. New Jersey sets guidelines for cleanup of 200 toxic waste sites.
55. Federal transportation legislation encourages people to consider the environmental and energy consequences of transportation choices.
56. The Draft Version of the "Yellowstone Vision Document."
57. Black bear listed for protection under CITES.
58. Increasing environmental concerns among religious leaders.
59. States increase efforts to control smog.
60. PG&E and Southern California Edison announce plans to reduce their emissions of CO_2 by 20 percent over the next 20 years.
61. Signing of treaty to block development in Antarctica.
62. Honda introduces a 50-plus-mpg car.
63. Designation of Monterey Bay National Marine Sanctuary—largest marine sanctuary ever.
64. Federal Trade Commission issues guidelines to assist consumers in determining validity of environmental terms on products, such as "biodegradable," "recycled" or "ozone friendly."
65. Congressman Gerry Studds of Massachusetts introduces Endangered Species Act Amendment of 1992.
66. Dolphin-safe tuna labeling regulations take effect.
67. *Beyond the Limits* published.
68. Eighty-two airlines and two pet store chains agree to stop importing rainforest parrots and tropical birds.
69. Introduction of Environmental Justice Act. First real step to addressing this issue.
70. Removing the California gray whale from endangered species list is proposed.

Q: *How much CO_2 does just one healthy tree remove from the air in one year?*

Eco-Forecasts for 1993

What will be the hot environmental issues in 1993? BUZZWORM *Senior Editor Elizabeth Darby Junkin offers insights on the emerging trends and issues that will command our attention over the next year.*

THE ELECTION OF 1992 had remarkable impact on the environmental issues that will become the big stories of 1993. Perhaps never before was there such a stark contrast between the presidential candidates and each one's approach to environmental solutions. That's not so much a statement about candidates George Bush, Bill Clinton and Ross Perot as it is a reflection of the importance the environment has come to have in American politics, foreign policy and economic welfare. The environmental stories of 1993 will be a clear reflection of the man who is President of the US in 1993.

Economics and Environment

The most dramatic issue of the year, the need for more jobs and a healthy ecosystem, is also the issue that will most reflect the leadership of the US. Following the lead of enlightened business in Rio de Janeiro in June 1992, the connection between a positive economic climate and a healthy environment will be made clearer in 1993. In Europe and Japan, business will seize the opportunity presented by our precarious global environment and begin to plan the 21st century of sustainable economic growth. Whether the US will lead, much less be part of, this international business effort depends on who is leading the nation—the Congress or the President.

With the election of a Bill Clinton/Al Gore Administration, the connection between the economy and the environment will be seen as the proactive business opportunity it is. Initiatives of the Clinton/Gore Administration will concentrate on business incentives. Understanding the mechanisms of environmentally benign growth and sustainable development—as dramatic an attitude shift for our current free market economy as the end of Communism is for the former Eastern-bloc nations—will become a centerpiece of the administration. A healthy ecosystem will be seen as the foundation of a healthy economy—a trend that will lead the way into the 21st century.

Population

Congress will begin legislation that will bring efforts to stabilize world population to the forefront in preparation for the World Population Conference to be hosted by the US in 1994.

With the election of a Clinton/Gore Administration, population will take the main stage, following the lead of Congress. Efforts to provide safe contraceptive technology to developing countries will be supported, as will efforts to place America in the lead of the global population discussion.

Population, the crisis of the multitudes that puts millions at risk after even moderate disruption in crop cycles, will continue to make headlines in 1993. The question is: What will be the caliber of discussion?

Energy

What to do with our decaying nuclear power plants will continue to be a hot issue in 1993. How to meet the nation's energy needs in 2000 will also emerge as one of the year's critical questions.

 Between 25 and 45 pounds, depending on the tree
Source: The Student Environmental Action Guide, *EarthWorks Press*

Nuclear power currently supplies 20 percent of our energy needs, but no new nuclear power plants are planned. We'll hear a lot of rhetoric about making America energy-independent, but the policy steps taken to do so will depend on the philosophy of the leader. Will we gain the willpower to unhook ourselves from the fossil fuel pipelines and enter the 21st century with other sources of energy?

With the election of a Clinton/Gore Administration, perhaps some new oil lands will be opened in the name of a secure oil supply, but only in addition to monetary and political support for energy alternatives that can jump-start American research and development sectors. The White House will increase support for new standards of energy efficiency. Tax incentives would likely make environmentally destructive energy policies more expensive for polluters and producers, giving tax breaks to environmentally beneficial energy sources.

Public Lands—Ending Giveaways and Benign Neglect

The public's lands will be the focus of heated debate during 1993. The public is growing increasingly disgruntled with perceived giveaways of grazing rights, mining lands and timber sales, especially as the nation's treasures—the national park system—begin to fray around the edges from too many people and too little money spent on protection. The year 1993 will see the public leading the charge to raise grazing fees, and to eliminate low-cost mining land leases and below-cost timber sales.

Environmental Law—America's Standards Questioned

The Endangered Species Act, the US gemstone of environmental standard, will be debated even more than it was in 1992. Perceived as getting in the way of jobs, especially in the regions where salmon and the spotted owl reside, the provisions of the Endangered Species Act will face fierce attacks, and its future will depend in large part on the results of the presidential election.

With the election of a Clinton/Gore Administration, the act will probably survive the year in its present form, and become an international standard for protection of endangered species in other nations as part of an international environmental foreign policy. Still, the new president will have to walk carefully through the rhetoric already stirred up over endangered species and perceived loss of jobs, remembering that perception *is* reality when it comes to putting people to work.

Nuke Disposal/Crisis in Europe

Regardless of who enters the White House in 1993, disposal of nuclear weapons components, most notably uranium and plutonium, will continue to be a growing issue in 1993. The international arms trade is lucrative business for those countries severely needing cash, and there are always willing buyers for the most powerful war chest in the world. The year will see more nuclear weapons components showing up in the hands of those who do not have the best interests of Mother Earth at heart—and see little debate about curbing the international arms trade.

International Environmental Policy

The Earth Summit in Rio de Janeiro in 1992 kicked off a new level of international diplomacy over the fate of the Earth. There will be new discussions in 1993, but the role of the US in these negotiations will be

 What was the British government's advice to nuclear plant workers to alleviate their fears of potential birth defects due to radiation exposure?

determined by its leader.

With the election of a Clinton/Gore Administration, the debate will increase at the global environmental diplomacy level. A change of administration would most affect the international ozone and global warming treaties, but it would also offer leadership in the international debt crisis and in environmental protection provisions in trade talks.

—*Elizabeth Darby Junkin*

What happened in 1992

*L*ast year, Earth Journal *predicted the hot topics of 1992. Here's a look at what happened.*

North/South Divide
The Earth Summit in Rio de Janeiro was held in June 1992. More governments than ever before gathered to discuss the fate of our planet from an environmental perspective. The meeting served to highlight the imbalance between the "North" and "South," the developed and underdeveloped countries. During the 13-day conference, an Earth Charter was signed, as were treaties on the Earth's forests and greenhouse gases. Although many hoped that the developed nations would commit to monetary assistance for developing countries, few specifics were forthcoming. Still, the problem of underdevelopment and the resulting destruction of the environment was discussed at the highest diplomatic level ever.

Yet the worst drought in this century, in Zimbabwe and southern Africa, forced Zimbabwe to resort to killing thousands of elephants and impalas to provide food for starving humans and to help other animals survive. War-exacerbated drought in Somalia caused tens of thousands to die from hunger later in the summer of 1992. The developed countries offered aid more slowly than in the 1984-1985 drought in Ethiopia, a disaster of equal caliber. By the end of summer 1992, an estimated one in four children under 5 years of age in Somalia died from starvation. Another face of underdevelopment, AIDS, has swept Africa's teeming population, too, leaving 1.5 million children orphaned in Uganda alone. Another 1.5 million children and adults have the HIV virus.

Pollution and Poverty: Whose Backyard Now?
The trend away from large, more bureaucratic environmental groups to smaller, community-based nongovernmental groups increased in 1992, creating a chorus of voices for environmental change from all quarters. At the Rio conference, some 15,000 participants representing nongovernmental organizations from 165 nations held a parallel forum that truly represented the global spectrum of backyard concerns.

New Environmentalists
A poll commissioned by the Times Mirror Magazines revealed that 64 percent of American adults interviewed said environmental protection was more important than economic gain. The survey, conducted by the Roper Organization, found that the environment remains a primary concern of Americans, even when environmentalism conflicts with economic growth. In another Roper poll, four out of five

A: *"Don't have children"*
Source: Greenpeace

Americans feel that the nation should "make a major effort to improve the quality of the environment." The American Association of Retired Persons, representing some 33 million Americans over age 50, has thrown its support behind efforts to protect the environment, noting that pollution increases could adversely affect aging Americans. The group also notes that its elderly constituents are more politically active than other age groups and more willing to involve themselves in strategies to preserve the natural habitat of the nation.

Nuclear Hot Cakes: Proliferation and Cleanup

Earthquakes in California and southern Nevada caused concern about the safety of nuclear waste that will be buried on Yucca Mountain, Nevada. It is tentatively scheduled to open in 2010 and will hold the nation's most radioactive wastes—supposedly isolated for tens of thousands of years.

New arms reduction agreements call for nuclear warheads to be disabled, but not to be destroyed. The components will be dismantled and the nuclear cores either stored or recycled into new weapons, raising the specter of black marketeers obtaining the material and selling it to willing nations or, in the case of some of the more volatile countries of the CIS, the weapons being reassembled and used as blackmail. Nothing in the arms agreements prevents the materials from being recycled into newer weapons for the signing parties, either. A former Reagan Administration official says the situation "could pose greater dangers" than the Cold War did.

Deteriorating nuclear plants in the former Eastern Bloc and Russia also pose hazards. Some 60 plants are believed to be operating without the safeguards required in most Western countries, among them 15 plants with the same design as Chernobyl.

The first criminal prosecution of a company that made nuclear weapons for the US government ended with the company, Rockwell International, pleading guilty to five felonies and five misdemeanors for mishandling poisonous waste at the Rocky Flats nuclear bomb plant near Denver. The company agreed to pay $18.5 million in fines.

Packaging

Although some 4,000 communities have curbside recycling programs, the stuff is piling up due to lack of markets to reuse it. The stacks are made worse by a glitch in the system that leaves industries that might be interested in using recycled materials in production not sure if the materials they need will always be available. One step forward on this front in 1992 was a proposal by the Senate Committee on Environmental and Public Works that would create federally mandated recycling targets for packaging and paper products to stimulate recycled material use.

Water

After six years, drought came to an end in California in 1992, but the reprioritizing of water allocation began. The lockhold California agriculture maintains on cheap water for water-intensive plants like alfalfa and cotton was eased during 1992, giving cities easier access to needed water. But providing more water to cities will likely accelerate the rapid urbanization of California lands that produce half the nation's fruits and vegetables. Plans were proposed for a new dam on California's American

 In what year did curbside recycling originate?

River for flood control, but more importantly for water supply and hydropower. The project threatens scarce remaining habitat for wildlife and whitewater enthusiasts.

A study released in 1992 revealed that chlorine used to purify drinking water is connected with bladder and rectal cancer. The study was released in the July 1992 issue of *American Journal of Public Health*.

Endangered Species Act

The federal "God Squad," a panel of government officials, met and voted to override the Endangered Species Act's protection of the northern spotted owl. The officials voted to allow logging in some of Oregon's old growth forests, even though Fish and Wildlife Department studies have found that such acreage is vital to the threatened owl. It was only the second time protections deemed necessary to protect a species from extinction were overridden. According to members of the Bush Administration, saving the owl was not worth the loss of jobs in the Northwest.

Some nearly extinct California condors were set free in 1992. The last remaining condors were taken from the wild nearly a decade ago to save the species from extinction. Intensive efforts were made to help the bird reproduce in captivity, yet remain wild enough to release. The effort seemed successful as two birds were released—but one was found dead in October, of unknown causes. Six more condors are scheduled to be released December 1, 1992.

In late 1991, 50 black-footed ferrets were returned to the wild in Wyoming. Another 350 ferrets bred in captivity have been released throughout the last year. The black-footed ferret was believed to be extinct in the wild.

Global Warming

In June, Mt. Pinatubo pumped enough gas and debris into the upper atmosphere to cool the planet slightly over the next two years. According to global climate models, the Earth's temperature should drop one degree for two years, enough to offset the global warming to date. Scientists also expect the blast from the Philippine volcano to further deplete the ozone layer. After Mt. Pinatubo's debris has dissipated, scientists predict, we will see even more warming of the planet.

A climate treaty was signed by all nations, including the US, at the Earth Summit. But the treaty was signed without actual deadlines for curtailing emissions of CO_2 and other greenhouse gases, due to pressure by the Bush Administration. Twelve European nations went beyond the terms of the treaty and agreed to stabilize emissions of CO_2 at 1990 levels by the year 2000.

Expense of the Environment

A 22-nation poll conducted by the George H. Gallup International Institute found that a majority of people in 16 nations, including some of the world's poorest, said they would willingly pay higher prices so that industry could better protect the environment. Only in the Philippines, Japan, Russia, Turkey, Poland and Hungary did a majority of people say they would not accept higher prices for a safer environment. In the US, 65 percent of Americans polled said they would accept higher prices. The majority of the people polled also said they would accept slower economic growth to protect the environment; only in India and Turkey did majorities give economic growth the priority.

—*Elizabeth Darby Junkin*

 1874, in Baltimore
Source: The Recycler's Handbook, *Earth Works Press*

PART 2
EARTH PULSE

EARTH ISSUES

Nuclear waste, famine, the ozone hole, endangered species—how can one person help solve these overwhelming problems? Start by reading this section, written by experts in the field, and choosing one area in which you want to get involved. The late Robert Rodale, author of Save Three Lives, suggests that the only way to "save the world" is to "save three lives." Whether you decide to start a beach clean-up, or grow food for homeless people, or help one neighborhood fight a toxic dump, this section will help you find your issue.

Life Out of Balance

- By Dr. Thomas E. Lovejoy -

Life is an exception to one of the basic laws of the universe: the second law of thermodynamics, which says energy runs down hill and order tends to chaos. Life is able to do this by harnessing energy, largely derived from the sun, and converting it into structures which are highly complex compared to anything else known to science. Generally, these structures can only be maintained by continued input of energy. For green plants, the energy source is sunlight and is converted through photosynthesis into organic molecules; plants, in turn, are the energy source for most other organisms.

The ability to create order and structure is the basis for the extraordinary product of organic evolution: a richness of organic form which science has only superficially cataloged or begun to understand. A *single* species has apparently been able to go even further, by devising new sources of energy or new ways to use old sources, and by building elaborate extensions of organic structure through information storage and retrieval—once cuneiform tablets or papyrus scrolls, later books, and today informatics. We, as a species, are the glorious new product of evolution, finding ways to break some of the previous limitations, and, as Ed Wilson points out, able to appreciate the wonders of organic creation at the very moment we are destroying it. It is as if we were just discovering how to read books as the Alexandria library goes up in flames.

What our species is discovering, however, is that this apparent escape from limits which have applied to all other life forms may not be quite the escape we thought it was. There is abundant evidence that we are affecting our environment in significant, sometimes essentially irreversible ways: toxic substances, ozone depletion, soaring extinction rates, urban pollution, declining fisheries. The list is long. And growing. The endless frontier has limits after all.

Out of all of this has grown the environmental movement. Like other political movements, it is as multifaceted and complex as a dragonfly's eye. The Wise Use movement likes to ridicule the "enviros" as, at best, uninterested in people, and probably at heart "pro" animals and "con" people. Yet the truth is that, ever since life began billions of years ago, it has been impossible for any organism not to affect its environment. That was true of trilobites. It is true of hummingbirds and lady slipper orchids.

The problem, therefore, is not whether to affect the environment or not, but rather how much and in what ways. The current environmental problem is complex and deeply problematical because we have only come to terms with it in small part and so late in the game. Now, with 5.4 billion of us, and growing at 0.1 billion per year, we have generated a set of problems that are so interrelated it is difficult to deal with any one facet effectively.

An example of complex linkages is tropical deforestation. Often measured nowadays in units of football field equivalents, deforestation causes biological diversity to go up in smoke and adds to greenhouse gas accumulation in the atmosphere. This leads to global climate change which, in turn, will negatively affect forests and biological diversity. Some tropical deforestation is driven by energy "needs," e.g. hydroelectric projects, as well as fuelwood needs, oil exploration and exploitation. The concomitant, if inadvertent, side effect is spontaneous colonization. In many places, indigenous people find the ability to maintain some control over their future is being wrested from their hands.

For the urban poor, these issues can seem terribly remote. Their concerns are with immediate problems of making some kind of living, and with air and water pollution so bad that epidemiological studies seem superfluous. It is easy to understand why such issues as animal rights, endangered species or deforestation seem remote or marginal at best. Yet, in the end, all people are dependent on the biology of the planet, and many medicines are derived from distant and esoteric natural sources. More important, the ability of society to come to grips with these staggeringly large problems requires the political will that lies, in part, in the hands of urban dwellers.

The most frightening part of the environmental crisis is the notion that caring about the environment means not caring about people. Caricatured in 1992 as spotted owls vs. people, this is a myth which springs up like dragons' teeth and often lurks hidden in the term "balanced approach." The truth is that, as much as the problems are generated by people, so too are the solutions emanated from people. That is why the Rio de Janeiro Earth Summit appropriately focused on environment and development and the relatively new, if yet vague, term of sustainable development.

Environmentalists have a love-hate relationship with science and technology. They have seen all too many examples of technology, such as CFCs, applied to the detriment of the environment. People are increasingly aware that the problem is not just one of numbers of people, but also of environmental impact per capita. Yet it is important to recognize that it is science which enabled us to detect or prove environmental problems such as the ozone hole or toxic substances.

The key will be to seek scientific solutions which fit the matrix of intertwined environmental problems—solutions which do not spawn new problems. An interesting example is global management of forests. So much carbon is stored in trees that just halting slash and burn can cut annual CO_2 emissions by a billion tons or more. A massive reforestation program could capture a billion tons a year of CO_2 by turning it into wood. This latter would work for three decades or so in the rapid growth phase of trees, and buy us serious time to work out a new and environment friendly energy scenario. Forests are also critical in watershed protection, as a home to indigenous people, and as the place where the bulk of biological diversity resides.

Most important of all is to recognize that we can spend too much time debating about which environmental problem is most important. The reality is that we must work in a reasonably coordinated way on many fronts at the same time. Prominent among them must be the growing human population and the understandable and justifiable development aspirations of the developing nations. Their ongoing drumbeat continually reminds us that no solution or set of solutions will possibly work unless population growth is halted.

Thomas E. Lovejoy is the assistant secretary for external affairs at the Smithsonian Institution.

EARTH JOURNAL

Shaded areas indicate greatest impact from air pollution.

What You Can Do

• Walk, ride a bicycle, take buses and trains. Act like gas isn't cheaper than people. One person generating about 0.1 hp paid $5/hour would cost about 1,700 times more than gasoline at $1.25/gallon.

• Pay attention to local government land-use policies. People are driving more and more, whether due to suburban sprawl, inconvenient child-care or low gas prices. Be sure you have realistic choices so you can choose not to drive.

• Shopping is to corporations as voting is to politicians. Corporations are as interested in your money as politicians are in your votes. So make sure they know you want environmentally friendly products, and investigate all claims.

• Contact your local Lung Association

Air Pollution

- By Muriel Strand and Ralph Propper -

Air is one of Earth's circulation methods. Along with water, it is a basic ingredient in Gaia's metabolism of photosynthesis and digestion. Biologists are beginning to think of the Earth as a living organism, and Gaia (the Greek Earth deity) is the name most often used for this creature. Contaminated air can be very unhealthy for Gaia and for her inhabitants.

Air pollution, like all pollution, is generally a by-product of human efforts to make life somehow easier, more comfortable or more amusing. While natural biosphere processes eventually recycle most pollutants, air pollutants are typically removed sooner than water contaminants or "Earth pollution," such as hazardous waste sites. However, some air pollutants are transformed into water and Earth pollutants (and vice versa); these transformation products can be even more toxic.

Most air pollutants stay aloft for days or months; chlorofluorocarbons (CFCs—refrigerants) and halons (used in fire prevention) will remain in the stratosphere chewing up ozone for decades. Each chlorine and bromine atom formed there from CFCs will destroy many of the ozone molecules that protect us from the sun's harmful ultraviolet radiation.

: *If only 10 percent of notebooks made in the US were made from recycled paper, how many trees could be saved annually?*

Stratospheric ozone depletion is already beginning to cause cataracts and skin cancer in Australia.

Another serious global air pollution problem is the greenhouse effect. While CO_2 can cause more heat retention, higher temperatures can evaporate more water to form clouds, which reflect more sunlight. Also, more turbulent weather will result from the larger temperature differences expected between the poles and the equator.

Motor vehicles are responsible for most of the tropospheric ozone formed by the reaction of nitrogen oxides (NOx) with volatile organic compounds (VOCs) on warm sunny days. (Unlike in the stratosphere, which is miles above us, we breathe the air in the troposphere. Unfortunately, a higher ozone concentration in this layer begins to harm our lungs long before it protects us from ultraviolet.) Busy intersections often have high levels of carbon monoxide, which degrades the blood's oxygen capacity. Vehicles are the source of diesel exhaust, which is more carcinogenic than all the rest of vehicular pollution. Automobiles enjoy massive subsidies of public money, yet are still costly to purchase, maintain, fuel, insure and drive. They require vast amounts of pavement, and isolate us from our community.

Tropospheric ozone and toxics from vehicles play a major role in health problems in California. Recent research indicates that lung function in children living in Los Angeles is 25 percent below normal. Diesel

chapter; most have an Air Conservation Committee. Contact the Bicycle Federation of America; they are currently forming local groups.

• Relax and enjoy life. Slow down and hug the trees. Reduce your consumption and increase your quality of life. Respect other species as you would have them respect you.

What's Being Done
• The Northeastern states are adopting California vehicle emissions standards, including requirements for increasing percentages of clean-fuel vehicles. Air quality agencies are defining ways to get emissions offsets from mobile source reductions to allow for population growth and industrial expansion.

• California is requiring 10 percent of new vehicles to be electric within

Pollutants spew from stacks at Pemex petroleum plant north of Guaymas, Mexico.

A: *45,000*
Source: The Student Environmental Action Guide, *EarthWorks Press*

10 years. Electricity from utilities causes far less pollution per mile than car motors powered by internal combustion, and overnight recharging fits utilities' load profile perfectly.

• CFC production will cease by the end of 1994; hydrochlorofluorocarbon (HCFC, with about one-tenth the hazard) production will continue. Safer alternatives may come later, but would require major engineering redesign. Reclamation and recycling of CFCs will permit their continued use.

• Market-based incentives are being discussed, although "free market" conservatives tend to resist removal of existing subsidies for vehicles, oil and roads.

• Voters in California have approved tax increases to pay for public transit as well as road construction. Congress recently approved the Intermodal Surface Transportation Efficiency Act (ISTEA) which designates significant funding for transit and bike facilities.

Resources
• Air Resources Information Clearinghouse, 46 Prince St., Rochester, NY 14607, (716) 271-3550.

• Center for Clean Air Policy, 444 N. Capitol St., Ste. 602, Washington, DC 20001, (202) 624-7109.

exhaust causes an estimated one extra cancer case for every few hundred people in California. Tropospheric ozone can lower agricultural yields by half, and costs California farmers up to $300 million annually.

Recent research has shown that cars are responsible for at least three-fourths of the air pollution in most American cities. And 10 percent of the cars cause more than half of the total auto pollution. Many of these "gross emitters" have been tampered with; they are not just older cars. New technology, called remote sensing, has recently been shown to spot gross emitters on the road. It is clear that remote sensing must be used for enforcement purposes, but the EPA has so far prevented this testing device from being employed.

In addition, the EPA has been holding up approval of California's plan to reduce motor vehicle pollution by requiring cleaner fuels and low-emission vehicles. The reason is that the new Clean Air Act allows other states to adopt California vehicle standards, and many have plans to do so—but the Bush Administration has been blocking them.

Acid rain, which results from NOx and SOx (sulfur oxides) emitted from cars or other combustion, threatens our natural and agricultural resources. The worst problems are in the northeastern US and nearby Canada, where the soil is especially susceptible to the effects of acid rain. In addition, acid rain and tropospheric ozone acting together make the problem far worse. Acid aerosol is a health threat, especially to children.

Tobacco smoke is one of the most noxious types of indoor air pollution. This kind of pollution is significant because we typically spend 80 percent to 90 percent of our time indoors. New energy-efficient buildings have little infiltration from outside air, resulting in higher concentrations of tobacco smoke, toxics from consumer products, and molds and dust.

Ralph Propper is an officer of the American Lung Association of Sacramento-Emigrant Trails, and Muriel Strand is a member of the Sacramento Environmental Commission. Both authors are staff scientists with the California Air Resources Board and board members of the Environmental Council of Sacramento, California.

: *According to the World Health Organization, how many people are injured or killed as a direct result of pesticide exposure or ingestion every year?*

EARTH ISSUES

Shaded areas indicate greatest abuse of animal rights.

Animal Rights

- By Ingrid E. Newkirk -

Animal rights were in the news throughout the summer of 1992. People for the Ethical Treatment of Animals (PETA) "slaughtered" a fake cow 60 times at the Earth Summit in Rio to spotlight the eco-consequences of meat eating. When "Batman Returns" hit the box office, activists hit the streets with signs reading, "Penguins belong in the Antarctic, not in Hollywood." In July, a baboon's liver was used for a doomed transplant experiment that relegated our fellow primates to spare parts. And by Labor Day, 114 civil disobedients were in jail in Pennsylvania for trying to stop a bloody bird shoot.

Unlike humans, animals don't need to be told that rivers, forests and icebergs are more than lumber and oil company playgrounds, all-terrain-vehicle riders' turf and military junk piles. To them, such places are home and life itself, although claws and teeth offer as little protection against human encroachment as tribal peoples' spears do against bulldozers. Most other-than-human beings don't even fight back: They scream, run or freeze in place as the nets fall and the knives come out.

The animal rights movement reminds us that we once openly performed experiments on human orphans. We slaughtered Native Americans (other-

What You Can Do

• In your workplace, ask the management to order cruelty-free office supplies (for example, Faber-Castell and International Rotex), and to provide vegetarian options in the cafeteria (request the Physicians Committee for Responsible Medicine's *Gold Plan*, P.O. Box 6322, Washington, DC 20015). Set up a magazine rack where employees can swap animal rights and environmental magazines and literature.

• Save an acre of trees a year as well as animals' lives, precious resources and your own health by adopting a vegan diet. Livestock agriculture is responsible for 85 percent of the loss of US topsoil, two-thirds of which is now gone; requires up to 240 times more water per pound of food than wheat; and produces 20 times more excrement (which

A: *Up to 1 million injured, and 10,000 killed annually*
Source: Greenpeace Action

contaminates our water, as there are no cattle sewer systems to manage it) than the total US population. Factory-farmed animals are surely among the most miserable in the world, and the links between animal products consumption and heart disease and cancer are now indisputable.

• Ask your local school's biology department to replace dissection, which has been linked to frog depletion and companion animal theft, with non-animal teaching aids. The animals will be spared, and studies have shown that students do equal or better work with models, computer simulations and books than with scalpels.

• Avoid entertainment and movies that involve animals, especially

than-Europeans) as casually as we slaughtered the North American continent's buffalo citizens. We described other-than-white beings as "subhuman savages," and other-than-male beings, considered the property of fathers and husbands, as having "as much sense as asses." Taking David Brower's "History of the Earth Compressed into One Year" concept, our exploitation of those "others" ended in only the last eight-tenths of a second.

Can we ever stop directing and start playing along with the other members of the great ecosystem orchestra? To bestow rights on animals and the Earth threatens furriers and bow hunters, just as racists feared emancipation, and women's lib gives some men the jitters.

At the 1992 International Whaling Commission meetings, Norwegians complained that the emphasis had changed from "acceptable kill quotas" for menke whales to questioning the morality of killing whales at all. True, all over the world, while many people still rationalize that they can deprive a "less-important" being of his or her life if there's even a remote possibility that there might be something in it for them (blubber, a xenograft or a ham sandwich), a growing number of humans are demanding that the Golden Rule be extended to animals—that we switch from humanocentric "preservation of the species"

Hair color was sprayed into this rabbit's eyes at Biosearch Lab. Reactions include ulceration, bleeding and massive deterioration.

Q: *In the last decade, how much has government funding for research and development of renewable energy technologies been cut?*

arguments to concern for the rights of individuals, regardless of herd, flock or pod size.

In the 1970s documentary "Primate," a Ph.D. (who shares 98 percent of his DNA with his victim) inserts electrical wires into a chimpanzee's penis. As he prepares to press the switch that forces an erection, he says: "Now we'll let Nature take its course." That out-of-touch remark was made less than a trillionth of a Brower second ago. But despite such ignorance and President Bush's disregard for the Endangered Species Act, times *are* changing. Over 420 cosmetics and household product companies have abandoned animal tests; General Motors is the only car company in the world that still crashes pigs into walls; and an estimated 20 percent of people who eat out choose a restaurant that offers vegetarian menu selections.

A 1992 survey showed that two-thirds of American high school students believe in animal rights. They realize that being amused by ever-smiling dolphins trapped in Sea World bathtubs and brown bears crammed into tutus at the circus is as unthinking as laughing at the disabled; that the true cost in habitat and eco-health of the rainforest isn't reflected on the "sale" tags on "cheap" leather jackets; that birds, whose bones are crunched between human teeth, and tree frogs, stolen by biology supply companies from Bangladeshi ponds, once enjoyed their tiny bird and frog lives.

In 1992 there were the first animal rights protests in Japan, France and Russia. It was the year Wella, Neutrogena and Fisher-Price banned animal tests forever; the US military was forced to stop shooting cats in the head; some universities were compelled to end secret meetings of animal care oversight committees; more medical schools dropped dog and pig labs from their curriculums; and more furriers shut up shop.

The animal rights movement believes that human compassion and tolerance needn't be tested in space somewhere, but here, among all the diverse lifeforms: among octopuses, foxes, mice, deer, pigs and bats. That's our goal and our challenge for 1993.

Ingrid E. Newkirk is national director of People for the Ethical Treatment of Animals. Her most recent book is Free the Animals! The Untold Story of the US Animals Liberation Front & Its Founder, "Valerie" *(Noble Press, Chicago, 1992).*

exotics. Circuses, rodeos, bullfights and zoos teach precisely the *wrong* lessons about compassion, respect for, and understanding of animals. Instead, rent video documentaries, watch public television and cable programs, take field trips and check out library books.

Resources
• *Animal Rights and You: A Guide to Sharing the Planet* (audiotape), by Ingrid Newkirk, Enhanced Audio Systems, 1160 Powell St., Oakland, CA, 94608, 1992.

• *Save the Animals! 101 Easy Things You Can Do*, by Ingrid Newkirk, Warner Books, New York, 1990.

• *Diet for a New America* (video program), Lifeguides, KCET Video, 4401 Sunset Blvd., Los Angeles, CA 90027. Addresses the environmental, ethical, and health implications of the American meat habit.

• *Shopping Guide for Caring Consumers (1993 Edition)*, PETA, P.O. Box 42400 Washington, DC 20015.

• EarthSave, 706 Frederick St., Santa Cruz, CA 95062. Environmental and vegetarian group.

• The Fund for Animals, 200 W. 57th St., New York, NY 10019. Wildlife and hunting issues group.

A: **By 80 percent**
Source: Union of Concerned Scientists

Shaded areas indicate greatest recent loss of biodiversity.

What You Can Do

Read

• *Global Biodiversity Strategy*, by World Resources Institute, UN Environment Programme and IUCN (The World Conservation Union), Washington, DC, 1992.

• *The Diversity of Life*, by Edward O. Wilson, Harvard University Press, Cambridge, MA, 1992.

Write

• The Endangered Species Act is one of the world's strongest environmental laws, but it is under attack from industry and conservative groups who seek to weaken it or prevent its reauthorization. Write to your congressperson and to the president to express your support for a strengthened—not weakened—Endangered Species Act. (See "How to Write Congress," page 322).

Biodiversity

- By John C. Ryan -

Complex beyond understanding and valuable beyond measure, biological diversity is the total variety of life on Earth. Scientists can only guess how many millions of species there might be in little-explored environments like the tropical forest canopy or the ocean floor. But nature retains its mystery in familiar places as well. Even a handful of soil from the US is likely to contain many species unknown to science.

One certainty is that biological diversity is collapsing at rates that can only be described as mindboggling. Biologist Edward O. Wilson of Harvard University estimates that 0.5 percent of tropical rainforest species are either extinguished or condemned to eventual extinction each year by destruction of their habitat. If there are even 10 million species, that equals 50,000 per year.

But species and genetic varieties are rapidly disappearing all over the globe, in freshwater lakes, oceanic islands, conifer forests and rice fields, at rates thousands of times greater than the natural "background" rate of extinction of one to ten species per year. Three-fourths of the world's bird species are declining in population or threatened with extinction. Amphibians (species like frogs and salamanders) are

Q: *Which US National Park contains 25 different keys and islands within its boundaries?*

declining worldwide. In one lake alone—Africa's Lake Victoria—roughly 200 fish species have disappeared in the past two decades.

The top priority for halting the loss of biodiversity will be the protection of wildlands, those areas so far minimally degraded by human activities. But the pervasive nature of the problem and the imminent threat of global warming mean that parks and reserves alone cannot do the job. Only if biodiversity becomes a central concern in our mainstream economic activities, as well as in our protected areas, will we avoid squandering our biological inheritance.

The direct causes of biodiversity loss are straightforward—habitat loss, overhunting, pollution and the introduction of exotic species. But the root causes are more complex and include maldistribution of farmland and population growth, both of which push poor farmers to clear wildlands for survival, and poor government policies and overconsumption, both of which encourage the destruction of natural areas for limited benefit.

Today, national parks and other protected areas cover nearly 5 percent of the Earth's land surface, and wilderness areas (many maintained by indigenous people) cover as much as one-third of the planet's land. But many indigenous management systems are

Support
- In many parts of the world, to speak up for the environment or biodiversity is literally to put one's life in danger. In Brazil, for example, more than 1,500 rural people (the most famous being rubber tapper Chico Mendes) have been killed in land disputes since 1965. By publicizing governmental harassment of indigenous people, the rural poor and environmental activists, human-rights groups make important contributions to biodiversity protection.

- Human Rights Watch, 485 Fifth Ave., Third Fl. New York, NY 10017.

- Amnesty International— USA, 322 Eighth Ave., New York, NY 10001.

Did You Know?
- The relatively small nation of Colombia is

Every species depends on a network of relationships with other species to survive.

A: *Biscayne National Park, on the eastern coast of Florida*
Source: The W.A.T.E.R. Foundation

home to more bird species—1,700, or nearly one-fifth of the world's total—than any other country on the planet. And only Brazil harbors more plant species than Colombia.

• On average, one Amazon tribe has disappeared each year since 1900. Many more have lost their lands or been assimilated into the mainstream culture.

• Brazil nut gatherers—who use tropical forests in a largely nondestructive manner and seek to stop deforestation by large landowners—receive about 4 cents a pound for their labors, just 2 percent to 3 percent of the New York wholesale price.

• Almost half of Australia's surviving mammals are threatened with extinction.

• One out of six plant species in Western Europe is threatened with extinction.

• The genetic diversity of food crops is one of the most important, and most threatened, types of biodiversity. In Indonesia, 1,500 local varieties of rice have disappeared in the past 15 years, and nearly three-fourths of the rice grown today descends from a single plant, leaving food supplies highly vulnerable to widespread pest or disease outbreaks.

unraveling, as cultures erode and national governments confiscate or privatize resources held by communities.

Restoring some degree of local control over resources is probably the only way that vast areas in the tropics can be "managed" at all. Governments claim ownership of 80 percent of the world's remaining mature tropical forests, but only by sharing management responsibility with the millions of people living near the forests do governments have any hope of controlling their use. Fortunately, since mid-1991, several governments (including those of Brazil, Canada and Venezuela) have recognized large indigenous homelands, giving several beleaguered cultures, and the biological diversity they have maintained, hope for survival.

But national governments are generally lagging behind local communities in their efforts to protect life's variety. At the United Nations Conference on Environment and Development in Brazil in June 1992, the US refused to sign an international biodiversity treaty signed by 150 other nations. The reasoning: It would allegedly cost American jobs in the pharmaceutical and biotechnology industries, which use genetic resources from the tropics.

What the US administration failed to recognize is that without a deal between poor nations (home to most of the world's biodiversity) and wealthy nations (home to most of the world's biotechnology), neither will prosper. Unless poor nations—and poor people within those nations—have the ability to benefit from biodiversity, they will not put much effort into conserving it. And without active conservation efforts, great portions of the world's living wealth will not survive to be tinkered with by biotechnology workers.

Signing the biodiversity treaty and supporting other countries' efforts to protect native habitats are crucial, but the most important action the US could take for biodiversity's sake is to reduce its profligate, and unparalleled, levels of consumption. No conservation effort, however ingenious, can get around the fact that the more resources one species consumes, the fewer are available for all the rest.

John C. Ryan is research associate at the Worldwatch Institute, a Washington, DC-based nonprofit research organization focusing on international environment and development issues.

 Q: *Worldwide, what is the annual market value of medicinal drugs containing active ingredients derived from plants?*

Shaded areas indicate forests most threatened.

Deforestation

- By Dr. Kenton R. Miller -

About 17 million hectares (42 million acres) of tropical forests—an area nearly twice the size of Ireland—are being felled each year, amounting to as much as 50 percent per annum since 1980. Because these forests are home to more than half the Earth's plant and animal species, tropical deforestation is the main force behind an extinction rate unmatched in 65 million years.

Forest destruction is not limited to the tropics. In the US, the old growth forests of the Pacific Northwest have already been cut back to just 13 percent of their original extent. Forests throughout the industrial world are being decimated by smog and acid rain. In 13 European countries, air pollution has damaged 10 percent to 20 percent of all trees. Clouds over some mountains in the eastern US are as acid as lemon juice, and 20 percent of the red spruce high in the mountains of New England have succumbed to smog over the past decade.

This loss of forests has profound ethical and aesthetic implications—and economic costs, as well. Most immediately, deforestation dims the developing world's prospects for making economic progress and relieving poverty, but it has other costs, too. As species die out, so do untold options for medical and

What You Can Do

• Forestry and development planners from many countries are working to redesign the *Tropical Forestry Action Plan* (TFAP) to incorporate sustainability into each country's forest management plan. For more information contact Matt Heering, TFAP Coordinator, Forestry Department, Food and Agriculture Organization, Via della Terme di Caracalla, 00100 Rome, Italy.

• Canada, Sweden and several other countries are working to establish a *World Forestry Commission* that would include scientists, government officials and representatives of nongovernmental organizations and be modeled after the Intergovernmental Panel on Climate Change (IPCC), under whose aegis the climate convention signed at Rio

A: *$40 billion each year*
Source: National Wildlife *magazine*

was negotiated. For more information contact Dr. George M. Woodwell, Director, Woods Hole Research Center, P.O. Box 296, Woods Hole, MA 02543.

• The final communiqué issued by the G-7 (US, Japan, Germany, France, Britain, Italy and Canada) economic summit in Houston in 1991 promised funding to Brazil for rainforest conservation. As a result, the *Rainforest Pilot Program* administered by the World Bank is investing up to $4 billion donated by G-7 nations in such projects as marking out the boundaries of indigenous reserves in Amazonia and developing a Brazilian national biodiversity conservation strategy. For more information contact Robert Kaplan, Rainforest Pilot Program Coordinator, The World Bank, 1818 H St. NW, Washington, DC 20433.

• Member nations of the *International Tropical Timber Organization* (ITTO) were scheduled to meet in the fall of 1992 to discuss re-authorizing the organization, which may depend on temperate countries' willingness to adopt the ITTO's "Target 2000" goal of ensuring that all timber is harvested in sustainable ways by the year 2000. For more information contact B.C.Y. Freezailah, Executive Director, ITTO, International Organizations Center, Fifth Fl., Pacifico-Yokohama, 1-1,

agricultural advances that humanity may someday desperately need. As just one biomedical example, consider the Pacific yew tree: Though treated as "trash" by loggers for generations, it has recently been found to contain a substance useful in treating certain cancers. Or weigh the agricultural costs of declining genetic diversity: The genetic similarity of Brazil's orange trees opened the way for a disastrous outbreak of citrus canker in 1991. Similarly, the Irish potato famine in 1846 and the failure of the Soviet wheat crop in 1972 both stemmed from loss of genetic diversity. Deforestation also wreaks havoc on climate—locally, by disrupting the hydrological cycle, and globally, through its effect on the carbon cycle.

For a time, hopes were high that a forestry convention, or treaty, would be ready for signature at the Earth Summit in Rio de Janeiro in June 1992, but it was derailed. The climate and biodiversity conventions that were signed at Rio have a bearing on forests, however, since forests provide habitat for wild

Photo: Sipa Press

Each year, tropical forests amounting to nearly twice the size of Ireland are felled, like this area in Brazil.

 : *What is the estimated cleanup bill for Gulf War oil well destruction?*

species and a carbon storage mechanism that can help offset the carbon dioxide given off by burning fossil fuels. But nations could not agree on a forest convention, and instead they settled for a non-binding "statement of principles" on the management, conservation and sustainable development of tropical, temperate and boreal forests—a statement significantly lacking any agreement on preserving key forest areas. Weak as they are, the forest principles that came out of Rio at least show that nations are finally acknowledging that deforestation is a grave problem.

The challenge now is for nations to show that they're serious about protecting the health of the world's forests—and some are. It would be premature to speculate on the prospects of various moves now afoot to curb deforestation, and it would be myopic to think that any one of them is sufficient. A problem as complex as deforestation can be mitigated only by an equally complex set of solutions. Nonetheless, everyone who is concerned about the fate of the forests can take heart from the fact that governments are exploring new options.

Dr. Kenton R. Miller is director of the program on biological resources and institutions at the World Resources Institute (WRI), a center for policy research on global resource and environmental issues, based in Washington, DC. Before joining WRI, he was director general of the International Union for Conservation of Nature and Natural Resources in Gland, Switzerland. He has also been a professor at the University of Michigan and a regional director of wildland management projects in Latin America for the Food and Agriculture Organization.

Minato-Nirai, Nishi-ku, Yokohama, 220 Japan.

• A *Center for International Forestry Research* (CIFOR) is being organized to conduct strategic research in tropical forest ecosystems and promote the transfer of new and appropriate forest technologies. For more information contact D.I. Bevege, Principal Advisor, CIFOR, Australian Centre for International Agricultural Research, Third Fl., Drake Centre, 10 Moore St., Canberra, ACT 2601, Australia.

• The *Forest Stewardship Council* is being established to certify whether timber is harvested sustainably and therefore merits "green seals of approval," such as those awarded by groups like Smart Wood and Green Cross. For more information contact Chris Elliott, World Wide Fund for Nature, Avenue Mont Blanc, CH-1196 Gland, Switzerland.

A: *Around $20 billion*
Source: Newsweek *magazine*

Shaded areas indicate regions most affected by desertification.

What You Can Do

Contact
• Chihuahuan Desert Research Institute, P.O. Box 1334, Alpine, TX 79831.

• Earth Island Institute, 300 Broadway, Ste. 28, San Francisco, CA 94133-3312.

• World Resources Institute, 1709 New York Ave. NW., Seventh Fl., Washington, DC 20006.

• United Nations Environment Programme, Room DC2-803, United Nations Plaza, New York, NY 10017.

Read
• "Expansion and Contraction of the Sahara Desert 1980 to 1990," Harold E. Dregne, Compton Tucker and Wilbur W. Newcomb, *Science Magazine,* July 19, 1991.

• "Desertification: A Review of the Concept,"

Desertification

- By Dr. Vivien Gornitz -

Desertification can be defined as the development of desert-like conditions in arid, semi-arid and some sub-humid zones, under the effects of drought and/or human activities. While a precise definition is elusive, the concept usually implies a process of land or soil degradation due to overgrazing, deforestation, poor agricultural practices or soil salinization. Adverse impacts include: an impoverishment or diminution of the natural vegetation cover; increased water and wind erosion; and ultimately reductions in crop yields and the land's carrying capacity for livestock, with deleterious consequences to humans.

The world's dry lands bordering deserts are inherently sensitive to water stress and drought. For example, droughts have occurred in the Sahel region of west Africa in the mid-18th century, the 1820s-1830s, the early 20th century, as well as during the last 20 years. During the drier-than-average period of the 1980s, the Sahara desert expanded by 15 percent within a four-year period (1980 to 1984), but then shrank by 8 percent in the following year. The ebb and flow of the Sahara during the 1980s corresponded to fluctuations in rainfall. This suggests that the popular conception of the desert marching south is grossly oversimplified, and underestimates the

Q: *What is the total amount of time Americans spend stuck in traffic each year?*

primary importance of rainfall on vegetation growth in arid to semi-arid climates.

Drought is a natural phenomenon, and arid and semi-arid ecosystems are well adapted to it. However, these regions are particularly vulnerable from unsustainable land-use practices. Desertification generally takes place in a sequence of interactive processes. Disruption of the natural vegetation cover caused by overgrazing and trampling by livestock, clearing of trees and shrubs, shortening fallow periods and other harmful agricultural activities degrade the soil. Then water and wind erode topsoil. Without replenishment by manure or chemical fertilizers, bare soil loses nutrients, suffers decomposition of humus (soil organic matter) and holds less water. Severely degraded soils form gullies and, in extreme cases, sandy soils break up into sheets and dunes. Even in moderately degraded terrain, less desirable shrubs replace beneficial vegetation, and crop productivity declines. These changes simulate more desert-like conditions even when rainfall has not decreased much—a process also called *xerotisation*.

Desertification, as an environmental problem, is not confined to the Sahel, or to desert margins. Vast semi-arid grasslands in East Africa, the Middle East

Michael H. Glantz and Nicolai Orlovski, *Desertification Control Bulletin*, United Nations Environment Programme, December 1983.

• *The Threatening Desert: Controlling Desertification*, Alan Grainger, Earthscan Publishers, London, 1990.

• "Dry Land Management: The Desertification Problem," Ridley Nelson, World Bank Technical Paper 116, April 1990.

• "Rethinking Desertification: What Do We Know and What Have We Learned," Steven L. Rhodes, *World Development*, September 1991.

• "Desertification Control Bulletin," United Nations Environment

Poor land management and climatic forces degrade soil, causing desert expansion.

A: *1 billion hours*
Source: Union of Concerned Scientists

- Programme, Nairobi, Kenya, 1988.

- "Assessment of Desertification in Arid and Semi-Arid Areas," Andrew Warren and Clive Agnew, International Institute for Environment and Development, London, 1988.

- "The Myth of the Marching Desert," Bill Forse, *New Scientist*, February 4, 1989.

- "Drought and Desertification: The Arid Earth," Charles Tyler, *Geographical Magazine*, Royal Geographical Society, May 1989.

Did You Know?
- The Sahel has been suffering from below-average rainfall over the last 24 years. This region has experienced climatic fluctuation lasting up to several decades.

- Over 1,200 million hectares (3,000 million acres) representing 11 percent of the Earth's vegetated land has been affected by soil degradation in the past 45 years.

- Africa and Asia contain nearly four-fifths of the world's desertified lands.

- Some 22 million acres of land have been rendered "unclaimable and beyond restoration" due to desertification.

and western Africa have been damaged by overgrazing. Desertification is not exclusively the problem of less-developed countries. Overgrazing in rangelands of Australia and Argentina has led to loss of productive plant species and topsoil. Poor agricultural practices coupled with drought created the "dust bowl" in the US Midwest in the 1930s. One of the worst desertified areas in the US is the Navajo Reservation of northern Arizona and New Mexico, where overgrazing by rapidly expanding sheep herds has created a denuded, gullied wasteland. Other seriously overgrazed areas in the western US include parts of eastern New Mexico, Colorado, southern Arizona and northwest Texas.

With proper conservation measures, the process of desertification can be halted and even reversed. For example, reforestation, innovative irrigation techniques and controlled grazing in the Negev Desert of Israel has made that region productive and prosperous. China has also pursued a program of reforestation and desert reclamation. However, more often than not, economic development programs do not give high priority to ecologically sound land use. Thus, insufficient funds are available for combatting desertification. Furthermore, there is a recent tendency to view desertification as a mostly climatological problem, therefore downplaying the role of nonsustainable land use. Unless the issue is addressed with a greater sense of urgency, more land resources will deteriorate and become unproductive.

If desertification is allowed to accelerate unabated, the end result, especially during droughts, will be more widespread famine, more human misery and major social dislocation, as environmental refugees migrate to other rural areas already stressed by overexploitation, or to overcrowded cities.

Since 1974, Dr. Vivien Gornitz has been an associate research scientist at the Lamont-Doherty Geology Observatory of Columbia University, working on climate studies and Earth resources. She is author of numerous works, most recently an essay, "Mean Sea-level Changes in the Recent Past," published in Climate and Sea Level Change, *edited by R.A. Warrick and T.M.L. Wigley, Cambridge University Press, New York, 1992.*

Q: *How many square miles of toilet paper are used in the US each year?*

EARTH ISSUES

Shaded areas indicate regions most affected by environmentally related disease.

Disease and the Environment

- By Robert S. Desowitz -

Infectious diseases, particularly those caused by vector-borne parasites, are in a real sense environmental diseases. Parasitism is a hazardous way of life, but evolution has conspired so that parasites exploit their environment and the behavior of their human hosts to insure their perpetuation. The hundreds of millions of people infected with vector-borne parasites such as the schistosomes (blood flukes), filaria (the cause of elephantiasis and river blindness) and malaria are testimony to the success of that exploitation.

Ecological "sanitization" has been applied to the control of some of these diseases—draining a swamp or other large body of water to deny the breeding of mosquitoes bearing malaria, yellow fever or viral encephalitis, for example. But more importantly, ecological perturbations have exacerbated the spread of these diseases. In the newly created lake behind the hydroelectric dam in tropical Africa the snails proliferate, and in the lakeside population the infection rate of the snail-transmitted urinary schistosome rises from 5 percent to 95 percent.

Malaria is an environmental disease aggravated by ecological abuse—it epitomizes human-made diseases.

What You Can Do

• Travelers to malaria-endemic zones should seek expert advice as to the most effective chemoprophylactic for that area. They should practice preventive defense by reducing contact with anopheline vectors by such measures as sleeping under a mosquito net; consider all fevers of sudden onset possibly malarial and seek expert medical advice promptly.

For more information on preventative medication, call the Centers for Disease Control at (404) 332-4555. Before traveling to an infected area, contact the World Health Organization, (202) 861-3200, for further information on boosters.

Other Environmental Diseases

Environmental diseases are infections spread from the environment to humans through air,

A: *22,627*
Source: Garbage *magazine*

water, food, puncture wounds and exchange of bodily fluids. These are a few examples.

• Cholera
Transmitted: through ingestion of food or water contaminated with feces, urine or vomit that contains bacteria known as *Vibro cholerae*.
Symptoms: severe diarrhea, vomiting and cramps in the arms, legs and abdomen. Temperature and blood pressure drop, and the skin and urine turn a dark purple shade. Can be fatal.
Prevention: Drink bottled water and avoid uncooked food, especially vegetables, in areas where the disease is likely.
Cure/treatment: A vaccine has been developed that produces immunity to cholera for a few months. Can be treated.

The statistics are stunning—200 million to 400 million cases each year, causing 1 million to 2 million deaths. Humans are heir to four species of malaria parasite, all found in *anopheles* mosquitoes. Of these, only one malaria is potentially fatal. However, the others can cause severe symptoms, including high fever, with rigor and sweats, headache and body aches.

In 1955, malariologists, led by the World Health Organization armed with the recently discovered powerful, inexpensive residual insecticide DDT, mounted a program to eradicate malaria globally by indoor spraying of houses throughout malaria-endemic regions for five years. Malarious nations throughout the world (with the exception of sub-Saharan Africa) joined the crusade. From 1956 to 1969, the US, through the Agency for International Development, gave $790 million to the Global Eradication of Malaria Program. By 1969 it had become apparent that the program had failed—that it had been an impossible dream.

Both behaviorally and physiologically, the anophelines had become resistant to DDT. By 1970 malariologists began looking toward a new strategy—antimalarial chemotherapy. This was a possibility because a marvelously effective, cheap and nontoxic drug, chloroquine, became available after World War II. By 1972, that avenue too began to close when

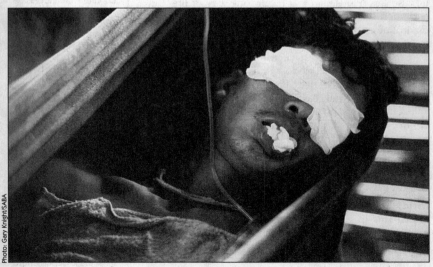

A comatose cerebral malaria victim at a refugee camp on the Thai-Myanmar border.

Q: *How much plutonium is the typical nuclear power plant capable of producing in one year?*

chloroquine-resistant strains of mosquitoes began their inexorable spread throughout the endemic world. There was no new antimalarial of similar qualities to replace chloroquine. The acute cases had to be treated with quinine—a 400-year-old drug. Pharmaceutical companies, faced with the $100 million, six-year cost to bring any new drug to market, could not justify to their stockholders drug research for poor people's diseases.

Meanwhile, the tropical world had physically changed, making it more malarial. Its population had exploded—in large part due to the good early years of the global malaria eradication scheme. Formerly, malaria would, in many hyperendemic regions, kill 40 percent of the children—it was a sad control of overpopulation. The malaria-spared, increased populations put new pressures on their habitats. Forests and other tropical ecosystems were destroyed to make way for agriculture. These altered environments too often created ideal habitats for anopheline vectors.

Thus by 1972 malaria was more widespread, more intractable and less controllable than it had been 20 years before. Few promising new antimalarial maneuvers exist. There is a mosquito net dipped in a pyrethrum analog, permethrin, with residual activity. There is a "new" 2,000-year-old Chinese antimalarial drug, qhinghaosu, derived from sweet wormwood (*Artemesia annua*), that appears to be highly effective; but it is not yet produced in quantity—and may never be. The environment continues to deteriorate, and malaria mosquitoes continue to breed in their human-made habitats. People sicken and die of malaria in unprecedented numbers. Fewer and fewer research resources are devoted to finding practical solutions. Only the malariologists have been brought to near-extinction.

Robert S. Desowitz has carried out research and teaching at the West African Institute of Trypanosomiasis Research in Nigeria, the University of Singapore, the US Army Medical Research Laboratory in Bangkok and the University of Hawaii John A. Burns School of Medicine, where he has been a professor of tropical medicine and medical microbiology since 1968. Author of many books, his most recent work is The Malaria Capers, *W.W. Norton, New York, 1991.*

• Giardia
Transmitted: through consumption of bacteria-contaminated water. The infectious bacteria, flagellate protozoan, is commonly found in streams and ponds. Contagious.
Symptoms: chronic diarrhea; can be fatal if dehydration occurs.
Prevention: Boil water at 180 degrees for one minute, use iodine tablets or a portable water filter (make sure the filter is clean), or drink bottled water when purified water is not available.
Cure/treatment: People recover with antibiotics and lots of fluids.

• Hepatitis
Transmitted: through ingestion of contaminated food or water. Some forms transmitted by sexual contact, injection or ingestion of infected blood or bodily fluids. Highly contagious.
Symptoms: nausea, gastrointestinal discomfort, fever and loss of appetite. The skin and whites of the eyes turn yellow and urine darkens. Weight loss, weakness and depression. Fatalities may result from liver failure.
Prevention: Hepatitis can be prevented by strict public health laws, personal hygiene and avoidance of unsanitary hypodermic needles.
Cure/treatment: Prolonged bed rest for at least four weeks. Medical supervision is imperative. May result in permanent liver damage if left untreated.

A: *About 550 pounds annually, enough for about 25 to 50 nuclear weapons*
Source: Union of Concerned Scientists

EARTH JOURNAL

Shaded areas indicate regions with the greatest amount of renewable water supply.

What You Can Do

• If your tap water comes from a public water system, get the no-cost annual water quality report from your water company and note all pollutants detected.

• One of the most lethal water pollutants is radon gas. Find out if your water company has tested for the presence of radon in the water. If they haven't, call your regional office of the Environmental Protection Agency (EPA) and ask them if any radon has been detected in your area. If yes, arrange to have your home tested for radon.

• If your tap water is from a private water system, you should test it annually for bacteria. In addition, get a comprehensive lab test of your water at least once a year, and stagger the times so that the water samples are taken at different times of the year.

Drinking Water

- By Colin Ingram -

The following general findings are based on a comprehensive, multi-year research program on all aspects of drinking water.

• While government health officials claim that our drinking water is safe, no one knows what "safe" is. Government standards for drinking water safety cover only a fraction of harmful pollutants that may be in the water.

• While our tap water is generally free of harmful microorganisms and other pollutants that can cause immediate health consequences, essentially all tap water in the US contains pollutants that can cause long-term health problems. At the present time there are cancer-causing chemicals in virtually every water supply in the US.

• The effects of water fluoridation are, over the long term, more harmful than beneficial.

• No one—regardless of his or her location—should drink tap water on a regular basis.

• There is a wide range of quality between different types and different brands of bottled water; most bottled water is superior to tap water, but some is worse.

• There is a wide range of effectiveness in the many water purifiers available for home use; some are excellent, most are fairly effective and some can actually

Q: *Due to commercial whaling, how many of the original 250,000 blue whales in Australia's Southern Ocean remain?*

add more pollutants to the water than they remove (see "Home Water Treatment Devices," page 243).

Since doing nothing about drinking water straight from your tap water is not a good option, your first decision should be whether to buy some kind of bottled water or make your own drinking water with a purifier. You can get high-quality drinking water both from bottled water and from water purifiers. By "high-quality," I mean drinking water that is *essentially free of most known pollutants* and is as safe as any of the other food we ingest.

If you decide you want the convenience of a home water purifier, your next decision is whether to rent or to buy. Purifier rentals are available where a water treatment dealer has a large customer base and has many units rented. If you are interested in renting a water purifier, call the dealers in your area to find out if they have rental programs. You must select a reputable dealer if you are going to rent a water purifier. Ask for several references and check them out before you sign a rental agreement.

If you decide to buy a water purifier for your home, you first need to find out what potential pollutants might be in the tap water in your area. A no-cost first step is to contact your water company and ask them for a copy of their annual water quality report (which they are required by law to supply). When you get it, it will probably be hard to interpret;

- Minerals in water are beneficial to health. Drinking hard (highly mineralized) water is better for you than drinking soft (demineralized) water.

- If you must drink unpurified tap water, turn the faucet on and wait 30 seconds to flush out loose pollutants before drinking.

- Buy bottled water from a store that sells a lot of it (so it doesn't remain on the shelf for a long time before you buy it).

- When you buy bottled water, to avoid growth of microorganisms, store it in a cool place away from direct sunlight.

- Store 5 to 10 gallons of water for emergencies. Fill your clean containers to the very top (the less air in the bottle, the less microorganism growth there will be) with tap water. Sterilize the water

Muddy waters of the Colorado River flow freely before being processed for drinking.

A: *Only about 200*
Source: Greenpeace Action

by adding iodine tablets or concentrate (from your drug store). Shake the bottles to mix the iodine and turn them on their sides for at least ten minutes to sterilize the caps. Then return them to an upright position and store them in a cool, dark place.

- In addition to considering the cost and the performance of a water purifier, make sure it is convenient to use. Many purifiers sit unused in closets and garages because their owners discovered, too late, that they were too much trouble to operate.

- If you own a water purifier that uses filter cartridges, change the cartridges every six months even if the manufacturer recommends a longer interval.

- If you need to have your water tested for chemical pollutants, don't use expensive, local test laboratories. Instead, use one of the several automated mail-order test labs that offer the best bargains in water testing. You can get their names and addresses from *The Drinking Water Book*.

- *The Drinking Water Book*, by Colin Ingram, published by Ten Speed Press, Berkely, CA, 1991, includes everything you need to know about getting safe drinking water for you and your family.

don't be shy about asking them to interpret it for you. The main thing you will want to know is whether their tests have shown the presence of any toxic pollutants in the water. If specific pollutants have been detected, you will need to select a type of water purifier that is effective at removing those types from water.

If you choose to buy bottled water for your drinking water, you have several options. First, you can buy bottles of drinking water at stores; if you live in an urban area you can have drinking water delivered to your home; or you can use your own containers and buy drinking water either from vending machines or from a store that specializes in selling bulk drinking water.

If you buy pre-bottled water from a store, choose a well-known brand that says "drinking water" on the label. Don't buy bottled water whose label says "natural spring" water or otherwise indicates that the water is from a natural (but untested) source. If you buy water from a vending machine, make sure your containers are clean and choose a vending machine that is used frequently; check the vending machine for a label that certifies that it has been inspected by your local health department. Water from vending machines, from companies that home-deliver and from specialty water stores is all of high quality and a big improvement over tap water.

Recent studies have confirmed that drinking chlorinated water for many years increases the risk of getting cancer—and essentially all public water supplies in the US are chlorinated. So even if your tap water is free of all other pollutants (very unlikely), you need to improve it with a high-quality purifier or you need to buy a good quality of bottled water.

Colin Ingram has been a scientific writer for more than 30 years. From 1982 to 1987, he conducted an extensive research program on all aspects of drinking water, including health-risk assessment, water testing, bottled water and product evaluation. The results of his research are in his latest book, The Drinking Water Book, *published by Ten Speed Press, Berkeley, CA, 1991.*

Q: *How much more motor oil is dumped by Americans in the US annually than was spilled by the* Exxon Valdez *oil tanker?*

EARTH ISSUES

Shaded areas indicate regions with the most ecofeminist involvement.

Ecofeminism

- By Gloria Feman Orenstein and Rebekah Cross -

While female representatives at the Earth Summit were few—25 out of more than 400 speakers were women—the most active tent at the Global Forum was the *Planeta Fêmea,* the "Women's Planet." After much lobbying for influence at the summit, a full chapter of Agenda 21 (a nonbinding agreement reached between participating countries) focuses on women's needs. This is no small feat—it shows a growing consciousness about ecofeminism.

Ecofeminism is both a philosophy and a political movement. As a philosophy, ecofeminism sees the interconnectedness between the oppression of women and the exploitation of nature within patriarchal civilization. Western industrial patriarchal societies have continually elevated males, humans, spirit and culture far above females, animals, plants, matter and all of nonhuman nature. Thus they have linked women to nature, and have consistently subordinated both women and nature to men and culture.

While women are not inherently more connected to nature than men, women around the world, especially in third world countries, are the major fuel gatherers, water collectors, agricultural workers and forest preservers. As childbearers, women's bodies are often the first to experience the effects of the

What You Can Do

Resources
• *The Ecofeminist Newsletter,* Noel Sturgeon, c/o Women Studies, Washington State University, Pullman, WA 99164-4032.

• Ecofeminist Visions Emerging (EVE), 402 W. 46th St., #3W, New York, NY 10036, (212) 315-3107. Meetings, newsletter and booklet, *What Is Ecofeminism Anyway?* by Cathleen and Colleen McGuire.

• Women and Environments Education and Development Foundation (WEED), 736 Bathurst St., Toronto, Ontario, M5S 2R4, Canada, (416) 516-2379. Magazine published quarterly; subscription $20 per year.

• Women's Environment and Development Organization (WEDO), 845 Third Ave., 15th Fl.

A: *10 to 20 times more*
Source: Protect Our Planet Calendar, Running Press Book Publishers

New York, NY 10022, (212) 759-7982. Helped to organize the Global Assembly and the World Women's Congress for a Healthy Planet. Offers "The Earth Spinners: Women Mend the Earth," video documentary.

• Global Assembly of Women and the Environment, WorldWIDE Network, 1331 H St. NW, Ste., 903, Washington, DC 20005, (202) 347-1514, Fax (202) 347-1524.

• Immaculate Heart College Center, 425 Shatto Place, #401, Los Angeles, CA 90020-1712, (213) 386-3116. Master's degree courses offered in feminist spirituality; course includes ecofeminist classes.

• Feminists for Animal Rights (FAR), Box 694

various forms of pollution in the environment. Thus, it is women who suffer the fate of the Earth first.

As a political movement, ecofeminism sees all life as part of one interconnected web, and seeks to transform the political and ideological systems that have led to the current global environmental and development crises through a variety of methods, among which are numerous grassroots projects.

A growing number of ecofeminists also reclaim the Goddess as the symbol of a Mother Earth-based spirituality, and study the history of the ancient goddess-revering civilizations in order to understand the non-sexist ecological and spiritual values of our ancestors, whose ideas and behavior did not lead to the destruction of our planet.

By studying the beliefs and practices of many indigenous people around the world today, ecofeminists are becoming conscious of the different ways time functions in various cultures. They have observed that in modern Western cultures consumption patterns reflect a world-view that indulges the present at the expense of the future. Humans take today what rightfully belongs to tomorrow. Scarcity of nonrenewable resources is the consequence.

Not only do indigenous people not steal the future's supply of resources, but they also offer a

Demonstrators at the Earth Summit, where ecofeminist issues were discussed, sing "We Shall Overcome."

Q: *According to the EPA, in 38 states, how many different kinds of pesticides have been found in groundwater?*

model of holistic consciousness worthy of our consideration. Ecofeminism seeks to reweave a vision of consciousness and of time that reconnects spirit and mind to matter, secular to sacred, and past to present to future, so that the links between all parts of the whole of life may be visible once more.

Ecofeminists see the challenge of the present as threefold.

1. Women must gain an important voice in the decision-making bodies that determine the policies that will create a sustainable future for life on our planet.

2. We must all create a model for living and thinking in harmony with all of nonhuman nature and, as far as possible, with an awareness of our human connectedness with the invisible dimensions of spiritual reality.

3. We must become conscious of the ways in that we have gendered nature as "feminine" and have thus connected the fate of women with that of the Earth. Knowing this, we must realize that our attitudes and treatment of both women and the Earth are mutually reinforcing, and that in order to save our planet we must immediately transform our concepts and actions regarding both women and nature.

Ecofeminists are both women and men who espouse these ideas and practices. Why, then, do we not simply say "ecohumanism"? The answer lies in the fact that the historical period of the Renaissance, known as the period of "humanism," brought about many of the changes in thinking that led to our conceiving of life as a machine, of the human mind as superior to and above nonhuman nature, and of women as subordinate to men. Thus, the word *humanism* carries with it actual historical connotations that, unfortunately, communicate those specific values that the ecofeminist movement both critiques and challenges.

Gloria Feman Orenstein is a professor of comparative literature and women's studies at the University of Southern California. She is author of The Reflowering of the Goddess *(Teachers College Press, Columbia, Athene series) and co-editor of* Reweaving the World: The Emergence of Ecofeminism *(Sierra Club Books, San Francisco, 1990). Rebekah Cross is a graduate student at Vermont College of Norwich University.*

Cathedral Station, New York, NY 10025. Feminist, vegetarian women who are dedicated to ending all forms of animal abuse.

• The Green Party, Women's Caucus, Regina Endrizzi, 725 Grove St., #4, San Francisco, CA 94102, (415) 621-3895.

• Ecofeminist Task Force, subgroup of the National Women's Studies Association, University of Maryland, Rm. 1121, Mill Bldg., College Park, MD 20742-4521.

• National Women's Mailing List, P.O. Box 68, Jenner, CA 95450, (707) 632-5763. Choose women's groups from which you want to receive mail about events, publications, services, etc.

• *Women and the Environment*, edited by Annabel Rodda, Zed Books, Atlantic Highlands, NJ, 1991, $15.95. Detailed sourcebook on women and the environment; covers the world.

Did You Know?
• Although women represent half the world's population and one-third of the world's labor force, they receive only 1 percent of the world's income and own less than 1 percent of the world's property.

A: *At least 74*
Source: The Student Environmental Action Guide, *EarthWorks Press*

EARTH JOURNAL

Shaded areas indicate where species are most endangered.

What You Can Do

Resources

• *The Endangered Species Act: A Commitment Worth Keeping* is a publication produced by the Endangered Species Coalition, a group of more than 40 environmental, professional and animal welfare organizations. It covers the rationale behind species conservation, the Endangered Species Act and strategies for strengthening the US commitment to endangered species protection. It is available from the Wilderness Society, 900 17th St. NW, Washington, DC 20006.

• *Green Fire* is a new newsletter devoted to providing timely information relating to the Endangered Species Act and the upcoming congressional debate over its reauthorization. To join the mailing list, write to the Wilderness Society, address above.

Endangered Species

- By Kathryn A. Kohm -

Despite overwhelming odds, proponents of endangered species conservation can celebrate several success stories from the past year. In fall 1991, California condors, once teetering on the edge of extinction, were released in the wild. The young birds that now soar over the mountains of the Los Padres National Forest in Southern California are the offspring of the last known individuals of the species taken into captivity as a last-ditch conservation measure in 1987. In Wyoming, 50 black-footed ferrets were returned to their native habitat in September 1991—just over a decade after they were thought to be extinct. During this last year, biologists have continued to release the remaining 350 ferrets bred in captivity at sites throughout the Rocky Mountain states. And in the oceans, hunting restrictions have provided enough breathing room for several species of whale to begin a promising recovery.

Yet, as important as these successes are, they are too often overshadowed by an ever-expanding list of endangered species both in the US and throughout the world. One of the harshest lessons of the past quarter century of experience in endangered species conservation is that we have grossly underestimated the magnitude of the extinction crisis—as well as the

Q: *In the US, what percentage of all raw material consumption is used solely for the production of meat, dairy and egg products?*

resources needed to deal with it. In its 1991 annual report, the Council on Environmental Quality, an advisory board to the president, quoted estimates that some 9,000 US plants and animals may be at risk. Since October 1991, 90 species have been added to the US endangered species list. Another 3,700 species remain stuck in the listing process because there are inadequate resources or incomplete data to complete official listing. Yet listing alone does not guarantee survival. To win philosophical support for endangered species protection as an abstract concept is easy—large mammals and majestic birds stir public sentiment. But to garner the necessary resources and political support to change our own behaviors is far more difficult. Stemming the tide of extinction means, among other things, taking a critical look at how we harvest timber, dredge ports, build housing developments and generate electricity.

In the Pacific Northwest, for example, the spotted owl has become a symbol for old growth forests. Listing the owl as a threatened species in June 1990 ignited one of the most heated controversies in the history of the Endangered Species Act. In spring 1992, a high-level political committee (nicknamed the "God Squad" because of its authority to explicitly allow a species to go extinct) was convened to consider a petition to exempt selected timber sales from Endangered Species Act regulations. The committee's

• An activist tool kit focusing on endangered species issues and the Endangered Species Act is available from the National Audubon Society, 666 Pennsylvania Ave. SE, Washington, DC 20003, (202) 547-9009. It contains fact sheets, brochures, bumper stickers and a button that says "Extinction . . . Not! Support the Endangered Species Act." The kit costs $6.00.

• To obtain recovery plans and other scientific documents relating to endangered species, contact the US Fish and Wildlife Reference Service, 5430 Grosvenor Lane, Ste.110, Bethesda, MD 20814, (800) 582-3421.

• *The Official World Wildlife Fund Guide to Endangered Species of North America*, Beacham Publishing Inc., Washington, DC, 1990 & 1992.

Some of the few remaining white rhinos roam the plains of South Africa.

A: *33 percent*
Source: 50 Simple Things You Can Do to Save the Earth, *EarthWorks Press*

Information on individual species on the US threatened and endangered species list. Each entry includes vital information on a species' habitat, distribution, conservation efforts and more.

Did You Know?

• More than half of all modern medicines can be traced to wild organisms. Chemicals derived from plants are the sole or major ingredient in one-quarter of all prescriptions written in the US each year.

• Fossil records indicate that we are in the midst of the largest extinction episode in 65 million years—the last such episode was when dinosaurs went extinct. Unlike other extinction episodes, the current crisis is unprecedented in the history of life on Earth. This is due to the current dramatically high rate of species loss and the fact that single species is causing it.

• Contrary to claims made by its opponents, the Endangered Species Act, as it currently stands, does allow for the consideration of social and economic factors. The decision to list a species under the act is made solely on the basis of biological information. However, once a species is listed, the act provides for consideration of social and economic factors through the recovery process, the designation of critical habitat and referral to the Endangered Species Committee.

decision to exempt 13 sales is currently under appeal by environmentalists. The debate, however, extends well beyond the fate of the owl. In question is the extent of timber harvesting throughout the region, traditional logging practices (such as clearcutting) and the loss of critical habitat to urbanization. Similarly, the listing of several species of wild salmon as threatened means that we will have to reexamine how we use water to generate electricity and irrigate crops.

Our focus must turn from racing to protect one species after another to protecting the larger context in which endangered plants and animals live—ecosystems, natural communities and regional landscapes.

This is increasingly the direction we are headed. In May 1992, the chief of the US Forest Service announced that ecosystem management would be the new guiding principle for research and management of national forests and grasslands. In Congress, at least two biodiversity protection bills have been introduced over the past year. Among their provisions is the creation of a national center for biodiversity research and the development of a national strategy for conserving biodiversity. And at the 1992 Earth Summit in Rio de Janeiro, an international treaty on biodiversity protection was signed by most attending nations—excluding the US.

Finally, the reauthorization of the Endangered Species Act will be a key political issue in 1993. Among the more positive amendments proposed is a move to develop integrated multi-species recovery plans for maintaining and restoring ecosystems and ecological communities. This sort of broad-based thinking is critical not only to the survival of endangered species, but ultimately to ourselves.

Kathryn A. Kohm is an editor and publications specialist working for the Olympic Natural Resources Center at the University of Washington and the USDA Forest Service Pacific Northwest Research Station in Seattle. She is also editor of Balancing on the Brink of Extinction: The Endangered Species Act *and* Lessons for the Future.

 What is the temperature in a lightning channel?

EARTH ISSUES

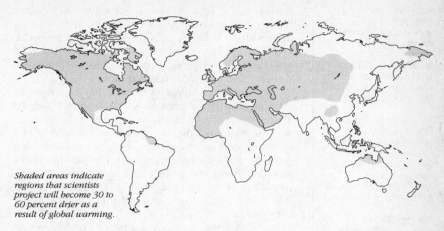

Shaded areas indicate regions that scientists project will become 30 to 60 percent drier as a result of global warming.

Global Warming

- By Andrew Revkin -

In 1992, the year of the Earth Summit, a critical shift occurred in global thinking about global warming. Previously, discussions centered on scientific questions about the greenhouse effect: How warm will it get as a result of the sharp rise in levels of carbon dioxide and other heat-trapping gases produced by burning fossil fuels and forests? Will seas rise or fall? Will clouds cool things off?

Last year was the first year most of the nations of the world acknowledged that the scientific evidence, although incomplete, was sufficient to justify action to reduce the risk that catastrophic warming might occur. It appeared that a global consensus had been reached on a key point: If emissions of carbon dioxide continue at the current rate, there is an unacceptable risk that disruptive climate shifts will occur within the lifetimes of children born today.

At the 1992 Earth Summit in Rio de Janeiro, 153 nations signed a draft climate convention that called for the creation of programs for reducing emissions of CO_2 and other greenhouse gases. The goal was to cut emissions so that by the year 2000, nations will have stabilized their output of heat-trapping gases at 1990 levels.

Most governments were amenable to going even

What You Can Do

Museum Exhibitions
• In 1992, a new source of information on global warming appeared as two science museums inaugurated the first exhibitions devoted to the subject. The American Museum of Natural History in New York City, in partnership with the Environmental Defense Fund, has assembled dozens of innovative, interactive displays that convey basic facts about climate change and the impact of human activities on the atmosphere. A similar show was put together by Philadelphia's Franklin Institute. Both exhibitions will travel to museums from coast to coast over the next few years.

Greenhouse Diet
• As an appendix to the book *Global Warming: Understanding the Forecast*, which is the companion volume to the American Museum of

: *Between 50,000 and 60,000 degrees Fahrenheit*
Source: The Colorado Daily, *Boulder, Colorado*

Natural History's new exhibition on global warming, the Environmental Defense Fund came up with a "Greenhouse Diet"—20 steps an individual can take to reduce his or her personal "greenhouse effect."

• The average American produces 40,000 pounds of CO_2 each year by driving a car, running air conditioning, lighting and appliances, and other actions that require the burning of fossil fuels.

Saving CO_2
Here is a sample of the significant CO_2 savings that can be made with simple lifestyle changes:

• Wrap your water heater in an insulating jacket, which costs just $10 to $20. It can save 1,100 pounds of CO_2 per year for an electric water heater, or 220 pounds for gas.

further—agreeing to a firm timetable for sharp cuts in these gases. But the US, alone among the wealthy industrialized nations, refused to go along. The reason? Bad timing. Because 1992 was a presidential election year, the political rhetoric surrounding global warming heated up to an extraordinary degree.

The Bush Administration insisted that the American economy would be imperiled if we agreed to sign a climate treaty containing hard commitments to reducing the human output of carbon dioxide. Conservative columnists and politicians began a campaign to discredit the science supporting the global warming theory. An effort was made to label environmentalists as the new Marxists. As columnist George Will wrote, some environmentalism is a "green tree with red roots."

Ironically, the scientific basis for concern about global warming became stronger than ever in 1992. A survey of climate research in the journal *Science* indicated that projections for warming in the coming century have not significantly changed in 15 years. There is still a broad consensus that a doubling of atmospheric carbon dioxide will lead to a 1.5 to 4.5 degrees Celsius rise in global mean temperature (3 to 8 degrees Fahrenheit).

Evidence that a nearly unprecedented warming is occurring became more robust, as well. By checking

The eruption of Mt. Pinatubo will cool the Earth's atmosphere for two years—then global warming will resume.

Q: *What is the yearly energy cost of a lack of insulation, drafty doors, and other energy leaks in American buildings, industry and transportation?*

the ratio of heavy and light oxygen isotopes locked in ancient glacial ice, scientists can construct a natural temperature record going back thousands of years. A sharp warming trend was indicated by tests of isotope ratios in glaciers in China, Khirgizia (in the former Soviet Union) and Peru. Glaciers around the world are also retreating at a record clip. In February 1992, scientists from the World Meteorological Organization and the University of Wisconsin, Madison, reported that the ice cap on Africa's Mt. Kenya shrank 40 percent between 1963 and 1987.

The 1991 eruption of Mt. Pinatubo in the Philippines spewed a vast amount of sun-blocking particles into the atmosphere. James Hansen, the director of NASA's Goddard Institute for Space Studies, said computer models predict a temporary cooling of the Earth from the volcanic aerosols. After a record warm winter, 1992 indeed began to exhibit a cooling trend. If the cooling persisted through the year and into 1993, Hansen said, that would increase confidence that the models are correct in predicting a sharp warming in coming decades.

Hansen cautioned, though, that we should not rely on volcanoes to cool off the greenhouse. The aerosols fall back to Earth within a few years. Projections are that global temperatures will resume their record climb around 1994.

"The road is now marked out," said United Nations Secretary General Boutros Boutros-Ghali, speaking at the Earth Summit. "Today we have agreed to hold to present levels the pollution we are guilty of. One day we will have to do better—clean up the planet."

An award-winning author and journalist, Andrew Revkin has spent the past decade reporting on the environment and the interplay of science and society. Revkin's latest book is Global Warming: Understanding the Forecast, *published by Abbeville Press in conjunction with the first major museum exhibition on global warming. His first book,* The Burning Season: The Murder of Chico Mendes and the Fight for the Amazon Rain Forest *(Houghton Mifflin, Boston, 1990), won a Robert F. Kennedy Book Award and the Sidney Hillman Foundation Book Prize. Revkin is formerly a senior editor of* Discover *magazine and a staff writer for the* Los Angeles Times.

• Buy energy-efficient compact fluorescent bulbs for your most-used lights. In a typical home, one compact fluorescent bulb can save 260 pounds of CO_2 per year.

• When you buy a car, choose one that gets good mileage. If your car gets 40 miles per gallon instead of 25, and you drive 10,000 miles per year, you'll cut your annual CO_2 emissions by 3,300 pounds.

Contact
• American Council for an Energy-Efficient Economy, 1001 Connecticut Ave. NW, Ste. 801, Washington, DC 20036, (202) 429-8873.

• League of Conservation Voters, 1707 L St. NW, Ste. 550, Washington, DC 20036, (202) 785-8683.

Read
• *Earth in the Balance: Ecology and the Human Spirit*, by Senator Al Gore, Houghton Mifflin Co., Boston, 1992. A well-reasoned proposal for improving the relationship between humans and the planet.

• *Global Warming: Understanding the Forecast*, by Andrew Revkin, Abbeville Press, New York, 1992. A large-format, illustrated guide to the global warming issue and possible solutions, written for the entire family.

A: *$300 billion annually*
Source: National Wildlife *magazine*

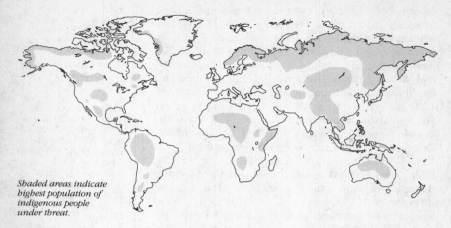

Shaded areas indicate highest population of indigenous people under threat.

Indigenous People

- By Ted Macdonald -

"Whenever a shaman dies, it's as if a library disappeared." Such statements have been repeated and expanded through films like "Medicine Man" and countless articles. The images illustrate how the Western world allocates space in the environmental landscape—indigenous people provide us with botanical knowledge and holistic world views. As most astute and sympathetic observers agree, indigenous people manage their resources in a sophisticated and sustainable manner.

Yet many would lump indigenous people in with the flora and fauna—another "endangered species" to be preserved and observed in the name of biological diversity. While for some small, isolated societies, protection is essential for survival, to extend that response to all forest people overlooks a critical distinction between science and history—people evolve socially and politically as well as biologically. Thus, about 20 years ago, many indigenous communities, particularly those of Latin America, began to accelerate their evolution through political mobilization. These new, grassroots organizations are positioned to stem forest destruction far more effectively than any Western pleas to preserve indigenous knowledge.

The human rights link to environmental issues was

What You Can Do

The situation of indigenous people in general remains very difficult. Articles, books and other sources of information on the problems faced by indigenous and tribal people are widely available from a variety of sources.

• Cultural Survival distributes books and pamphlets. Write for a free catalog: Publications, Cultural Survival, 215 First St., Cambridge, MA 02142.

Additional information on recent indigenous initiatives to secure land and resource rights (and programs to manage those resources) is available from a variety of sources. These include:

• World Rainforest Movement (WRM), International Secretariat, 87 Contonment Rd., 10250, Penang, Malaysia;

Q: *How many tourists visit America's national parks each year?*

dramatically illustrated by the 1988 assassination of Francisco "Chico" Mendes, a Brazilian rubber tapper who promoted the sustained use of standing forests rather than clearcutting for cattle pasture lands. In the eyes of many in the US and Europe, he was a conservationist. But to the local elites of the Brazilian Amazon, he was a labor organizer who threatened the status quo.

On a far greater scale, indigenous communities throughout the Americas and many other areas of the world have now organized themselves into local, regional, national and international ethnic federations. Indigenous organizations in Bolivia and Ecuador recently challenged local powers through extended protest marches, and many of their leaders, like Mendes, risk violent consequences.

Where are the environmentalists in all this? Indigenous people form part of the forest landscape, and their culture—indeed their fate—is linked closely to that forest's future. Yet while some convergence of interests exists between long-term forest residents and those concerned with environment and development, the links are indirect and alliances are fragile. Most indigenous and other forest people recognize that any outsider can put their land and its resources—their present and future capital—at risk. So indigenous people are as wary of the term "conservation" as they

or: 8 Chapel Row, Chadlington OX7 3NA, England. In early 1992 the WRM convened an international meeting of indigenous and tribal people to prepare the final draft of the "Charter of the Indigenous-Tribal Peoples of the Tropical Forests." The document highlights the relationship between environment and culture. Copies can be requested from WRM.

• COICA (Coordinating Group for Indigenous Organizations of the Amazon Basin), Jiron Larco Herrera, 1057, Magdelena del Mar Lima 17, Peru. In collaboration with Oxfam-America, this international indigenous organization recently published a study on indigenous land and land rights initiatives in South America—"El indigena y su territorio" (by Richard Smith, Pedro

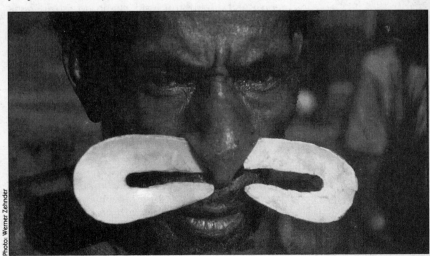

A tribesman from the remote New Guinean lowland region of Asmat.

A: *250 million*
Source: "The World of Audubon Special"

Garcia and Alberto Chirif).

• World Wildlife Fund, 1250 24th St. NW, Ste. 500, Washington, DC 20030. Request literature on their Tropical Forestry Program and the Wildlands and Human Needs Program.

• National Geographic Society, 17th and M Sts. NW, Washington, DC 20036, (202) 857-7000. The society recently published an excellent map, "The Coexistence of Indigenous Peoples and the Natural Environment in Central America," a special supplement to "Research and Exploration: A Scholarly Publication of the National Geographic Society" (Spring 1992).

• Dene Culture Institute, 35 Morrison Dr., Yellowknife, NWT X1A 1Z3, Canada. The institute has prepared a broad, detailed directory of indigenous projects and programs, entitled "Amerindian Initiatives in Environmental Protection and Natural Resource Management."

• Second InterAmerican Indian Congress on Conservation and Natural Resources, Casilla 14259, La Paz, Bolivia. This congress was a December 1991 meeting of indigenous people involved in a variety of conservation and natural resource management programs. Request copies of the proceedings.

are of "development," always asking: "For whom?"

In other words, indigenous people's primary concern lies with gaining secure land and resource rights, not just the complex management of that environment. They are often as suspicious of environmentalists as they are of development agencies—both groups often design plans for the use of land and resources claimed by forest people. Park planners and managers in many countries of the developed as well as the less-developed world still cast indigenous people's rights alongside those of forest exploiters who must be restrained or expelled from protected areas.

Indigenous ethnic federations initially focused solely on political mobilization, an essential first step toward broader empowerment. Now many indigenous communities ask their organizations: "OK, we're organized. Now what?" These people are asking their leaders to pursue land claims and obtain technical assistance for programs of resource management.

Such requests enable those concerned with the environment and sustainable development to work directly with indigenous people as they draw from traditional technologies and create new ones. Support, however, must include respect for rights to land and resources. As illustrated in indigenous resource managed projects such as Panama's Project PEMASKY and Ecuador's Awa Project, indigenous people and other forest residents do not isolate technology, politics and culture.

Through their organizations, indigenous people have raised their heads and are demanding attention, not simply as objects of research, largesse or sympathy, but as legitimate actors on a broad political landscape. Given that attention, indigenous people will play a vital role in maintaining the world's resources.

Ted Macdonald, an anthropologist, has been projects director since 1979 for Cultural Survival, a Massachusetts-based organization that undertakes advocacy, research, publications and field projects on behalf of indigenous people and ethnic minorities. During that time he has also been a research associate in the Department of Anthropology at Harvard University.

Q: *Growing grains and vegetables uses how much raw material as compared with meat production?*

EARTH ISSUES

Shaded areas indicate presence of minorities involved in environmental activism.

Minorities and the Environment

- By Richard Moore -

More than two busy years have passed since people of color began to raise the call for environmental justice to mainstream environmental organizations and others. During 1990 alone, the Gulf Coast Tenants Organization and SouthWest Organizing Project initiated open letters to the Group of Ten (a coalition of environmental groups), calling for equitable distribution of resources and board and staff representation for people of color; the Panos Institute published *We Speak for Ourselves*, which provided African, Asian, Latino and Native Americans with a voice on the environment; and Earth Day activities were conducted in communities of color in various parts of the country. Some EPA staff went public with concerns regarding institutional racism at the agency, concerns they had been raising internally for years. The newly formed Southwest Network for Environmental and Economic Justice called on environmentalists organizing in communities of color to provide resources to and accept direction from those affected workers and communities. Even *Time*, *Newsweek*, the "Today Show" and other major media outlets covered, for

What You Can Do

• If you know or suspect that industrial, military, agricultural, governmental or any other type of pollution is impacting your health or safety, the first thing to do is to talk to your neighbors to see if they also feel affected by the problem. You may have to go door to door in order to find this out, and spend a lot of time with others in your community.

• If the source of the pollution is an employer in your community, talk to neighbors who may work there. They can give you information about any concerns they may have inside the facility, as well as an idea of what kinds of chemicals are being used there.

• Research the problem. Contact local or state environmental health agencies or the EPA. These are supposed to give you important

A: *95 percent less*
Source: The Student Environmental Action Guide, *EarthWorks Press*

information about what types of chemicals the polluter is discharging into the water and air.

• Talk to other groups in your area, region or in other parts of the country that have addressed the types of problems you face.

• Recruit potential allies in your community who are either directly or morally affected by the problem. These can include civic groups, civil rights organizations, neighborhood associations, churches, student organizations and trade unions.

• Conduct a community tour of your neighborhood to publicly expose the problem. Invite representatives of local organizations, the media and elected officials to go on the tour, and then discuss the issues of concern so that everyone can voice opinions.

• Conduct accountability sessions with elected

better or worse, efforts to fight the poisoning of communities of color.

Two years later, things have definitely changed, though perhaps not as much within the traditional environmental movement as might have been expected. True, some diversification has taken place within the major organizations. True, there have been a few partnership efforts between grassroots communities and Group of Ten organizations. More significant changes have certainly occurred within two organizations outside the Group of Ten—the National Toxics Campaign and, to a lesser extent, Greenpeace—both of which had shown a greater commitment previously to working with grassroots communities of color. At the National Toxics Campaign, for example, nearly one-half of the board of directors is now composed of people of color with concrete experience in efforts to address environmental racism and injustice. Others, such as the Environmental Careers Organization, which formerly catered exclusively to the needs of the Group of Ten and corporations, have begun to place interns with grassroots groups and progressive policy institutes like Panos.

But what is really different is that people of color have come together across the country to build our own movement, one deeply rooted in past and continuing struggles for human and civil rights, safe and affordable housing, cultural preservation, employment rights, healthy working conditions, and accountability among our elected officials. Our

Manuelito, Arizona, residents pump water at the only well for 1,500 people since the Puerco River on the Navajo Reservation was contaminated with uranium.

Q: *Swedish municipal incineration is responsible for what concentration of mercury emission in that country?*

movement defines *environment* as where we live, work and play, and therefore is integrally linked with all other issues that confront us day to day. Our movement is inclusive. It recognizes the urgent need for people of color to come together so that we may define our own issues and needs. But we also recognize the need to reach out and develop alliances with whites and white organizations, and the importance of ultimately developing a fully multiracial movement for justice.

The 1991 First National People of Color Environmental Leadership Summit reflected the groundswell of activity that has led to the development of this movement. More than just a high-profile "national event," the summit epitomized the growth of local grassroots efforts for environmental justice. Community leaders with many years of experience came from throughout the US (from as far away as Hawaii and Puerto Rico) to join with one another to celebrate the fact that we are not alone in our efforts, and that we are growing stronger every day.

The network reflects the fact that there is as much or more local organizing taking place among people of color today as in the 1960s and early 1970s. The network is serving as an attractive model for others, particularly those in the southeastern US, who are looking for ways to strengthen local work and impact national policy without having to rely on national "advocacy" organizations to speak on our behalf.

After surviving a decade of Reaganism, people are looking for mechanisms through which they can mutually strengthen their efforts. Clearly, the environment as defined in its broadest sense is the thread that has the potential to bring us together. We look forward to making this happen, understanding that there will no doubt continue to be a lot of struggle and contention in the process. But, as Frederick Douglass once said, without struggle there is no progress.

Richard Moore is coordinator of the Southwest Network for Environmental and Economic Justice and a founder of the Albuquerque-based SouthWest Organizing Project. Moore served as a member of the Planning Committee for the First National People of Color Environmental Leadership Summit and is a longtime member of the Eco-Justice Working Group of the National Council of Churches.

officials. Obtain concrete commitments to take action to enforce environmental regulations.

• Confront the polluters to learn what they may do to address the problem. You will have to decide on the best time to do this (e.g., before or after you go public with your concerns). It is important to have a committed base of support when you do this.

• Keep your community informed. Develop brief fact sheets periodically that inform the community of the progress in the fight, and distribute them throughout your neighborhood. Appear on radio and TV talk shows.

Contact
• Southwest Network for Environmental and Economic Justice, P.O. Box 7399, Albuquerque, NM 87194, (505) 242-0416.

• SouthWest Organizing Project, 211 10th St. SW, Albuquerque, NM 87102, (505) 247-8832.

• Gulf Coast Tenants Organization, P.O. Box 56101, New Orleans, LA 70156, (504) 949-4919.

• Southern Organizing Committee for Economic and Social Justice, P.O. Box 811, Birmingham, AL 35201, (205) 781-1781.

• United Church of Christ Commission for Racial Justice, 475 Riverside Dr., Ste. 1950, New York, NY 10115, (212) 870-2077.

A: *60 percent*

Source: Greenpeace Action

EARTH JOURNAL

Shaded areas indicate nuclear power, weapons, suspected nuclear weapons, nuclear programs, uranium mining, or nuclear test sites.

What You Can Do

• Help to declare your local community a Nuclear Free Zone (NFZ), banning the production of nuclear weapons, nuclear power and the disposal of radioactive wastes. Join a global movement of over 4,500 locally declared NFZs in 43 countries, including 188 in the US in cities as large as Chicago and as small as Homer, Alaska.

• Let your local, state and federal officials know that you do not support the rush to site radioactive waste facilities. Tell them you support ending nuclear power generation and nuclear weapons production, which generate the vast majority of radioactive wastes. Wastes can be temporarily stored on-site until thorough, scientific research can be done, without the pressure for a "quick fix" answer.

Nuclear Weapons, Power and Waste

- By Charles K. Johnson -

Despite the end of the Cold War, the dismantling of the Soviet Union as a superpower and a tentative agreement with Russia to reduce our respective nuclear stockpiles of between 3,000 and 3,500 warheads, US military policy continues to emphasize building more nuclear weapons and large-scale nuclear weapons systems like the Trident submarine and the B-2 stealth bomber.

And, despite increasing evidence of the hazards of nuclear power and the insurmountable environmental problems caused by a massive volume of nuclear waste, US energy policy ignores safer alternatives and continues to promote a so-called "new generation" of nuclear power plants as the major new source of energy into the 21st century.

These two seemingly unrelated activities are presided over by the US Department of Energy (DOE) in its twin role as atom bomb maker and nuclear power researcher and advocate.

Sometimes the lengths to which nuclear planners go to advocate the nuclear industry are astounding. When earthquakes heavily cracked the test mine shafts of the proposed Yucca Mountain High-Level

Q: *For every degree decrease in a heating system, by what percentage are heating costs reduced?*

Radioactive Waste Repository in Nevada in August 1992, DOE officials claimed they were pleased to have all the new seismic data that resulted from the quakes. One can only marvel at their optimism.

But to illustrate how the determination to continue these misguided policies translates into cold, hard cash, this year federal funding for nuclear weapons and weapons systems remains at $61.9 billion, roughly the same level as for each of the last five years. To put this in perspective, that $61.9 billion represents more federal money spent in one year than was spent on housing during the entire 12-year Reagan-Bush Administrations. The Bush Administration is still planning to modernize its aging atomic bomb plants to extend their capability to make nuclear weapons well into the next century.

At the same time, the administration, influential members of Congress and the nuclear power industry are gearing up for another attempt to sell Wall Street and the American public on a supposed new generation of "inherently safe" nuclear reactors. Funding for nuclear fission and fusion research is rising and will be nearly $1.5 billion for fiscal year 1993, dwarfing conservation and renewable energy research. In addition, the industry is pushing for taxpayer forgiveness of $10 to $14 billion of debt by nuclear facilities for uranium fuel and fuel-processing services. This is on

- Lobby your members of congress and the president to support a permanent end to nuclear testing. Congress recently passed a nine-month moratorium on testing, which President Bush signed as part of other legislation. Russia and France already have temporary nuclear test bans. A permanent ban on nuclear testing will make it more difficult to develop new nuclear weapons; now is the time to push for a ban.

- Promote energy efficiency and renewable energy in your home and business. Support the development in your state and in your community of energy conservation and safer, cleaner sources of energy.

- Boycott investment or purchase of goods and services from companies that make nuclear

An abandoned commercial nuclear power reactor in Satsop, Washington.

A: *2 percent*
Source: Protect Our Planet Calendar, Running Press Book Publishers

weapons or generate nuclear power. Avoid radioactive materials in consumer products. Get your place of worship, civic group or local community to do the same.

Resources

• Nuclear Weapons Production and Rad-Waste:
The Military Production Network, 218 D St. SE, Second Fl., Washington, DC 20003, (202) 544-8166.

• Nuclear Power Generation and Rad-Waste:
Nuclear Information and Research Service, 1424 16th St. NW, #601, Washington, DC 20036, (202) 328-0002.

• Nuclear Weapons Testing:
The Plutonium Challenge, c/o Natural Resources Defense Council, 1350 New York Ave. NW, #300, Washington, DC 20005, (202) 783-7800.

• Renewable Energy:
Rocky Mountain Institute, 1739 Snowmass Creek Rd., Old Snowmass, CO 81654, (303) 927-3851.

• Nuclear Free Zones:
Boycott Information—Nuclear Free America, 325 E. 25th St., Second Fl., Baltimore, MD 21218, (410) 235-3575.

Read

• *Trinity's Children, Living Along America's Nuclear Highway,* Tad Bartimus and Scott McCartney, Harcourt Brace Jovanovich, New York, 1991.

top of such existing subsidies as the Price-Anderson Act, which limits nuclear utilities' liability to $500 million in the event of a core meltdown, a tiny fraction of the estimated cost.

Yet all this pro-nuclear momentum faces a major stumbling block—the question of what to do with mass quantities of toxic radioactive waste.

Since the dawn of the atomic age, millions of gallons of radioactive waste have been accumulating in storage facilities around the country. Much of the waste is being held in disposal sites where safety is not assured. Improper disposal methods and leaking tanks are being discovered with alarming regularity. At the high-level radioactive waste facility at Hanford, Washington, for example, some of these nuclear wastes could explode, spewing radioactive particles over a wide area of eastern Washington and Oregon, as a waste dump at Kyshtym, Siberia, did in 1957, contaminating hundreds of square miles and poisoning thousands of people.

The need for additional high-level and low-level radioactive waste storage sites grows increasingly dire, as we continue to generate waste without finding the solutions for its disposal. Citizens and, in some cases, local and state governments are opposing locating waste sites in their communities. Meanwhile, as the nuclear industry and the federal government become more desperate in their search for waste sites, their tactics have become more high-pressure and coercive. Increasingly, they resort to undemocratic methods, such as pre-empting local siting authority and state policy.

Citizen organizing and community collective action have brought about major victories for the anti-nuclear movement. A grassroots chorus of communities saying "no" to nuclear waste dumps has emerged. The nuclear industry continues to pressure Congress, and state and local governments, to site rad-waste facilities of all types. They know that the failure to site new dumping grounds for their radioactive materials would mean the death knell for their industry.

Charles K. Johnson is the director of Nuclear Free America, an international clearinghouse for nuclear free zones and US military contract information, based in Baltimore.

: *What percentage of the world's projected 1-billion population growth (between 1990 and 2000) will be in developing countries?*

EARTH ISSUES

Shaded areas indicate coastal areas most polluted.

Ocean Pollution

- By Thomas Miller -

There was a time, not so long ago, when the world's oceans were viewed as little more than a barrier to people's conquest of the terrestrial environment. Except for a few high seas fishing fleets, most interaction between the marine environment and humans was, and remains, at the water's edge. And here is where human impact on the marine environment is most visible.

Pollution along our coasts and beaches appears to have become the rule rather than the exception. Humankind's impact on the coastal zone has been documented for many decades, from urban harbors to remote bays, estuaries and gulfs, where the human waste stream comes to rest as debris.

The International Coastal Cleanup, the world's largest, collected 3.7 million pounds of trash along 4,743 miles of beach in 1991. Coordinated by the Center for Marine Conservation, the 1991 cleanup enlisted more than 145,000 volunteers in 34 US states and territories and 12 countries.

Plastics were the most common debris item reported, accounting for approximately 59 percent of all debris collected. While federal and international statutes and treaties have been in force for some time, they don't appear to steam the tide of trash

What You Can Do

Get Involved
• Join the Center for Marine Conservation, 1725 DeSales St. NW, Ste. 500, Washington, DC 20036, (202) 429-5609. Basic membership is $20 annually and includes a subscription to the quarterly newsletter *Marine Conservation News*. The center also publishes *Coastal Connection*, for those interested in keeping informed on the International Coastal Cleanup.

Stencil Program
• Join a local group involved with stenciling clean water messages on storm drains throughout the US. For more information on how to get involved, contact the Center for Marine Conservation's Pollution Prevention Program.

Beach Cleanup
• Be a part of the world's largest beach

A: *93 percent*
  Source: International Herald Tribune

cleanup. Each year in September, the Center for Marine Conservation's International Coastal Cleanup takes place throughout the US and in other countries. Volunteers spend a Saturday morning picking up and cataloging beach trash. This information is then compiled by the Center for Marine Conservation and released in a report to the media, policymakers and the public.

Monofilament Line Recycling
• Monofilament fishing line left in the water can prove deadly to birds and marine life. Properly dispose discarded fishing line and, where possible, recycle. Berkley, a fishing tackle manufacturer in Spirit Lake, Iowa, has a nationwide recycling program to reduce this threat to wildlife. Some 38 cases of

along the world's coasts.

The shipping industry, offshore oil interests, fishing fleets, passenger cruise lines, merchant vessels and the navies of the world are just a few of the culprits that need to take a greater, more visible role in ending this modern-day trashing of coasts.

But ocean-based sources are only part of the problem. What begins well upstream of the coast also has a direct and often deadly impact on coastal eco- systems and their wildlife. More than 1,100 US cities continue to discharge billions of gallons of raw sewage into some of the country's most important estuaries and coastal waters. Every year, sewage treatment facilities discharge 5.9 trillion gallons of sewage wastewater into coastal waters. The Environmental Protection Agency (EPA) estimates that 160,000 factories dump between 41,000 to 57,000 tons of toxic organic chemicals and about 68,000 tons of toxic metals into these systems each year. Both the Center for Marine Conservation and the Natural Resources Defense Council (NRDC) have documented the problems associated with combined sewer overflows (CSOs). CSOs occur when rains and storms swell the volume at local sewage treatment plants so that, together, the storm water and household waste exceed the capacity of the system to treat the flow, and it is simply discharged directly into rivers, lakes and coastal waters.

Long Island Sound illustrates the problem of pollution in coastal waters.

Q: *During a "normal" summer, how many Americans die from excessive heat and sun?*

CSOs impair 2,997 miles of river, 3,569 acres of lakes and 228 miles of estuary in the US. Seven of 17 estuaries in the EPA's National Estuary Program are adversely affected by CSOs. In 1990, 2,400 public beaches and 597,000 acres of shellfish beds were closed due to CSOs. The economic losses amount to countless millions of dollars annually.

In 1988, the state of New Jersey lost an estimated $1 billion from beach closures. NRDC has documented CSOs as a major factor in 1,467 days of beach closures or advisories in 1990. The Washington State Department of Fisheries estimates that more than $3 million is lost annually because of pollution-related shellfish bed restrictions.

CSOs, like marine debris, affect not only the livelihoods and quality of life for millions of people, but also the wildlife that depends on a healthy coastal ecosystem. Polluted beaches, closed fisheries, contaminated seafood and dead and dying wildlife are the price we pay for failing to spend the money necessary to improve antiquated sewer systems.

While much debate has focused on coastal zones, nuclear waste, munitions, dredge spoils and other toxics have found their way into waters far from the coast. Scientists continue to debate whether the ocean is an appropriate dumping ground for millions of tons of toxic waste and sewage sludge.

Scientists are just beginning to unravel the mysteries of deep sea ecosystems and the marine life that lives there. Who can say with certainty what short- and long-term effects the dumping of millions of gallons of sewage sludge and toxics into deep oceans will cause?

This year marks the 20th anniversary of the Clean Water Act. But we have yet to achieve the act's goal for all the nation's waters to be "fishable and swimmable." Let us all renew our personal and institutional commitment to continue the fight for clean water—both at land and at sea.

Thomas Miller is the director of press relations for the Center for Marine Conservation, an international conservation organization. He works on issues such as the Endangered Species Act, offshore oil, coastal pollution, and sea turtle, marine mammal and fisheries conservation.

wildlife entangled in monofilament line were reported during the 1991 International Coastal Cleanup, accounting for over one-third of all wildlife entanglement incidents reported.

Six-pack Holders
• Seemingly harmless, six-pack holders can pose a serious, possibly deadly, hazard to wildlife. During the Center for Marine Conservation's 1991 Cleanup, 34,492 six-pack holders were reported by cleanup volunteers. As with all items you take to the beach, what you take in, you should take out. Cut up six-pack holders prior to disposal.

Massive Balloon Launches
• During the 1991 Cleanup, 36,164 balloons were collected. Outdoor balloon launches pose a potential hazard to birds and other wildlife, while littering our waterways and beaches. Balloons themselves are not the problem. Responsible use and disposal of balloons is the key to reducing litter and protecting wildlife.

Get Involved
• By far the most important thing you can do for the marine environment is to become involved in coastal and ocean issues in your community. Write your elected officials and vote for candidates who support clean water initiatives.

A: *175 to 200 each summer*
Source: The Colorado Daily, Boulder, Colorado

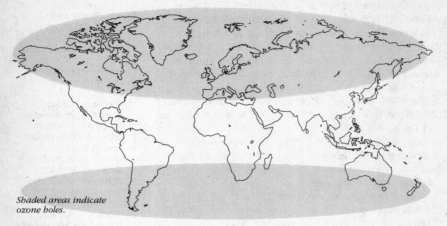

Shaded areas indicate ozone holes.

Ozone Depletion

- By Sasha Madronich -

What You Can Do

The Damaging Effects of UV

• Exposure to ultraviolet (UV) radiation is the main cause of skin cancer. A recent UN panel concluded that a 10 percent ozone depletion, if sustained over several decades, would result each year in 300,000 new cases of nonmelanoma skin cancers and 4,500 new cases of melanoma, the deadliest form of skin cancer. Because the induction time of skin cancer is long, the effects will not appear for several decades. Some other cancers, including those of the lip and the salivary glands, may also be related to UV exposure.

• UV causes a number of eye disorders, including cataracts, snow-blindness, retinal damage, intra-ocular melanoma and the onset of farsightedness at an earlier age.

In the early 1970s, scientists began suspecting that stratospheric ozone may be destroyed effectively by human-made chlorofluorocarbons (CFCs) used for refrigeration and as solvents, aerosol propellants and foam-blowing agents. The CFC gases released at the surface waft upward to the stratosphere, where they are decomposed by intense ultraviolet (UV) radiation. Each liberated chlorine atom can destroy thousands of ozone molecules before being removed from the stratosphere. Although ozone is regenerated continuously by the sun's rays, it remains lower than its natural amount as long as chlorine atoms are present. Consequently, more UV radiation penetrates to the surface of the Earth, threatening all forms of life.

The ozone depletion theory (as it was then known) was at first quite controversial. Measurements showed that ozone amounts vary from day to day, not unlike temperature or humidity. Most scientists predicted that reduction in ozone due to the CFCs would be minor in comparison to the natural fluctuations. The concerns led some countries to ban the use of CFCs as aerosol propellants for personal products such as deodorant and hair sprays, but most CFC uses remained unabated.

Then, in 1985, an unexpected discovery changed

 What is the price for one minute of fresh air from a sidewalk oxygen vendor in Mexico City?

everything. British scientists had been monitoring ozone in the Antarctic continent continuously since 1957. The springtime ozone had been relatively constant from 1957 until 1980, but then began to decline until, in 1985, it fell to about half of the average value from the earlier decades. Scientists quickly realized that they had underestimated the problem, and that the effectiveness of ozone-eating chlorine atoms is greatly increased by stratospheric particles such as the high-altitude clouds normally present in the cold temperatures of the polar winter. Policymakers began a historic process of negotiations culminating with the 1987 Montreal Protocol, strengthened by the 1990 London amendments, in which the members of the United Nations agreed to phase out CFC production by the year 2000. The US has recently accelerated its own schedule to 1995.

Today, stratospheric ozone depletion is a reality, not just a theory. Many more ozone measurements have been collected, from ground-based stations looking up and from satellites looking down through the atmosphere. Substantial ozone reductions have occurred over many parts of the globe. The strongest reductions are still over the Antarctic continent and its surrounding ocean. The last three years (1989, 1990 and 1991) have been particularly severe, with 1991 values lower than ever before. Lesser but still significant reductions in ozone have occurred at

• UV has been shown to suppress the immune system and to activate the herpes simplex, HIV-1 and papilloma viruses. Resistance to bacterial infections, such as malaria, tuberculosis and leprosy, may also be diminished. The evidence so far suggests that UV increases the severity of these diseases, rather than the rate of infection. UV may also interfere with the effectiveness of vaccination programs. In contrast to skin cancer, UV-induced immune suppression seems to affect all races, regardless of skin pigmentation.

• UV reduces the growth of many plants. UV-sensitive species include economically important varieties of conifer trees and food crops such as wheat, rice and soybeans. Seedlings appear to be more vulnerable than mature plants.

Scientists in Sweden prepare to launch a balloon to study ozone depletion.

A: $1.15

Source: Harper's magazine

- UV penetrates the ocean and impairs the growth of marine phytoplankton, zooplankton, shrimp and crab larvae, and juvenile fish. The effects on phytoplankton are of particular concern, as these microscopic plants form the basis of the marine food chain. Biologists have recently found that phytoplankton is reduced by 10 percent to 15 percent under the Antarctic "ozone hole."

- Together with hydrocarbons and nitrogen oxides pollutants, UV is one of the main ingredients in the formation of photochemical smog, which affects cities worldwide. Increase in UV radiation will make it more difficult to achieve clean air standards.

Resources
- *Environmental Effects Panel Report* and *Environmental Effects of Ozone Depletion: 1991 Update*, available from the US EPA, 401 M St. SW (RD-682), Washington, DC 20460.

- Practical suggestions for reducing the use of ozone-destroying chemicals may be found in: *Protecting the Ozone Layer: What You Can Do*, Environmental Defense Fund, 257 Park Ave. S., New York, NY 10010. Also, EarthWorks Press, 1400 Shattuck Ave., Box 25, Berkeley, CA, 94709, publishes *50 More Simple Things You Can Do to Save the Earth* and *50 Simple Things Your Business Can Do to Save the Earth*.

mid-latitudes of the Southern Hemisphere, which cover large portions of the oceans as well as Australia, New Zealand, South Africa and parts of South America. The reductions range from 3 percent to 10 percent between 1979 and 1989. The mid-latitudes of the Northern Hemisphere (North America, Europe and much of Asia) have experienced spring and summer ozone reductions of 2 percent to 6 percent, and winter decreases of 4 percent to 8 percent. For now, no ozone depletion has been observed over equatorial regions.

The next few years will be crucial to understanding the full severity of ozone depletion. In January 1992, chlorine amounts in the northern polar regions were higher than ever observed before. According to current scientific understanding, large ozone reductions could have occurred if temperatures in the stratosphere had remained low into early spring. Fortunately, temperatures rose more rapidly than usual during February and no northern ozone hole was observed, but a more typical polar winter in the next few years might result in large ozone depletions.

An insidious aspect of the problem is that a long time lag exists between the industrial production of CFCs and their ultimate conversion to ozone-depleting chlorine atoms. There are no practical ways to accelerate the removal of CFCs and chlorine already in the atmosphere. Even with the Montreal Protocol in effect, chlorine amounts are expected to increase during the 1990s, peaking near the year 2000, then decreasing gradually back to 1990 levels by the year 2030.

The Montreal Protocol is an unprecedented human achievement. But its history is also a warning that, even with our best intentions, damage already done to the environment cannot be repaired instantly, and may be passed on to future generations.

Sasha Madronich is a scientist with the Atmospheric Chemistry Division of the National Center for Atmospheric Research in Boulder, Colorado. He has served on United Nations Environment Programme panels assessing stratospheric ozone depletion and the effects of increased UV radiation on the biosphere, and on international commissions for the monitoring of UV radiation changes.

: **What is found in the bark of the Pacific yew tree?**

EARTH ISSUES

Shaded areas indicate most extreme over-population.

Population

- By Dianne Sherman -

Every day we share the Earth and its resources with 250,000 more people than the day before. Every year there are another 95 million more mouths to feed and, due to environmental degradation, the world's farmers must do it with an annual loss of 24 billion tons of topsoil.

For most of history, human population had an almost imperceptible impact on the Earth. In fact, the vast majority of the population explosion has taken place in less than one-tenth of 1 percent of human history.

When Columbus "discovered" the Americas 500 years ago, global population was small, numbering only 425 million. The planet seemed an infinite resource—its air, water and land at our disposal for the creation of our wealth and the deposit of our waste.

Much has changed. In just the past 40 years, the world economy has quadrupled in size, and world population has tripled to 5.4 billion. There is mounting evidence that this massive "growth machine" may have already gone beyond its limits. Global warming, ozone depletion, deforestation, disappearing species and water and food shortages are but a few of the warning signs.

The population crisis is not just caused by

What You Can Do

• Communities across the nation are grappling with shrinking open space, traffic-choked highways, worsening air pollution, lack of landfill space, and diminishing quality and availability of services. Check on local zoning, planning and growth management laws to see if your community has an ordinance to regulate population growth and development, and to ensure the environment is fully protected. If not, pressure local officials to introduce such legislation.

• Write the president, and your local and state elected officials, encouraging the US to take the lead in promoting sustainable population policies in the US and worldwide.

• Over the next two decades, 3 billion young people will be entering their reproductive years.

 Taxol, a potent drug used in the treatment of breast, ovarian and lung cancers
Source: National Wildlife magazine

At current levels of distribution, only about half will have access to contraception. Support family planning programs and contraception research, both in the US and worldwide.

• The phenomenal growth in human population is taking its toll, as wildlife habitats are paved over, built on, polluted or mined—all to "benefit" encroaching civilization. Work to protect local wetlands, forests and other biologically rich habitats.

• The US government spent at least $500 million daily during the Persian Gulf war. By comparison, the 1991 federal budget allocations for domestic family planning programs could pay for only 7.6 hours of that war. International family planning funds

expanding human numbers. In many ways, the 1.2 billion people in developed countries are doing more damage than the 4.2 billion in developing countries. Burgeoning consumption among the rich and increasing dependence on ecologically unsound technologies to supply that consumption play major roles. This is especially true in the US, where we consume more energy, food and water and produce more solid waste than any other nation's population.

But as incomes and technologies in the third world increasingly resemble those of the North, developing nations will have average consumption levels on a par with developed countries. Thus it will make a very big difference to almost every form of environmental degradation how quickly—and at what size—the world population eventually stabilizes. A new set of United Nations projections for the next century dramatically illustrates this point—global population could stabilize at 8 billion, or reach 19 billion and continue to grow!

Because of the political explosiveness of the issue, population has often been conspicuously absent from environmental discussion. The Earth Summit in June 1992 was a clear example. Despite much commitment and perseverance from organizations that worked hard to incorporate overpopulation in the

A crowded street on a Saturday afternoon in Tokyo. To keep its economy growing, Japan is encouraging its women to have more children.

Q: *How much food is dumped into landfills by Americans every year?*

summit agenda, politics prevailed, and population was essentially glossed over in the official proceedings. One can only hope that public pressure will prevent this from happening again in the proceedings leading up to the United Nations Conference on Population and Development, scheduled for 1994.

In spring 1992 the National Academy of Sciences and the Royal Society of London, in a first-ever joint statement, warned that "if current predictions of population growth prove accurate and patterns of human activity on the planet remain unchanged, science and technology may not be able to prevent either irreversible degradation of the environment or continued poverty for much of the world."

This unprecedented statement strongly counters the many "techno-fix" advocates who have long argued that technological advancements will allow humankind to keep up with the needs and demands of growing populations.

So what are the chances for success, especially at a time when the anti-choice, anti-environmental movements continue to target our nation, and when these reactionary campaigns are actively spreading to other parts of the globe as well?

Most agree that stopping overpopulation will take a combination of personal choices and public policies. These include making family planning services universally available; developing safer and more effective contraceptives; affording women legal, educational and social equality; improving infant and child survival rates; breaking the cycle of poverty and resource disparity; implementing better technologies; and curbing overconsumption.

Of course, solving the population crisis will not alone put an end to environmental degradation and social injustice. But it will buy time to improve the quality of life for present and future generations. The alternative might put a livable and just world permanently out of reach.

Dianne Sherman is the director of communications of Zero Population Growth the nation's largest grassroots advocacy organization working to stop overpopulation both in the US and worldwide. The Natural Resources Council of America awarded Sherman the 1991 award of achievement in the field of environmental and conservation publications.

could defray only 12.5 hours of war expenses.

• By the time an American reaches age 75, he or she will have produced 52 tons of garbage, consumed 42 million gallons of water and used five times as much energy as the world average. If you decide to have children, consider having only one and definitely stop at two—you might also want to consider adoption.

Resources
• Zero Population Growth, 1400 16th St. NW, Ste. 320, Washington, DC 20036, (202) 332-2200.

• Population Crisis Committee, 1120 19th St. NW, Ste. 550, Washington, DC 20036, (202) 659-1833.

• Population Institute, 107 Second St. NE, Washington, DC 20002, (202) 544-3300.

• Population Reference Bureau, 1875 Connecticut Ave. NW, Ste. 520, Washington, DC 20009, (202) 483-1100.

• United Nations Population Fund, 220 E. 42nd St., New York, NY 10017, (212) 297-5000.

Did You Know?
• It is estimated that 80 percent of deforestation is caused by population growth.

• In the US, a teenager has a baby every 67 seconds.

: *The equivalent of about 21 million full shopping bags*
Source: The Student Environmental Action Guide, *Earth Works Press*

Shaded areas indicate highest volume of waste per capita.

What You Can Do

• Buy recycled—if you don't, you're not recycling.

• Use cloth napkins and towels instead of paper ones. Using paper napkins generates 15 times as much solid waste as cloth. Buy a set of reusable dishes for picnics and barbecues to replace disposable ones.

• Contact the Loading Dock, 2523 Gwynns Falls Parkway, Baltimore, MD 21216, a nonprofit group that collects and distributes surplus building supplies. They sell them at one-third their retail value to nonprofit organizations for use in low-income housing. Consider replicating this operation in your community.

• Influence the publishing office of your local newspaper to use recycled newsprint. Over 60 percent of a newspaper's value is added at the

Recycling

- By Neil Seldman and Hugh Stevenson -

The growth of recycling programs throughout the US is a strong indicator that recycling has captured the attention and support of many Americans and their public servants. However, waste prevention and market development for recyclables have emerged as major challenges for solid waste managers everywhere. The creation of markets for recyclables depends upon the efficiency of collection programs, the quality of the materials collected and on communicating the availability and benefits of recyclables as feedstock for industry. The continued success of recycling requires that all Americans—businesses, government agencies and policymakers, and consumers—see recycling as an effective economic development strategy.

As 1992 began, there were almost 4,000 curbside recycling programs in the US—an increase of almost 250 percent since 1988. Much attention has been given to the costs of curbside recycling in 1992, but few Americans are aware that a dozen US communities are recovering 50 percent and more of their municipal solid waste at costs far below traditional collection and disposal methods.

Even cutting-edge collection programs like Seattle's, which charges residents according to the

Q: *Which US national park is part of the only mountain belt in the continental US that runs east and west?*

volume of trash they produce, can top their performance. To reach its 1995 goal of a 60 percent recovery rate, Seattle plans to expand its collection to include more from the commercial sector and apartment buildings, and to target food waste for composting. The city will also upgrade its handling of glass to enhance marketability.

To be a cost-effective waste management strategy, recycling does not have to pay for itself from the sale of collected materials. In densely populated areas, avoiding high tipping fees at landfills is an obvious money saver for many communities. However, evidence suggests that the real payoff comes when recycling is implemented as a strategy for economic development. When used locally to produce new goods in scrap-based enterprises, recyclables can contribute significantly to job creation, business expansion and the vitality of local economies. The operating experience of existing plants indicates that a city of 1 million recovering half its waste stream can support 60 small-scale manufacturing plants, which can create 1,500 jobs and add $260 million to the local economy each year.

For example, New Jersey currently realizes $41 billion in sales from its largest scrap-based manufacturers—more than the total farm receipts in the Garden

publishing stage; it is the publisher, not the pulp or newsprint producer, who will dictate any change in a newspaper's recycled content.

• Influence your local government to buy products with high scrap content. Contact the city procurement office.

Resources
• *Worms Eat My Garbage*, by Mary Appelhof, 10332 Shaver Rd., Kalamazoo, MI 49002. A practical guide to setting up a worm composting system to reuse kitchen scraps to produce fertilizer for plants and gardens, and to grow fishing worms. An excellent project for schoolchildren.

• Take It Back Foundation, 111 N. Hollywood Way, Burbank, CA 91505. Promotes recycling education in

Safe Recycling, Inc., based in New Jersey, receives local recyclables to sort and sell to major corporations.

A: *The Channel Islands National Park, on the coast of California*
Source: The W.A.T.E.R. Foundation

schools. Has produced a music video in which rock stars sing about recycling, and other education materials for young people.

• National Toxics Campaign, 1168 Commonwealth Ave., Reston, MA 02134. Advocates citizen-based preventive solutions to the environmental crisis. National Toxics Campaign provides organizing assistance, technical and legal help, leadership training and environmental testing to grassroots groups.

• *Beyond 40 Percent: Record-Setting Recycling and Composting Programs*, Brenda Platt, et al., Institute for Local Self-Reliance, 2425 18th St. NW, Washington, DC 20009. Documents the operating experience of 176 communities—14 of which are recovering over 40 percent of their waste stream. Establishes materials recovery as a cost-effective primary waste management strategy.

• National Association of Recycling Industries, 330 Madison Ave., New York, NY 10017.

• National Recycling Coalition, 1101 30th St. NW, Ste. 305, Washington DC 20007.

• National Solid Waste Management Association, 1730 Rhode Island Ave. NW, Washington, DC 20036, (202) 659-4613.

• *Taking Out the Trash*, Jennifer Carless, Island Press, Washington, DC, 1992.

State—and yet there remains great potential for expansion in New Jersey and throughout the country. Governments can cultivate scrap-based manufacturing through grant and loan programs, investment tax credit programs, technical assistance, recycled-content laws, sales tax exemptions, aggressive public education and recycled product procurement policies and legislation.

A developing trend is joint ventures among cities, environmental and community groups, and manufacturers that use recyclable materials as the feedstock for their products. In these joint ventures, citizen groups, for the first time, have a role in the management of industry. In Los Angeles, an entire industrial park has been dedicated to "joint venture" remanufacturing. The city has provided incentives for manufacturers to build scrap-based manufacturing plants; the community groups, which will supply the scrap materials, will receive percentages of the profits.

The quality and cost-effectiveness of scrap-based products continue to increase with demand. For example, Coon Manufacturing, of Spickard, Missouri, illustrates the value added to products through remanufacturing. Coon buys plastic milk jugs at $.10 per pound, and sells goods remanufactured from them at $4 per pound. Elsewhere, 13 US mills currently produce approximately 4 million tons of recycled newsprint per year; at least 17 other such mills are planned for US construction.

Recycling, combined with waste prevention, enters the 21st century as the waste management strategy of the future. New partnerships, born of recycling, involving environmentalists, community development groups, government and industry, can do much to improve the environment and economy of US communities. Using the interconnectedness of the environment as our guide, we will all have to connect, communicate and cooperate if we are to achieve a sustainable economy.

Neil Seldman is president and co-founder of the Institute for Local Self-Reliance. As economic development planner, Seldman emphasizes environmental quality in municipal and solid waste management and the creation of economic development opportunities for minority youths, community groups and small businesses. Hugh Stevenson is a staff member of the Institute for Local Self-Reliance. He writes on recycling for the institute.

: **Which country has recently implemented a tree planting program designed to forest 75 million acres by the year 2000?**

Shaded areas indicate greatest energy consumption per capita.

Renewable Energy

- By Dr. Warren Leon -

To address a wide range of environmental problems, the US must make the switch from polluting fossil fuels to renewable energy sources. Expanded use of renewables—those energy sources like sunlight and wind that are regenerated at the same rate they are used—would reduce air pollution, acid rain and oil spills. Because fossil fuels emit the greenhouse gas carbon dioxide when they are burned, a transition to renewables worldwide would also curb global warming.

Most Americans understand the environmental and economic problems caused by excessive reliance on fossil fuels. They express enthusiasm for renewable energy and support policies to spur its use. But, unfortunately, most of them do not believe that renewable energy can be deployed on a large enough scale to displace significant quantities of oil, coal or natural gas.

Renewable energy's brief period of popularity in the late-1970s did not dispel this skepticism. At that time, annual funding for research and development of renewables grew from almost nothing to more than $700 million per year. Tax credits and other programs made solar collectors, wind turbines and other devices attractive business investments. So many

What You Can Do

Write
• Write to your elected officials to let them know you want an energy future based on renewable energy rather than fossil fuels. Encourage them to support policies to eliminate the barriers to renewable energy development. (See "How to Write Congress," page 322)

• Write your local electric company officials to find out if they have plans to increase the use of renewable energy. If not, encourage them to include renewables in their plans.

Educate
• Help change the public perception of renewable energy by organizing renewable energy fairs, forums, demonstrations and workshops.

• Organize tours of homes in your community that use solar design.

A: *China*
Source: Protect Our Planet Calendar, Running Press Book Publishers

- Join a national educational campaign on renewable energy, such as "Renewables Are Ready" sponsored by the Union of Concerned Scientists, 26 Church St., Cambridge, MA 02238, or "Sun Day 1992" sponsored by Public Citizen, 215 Pennsylvania Ave. SE, Washington, DC 20003.

Use Renewable Energy
- The most complete catalog of renewable energy products for homes and small businesses is the annual *Alternative Energy Sourcebook* published by Real Goods Trading Corp., 966 Mazzoni St., Ukiah, CA 95482.

Read
- *Cool Energy: Renewable Solutions to Environmental Problems*, by Michael Brower, MIT Press, Cambridge, MA, 1992.

families—from President Carter's on down—placed solar collectors on the roofs of their houses that sales increased fivefold between 1975 and 1980.

But the boom did not last long enough to convince most people that our energy future lies in the widespread use of renewable energy. In fact, because some of the crash implementation projects of the late 1970s were poorly conceived—and in some cases the technologies simply were not ready for deployment—renewable energy earned a reputation for high cost and unreliability.

Nevertheless, during the 1980s, while renewables were ignored by the public and dismissed by the federal government, the technology advanced. The reliability and efficiency of renewable energy equipment improved, and the cost of installing, maintaining and running it declined. In the case of wind turbines, for example, more advanced designs, better choice of materials and careful siting have made the cost of generating electricity from wind one-fourth of what it was a decade ago. In many locations, a utility company can now build a wind power facility that will produce electricity at a cost comparable to a new fossil-fuel-powered plant.

A 1991 study by four national energy and environmental organizations demonstrated that, if coupled

Careful siting and advanced designs have made wind energy a competitive alternative to fossil-fuel-based power.

Q: *Worldwide, how many reported nuclear tests have been conducted in the period between 1944 and 1990?*

with strong measures to use energy more efficiently, renewable energy sources could provide more than half of the US energy supply by the year 2030, and at a net savings to consumers. Fossil-fuel use would not be eliminated completely, but, with the right policies, oil consumption could be reduced to just one-third of current levels by 2030.

Despite the logic of pursuing this course, it will not happen automatically. Current government policies and the marketplace are structured in a way that encourages the wasteful use of fossil fuels, not the efficient use of all available energy sources. Tax laws, for example, favor fuel-intensive technologies that use coal and oil, rather than those such as solar and wind that have high initial construction and start-up costs but low ongoing fuel costs.

If the amount of money society must pay to deal with the environmental and health impacts were included in the price of fossil fuels, renewables would be more competitive. If the federal government spent as much on research and development of renewable energy technologies as it does on fossil fuels or nuclear power, progress would be faster.

But a level playing field will not be enough, as long as the public and policymakers retain an outdated image of renewable energy technologies and have insufficient information about them. For example, new homes can be built cost-effectively with passive solar designs that use the sun to minimize heating and air-conditioning costs, but neither home buyers nor most architects or builders demand or build such houses. In some parts of the country, a new generation of solar water heaters could reduce fossil-fuel use and save money for the purchasers of new homes, but they are rarely installed.

Because additional information and new public perceptions of renewables are so essential, environmentalists can play a useful role by helping to spread the word that renewable energy technologies are ready to be put in place to help solve many of the world's most critical environmental problems.

Dr. Warren Leon is director of public education for the Union of Concerned Scientists, a nonprofit organization focused on energy policy, global environmental problems and arms control. He supervises the organization's publications program and outreach efforts.

Resources

• The American Solar Energy Society, 2400 Central Ave., G-1, Boulder, CO 80301, publishes reports and conference proceedings as well as the magazine *Solar Today*.

• Renew America, 1400 16th St. NW, Ste. 710, Washington, DC 20036, a nonprofit advocacy group, produces informative materials and sponsors the "Searching for Success" program, which recognizes outstanding environmental projects, including those that promote renewables.

• The Union of Concerned Scientists produces a wide range of educational materials, posters and reports on renewable energy.

• The various renewable energy industry trade groups based in Washington (the American Wind Energy Association, the Solar Energy Industries Association, the National Wood Energy Association, the National Hydropower Association, the Renewable Fuels Association and the US Export Council for Renewable Energy) are storehouses of valuable (if one-sided) information and also publish newsletters and journals. The Geothermal Resources Council, P.O. Box 1350, Davis, CA 95617-1350, serves the geothermal industry.

A: *1,909*
Source: Greenpeace

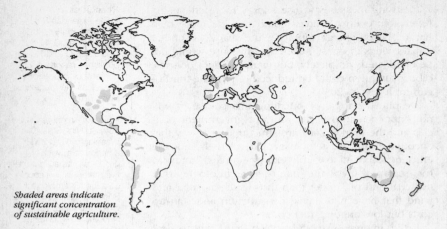

Shaded areas indicate significant concentration of sustainable agriculture.

Sustainable Organic Agriculture

- By Thomas B. Harding -

To get right to the point, sustainable organic agriculture—and the food and fiber products it delivers—cannot be realized until consumers prove their commitment at the retail food cash register.

There must be changes in the current state of worldwide food production and its impact on the environment, and on the general social and economic well-being of food producers, particularly those in less developed countries.

Can we establish, as is needed, a new "corporate structure," with a new ethic and a new sense of responsibility? Can industry realize that protecting our Earth environment and our natural systems is just as important as protecting our bottom line and our marketplace?

Most of the world's food was grown organically until about 150 years ago, when synthetic materials began to be developed and used for agriculture. The organic system of the past, and today, means: managing the farm system in a holistic, integrated way; using no harmful chemicals; using appropriate livestock and regional crop biodiversification (raising several crops); utilizing on-farm resources for

What You Can Do

• Support local, regional organic producers. Develop mutually beneficial partnerships—adopt a farm.

• Demonstrate your commitment to the environment, get involved and buy organic and environmentally beneficial products. Walk your talk!

• Support businesses that are committed to protecting the environment—tell them the type of products you want and will buy, and make sure they practice what they preach. There is green (profit) in protecting the environment.

• Educate someone every day about the benefits of organic gardening and farming, and about the other environmentally friendly actions they can personally adopt.

• Take time to educate our children about the

 Over a lifetime, the average American will use how many gallons of water?

fertilizers (livestock manures, compost and green plant manures); controlling insect pests and weeds through natural and biological means; and working within the carrying capacity of the land and resource base, and in harmony with the natural system.

Today, certified organic producers must utilize these same age-old principles—including not having used any prohibited materials or practices for at least three years—if they are to comply with the proposed "Organic Food Production Act of 1990." This act, a portion of the 1990 farm bill, was designed to create national standards for organic food production and labeling, but has yet to receive funding. In order for organic growers to compete in a world market, the US should have national standards in place by October 1993. Whether funding will be appropriated by Congress, however, remains to be seen.

Throughout the global community, smaller family farmers, including very poor farmers in the third world, are returning by the thousands to the sound ecological, economic and social principles of sustainable organic agriculture. Why? Because they have no choice. They continue to be hurt and neglected by the conventional production system. This system gives them low farm-gate prices for their products, is chemically intensive, and is expensive. Most farm credits and subsidy programs favor large, corporate, farm monocultures, which natural system. Visit an organic farm, and spend the day with a farm family. Know where our food comes from. It's not grown at a grocery store.

• Be politically active. Help change food and agriculture policy that keeps high-quality third world organic products from being traded in the international marketplace. These and other barriers keep much of the third world in debt, which forces them to sell their natural resources to maintain the debt-merry-go-round controlled by a few.

• Use only the principles of organic gardening in your family vegetable garden, lawn, flower beds, and herb garden. Grow much of your own food: It's fresher, safer and tastier. It is also great exercise and fun!

Organic sesame is harvested in Mexico.

A: *7.5 million*
Source: *Children's Television Workshop*

- Be inquisitive—make sure that your organic products are certified. Ask the retailer about the certification, the seal of approval and how he/she verifies the product guarantee.

- Visit another country. Learn about the people, their language and their culture firsthand—and recognize your responsibility to be a good neighbor. You may also see how unbalanced the world has become.

- Believe that it is not too late to change. Remember that real change doesn't come about until you have made a personal commitment.

Resources
- Organic Crop Improvement Association International, 3185 Township Rd. 179, Bellefontaine, OH 43311, (513) 592-4983.

- OFPANA, P.O. Box 1078, Greenfield, MA 01301, (413) 774-7511.

- International Federation of Organic Agriculture Movements (IFOAM), c/o Okozentrum Imebach, D-6695 Tholey-Theley, Germany, (49) 68 53 5190.

- Organic Food Foundation, 125 W. Seventh St., Wind Gap, PA 18091, (215) 863-8050.

produce large quantities of one crop that sells at cheap, marginal farmgate prices to a fixed market. Most crops (including all major commodities and a lot of fresh and processed foods) are controlled by less than ten multinational companies. This practice displaces hundreds of thousands of farmers, families and rural communities throughout the world.

Why should you as a food consumer care about saving farms in first and third world communities, as long as your food is abundant and cheap? The agriculture dollar multiplies several times throughout the rural communities where it is spent, creating many non-farm jobs that keep people out of our crowded cities, help build a more sustainable social and economic system, and improve the overall quality of life.

The present US food system is cheaper than any place else on Earth, mostly for all the wrong reasons: chemically-intense, monoculture crops; non-sustainable, large, corporate farms; marketplaces controlled by a few multinationals; tax-supported farm credit and subsidy programs; limited environmental and natural resource accountability.

If the present system were run like a responsible business, it would not be allowed such high, government tax-supported programs. It is clear that the present system is unsustainable and nearly bankrupt.

We can make a great difference, with our cash register vote, in how food and agriculture policy is developed and carried out in the future. By buying certified organic foods and other enviromentally beneficial products, we can send a loud, clear, economic message to the food production cartels, governments and the politicians of the world.

Organic producers are the true "environmental stewards"—but their efforts will be lost if green consumers refuse to buy their products, and simply continue to give lip service to "saving the environment."

Thomas B. Harding is president of AgriSystems International, a company providing safe, profitable alternatives for food and agriculture industries. Harding also assists conventional and organic farmers to improve farming systems, is an internationally approved independent third-party certification inspector for the Organic Crop Improvement Association and IFOAM vice president.

 What is the estimated number of living species on Earth?

EARTH ISSUES

Shaded areas indicate most acute toxic pollution.

Toxic Pollution

- By Michael S. Brown and Michael R. Reich -

Toxic chemicals seem to be everywhere in our environment. We have more information than ever before about the hazards in the building materials of our homes, the products that we buy, our nearby industries and the air, soil and water of our neighborhoods. More people are becoming motivated to reduce exposure and remove toxic chemicals from our communities. They soon discover that toxic pollution is not simply a technical problem, not simply a problem of information, but is a deeply rooted political problem involving the distribution of power in society.

What seems like a simple task with straightforward solutions—correcting and preventing toxic pollution—quickly leads to intense conflicts with powerful organizations. Whether it is a contaminated waste site, an industry spewing toxic metals into a local neighborhood or a proposal for a hazardous waste incinerator, community groups soon realize that it is through politics that they must address the toxic problem. Obtaining redress for harms suffered and reducing the likelihood of future harms rarely can be achieved through scientific discussion or even legal action alone.

The political processes around toxic pollution follow certain predictable patterns. Polite requests for

What You Can Do

Collect Toxics Information
• Contact your state agency that regulates hazardous wastes and ask for a list of known contaminated sites to find out if there is a site in your community. Check with the EPA, (800) 535-0202, to see if there is a federal site listed in your area. Ask specifically about sites you suspect are contaminated.

• Contact your local planning department for available information about planned facilities, such as a solid or hazardous waste incinerator. Investigate whether the developer submitted an application for federal, state or local permits, and obtain the information in the application.

• Contact businesses directly to request information about toxic pollution.

A: *80 million*
Source: International Herald Tribune

- Call the EPA Emergency Planning and Community Right-to-Know Information Hotline at (800) 535-0202, to inquire about the Toxics Release Inventory and request information for specific facilities.

- Use the RTKNET Pilot Project, (202) 234-8494, for computer access. The Working Group on Community Right-to-Know, (202) 546-9707, can help you figure out how to use the data, which can be provided by city, county, state, type of industry, type of chemical, etc.

- Contact your state regulatory programs for air, water and hazardous waste, to obtain the information they collect from businesses.

Contact
Contact national and local organizations to

action rarely result in an adequate response. Solutions that address the core of the problem, whether it is prevention, cleanup or compensation, depend on identifying the problem as a collective issue, mobilizing for group action, and joining with other groups to persuade the political system to respond.

Toxic victims and those at risk of becoming victims begin at a distinct disadvantage compared to the private corporations and government agencies that are the source of the problem. Community groups are less prepared to fight, possess fewer financial and scientific resources, and rarely have easy access to powerful decision-makers. Moreover, they don't begin with an understanding of the political nature of toxic pollution or how the system works. Community groups need to create their own resources and learn about the system on their own.

Many chemical disasters occur in places where people are socially marginal and politically powerless. Poor communities and communities with more minorities are at greater risk of harm from toxic pollution than communities with high concentrations of professionals and high incomes. The relative powerlessness of a community increases its vulnerability to toxic pollution.

Effective political strategies for fighting toxic pollution can be learned. Indeed, private corporations and

Toxic waste is dumped in Chaborovice, Czechoslovakia.

Q: *During the next 25 to 30 years, how many of today's current living species face possible extinction?*

government agencies spend huge sums, both before and after toxic disasters, to learn about methods to protect their interests. Community groups need to develop their own successful political strategies to protect their communities from toxic pollution.

Community groups should also be aware of transportation of hazardous materials. Several large-scale chemical spills erupted in the US during 1991 and 1992, with severe effects on the local environment. A train wreck in northern California resulted in the release of a pesticide, meta-sodium, into the upper Sacramento River, creating an ecosystem disaster that is expected to take a decade or more to recover. That same month, another train derailment, this time in Southern California, caused the evacuation of hundreds of residents due to the presence of aqueous hydrazine.

Citizens should also be aware of some progress being achieved in the legal arena. The EPA negotiated the payment of cleanup costs by responsible companies at several federal Superfund sites, which should provide some relief for long-suffering residents. In the past year, several firms, including Chevron and Rockwell, were required to pay record fines for environmental harms.

From a global perspective, the most significant story of toxic pollution has been the disclosure of the extent of environmental degradation in Eastern Europe and the former Soviet Union. Toxic pollution in those countries reached extraordinary levels, with high costs to human health and the natural environment. Achieving reasonable levels of pollution control will require substantial overhaul of the outdated production infrastructure, with serious questions remaining about who should pay for the cleanup and rebuilding, and how.

Michael S. Brown is manager of environmental affairs for the city of Irvine, California. Michael R. Reich is associate professor of international health at the Harvard School of Public Health and author of Toxic Politics: Responding to Chemical Disasters *(Cornell University Press, Ithaca, NY, 1991).*

request help in collecting information and planning your action on toxic pollution, including the following organizations:

• The Citizen's Clearinghouse for Hazardous Wastes, 119 Rowell Ct., Falls Church, VA 22046, (703) 237-2249.

• The US Public Interest Research Group, 215 Pennsylvania Ave. SE, Washington, DC 20003, (202) 546-9707, and its state PIRGs.

• Greenpeace, 1436 U St. NW, Washington, DC 20009, (202) 462-1177.

• The National Toxics Campaign, 1168 Commonwealth Ave., Boston, MA 02134, (617) 232-0327.

• The Sierra Club, 730 Polk St., San Francisco, CA 94109, (415) 776-2211.

• A number of labor unions, including the United Auto Workers and the Oil, Chemical, and Atomic Workers, have active health and safety programs. Union locals can be important allies in addressing toxic pollution in the community, since their workers are often directly affected in the facility.

A: *25 percent, potentially 20 million species*
Source: International Herald Tribune

Shaded areas indicate most cars per capita.

What You Can Do

• Ride a bus or bicycle for your errands whenever you can. Carpool when possible.

• When driving at highway speeds, an air-conditioned car with the windows rolled up and the vents closed gets better mileage than a car with its windows down and no air-conditioning. Have your air conditioner checked periodically for leaks, to avoid dumping ozone-depleting chlorofluorocarbons (CFCs) into the atmosphere.

• Reduce your car's rolling resistance. Keep the tires inflated to the level specified by the manufacturer. This increases the mileage of the car and extends the life of the tires. Consider buying special low-rolling-resistance tires, manufactured by Goodyear and other companies.

Transportation

- By Philip Terpstra -

The greatest single cause of the air pollution problem in US cities is the large number of automobiles that are powered by internal-combustion engines (ICEs). Road vehicles account for 60 percent of ozone emissions, 80 percent of carbon monoxide emissions and 63 percent of petroleum consumption. The US ranks number one in the world in both per-capita and total number of automobiles in use. The US has more than one car for every two people. With nearly 150 million cars (in addition to buses and commercial trucks) clogging the roads and dumping pollutants into the sky, we have some real problems to solve.

What can be done? A big improvement in public transportation services would certainly help. The benefits of mass transit include less pollution, less traffic congestion, greater safety and lower total transportation costs. If we had buses and other forms of mass transit running more regularly and with more convenient destinations, more people would use them. Many African nations, though plagued with myriad political and economic difficulties, have available to the common citizen bus and taxi systems that are far more convenient and financially feasible than those offered in the US.

Q: *How many endangered sea turtles die in the US every year as a result of incidental capture and drowning in shrimp nets?*

The big automakers and tire manufacturers were largely responsible for ruining the functional public transportation systems that many large US cities had 50 years ago. They worked together to create dependence on personal cars, thus ensuring a large market for their products. Now public bus systems don't have enough paying customers to expand their services, so many people continue to need cars. It will take time for us to get out of this vicious cycle. Now 70 percent of the land area in Los Angeles is devoted to the use of cars—roads, gasoline stations, auto parts stores, parking lots, etc.

Widespread replacement of petroleum-burning cars with electric vehicles (EVs) would greatly reduce air pollution and some other environmental ills. If recharged from the present mix of electric power generation facilities (hydroelectric, coal, nuclear, solar-thermal, etc.), EVs would be responsible for the production of only about one-tenth of the air pollution emitted by petroleum-fueled cars. This is because grid utilities are much more efficient at producing electric power than a car's internal combustion engine is at producing mechanical energy. Also, EVs do not waste energy while stopped in traffic—a huge advantage in city driving, for which most EVs are used. Of course, not all EVs have to be charged from polluting power stations.

• When shopping for replacement parts to repair your car, buy used parts. Many independent auto parts stores now stock plenty of fully functioning used parts from auto salvage yards.

• Don't throw worn-out tires away. Drop them off at a tire dealer. Some old tires can be recapped, others are shredded and used for other purposes, such as additives to paving materials for roads, and porous soaker-hoses for gardens and landscaping.

• Take in your old oil and antifreeze for recycling. Used antifreeze can be restored to the same specifications as the new liquid. Recycling oil is a bit more difficult, but it is possible with special equipment now owned by many facilities. Take the old oil into a local

Alternative transportation could free cities such as Los Angeles from ensnaring nets of asphalt.

A: *Approximately 11,000*
Source: Greenpeace Action

gasoline station or auto parts retailer that will accept it. Check the phone book for suitable drop-off places if you're unsure. If that doesn't help, contact your local or regional office of the Environmental Protection Agency. Don't pollute the earth by dumping the old fluids down a drain or onto the ground. If you have someone else take care of your vehicle's maintenance, make sure that shop recycles its used oil

• Start a business to convert used cars to electric power (see resources, below).

Resources
• *The Electric Vehicle Directory*, updated annually. Lists ready-to-drive electric cars and over 100 manufacturers and suppliers of EV components for those who wish to convert their own cars to electric power. Spirit Publications, Box 23417, Tucson, AZ 85734, (602) 822-2030.

• *Convert It* is a step-by-step guide to converting a used ICE car into an electric vehicle. Published by Electro Automotive, Box 1113, Felton, CA 95018-1113, (408) 429-1989.

• Solar Electric Engineering sells solar-assisted electric vehicles capable of 65 to 70 mph, and markets conversion kits for do-it-yourselfers. Contact them at 116 Fourth St., Santa Rosa, CA 95401-6212, (707) 542-1990.

Renewable and nonpolluting energy sources, such as solar and wind power, are becoming more widespread.

EVs use no engine oil, so oil doesn't leak onto roads or get poured down drains. They use no antifreeze (about 250 million gallons of antifreeze are sold annually in the US; much of it is disposed of improperly). An electric motor lasts about 20 times longer than an internal combustion engine. The main hindrance to widespread use of EVs at this time is their high initial price; their operating cost, including periodic replacement of batteries, is less than that of a petroleum-fueled vehicle.

An excellent system currently under development for city dwellers is the dovetailing of personal EVs with public transportation. Trains and other forms of mass transit are more efficient than millions of individual cars. But how do you get from your house to the train? A firm called Packer Engineering, based in Naperville, Illinois, is developing Stackable Electric Rental Cars (SERC). The concept is to have special electric cars that commuters can rent from a public transportation drop-off point to drive to their houses each evening. In the morning, the rental car is driven back to the terminal. The rental agency takes care of charging the vehicle and handling other maintenance. The driver still has the flexibility of a personal-use vehicle, but doesn't have the parking and maintenance hassles or the high cost of ownership.

The stackable feature of the SERC allows for space-saving at the rental agency's parking lot. The nose and tail sections of the vehicle swing up, so that many cars can be parked in a small space. These cars may be available in late 1993.

In December 1991, President Bush approved a $151 billion transportation bill that allows states to allocate funds to other forms of transportation besides cars. The bill left many in the industry hopeful for the integration of different types of transportation.

Philip Terpstra is the owner of Spirit Publications, which publishes technical books. He is also a commercial pilot and aircraft mechanic. Born in Africa, he has traveled widely and studies international environmental concerns. He is a founding member of the Tucson Electric Vehicle Association and is an active participant in its EV Technology Research Committee.

: *Since 1986, how much has the US demand for natural gas risen?*

EARTH ISSUES

Shaded areas indicate major wilderness areas.

Wilderness

- By Judy Anderson -

Wilderness . . . a mystique has been attached to that word. To retain for our heirs the vision we have—of exuberant vistas, abundant wildlife, nature interacting without human intervention, unfettered and free—we have a prosaic official government certificate. Wilderness lands additions in the first decade after the Wilderness Act of 1964 averaged less than a half million acres annually. It rose to 2 million acres per year in the late 1970s, bulged to almost 70 million acres with the Alaska Lands Act, and then dropped to less than half a million acres annually the past eight years.

The slow pace of recent additions stems not from exhaustion of suitable areas, but from the political infighting that causes them to languish for decades, while competing users nibble at their edges or eat away whole areas.

On the table again in 1992 was a Montana Wilderness bill for 1.2 million acres of US Forest Service land. The bill has been around for more than 12 years, changing from one year to the next. Colorado proposals range from 600,000 to 800,000 acres. Both the Montana bill and the smaller Colorado bill were passed by the Senate. An old Idaho proposal wasn't introduced this time; people are exhausted from previous battles and see no

What You Can Do

• Plan a visit to the public lands. Nothing can replace on-the-ground knowledge. You can write for information to the Forest Supervisor for the particular national forest, national park, wildlife or refuge, or the state director for the Bureau of Land Management (BLM). Parks, forests and refuges are found throughout the country, BLM land only in the West.

• While most environmental organizations have some concern for wilderness, those with an emphasis on wilderness are:

The Wilderness Society, 900 17th St. NW, Washington, DC 20006.

Sierra Club, 730 Polk St., San Francisco, CA 94109.

Most states have a local wilderness coalition or

  4 trillion cubic feet
Source: Protect Our Planet Calendar, Running Press Book Publishers

alliance of groups that are advocates for particular areas. To contact one of these, the national groups or their field offices in various parts of the country can be of assistance.

• "Adopt" a wilderness or proposed wilderness area. Federal government personnel are frequently reassigned and have no long-term perspective. You can provide that continuity by visiting the area on a regular basis so that you can see changes—both degradations and recovery. Get on the right mailing list so that you are notified whenever something that affects your ex-orphan is planned. Defend it from incursions, make sure it isn't neglected by the federal land managers and let them know that you're out there looking over their shoulder as they do the work.

prospect for meaningful progress. These are the last three states with major forest wilderness—left over from forest planning done in the 1970s.

"Public lands" managed by the Bureau of Land Management (BLM) in 12 Western states are just moving into the spotlight. The Department of Interior is announcing its wilderness recommendations in 1992, state by state. In California, the first disclosed, BLM lists about 2 million acres statewide; the environmental community supports a proposal for 4.5 million acres in just the desert, and over 4 million acres of national park wilderness.

One wilderness act has been signed in 1992: the little-known Los Padres Condor Range and Rivers Protection Act, raising the wilderness total to almost half of the 1.8 million-acre Los Padres National Forest and designating three wild rivers and mandating study of six others in this area northwest of Los Angeles. And some 56,000 acres of wilderness, conservation areas and scenic areas have been protected in North Carolina.

Ah, wilderness, if only the problems of maintaining the vision were solved by making it *Wilderness*. In 1992, wilderness legislation took a new twist with the first bill designed to improve the management of existing areas.

What can go wrong? Can't one just leave the

Cascades Range peaks rise high above the threat of land development.

Q: *If you install a low-flow aerator on faucets, you can cut water flow by how much, without feeling any difference?*

wilderness as it is? Manage with benign neglect? No, not even if the decision were made to make it a "human exclusion zone"—prohibiting all entry. But wilderness areas are not excluded from all human entry, and using surrounding areas also has an impact. Human beings bring problems with them, as do their horses, cattle, sheep, dogs and other pets. Obviously, trenches dug across sensitive meadows by lug-soled feet are undesirable, but subtle stress on wildlife by interfering with their breeding, feeding and resting cycles also has to be monitored.

Wilderness designation is caught up in other arguments concerning the management of public resources: the Endangered Species Act/spotted owl/timber controversy, mining law reform and grazing reforms, Western water law and protection of biodiversity.

While traditional wilderness designations used scenic values as a major argument, new wildernesses will have major biological-values components. In contrast to the anti-wilderness arguments, biological arguments have become stronger and more sophisticated as botanists and zoologists have learned more about what physical arrangements of wilderness and other protective categories best help species survive, especially species that require large territories. This analysis of the physical arrangements and the breaks that cause problems, like loss of migration routes for elk and deer, is called "gap analysis."

Biologists looking at natural ecosystems are dealing with larger and larger units, bioregions of geographically related areas such as the Rocky Mountains. Gaps identified in the analysis of existing and potential parks, wilderness and other habitat protection units will provide the ammunition that wilderness activists need to return to Congress for boundary expansions or new areas.

Even if all the possible wilderness currently before Congress is designated, wilderness advocates will continue to work hard but refocus their efforts—to assure wilderness habitat so that the mountain lions, bears and condors that give life to the spectacular scenery will still be there for the next century.

Judy Anderson is chair of the Sierra Club's National Parks and Wilderness Campaign Steering Committee, and co-director of the California Desert Protection League.

• Learn more about wilderness. The Department of Agriculture, the Department of Interior and many universities and colleges sponsor colloquiums and discussions on wilderness issues. These are nearly all open to the public, but not generally announced. Your nearby university or college may be involved in wildlands issues; ask to be notified of future wilderness-related activities.

• Congressional committees are assembled from across the country. A senator from Maine or Vermont may be key to wilderness in Montana or Texas. Volunteer to assist with the Public Lands Defense Action Network, c/o Sierra Club (address on page 125). As a member of the network, you write or call congresspeople about a particular wilderness proposal, visit his or her office with the same message and get others to do the same.

Read
• *Wilderness at the Edge, a Citizen Proposal to Protect Utah's Canyons and Deserts*, by the Utah Wilderness Coalition, P.O. Box 11446, Salt Lake City, UT 84147. Distributed by Peregrine Smith Books.

• *Colorado BLM Wildlands, a Guide to Hiking and Floating Colorado's Canyon Country*, by Mark Pearson and John Fielder, Westcliffe Publishers, Inc., Englewood, CO 80110, 1992.

: *25 percent to 50 percent*
Source: The Student Environmental Action Guide, *EarthWorks Press*

III

REGIONAL REPORTS

A *child can shatter the world into a hundred bits of colored cardboard—and then piece them together again to form a jigsaw puzzle of our world. Regional Reports breaks the world into "puzzle" pieces of environmental news, statistics, maps and other data—and then pieces them together to form a clear picture of our global environment.*

KEY TO GREEN STATISTICS

CONSERVATION ISSUES HAVE BECOME major popular and political concerns. National and international conservation organizations have gained support from governments, international development agencies, corporations and commercial organizations in the creation of international conservation treaties. *Earth Journal* uses membership in these environmental conventions as a measure of a region's environmental commitment.

Antarctic Treaty and Convention ensures that Antarctica is used for peaceful purposes and that international cooperation in scientific research continues.

CITES: The Convention on International Trade in Endangered Species of Wild Fauna and Flora monitors trade and declares species endangered and threatened internationally.

Law of Sea: Signatories to the United Nations Convention on the Law of the Sea, which establishes environmental standards and enforcement provisions.

MARPOL: The Protocol of 1978 Relating to the International Convention for the Prevention of Pollution from Ships is intended to eliminate international pollution by oil (and other harmful substances) at sea.

Migratory Species: Convention on the Conservation of Migratory Species of Wild Animals protects wild animal species that migrate across international borders, by promoting international agreements.

Ocean Dumping: The Convention on the Prevention of Marine Pollution by Dumping of Wastes and Other Matter regulates ocean disposal of materials, encouraging regional agreements, and establishing a mechanism for settling disputes.

RAMSAR: The Convention on Wetlands of International Importance, Especially as Waterfowl Habitat, stems encroachment on and loss of wetlands.

World Heritage: The Convention Concerning the Protection of the World Cultural and Natural Heritage establishes a system of collective protection sites of outstanding universal value.

KEY TO GREEN RATINGS

1 Nations in the region are undertaking conservation initiatives of all types and have good policies in place.

2 Nations in the region are beginning to undertake conservation initiatives, but areas are threatened and countries need more encouragement and/or assistance to achieve their conservation goals.

3 Nations in the region may have joined international initiatives and have followed up with some actions, but the region is in danger and much more needs to be done to aid or improve conservation.

4 Nations have shown a green front, but it is mainly lip service to the environment, with little or no action.

5 Environmental impacts are severe, and nations in the region are not actively responding with conservation initiatives.

Q: *In Costa Rica, what was responsible for dropping the fertility rate from 7 to 3.5 children per woman in only 25 years?*

REGIONAL REPORTS

Antarctica

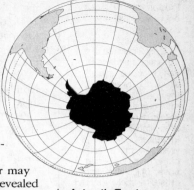

Antarctica is the world's driest, coldest and windiest continent. Yet in this solitary land emerge the first signs of many of the world's major environmental problems. The first hole in the ozone layer appeared over Antarctica. Evidence for a theory of global warming has also emerged: Core samples from an Antarctic ice sheet revealed the accumulation of carbon dioxide in the atmosphere.

Two major scientific finds in the past year may provide a hint at the Earth's future. One study revealed how severely the Antarctic food chain is disrupted by the ozone hole. When the Antarctic hole is open between January and March, the growth of phytoplankton is reduced by 6 percent to 12 percent, scientists have discovered. When the hole closes, production of phytoplankton resumes so that the overall reduction in growth is about 4 percent. Phytoplankton is at the bottom of the food chain; the krill that feed on it are reduced in number, leaving less food for the penguins and whales that feed upon the krill.

Scientists have also found that warmer global temperatures aren't likely to cause the Antarctic ice cap to melt. Global temperatures were higher than normal during the mid-1980s, yet the average annual snowfall in Antarctica was the highest ever recorded (record-keeping began in 1806). The ice cap actually increased in size, because snow accumulates in Antarctica faster than it melts, and ocean levels were four to six one-hundredths of an inch lower than expected. In a separate study, scientists concluded that the level of Antarctic glaciers was highest between 7,000 and 4,000 years ago, when global temperatures were one to two degrees Celsius warmer. Higher atmospheric temperatures allow the air to hold more water vapor, leading to greater precipitation, scientists reason. But what happens in Antarctica won't necessarily happen elsewhere, they warn. Glaciers in warmer areas of the globe are likely to melt if the atmosphere's temperature climbs higher.

While Antarctica is rich in minerals, and many want to mine there, no mining or oil exploration will be allowed on the continent for the next 50 years, as guaranteed by a treaty signed on October 5, 1991, by

Antarctic Treaty Nations with Voting Rights: Argentina, Australia, Belgium, Brazil, Chile, China, Finland, France, Germany, India, Italy, Japan, New Zealand, Norway, Peru, Poland, South Africa, South Korea, Spain, Sweden, United Kingdom, US, CIS, Uruguay

Non-Consultative Parties: Austria, Bulgaria, Canada, Colombia, Cuba, Czechoslovakia, Denmark, Ecuador, Greece, Hungary, Netherlands, North Korea, Papua New Guinea, Romania

Regional Statistics

Land Area: 5.4 million square miles, expanding to some 8 million square miles, including ice pack, in winter.

Geography: While 70 percent to 75 percent of the world's fresh water is stored in the ice of the Antarctic, it is considered a desert because the water, in its frozen state, is mostly unavailable as a life support.

 Government policies providing access to contraceptives
Source: Newsweek magazine

Precipitation averages only a little over one inch a year, similar to the Sahara Desert.

Only 2 percent of the continent's surface is free of ice, including about one-twentieth of the coast, the peaks of mountains and the dry valleys, where virtually no precipitation has fallen for at least 2 million years. Over nine-tenths of the coastline is comprised of ice shelves or cliffs. The shelves stretch out over the sea and break off into often-immense icebergs; one, sighted in 1956, was the size of Belgium. Another iceberg sighted in spring 1990 was the size of Rhode Island.

Climate: Few plants and animals inhabit this region because it is so cold.

Resources: Coal, metal ores, fossil fuels

Green Rating

2 Antarctica is in danger of being "loved to death," as tourists and scientists put unprecedented pressure on the region's fragile ecosystem. Antarctica belongs to everyone and to no one, which can be its salvation—or its downfall.

the 1959 Antarctic Treaty's 24 voting members. After 2041, the ban will continue, unless three-quarters of the voting members vote to end it. This condition was a compromise to the US, which wanted to end the agreement after the 50-year period.

A US research station received stinging criticism from Greenpeace after it detonated a load of hazardous waste on the Antarctic continent in December 1991. The waste could not be shipped because the slightest bump would cause it to explode. The blast left a crater 40 feet wide and 10 feet deep and could be heard 10 miles away. Though the staff of the American base insisted that the detonation site contained no plant or animal life, it did admit that it did not consider the environmental impacts of the explosion. "It's appalling that they damage the environment, even though they have a firm mandate to protect it," said Dana Harmon, a spokeswoman for Greenpeace, which has long advocated the establishment of Antarctica as a wild park. Greenpeace has lobbied Antarctic Treaty members to preserve the unspoiled nature of the once-isolated continent.

Although Japan continues to hunt whales in Antarctic waters, the continent is now largely protected, by world consensus, as a pristine land of ice.

HOT SPOT

TODAY, THE ANTARCTIC TREATY NATIONS operate 69 research stations on the continent. This high level of habitation could impact a fragile environment where one footprint can remain for 200 years. Potentially more damaging is the influx of tourists to Antarctica: Some 7,000 visited Palmer Station in one year alone. While these tourists may be ecologically conscientious, their sheer numbers could seriously affect the area and also disrupt scientific research.

Limiting this impact was one topic of discussion at the November 1992 meeting of Antarctic Treaty nations. Italy and France have asked that the treaty include regulations on tourism. The National Science Foundation, the government agency that sponsors all US Antarctic research, now works with US tour groups to regulate visits to research bases.

Sources: *Washington Post, New York Times, Los Angeles Times, USA Today, Nature, Science News.*

: *According to the EPA, for every 1 percent of ozone depletion, how many new cases of skin cancer deaths occur in the US each year?*

REGIONAL REPORTS

Southern and Central Africa

Southern and Central Africa's geography consists of desert and savanna lands. The climate is dry and hot in some places, but other areas are lush with tropical rainforests. Resources include diamonds, gold, nickel, iron and ore.

Africa has the lowest life-expectancy and highest child mortality rates in the world. Overpopulation results in a rapid drain of the region's natural resources—land is managed poorly in order to feed people quickly. Cattle outnumber people two-to-one in the many areas where people depend on livestock to make their living. Erosion due to overgrazing is a serious problem.

Wetlands all over Africa are drained to decrease the number of malaria-carrying mosquitoes. Although Zaire has one of the largest remaining rainforests in the world, most of Southern and Central Africa's forests have been depleted for fuel and farming. In Zaire, however, the government has declared a combined total of 5,000 square miles as a tropical rainforest reserve.

The Southern and Central African region is suffering the most severe drought in the area's history. This catastrophe has brought devastation to an estimated 115 million people. There had been almost no rain for eight months as of November 1992, and shortages of food and water are at a critical stage. Experts have urged that 10 million tons of food be imported over the next 12 months to prevent mass starvation. Malawi, Zimbabwe and Zambia have been hardest hit. These countries also have the biggest transportation problems, which make food delivery difficult.

Crops in South Africa and Zimbabwe have been destroyed by the drought. Zimbabwe has suffered nationwide power cuts, due to a lack of water to fuel hydroelectric power, prompting Zaire to contribute some of its water supply in aid. Underground pipes have burst across the region, because there is no water to prevent the buildup of gas and sludge. Some 30 million people are starving in 10 nations, and crop

Angola, Botswana, Burundi, Central African Republic (CAR), Congo, Lesotho, Malawi, Mauritius, Namibia, Rwanda, South Africa, Swaziland, Uganda, Zaire, Zambia, Zimbabwe

Regional Statistics

Population: 151,310,000

Land Area: 3,392,075 square miles

Geography: Countries of this region bordering the Atlantic Ocean are vast savanna and desert lands. The high-elevation interior plateau stretches across the central part of the region, with rocky, arid areas, notably the Kalahari Desert, at 4,000 feet. One of the world's largest dams, the Kariba, is located in landlocked Zimbabwe, irrigating a vast, high-elevation veld for ranching and agriculture.

Climate: Dry and hot in the desert region and tropical rainy conditions

20,000
Source: The Campaign for Safe Alternatives to Protect the Ozone Layer

in the scattered rainforests. Cooler in the mountains and high plateaus, hotter in the lowlands.

Resources: Diamonds, gold, iron, oil, fish, copper, salt, gas, timber, uranium, limestone, zinc, platinum, asbestos, cobalt, silver, bauxite, coal, chrome, nickel, tin and potash

Green Statistics

Major Conservation Initiatives:
• ANTARCTIC TREATY AND CONVENTION: South Africa
• CITES: Botswana, Burundi, Central African Republic, Congo, Lesotho, Malawi, Mauritius, Namibia, Rwanda, South Africa, Uganda, Zaire, Zambia, Zimbabwe
• LAW OF THE SEA: Angola, Botswana, Burundi, Central African Republic, Congo, Lesotho, Malawi, Mauritius, Namibia, Rwanda, South Africa, Swaziland, Uganda, Zaire, Zambia, Zimbabwe
• MARPOL: South Africa
• MIGRATORY SPECIES: Central African Republic, Uganda, Zaire
• OCEAN DUMPING: Lesotho, South Africa, Zaire
• RAMSAR: Lesotho, South Africa, Uganda
• WORLD HERITAGE: Burundi, Central African Republic, Congo, Malawi, Uganda, Zaire Zambia, Zimbabwe

Protected Areas: Angola: 1%, Botswana: 18%, Central African Republic: 6%, Congo: 4%,

losses could reach 80 percent.

Africans are clearing forests to make room for new crops. This leads to erosion and prevents soil regeneration, a problem that cannot be solved by much-needed rains.

Millions of people are suffering from the drought, and so are animals. Some 90,000 cattle have died since December 1991. Zimbabwe's wildlife reserve (one of the largest and most diverse in the region) and wildlife in Namibia are threatened, due to a lack of grazing areas. Overgrazing by livestock in the past has led to soil erosion and ultimately desertification, making the land unable to support crops or grasslands. Officials are shooting wild animals to feed the population. Meat from these animals goes mainly to children.

Various drought-stricken countries established the Special Aid Coordination Organization (SADCC) to properly disperse water and food. Although $526 million in aid has been sent to the region, there is concern that the money is not properly distributed.

There is speculation the drought may be caused by a periodic climate disruption known as El Niño (see "El Niño," page 7).

In addition to the drought, health is a perpetual concern in this region. At present rates, the sub-Saharan population will double in the next 23 years, but scientists predict some African populations severely afflicted by the AIDS epidemic will decrease in that time.

In Uganda, for example, 1.5 million people out of a population of 16 million are infected with HIV. Malawi and Rwanda are similarly affected. Some 140,000 people in Zimbabwe, out of a population of 10 million, have tested positive for AIDS. Critics claim that many cases go unreported and project this number to be closer to 500,000. Health programs have made free condoms available in all government clinics.

Controversy regarding the hunting and selling of ivory persists. The government of Zambia burned a nine-ton stockpile of ivory tusks and weapons from poachers and smugglers on February 14, 1992, as a symbolic gesture of the government's commitment to conservation. An international ban on elephant products (including meat, hides and ivory), implemented in 1989 by CITES, still exists. However, the smuggling of ivory continues, and hunters and poachers have

 Worldwide, how many more cars are there today as compared to 20 years ago?

reduced the elephant population from 100,000 in 1980 to less than 20,000 in 1992.

Zimbabwe, Malawi, Namibia, Botswana and South Africa want the ivory trade resumed. These countries claim their elephant populations are not endangered, and that the export of ivory is vital to their economic survival. Money earned from the ivory trade funds conservation projects in these countries.

PROMISING INITIATIVES

IN THE WAKE OF OUTCRY FROM LOCAL COMMUNITIES as well as environmentalists all over the world, the government of Botswana has dropped plans to dredge part of the Boro River, one of the major waterways of the wildlife-rich Okavango Delta. The original plan called for dredging and building embankments along 26 miles of the river in order to ensure a water supply for the fast-growing town of Maun near the delta and for diamond mines 160 miles away in Orapa. Opponents feared dredging would irreparably damage the Okavango, an inland delta which is gaining worldwide stature as a reserve for game and birds.

Local activists appealed to the international environmental community for a second opinion. When the International Union for the Conservation of Nature came out against the proposal, the government halted its plans.

Malawi has strengthened its wildlife protection laws, with convictions resulting in large fines and mandatory prison terms. A Wildlife Research and Management Board has also been developed to discuss policies and changes in wildlife policy. In Botswana, for the first time, the parliament is considering strengthening its wildlife laws.

Sources: *African Wildlife Update, 1992 Information Please Environmental Almanac* (Houghton Mifflin), *International Herald Tribune, New York Times, Wall Street Journal, Washington Post.*

Lesotho: less than 1%, Malawi: 11%, Mauritius: 2%, Rwanda: 11%, South Africa: 5%, Swaziland: 2%, Uganda: 7%, Zaire: 4%, Zambia: 9%, Zimbabwe: 7%, Burundi, Namibia: 0%

Green Rating

3 The horrors of war, famine, drought and environmental destruction are inescapable in this region. Hungry people do not have the luxury of thinking about the future. Outside aid is desperately needed— before it's too late.

"Why is it that we can find domestic cattle everywhere in abundance, but the black rhino is an endangered species? Both animals have commercial value and both are easily tamed. Black rhinos are not farmed in the same way as any other species is because the government doesn't allow it. If people were allowed to own rhinos privately, and were allowed to use them for the production of any goods and services, rhinos would be no more endangered than cattle, goats, donkeys, ostriches and crocodiles."
—*Mike T'Sas-Rolfes,*
Endangered Wildlife

A: *Twice as many*
Source: Protect Our Planet Calendar, Running Press Book Publishers

EARTH JOURNAL

East Africa

Long years of drought, civil war and burgeoning populations have transformed East Africa from a fertile region to an arena of unparalleled human misery. Starving Somalis, their anguish captured vividly by magazine photographs and television cameras, were the overwhelming image from the region in 1992. Before the current cycle of war and famine ends, millions of Somalis will die. Mozambique, Ethiopia and Kenya have also been plagued by war and drought.

Forests once flourished in parts of East Africa, particularly along the highlands of the Great Rift Valley in Ethiopia and Kenya. Those forests are nearly gone, cut to provide fuel wood and to create cropland and livestock range. Only 3 percent of Kenya remains forested and even those reserves are fast disappearing; in Ethiopia, forest remains on just 4 percent of the land.

All of East Africa suffers from periodic hard-hitting droughts that wipe out crops and essentially degrade farmland to desert. A massive drought between 1983 and 1985 caused the starvation of 1 million Ethiopians. Drought still plagues Ethiopia, as well as Somalia and Mozambique. In 1992, one-fifth of all Mozambicans relied on foreign food donations because Mozambique could not grow enough itself.

A 17-year civil war is also to blame for this lack of self-sufficiency in Mozambique. From 1976 until a ceasefire in October 1992, the war caused 1 million violent deaths and stunted Mozambique's development. The country is one of the poorest in the world and was rated the world's worst nation in human suffering by the Population Crisis Committee.

In Somalia, food donations often cannot reach millions of starving people because the military seizes them in an effort to starve a rebel army. The scarcity of food in refugee camps has granted nurses unwanted power: They must decide which children will be fed and which will die.

To flee civil unrest in Somalia, Mozambique and Ethiopia, hundreds of thousands of people have crossed borders in search of new homes. However, these environmental refugees find only rough living

Comoros, Djibouti, Ethiopia, Kenya, Madagascar, Mozambique, Seychelles, Somalia, Tanzania

Regional Statistics

Population: 140,200,000

Land Area: 1,853,489 square miles

Geography: Slashing the eastern side of Africa for 3,500 miles, the Great Rift furrows north from Mozambique to the Red Sea, forking into two branches. Deep lakes, such as Lake Tanganyika, the world's second-deepest lake, trace the course of the western branch. The eastern branch is traced by shallow alkaline lakes and volcanoes such as Kilimanjaro, Africa's highest peak, located in northern Tanzania.

Climate: Hot and humid in most regions. The highlands in Kenya and Ethiopia are cooler and more moist than surrounding lowlands.

: *Every year, how much wood and paper is thrown out by Americans?*

conditions in areas sparse in vegetation. For example, 500,000 Somalians and 200,000 returning Ethiopians have crossed into southeast Ethiopia in the past year, overcrowding the region and potentially causing the starvation deaths of another 1 million people.

Populations are growing in East African countries faster than most countries in the world. Kenya's growth rate of 3.6 percent is extremely high; if that trend continues, Kenya's population will double within 20 years. Somalia's rate of 3.3 percent is also high. As the population grows, land becomes scarce for the 85 percent of East Africans who are farmers. Many must till marginal lands that often support crops for only one or two growing seasons. After that land is depleted, farmers are forced to seek even more marginal land.

To ease the pressure on marginal lands in Eritrea, newly liberated from Ethiopia, the government has tried to convince nomads to give up their traditional way of life and instead live in cities. Kenya has tried to slow its growth rate through education. One reason for Kenya's high population rate is that a woman's worth traditionally has been measured by how many babies she can produce. Schools have been teaching young girls to value instead their ability to achieve outside of the family. Sex education has also had an impact: 20 percent of Kenyan women now use birth control, a rate twice as high as in other parts of the continent.

PROMISING INITIATIVES

IN KENYA AND TANZANIA, tourism has long been an important source of revenue. African game animals such as rhinoceroses and elephants attract thousands of camera-toting tourists and thrill-seeking safari-goers each year. Yet these animals are threatened by poachers.

In 1990, Robin Hurt, a longtime safari leader, started a project that makes it more profitable for Tanzanian natives to bag poachers than wildlife. In two years, the Cullman Project has had these results: 253 poachers prosecuted, 4,155 traps destroyed, and 20,775 animals saved. Many of the project's rangers are ex-poachers who are adept at tracking the movements of their former peers.

Sources: *New York Times, In These Times, Washington Post, Outdoor Life, 1992 Information Please Environmental Almanac* (Houghton Mifflin).

Resources: Gold, limestone, minerals, salt, copper, platinum, chromium, graphite, semiprecious stones, uranium, diamonds

Green Statistics

Major Conservation Initiatives:
• ANTARCTIC TREATY AND CONVENTION: None
• CITES: Ethiopia, Kenya, Madagascar, Mozambique, Somalia, Tanzania
• LAW OF THE SEA: Comoros, Djibouti, Ethiopia, Kenya, Madagascar, Mozambique, Somalia, Tanzania
• MARPOL: Djibouti
• MIGRATORY SPECIES: Madagascar, Somalia
• OCEAN DUMPING: Kenya, Somalia
• RAMSAR: Kenya
• WORLD HERITAGE: Ethiopia, Madagascar, Mozambique, Tanzania

Protected Areas: Comoros: less than 1%, Ethiopia: 6%, Kenya: 5%, Madagascar: 2%, Seychelles: figures unavailable, Tanzania: 13%, Djibouti, Mozambique, Somalia: 0%

Green Rating

3 Although this region has some good environmental policies, overpopulation, civil unrest and drought have overburdened this once-lush area. If population growth can be limited, the area may be able to return to its natural beauty and abundance.

A: *Enough to heat 5 million homes for 200 years*
Source: Greenpeace

EARTH JOURNAL

Ivory Coast

Benin, Cameroon, Cape Verde, Côte d'Ivoire, Equatorial Guinea, Gabon, Gambia, Ghana, Guinea, Guinea-Bissau, Liberia, Nigeria, Sao Tome and Principe, Senegal, Sierra Leone, Togo

Regional Statistics

Population: 179,270,000

Land Area: 1,198,215 square miles

Geography: Forested plateaus, rainforests and mangrove swamps are found along the coastal areas. Savanna-type areas cover the inland regions. Some semi-arid desert in the hilly northern interior. All countries have coasts on the Atlantic Ocean.

Climate: Ranges from subtropical and rainy to semi-arid desert. Winter is dry; summer is rainy.

Resources: Timber, low-grade iron ore, bauxite, diamonds, rubber, tin, coal, marble, fossil fuels

Deforestation, contaminated water, toxic wastes and wildlife extinction are ongoing problems faced by most countries on the Ivory Coast. Trees are cleared for agricultural purposes, with Benin losing its forest at an average annual rate of approximately 1.7 percent. Some 91 percent of the forest in Gambia has been cleared for fuel wood and agriculture. Logging is a large part of the economy of Cameroon and the Côte d'Ivoire, a country that suffers from a deforestation rate of 1,120 square miles per year, the highest in the world. Population density and forest fires contribute to the land degradation.

There are few land management programs in the area, although Guinea maintains biosphere reserves and has recently opened its first national park. However, lack of land management in the region has led to soil erosion, desertification, water depletion and groundwater pollution. Wildlife poaching, illegal farming and grazing increase land degradation in most areas.

Approximately 65 percent of the rural population of Benin lacks safe drinking water, and most incidents of disease are related to unsanitary drinking conditions. Malaria strikes more than half the population of Cameroon, and water drainage to minimize mosquito breeding grounds is vital to the population. The government of Gambia plans to build a dam that will increase the country's hydroelectric power, but the construction will destroy 5 percent of the mangrove forest. The dam will also alter the saline content of the water that many shrimp and shad depend upon for food. Another dam project in the works in Senegal threatens to damage the ecology of the area and alter the breeding cycles of fish.

Since 1987, toxic waste dumping in West Africa by American and European companies has increased. Nine shipments (totaling over 19 million tons of industrial waste) have been imported into Ghana, Guinea, Guinea-Bissau, Congo, Nigeria, Sierra Leone, Liberia and Cape Verde. Illegal dumping is apparently common, though difficult to detect when it is done offshore.

: *How much aluminum is thrown away by American consumers and industries?*

Illegal hunting is a large problem in the Ivory Coast region and many believe certain species have been given inappropriate protection under CITES. Although natural bushfires and farming contribute to the loss of many species, advocates of trade bans stress that trade is the biggest threat to wild populations, and relaxation of laws may seriously endanger species' populations. The ban on ivory, however, has not succeeded in quelling poachers, and the funds needed to enforce these laws are scarce. Poached ivory is frequently smuggled through various countries to South Africa, where it is shipped to the Far East.

Elephants and parrots are the most rapidly disappearing species in the Ivory Coast region, due to the wildlife trade. Senegal exports more parrots and Gabon sells more ivory than any other country. Overfishing has depleted fish populations and dam construction has reduced river sediment necessary for nutrition of shrimp and fish. Hunters in Gabon collect approximately four tons of meat every month for export as well as domestic use.

HOT SPOT

DEFORESTATION IS THE BIGGEST PROBLEM plaguing the Ivory Coast. These countries depend on agriculture for their subsistence, specifically the export of two crops—cocoa and coffee. For example, the Ivory Coast had produced 75,000 tons of cocoa and 147,000 tons of coffee annually before it acquired independence in the mid-1970s, when these quantities expanded to 228,000 and 305,000, respectively. This increased production pressure has led to soil damage and loss. Inadequate implementation, forest fires and poor maintenance have inhibited the region's attempts to reduce deforestation rates.

Sources: *A Green History of the World* (St. Martin's Press), *1992 Information Please Environmental Almanac* (Houghton Mifflin), *African Wildlife Update.*

Green Statistics

Major Conservation Initiatives:
• ANTARCTIC TREATY AND CONVENTION: None
• CITES: Benin, Cameroon, Gabon, Gambia, Guinea, Guinea Bissau, Liberia, Nigeria, Senegal, Togo
• LAW OF THE SEA: Benin, Cameroon, Côte d'Ivoire, Equatorial Guinea, Gabon, Gambia, Ghana, Guinea, Guinea-Bissau, Liberia, Nigeria, Senegal, Sierra Leone, Togo
• MARPOL: Côte d'Ivoire, Gabon, Liberia, Togo
• MIGRATORY SPECIES: Benin, Cameroon, Côte d'Ivoire, Ghana, Nigeria, Senegal, Togo
• OCEAN DUMPING: Cape Verde, Côte d'Ivoire, Gabon, Liberia, Nigeria, Senegal, Togo
• RAMSAR: Benin, Gabon, Ghana, Senegal
• WORLD HERITAGE: Benin, Cameroon, Cape Verde, Côte d'Ivoire, Gabon, Gambia, Ghana, Guinea, Nigeria, Senegal

Protected Areas:
Benin: 8%, Cameroon: 4%, Côte d'Ivoire: 6%, Gabon: 7%, Ghana: 5%, Guinea: less than 1%, Liberia: 1%, Nigeria: 1%, Senegal: 11%, Sierra Leone: 1%, Togo: 9%, Cape Verde, Gambia, Guinea-Bissau: 0%, Equatorial Guinea, Sao Tome and Principe: figures unavailable

Green Rating

4•5 Nations here pay lip service to conservaton.

A: *Enough to rebuild the US commercial airfleet every 90 days*
Source: Environmental Defense Fund

EARTH JOURNAL

North Africa

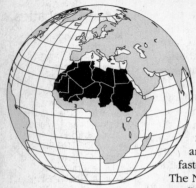

Much of North Africa is already desert, and that acreage increases each year. The great Sahara Desert has expanded southward over the last 20 years, claiming 270 new square miles in Mali and Sudan. Throughout North Africa, soil that was once fertile has turned to desert, sapped by drought and the pressures to sustain one of the world's fastest-growing populations.

The North African countries along the Mediterranean coast (Morocco, Algeria, Libya, Tunisia and Egypt) are relatively wealthy because of petroleum and mineral deposits. Burgeoning populations, however, are causing problems. Because the Sahara Desert takes up the southern portion of these countries, most of their populations live on the fertile northern coastal plains. Some 90 percent of Libya's people, for example, live in 10 percent of the country's land area.

Such extremely high population density has two effects. First, it forces many farmers onto lands with sparse vegetation. There, the land is quickly depleted through overfarming, overgrazing and fuel wood gathering. Topsoil is scattered to the winds as the farmers move on to farm other marginal lands. Second, the concentration of people, especially in urban areas, leads to water pollution. What little water there is has been contaminated by untreated urban sewage, industrial waste, oil refinery discharge, pesticides and topsoil runoff. Citizens of dense Mediterranean coastal cities suffer health problems. Rampant algae growth, spurred by fertilizer nitrates, has turned the Mediterranean from blue to green and upset the ecological balance. Also, dolphins, whales and seals have been found with PCB contamination levels that are 50 times higher than what is considered dangerous for humans.

For countries along the Sahara (Mauritania, Mali, Niger, Chad, Burkina Faso and Sudan), drought combined with a growing population have made life a struggle. Agriculture is the traditional economic activity in these countries; Chad and Niger, in fact, were self-sufficient in grain production until a six-year drought that began in 1967. Drought continues to ravage these countries. Subsistence farmers are still

Algeria, Burkina Faso, Chad, Egypt, Libya, Mali, Mauritania, Morocco, Niger, Sudan, Tunisia

Regional Statistics

Population: 364,343,000

Land Area: 8,154,348 square miles

Geography: The Atlas Sahrien mountains in the north and the Sahara Desert in the south bracket areas of plateau, savanna and steppe. Fertile land exists along the coastlines on the northern, eastern and western borders of the region, as well as the river valleys of Mauritania and the Sudan, but the majority of the area is barren and nonarable.

Climate: Hot and dry in most of the region; moderate temperatures along the coast.

Resources: Crude oil, natural gas, iron ore, gypsum, uranium, manganese

Q: *What percentage of Chernobyl's radioactive materials were released in its 1986 nuclear disaster?*

the base of the economy and continue to farm untouched lands. These lands are sparsely vegetated, however, and are quickly and inevitably rendered to waste, forcing the farmers to seek new farmland. The countries remain poor: Burkina Faso's per capita gross national product, for example, is $210 a year.

Poverty has forced populations to migrate en masse to urbanized areas and to other countries. It has also made wildlife poaching a lucrative trade. The problem is acute in Niger, Sudan and Mali, where elephant, rhinoceros, oryx, cheetah and gazelle populations are quickly dwindling. Only three countries are fulfilling their duties as members of CITES, a worldwide consortium of nations that regulates the illegal trade of wildlife. The rest opt either for token membership or no membership at all.

Even the large public works projects that are supposed to empower these nations are of questionable merit. Irrigation projects along the Nile, such as the Aswan Dam, have allowed Egypt to double its actual crop area. However, 28 percent of Egypt's irrigated soils have become salinized in the process, making uncertain the sustainability of this practice. In Mauritania and Mali, plans to dam the Senegal River bring promises of increased agricultural production. But there are environmental costs: The dams would disrupt fish spawnings and the wintering grounds of migratory birds, as well as limit traditional, sustainable methods of agriculture and livestock tending.

Other projects are in the works to more effectively use the precious fresh water in the region. Libya, for example, has a $25 billion plan to pump water from underground sources below the Sahara. No one knows the extent of these aquifers or how long they might yield water. But, in North Africa, a region dominated by sweeping desert, any water plan will be considered.

HOT SPOT

LONG-RUNNING CIVIL WARS in Chad and Sudan have worsened poverty and compelled entire populations to migrate en masse into sparse deserts. In 1992, the Islamic Sudanese government, fearing an uprising, forced 400,000 Christians and others out of their homes in southern Sudan and into the desert.

Sources: *Washington Post*, *New York Times*, *State of the World 1992* (W.W. Norton), *1992 Information Please Environmental Almanac* (Houghton Mifflin).

A: *About 7 percent*
   Source: Nuclear Information and Resource Service

Green Statistics

Major Conservation Initiatives:
- ANTARCTIC TREATY AND CONVENTION: None
- CITES: Algeria, Burkina Faso, Chad, Egypt, Mali, Mauritania, Morocco, Niger, Tunisia
- LAW OF THE SEA: Algeria, Burkina Faso, Chad, Egypt, Libya, Mali, Mauritania, Morocco, Niger, Sudan, Tunisia
- MARPOL: Algeria, Egypt, Tunisia
- MIGRATORY SPECIES: Burkina Faso, Chad, Egypt, Morocco, Niger, Tunisia
- OCEAN DUMPING: Chad, Libya, Morocco, Tunisia
- RAMSAR: Algeria, Burkina Faso, Chad, Egypt, Mali, Mauritania, Morocco
- WORLD HERITAGE: Algeria, Burkina Faso, Egypt, Libya, Mali, Mauritania, Morocco, Niger, Sudan, Tunisia

Protected Areas:
Algeria: 1%, Burkina Faso: 3%, Egypt: 1%, Mali: 1%, Mauritania: 1%, Morocco: 1%, Niger: 1%, Sudan: 3%, Chad, Libya, Tunisia: less than 1%

Green Rating

4 North Africa has been depleted by drought, hunger and civil war. As people leave the plundered and unproductive countryside for the cities, they find even less chance of making a living. This region pays lip service to the environment, but little action is being taken.

EARTH JOURNAL

Middle East

In the Middle East, coastlines are endangered by erosion and pollution. There is a great loss of wildlife due to industrialization, and the 1991 Persian Gulf war devastated what was already one of the most-polluted bodies of water in the world. Meanwhile, many Persian Gulf countries depend on the gulf for fish. The war severely damaged the region's soil, and a shortage of clean water is a major cause of health problems. Oil drilling and production also pollute in the area.

Bahrain, Cyprus, Iran, Iraq, Israel, Jordan, Kuwait, Lebanon, Oman, Qatar, Saudi Arabia, Syria, Turkey, United Arab Emirates (UAE), Yemen

Regional Statistics

Population: 197,370,000

Land Area: 2,427,414 square miles

Geography: This region is composed mainly of desert and semi-desert areas. The high Zagros mountain range is a natural border dividing Iran and Iraq. The Saudi Arabian Desert is the world's largest sand desert. Natural meadows exist in the more humid areas of Israel.

Climate: Hot and dry throughout the region, with more moderate temperatures in coastal and montane areas.

Resources: Oil, fish, copper, asbestos, gypsum, clay, salt, building stone, iron, natural gas, sulfur, potash, bromine,

Oil lakes half a mile wide and up to three feet deep are still forming in the desert, due to the war. Oil is killing plants, birds and insects, and creates more pollution than did the smoke from burning wells. Oil in some parts is hard enough that stones float on the surface. Tests show that oil has seeped into the ground eight inches deep. Saudi Arabia is considering using hoses to wash away some of the oil on its beaches, but there is concern that this may cause even further damage to marine life. Some 6.2 million acres of forest in Iran along the Persian Gulf and the Sea of Oman suffer from acid rain caused by the burning oil wells in Kuwait. An estimated $10 billion is needed to clean and restore the Persian Gulf area.

The Gulf Cooperation Council (GCC)—Saudi Arabia, United Arab Emirates, Oman, Qatar, Bahrain and Kuwait—is more concerned about illegal waste dumping since the war. These countries depend on the gulf to fill 70 percent of their desalinated water needs. Sanctions have never been imposed to deter polluters, and fees for disposing of oil have increased, encouraging increased illegal dumpings. As discussed at the 1992 Middle East Environmental Conference, the region's leaders are contemplating an emergency environmental fund.

The GCC hosted an international research project for the study of its ecosystems. A research vessel spent 100 days surveying the waters of the gulf, determining that although subtidally everything appears healthy, life in the intertidal zone has been eliminated. Coral is bleached and badly eroded. Environmentalists fear complacency now that the smoke has cleared.

Water is a divisive issue in the Middle East, and

Q: *What percentage of US greenhouse gas emissions are caused by autos?*

regional water plans have failed. Jordan and Syria accuse each other of stealing water. Syria and Iraq charge Turkey with hijacking water from the Tigris and Euphrates rivers. Turkish president Turgut Ozal proposed a $21 billion network of crisscross aqueducts, but Iraq and Syria feel this would only give Turkey more water. Syria and Iraq are aiding Kurdish separatists in their fight against Turkey, to punish Turkey for its use of Euphrates water.

Palestinians, Jordanians, Lebanese and Syrians claim water that is controlled by Israel. Jordan's West Bank and Syria's Golan Heights contribute two-thirds of Israel's water supply. Without major cutbacks in farming, Israel and Jordan will soon have to import water from Turkey or develop desalination plants. The United Arab Emirates will spend $137 million on a new desalination plant, and Bahrain will invest $196 million for its own.

In the late 1980s, Jordan drained the wetlands upon which 300 species depended, pumping the water 60 miles to the capital city of Amman. Jordan's water demands rise 10 percent per year, and the population grows 3.5 percent annually. Jordan must double its water supply in 20 years to keep pace with the population. The country cannot afford desalination plants, and its main hope lies in regional water sharing. World Bank officials predict Jordan and Israel will exhaust their natural water supplies by 1995.

HOT SPOT

IN THE MID-1980s, Turkey initiated a huge hydroelectric and irrigation project on the Euphrates without negotiating with water-sharing countries in the region. The finished product, the Ataturk Dam system, includes 19 dams and is the world's fifth-largest irrigation system. Critics say the investment will never pay off. Environmentalists worry that irrigated water will carry salts, fertilizers and pesticides back into the river, which is Syria's main source of drinking and industrial water. Turkey says Syria and Iraq are getting more water, due to regulated deliveries, than they would if they depended on the unpredictable flow of the Euphrates.

Sources: Agence France Presse, *The Atlanta Journal and Constitution, 1992 Information Please Environmental Almanac* (Houghton Mifflin), Greenwire, Inter Press Service, *New York Times*, Proprietary to the United Press International 1992, The Reuter Library Report, *Washington Post*, Xinhau General Overseas News Service.

limestone, chrome, asphalt, magnesium, chromite, boron

Green Statistics

Major Conservation Initiatives:
• ANTARCTIC TREATY AND CONVENTION: None
• CITES: Cyprus, Iran, Israel, Jordan, Kuwait, United Arab Emirates
• LAW OF THE SEA: Bahrain, Cyprus, Iran, Iraq, Kuwait, Lebanon, Oman, Qatar, Saudi Arabia, United Arab Emirates, Yemen
• MARPOL: Cyprus, Israel, Lebanon, Oman, Syria, Turkey
• MIGRATORY SPECIES: Israel, Saudi Arabia
• OCEAN DUMPING: Cyprus, Jordan, Kuwait, Lebanon, Oman, United Arab Emirates
• RAMSAR: Iran, Jordan
• WORLD HERITAGE: Cyprus, Iran, Iraq, Jordan, Lebanon, Oman, Qatar, Saudi Arabia, Syria, Turkey, Yemen

Protected Areas:
Qatar, Yemen: 0%, Bahrain, Iraq, Kuwait, Lebanon, Oman, Saudi Arabia, Syria, Turkey, United Arab Emirates: less than 1%, Cyprus, Jordan: 1%, Iran: 2%, Israel: 12%

Green Rating

3 Oil development has been the order of the day in this region—at the expense of the environment. Countries here are in danger of making matters worse in their search for water and other resources.

A: ***Approximately 30 percent***
Source: Campaign for New Transportation Priorities

EARTH JOURNAL

Western Europe

The proposed formal European Community (EC), currently an informal 12-nation coalition, would have a primary economic focus: to encourage free trade within Europe and to provide some collective clout in the world marketplace. But sustaining economic growth without seriously disrupting the environment is another concern of the EC. Its top environmental officials have struggled to create strict binding agreements on pollution that don't hinder any one country's opportunities for wealth.

With European countries engaging in free trade of pollution as well as goods, regulation is clearly needed. By the time the Meuse River reaches the Netherlands to empty into the North Sea, for example, it has already been contaminated with human sewage and factory chemical waste from France and Belgium. The Netherlands has complained to its neighbors for 25 years, to no avail.

Because European states are relatively small and close together, one country's pollution is likely to be every country's problem. This is clear to the Scandinavian countries of Norway, Sweden and Finland, which all have serious acid rain problems because of pollution drifting northward from the continent. In southern Norway, where rain falls as acidic as lemon juice, over 7,000 square miles of lakes can no longer support fish. In southern Sweden, acid rain has seriously affected 16,000 lakes. Of the total amount of acid rain that falls on their countries, Sweden and Norway contribute only about 10 percent.

The problem of acid rain is not news to Europe, where coal and oil burning has churned sulfur dioxide into the atmosphere for years. Acid rain has eaten away at the marble of Athens and the spires of Notre Dame Cathedral in Paris. Acid rain and other pollutants have also severely damaged almost 15 percent of all trees in the European Community, according to a 1991 EC report. The damage is even higher in Austria and Switzerland, where heavy truck traffic passes every day through what is known as the crossroads of Europe. There, over one-third of all trees have been affected.

Andorra, Austria, Belgium, Denmark, Finland, France, Germany, Greece, Iceland, Ireland, Italy, Liechtenstein, Luxembourg, Malta, Monaco, Netherlands, Norway, Portugal, San Marino, Spain, Sweden, Switzerland, United Kingdom, Vatican City

Regional Statistics

Population: 378,242,000

Land Area: 2,270,678 square miles

Geography: The Alps transverse Austria, France, Germany, Switzerland, Liechtenstein and Italy. The central and southern part of the region is largely fertile plains and rolling hills, intercepted by rivers and mountain ranges. The northern section, Iceland, Sweden, Norway and Finland, are low plain regions with snow and ice-covered lands reaching the Arctic Circle. Iceland is one of the most volcanic regions on Earth. Western Europe

: *Since 1983, nine out of ten cars manufactured in Brazil run on what type of fuel?*

The EC in 1992 ordered refineries to reduce the amount of sulfur in diesel fuel and heating oil by one-third by 1994, and then by 80 percent more by 1996. This should improve the situation, but will hardly save the forests, according to Sten Nilsson, an Austria-based pollution expert.

A tougher plan to reduce energy consumption by imposing high taxes on fossil fuels met great resistance this year. The EC-proposed plan called for oil to be taxed $3 per barrel for 1993, with the tax to increase each year to $10 in the year 2000. It was, in effect, an attempt to tie the price of environmental damage to the price of a barrel of oil. High energy prices in Japan have spurred great gains in energy conservation and efficiency, proponents of the measure argued. They also proposed a similar tax for coal and natural gas. Although the plan was approved by the EC executive commission, it has not met the needed approval of the 12 member nations.

A step toward a common ecological vision for the EC would be the establishment of a European Environment Agency. Given effective powers, it could resolve disputes such as that over the Meuse River. Yet France was the only country that voted against the agency's formation.

France's vote appears ill-conceived, partly because of the grave hazardous waste situation in Europe. The former West Germany had 50,000 dump sites in need of cleanup, and the cost of a proper cleanup for all of Western Europe is estimated at $200 billion to $500 billion. As of 1991, the EC had no overall dumping policy and, instead, was letting individual countries and the free market decide where waste should be disposed. This allowed Dutch polluters to skirt their country's stiff regulations by transporting their waste to Belgium. When Wallonia, the southern region of Belgium, banned the import of waste, the EC balked, charging that the law violated a free trade agreement. With the free market ruling the handling of hazardous waste, Greenpeace's Dutch waste trade coordinator in Amsterdam, Jim Puckett, fears that waste will pile up in Europe's poor countries: Spain, Portugal and Greece.

Western Europe's popularity with third world nations rose somewhat during the 1992 Earth Summit. The EC, which has promised to keep its emissions of global warming gases at 1990 levels, supported a strong global warming treaty and a biological diversity

was entirely forested before human habitation and is still 35 percent woodland.

Climate: A region of four distinct seasons, humid and warm in the summer and cold in the winter.

Resources: Mineral water, coal, iron, timber, brown coal, potash, marble (Italy), oil, salt, cryolite, fish (Italy and Iceland), lead, barite, peat, silver, gypsum, dolomite, copper, mercury, petroleum, pyrite, tungsten

Green Statistics

Major Conservation Initiatives:
• ANTARCTIC TREATY AND CONVENTION: Austria, Belgium, Denmark, Finland, France, Germany, Greece, Italy, Netherlands, Norway, Spain, Sweden, United Kingdom
• CITES: Austria, Belgium, Denmark, Finland, France, Germany, Ireland, Italy, Luxembourg, Malta, Netherlands, Norway, Portugal, Spain, Sweden, Switzerland, United Kingdom
• LAW OF THE SEA: Austria, Belgium, Denmark, Finland, France, Greece, Iceland, Ireland, Italy, Luxembourg, Malta, Netherlands, Norway, Portugal, Spain, Sweden, Switzerland, United Kingdom
• MARPOL: Austria, Belgium, Denmark, Finland, Germany, Greece, Iceland, Italy, Netherlands, Norway, Portugal, Spain, Sweden, Switzerland, United Kingdom
• MIGRATORY

A: *Ethanol*
Source: Union of Concerned Scientists

SPECIES: Belgium, Denmark, France, Germany, Ireland, Greece
• OCEAN DUMPING: Belgium, Denmark, Finland, France, Germany, Greece, Iceland, Ireland, Italy, Luxembourg, Malta, Netherlands, Norway, Portugal, Spain, Sweden, Switzerland, United Kingdom
• RAMSAR: Austria, Belgium, Denmark, Finland, France, Germany, Greece, Iceland, Ireland, Italy, Liechtenstein, Luxembourg, Malta, Netherlands, Norway, Portugal, Spain, Sweden, Switzerland, United Kingdom
• WORLD HERITAGE: Austria, Belgium, Denmark, Finland, France, Germany, Greece, Iceland, Ireland, Italy, Luxembourg, Malta, Netherlands, Norway, Portugal, Spain, Sweden, Switzerland, United Kingdom

Protected areas:
Austria: 19%, Belgium, Finland, Switzerland: 3%, Denmark, Portugal: 7%, France, Germany, Iceland: 8%, Ireland: less than 1%, Italy, Netherlands, Sweden: 4%, Norway: 16%, Spain: 5%, United Kingdom: 11%, Andorra, Greece, Liechtenstein, Luxembourg, Monaco: 0%, Malta, San Marino, Vatican City: figures unavailable

Green Rating

1•2 Western Europe has good environmental policies—and big environmental problems. It's an uphill battle, but there is hope that they head uphill, not down.

treaty, while the US rejected both. The EC also pledged $4 billion to fund environmental programs introduced at the conference. Germany made aggressive political strides in Rio, cutting compromises and back-room deals to revitalize stalled talks, and dwarfing the US pledge of $150 million to protect forests worldwide. Germany pledged $165 million to Brazil alone. Germany also announced it would triple its contribution (to $487 million) to a World Bank fund designed to promote sustainable development in the third world. German Chancellor Helmut Kohl called for the industrialized nations to contribute $4 billion total.

Many third world nations argue, however, that World Bank-funded projects in the past degraded the environment and benefited mainly the industrial elite, who have close economic ties to industrialized nations. Critics also say that $3 billion to $4 billion is simply a pine needle in the forest compared to the aid Western Europe will give Eastern Europe.

For Germany, reunification in 1990 meant inheriting the environmental problems of the former East Germany, such as the Schoenberg dump. Schoenberg accepted any waste from any country for low fees. Environmentalists who tested the site after the borders were opened found that the so-called "impermeable clay floor" of Schoenberg is really porous sand that allows toxic waste to seep into groundwater. Besides Schoenberg, Germany needs to clean up 15,000 to 20,000 other East German dumps.

Germany will also have to spend an estimated $11.2 billion in the East German central industrial belt, to make industry there meet environmental standards. Factory pollution in the region is currently 100 times worse than Germany will allow. Yet, as Germany declared this crackdown, it also said it would loosen environmental regulations to allow East German factories to compete for survival in the unified state. It is a balance of priorities that typifies Western Europe. As nations jockey for economic bargaining power within the European Community, it is uncertain whether any of them will heartily endorse any hard-hitting environmental policy.

Sources: *New York Times, Los Angeles Times, Washington Post, International Herald Tribune.*

: *What has been found in the breast milk of nursing mothers living in industrialized parts of the US?*

REGIONAL REPORTS

Eastern Europe

The repercussions of pollution on individual health and the general quality of life for an average Eastern European are terrifying. They include lung disease, astoundingly high infant mortality rates and high lead levels in children. This pollution has resulted from decades of mismanagement, lack of technology and poor wages under Communist rule.

In the past, highly toxic chemicals were consistently dumped into the soil, water and air in this region, and some of this dumping continues today. Poland emits as much sulfur dioxide as the entire 12-nation European Community, and Prague's sulfur dioxide concentrations average twice the World Health Organization standard. Chemicals throughout the region are poorly stored, underground tanks leak, and backyards pose as dumping grounds. Companies choose to purchase new, cleaner sites, rather than incurring the cost of cleaning contaminated areas. Other companies have assessed the environmental quality of the region and are deterred from investment altogether. Hungary's government has been able to sell only one industrial site, thus far.

However, 1992 saw some positive actions. Poland published a list of 80 polluting companies, in order to force them into cleanup programs. Czechoslovakia canceled plans to build a power plant, in order to allocate the funds to the environment. Walcho Environmental Systems, Inc., a US company, sold $2 million worth of pollution control devices to a Polish utility company and intends to sell similar systems in Czechoslovakia and Hungary. In the past two years, various governments have given industries incentives for cleanup. The governments have agreed to financially assist companies in their cleanup programs, but the money provided for treatment will be rescinded in five years if it is not properly used.

Admittance to the European Community (EC) means former Eastern-bloc countries must promise to meet specific environmental standards, including protection and quality codes. Most Eastern European countries lack the funding necessary to meet these standards. Hungary was granted associate status in

Albania, Bosnia-Herzegovina, Bulgaria, Croatia, Czechoslovakia, Federal Republic of Yugoslavia, Hungary, Poland, Romania, Slovenia

Regional Statistics

Population:
123,740,000

Land Area: 450,409 square miles

Geography: This region is on average 50% cultivated lands; the rest is rolling plains and forests. The Balkan Mountains run south and the Carpathian Mountains extend east-west; the fabled Transylvanian Alps in Romania extend east-west. Seas surround the countries, with the Baltic Sea in the north, Adriatic to the southwest, and the Black Sea to the southeast. Yugoslavia has over 2,000 miles of coastline.

Climate: This region has a cold winter and a warm summer, marked by few extremes of temperature.

: *More than 100 chemical contaminants from industrial pollution*
Source: Greenpeace

Resources: Petroleum, minerals, timber, metals, coal, lignite, natural gas

Green Statistics

"Yugoslavia" below refers to affiliations before the war.

Major Conservation Initiatives:
• ANTARCTIC TREATY AND CONVENTION: Bulgaria, Czechoslovakia, Hungary, Poland, Romania
• CITES: Bulgaria, Hungary, Poland
• LAW OF THE SEA: Bulgaria, Czechoslovakia, Hungary, Poland, Romania, Yugoslavia
• MARPOL: Bulgaria, Czechoslovakia, Hungary, Poland, Yugoslavia
• MIGRATORY SPECIES: None
• OCEAN DUMPING: Hungary, Poland, Romania, Yugoslavia
• RAMSAR: Bulgaria, Czechoslovakia, Hungary, Poland, Yugoslavia
• WORLD HERITAGE: Albania, Bulgaria, Czechoslovakia, Hungary, Poland, Romania, Yugoslavia

Protected Areas: Albania: 2%, Bulgaria: 1%, Czechoslovakia: 16%, Hungary: 6%, Poland: 7%, Romania: 1%, Yugoslavia: 4%

Green Rating

3•5 Eastern Europe's people are suffering the consequences of industrialization without environmental protection. People are working hard with what little they have, but the problems are

the EC, and it hopes to comply with regulations within three to five years. Poland fears it is incapable of receiving this status before the year 2000. The EC is working on an environmental code of conduct for those countries that intend to seek membership in the community.

For a fee, US companies have offered to help Eastern European countries develop new methods of cleanup and operation. They promise to track down funding, and predict results within a year. More than a dozen environmental firms have established themselves in Budapest, Prague and Warsaw, with lawyers counseling ministries on waste management and cost-effectiveness.

Amid these efforts operates a nuclear power plant in Kozloduy, Bulgaria, which some experts view as the most poorly run in the world. Critics urge that four of the six reactors at the plant be shut down. However, the plant generates 40 percent of Bulgaria's electricity, and there is no money to obtain alternate sources of energy. Power outages are common, and there is fear that some people may die of cold.

Using nuclear energy in this region has resulted in poorly stored radioactive waste, deficient controls and insufficient fire protection. Soviet-designed water reactors are considered the least safe—there are four in Bulgaria and two in Czechoslovakia. Four reactors in eastern Germany were shut down when plants in western Germany offered to compensate for the power loss. A faulty plant in Romania was ordered closed. Many plants remain open because of the region's heavy reliance on nuclear power. Hungary receives 50 percent of its electricity from nuclear reactors, and Czechoslovakia depends on nuclear reactors for 28 percent.

Leaders of major democracies have expressed an interest in contributing to a multimillion-dollar fund to improve the safety in nuclear power plants. Group 24, an organization of industrial nations, is working together to examine nuclear plants in the region. These plans involve refitting some reactors, closing unsafe plants, and offering alternate forms of electricity.

Years of preoccupation with economic survival in Eastern Europe have also led to the neglect of and cruelty toward this region's animal population. In the war in what was Yugoslavia, more than 1 million farm animals were killed by federal forces in the last

: *What percentage of junk mail is neither opened nor read?*

half of 1991. Bulgaria, Hungary and Czechoslovakia maintain no animal welfare laws, and zoos have cramped cages. Poland's animal shelters are overcrowded; however their environmental protection laws are stronger than most other areas. In Hungary, it is reported that over 300,000 stray dogs and cats roam the streets.

PROMISING INITIATIVES

POLAND HAS DEVELOPED A DEBT-FOR-ENVIRONMENT SCHEME. The "environment fund" will pay for and monitor a program supporting cleanup initiatives. Under the scheme, Poland will spend money on environmental initiatives instead of paying back debts owed to various countries. The US has formally agreed to this plan, which will add $350 million to the fund in the coming years. France and Scandinavia have expressed interest. Funds will be used to reduce pollution in the Baltic Sea, limit greenhouse gas emissions, and protect forests and wetlands.

HOT SPOT

THE WAR IN WHAT WAS YUGOSLAVIA, between the Serbians and Croatians, two factions fighting for control of the Herzegovina/Bosnia area, had left more than 8,000 people dead and some 2 million others displaced as of July 1992. Many cities were destroyed by warfare and abandoned. Croatian and Serbian concentration camps reminiscent of the Nazi era are a horrifying example of the extent to which the violence and destruction have escalated. Prisoners on both sides suffer from torture and starvation.

Sources: *Earthwatch, 1992 Information Please Environmental Almanac* (Houghton Mifflin), *Los Angeles Times, New York Times, Subtext.*

nearly insurmountable, as new menaces appear regularly.

"I do not like noise unless I make it myself."
—French proverb

"The only sustainable city—and this, to me, is the indispensable ideal and goal—is a city in balance with its countryside: a city, that is, that would live off the net ecological income of its supporting region, paying as it goes all its ecological and human debts.... Some cities can never be sustainable, because they do not have a countryside around them, or near them, from which they can be sustained."
—Wendell Berry, Atlantic Monthly

"The law is no panacea for anything, let alone the enormous and complex problems of environmental degradation. It cannot single-handedly take on the economic forces which have produced nuclear fall-out so severe that life expectancy has been reduced in parts of East Europe; and which have produced such overconsumption elsewhere that the global commons have been hideously fouled. Nor can it withstand the social and economic pressures that are eroding the resource base in many developing countries. But it can help."
—Daniel Nelson, Panos

A: *44 percent*
Source: 50 Simple Things You Can Do to Save the Earth, *Earth Works Press*

EARTH JOURNAL

Former Soviet Union

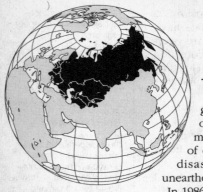

With the Iron Curtain lifted, the world now knows that the former Soviet government had much to hide. Its treatment of the environment was dismal. The government lied to the world about the true extent of damage wreaked by the Chernobyl nuclear disaster, according to documents recently unearthed by the newspaper *Izvestyia*.

In 1986, Core No. 4 of the Chernobyl nuclear power plant exploded, thrusting radiation 50 times as intense as that from the Hiroshima and Nagasaki explosions into neighboring Ukrainian and Belorussian villages. In the wake of this accident, the ruling Politburo attempted to hide any signs of disaster. To keep the number of hospitalized people low, it raised the standard of safe radiation tenfold. To maintain the region's level of farm productivity, it mixed contaminated meat and milk with food from other regions and shipped it across the country. Finally, Soviet officials claimed that only 100 people had suffered large initial doses of radiation. But the number was 10,000 to 12,000 for one Belrussian district alone, said one Russian physician.

Chernobyl's impact on public health is visible today. Many children, especially, show signs of chronic radiation sickness such as lymph gland inflammation, digestive tract and kidney failure, and anemia. Thyroid cancer and leukemia rates are much higher than normal. In Ovruch, Ukraine, the number of youngsters classified as "very ill" is at least six times greater than in 1986.

The Chernobyl disaster galvanized the Russian people not only against nuclear power, but also against the central Soviet government. The green movement became closely allied with the democratic movement, especially in the Balkan states.

The green movement is not, however, a concerted nationwide effort. It is driven by local groups protesting violations against the environment in their communities. These local groups have essentially stopped the construction of new nuclear facilities. At Krasnoyarsk, Russia, a 60,000-signature petition has blocked construction of a major reprocessing plant. In

Commonwealth of Independent States
(Armenia, Azerbaijan, Belarus, Georgia, Kazakhstan, Kyrgyzstan, Moldova, Russia, Tajikistan, Turkmenistan, Ukraine, Uzbekistan), **Estonia, Lithuania, Latvia**

Regional Statistics

Population:
426,900,000

Land Area: 8,571,805 square miles

Geography: Most of this region is a vast plain stretching from Eastern Europe to the Pacific Ocean, separately called the North European Plain in the east, the West Siberian Plain in the center, the steppes in the south and the Central Siberian Plain in the east. The north consists of frozen tundra, the central region contains a belt of temperate forests and grasslands, and the south has steppes or prairies.

Climate: Ranges from cold, dry, arctic conditions in the north to

 How many recycled plastic bottles does it take to create enough fiberfill for a sleeping bag?

Russia, the parliament passed a law in 1990 prohibiting the burial of radioactive wastes from outside the republic. And near the Chelyabinsk nuclear weapons plant, where Lake Karachay is so contaminated with radioactive waste that standing on its shores for an hour would be fatal, residents have actively opposed nuclear waste burial sites.

The former Soviet citizens have had more to protest than nuclear plants. For years, the Soviet Union fueled its rapid industrialization with coal-burning power plants. Since coal prices were set unrealistically low by the central government to encourage growth, Soviet factories have been extremely inefficient (using 1.5 to 2.5 times more energy than the US to produce the equivalent gross domestic product). Some 103 cities in the commonwealth exceed pollution standards (which are stricter than in the US, but seldom enforced) by at least 10 times. Sixteen towns have pollution that exceeds the standards by 50 times. The life expectancy of a Russian is 69.3 years, lower than the US and Japan's life expectancies of 75.3 years and 77.3 years, respectively. In many factory towns, the average life expectancy is lower than the average retirement age of 55. A widespread health crisis is partly to blame for the low life expectancy. But air pollution levels have been clearly linked to some diseases.

Environmental problems also affect the agricultural areas of the old Soviet republics. About 1.5 billion tons of fertile topsoil is lost to erosion each year. Some 25 percent to 30 percent of the rich topsoil has been lost in the Black Soil Belt in the western areas, partly because of reliance on fertilizer and pesticides. Efforts to restore soil fertility, however, have been lacking.

At Lake Baikal, which contains one-fifth of the world's freshwater and is home to 2,400 species, two-thirds of which exist nowhere else on Earth, the ecosystem is severely threatened by industry bordering the lake's southern rim. The most notorious of these polluters, a state-run paper factory, dumps its waste into the lake—even in winter, when the lake is covered by three feet of ice. But local citizens have had great success in fighting the state's pollution of the lake, blocking a plan to pipe waste into the Irkutsk River in 1988 and getting a verbal agreement from the state to shut down the paper factory.

Though Boris Yeltsin is amenable to environmental

temperate and humid conditions in the south.

Resources: Fossil fuels, lead, zinc, timber, mercury, potash, phosphate, nickel

Green Statistics

Major Conservation Initiatives:
• ANTARCTIC TREATY AND CONVENTION: Armenia, Azerbaijan, Belarus, Georgia, Kazakhstan, Kyrgyzstan, Moldova, Russia, Tajikistan, Turkmenistan, Ukraine, Uzbekistan
• CITES: Armenia, Azerbaijan, Belarus, Georgia, Kazakhstan, Kyrgyzstan, Moldova, Russia, Tajikistan, Turkmenistan, Ukraine, Uzbekistan
• LAW OF THE SEA: Azerbaijan, Kazakhstan, Russia, Turkmenistan, Ukraine, Uzbekistan
• MARPOL: Armenia, Azerbaijan, Belarus, Georgia, Kazakhstan, Kyrgyzstan, Moldova, Russia, Tajikistan, Turkmenistan, Ukraine, Uzbekistan
• MIGRATING SPECIES: None
• OCEAN DUMPING: None
• RAMSAR: Armenia, Azerbaijan, Belarus, Georgia, Kazakhstan, Kyrgyzstan, Moldova, Russia, Tajikistan, Turkmenistan, Ukraine, Uzbekistan
• WORLD HERITAGE: None

All CIS states have agreed to honor previous conservation commitments, excepting those that do not apply, i.e., Law of the Sea for

Source: The Denver Post *1991 Colorado Recycling Guide*

landlocked countries. Estonia, Lithuania and Latvia are still negotiating.

Protected Areas:
CIS: less than 1%, Estonia, Lithuania, Latvia: figures unavailable

Green Rating

3 The area is in severe danger. Local governments are having trouble organizing environmental efforts in increasingly nationalistic states. Yet the people have overcome seemingly insurmountable political suppression, and may yet overcome seemingly insurmountable environmental problems.

"Coal has always cursed the land in which it lies. When men begin to wrest it from the earth it leaves a legacy of foul streams, hideous slag heaps and polluted air. It peoples this transformed land with blind and crippled men and with widows and orphans. It is an extractive industry which takes all away and restores nothing. It mars but never beautifies. It corrupts but never purifies."
—Harry M. Caudill, Night Comes to the Cumberlands

"From the standpoint of pollution, the market will be no kinder than the old command system. Profit is the top priority, and the environment is considered an investment with no return."
—Spartak G. Akhmetov, New York Times

causes, according to Russia's environmental minister, and called for the contribution of 5 billion rubles to environmental concerns, the new government would rather be economically competitive than environmentally clean as it enters the world market economy. Nickel-smelting plants in Nikel continue to belch sulfur-laden smoke, which has killed trees around the factory and sent acid rain into nearby Norway. The life expectancy of a Nikel worker is 44 years. Children in Nikel have central nervous system disorders and respiratory disease. Factory officials refuse to consider stricter pollution control.

The paper plant along Lake Baikal continues to operate. The state cannot promise when it will close. And near Chernobyl, it is said that 1.5 million to 1.8 million people should be moved for health reasons. However, there is no place to which they can move. State radiation testing is underfunded. The machinery is often broken, and tests usually take ten months to return. The government tries to ship in food from other regions, and provides citizens in the Chernobyl area with a stipend for its purchase. The stipend is too small, however, and people continue to ingest food they grow themselves in the region, food likely contaminated with radiation.

HOT SPOT

THE ARAL SEA HAS BECOME THE SYMBOL of the old Soviet Union's environmental neglect. Located in central Russia, the sea was once as large as West Virginia, but in 30 years has lost 66 percent of its volume and 40 percent of its area, and has seen its water level drop 13 feet. The rivers that feed it, the Amu and the Syr, were diverted to irrigate cotton farms in the 1950s. However, the irrigation did not allow for proper drainage, causing salt to rise from water tables and salinate the soils, decreasing their fertility. Also, the exposed dried bed of the Aral Sea turned to a salty dust that winds blew over the farmland, further decreasing agricultural productivity.

Sources: *Los Angeles Times*, *The Times* (London), *Washington Post*, *New York Times*, International Center for Information, *State of the World 1992* (W.W. Norton).

: ***How much new garbage does the Fresh Kills, New York, landfill absorb each day?***

REGIONAL REPORTS

South Asia

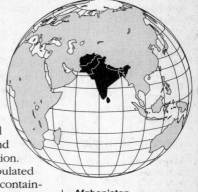

South Asia is bounded by the Himalaya Mountains to the north and the Peninsula of India to the south, with the Indo-Gangetic Plains in between. Although some of the wettest areas on Earth are in this region, it also includes hot and cold deserts. The climate is dominated by monsoons (seasonal wind reversals) that create thick, humid air and are the prime source of precipitation in the region.

South Asia is the second most densely populated region on Earth (after East Asia), with India containing one-sixth of the world's population. Some 74 percent of South Asians live in rural areas. Poverty and rising population are the largest contributors to ecological destruction.

Tourism often adds to the damage. The litter left on Mt. Everest by climbing expeditions continues to prove that no area on Earth is truly pristine. At 26,000 feet, a campsite is strewn with 50 to 100 empty oxygen bottles, empty food cans, torn tents and plastic food wrappings. While the path to the summit is hardly a freeway (just 386 people have duplicated Sir Edmund Hillary's historic feat), concern that Everest is enduring too much traffic grew after 32 people reached the summit at the same time in May 1992.

The population of India is expected to overtake China's population by the year 2000. India's population grows by over 2 percent per year, and is presently 684 million. Pakistan sustains one of the highest population growth rates, 3.4 percent, and only half the people have access to clean water. Afghanistan has one of the lowest literacy rates in the world (33 percent for men and 6 percent for women), and its average life expectancy is slightly over 34 years. Its annual population growth rate is 3.8 percent. Roughly half the citizens of Sri Lanka live in poverty, but the country's annual growth rate is just 1.5 percent. Bangladesh's rate of population growth is 2.7 percent, and many people there suffer from malnutrition.

Forest is the natural cover in this region, but nearly all of it has been logged or cleared for cultivation. Over 55 percent of the land in India and over 75 percent in Bangladesh is cultivated. In some areas of South Asia, only remnants of forest cover remain.

Afghanistan, Bangladesh, Bhutan, India, Maldives, Nepal, Pakistan, Sri Lanka

Regional Statistics

Population:
1,153,723,000

Land Area: 2,108,465 square miles

Geography: This region lies on the Arabian Sea just east of the Middle East. The Himalaya Mountains stretch across the North of Nepal, home to the world's highest mountain. Much of Afghanistan is covered by the snowcapped mountains of the Hindu Kush and by deep valleys. Pakistan, India and Sri Lanka have large regions of fertile plains, intersected with various river systems. Bangladesh is mostly low-lying ravines frequented by tropical monsoons and floods. Most of eastern Pakistan is desert.

Climate: Ranges dramatically from hot arid

A: *More than 34 million pounds*
Source: Garbage *magazine*

deserts to wet humid valleys and cool mountain lands.

Resources: Natural gas, oil, coal, copper, sulfur, lead, zinc, iron ore, salt, precious and semi-precious stones, limestone, manganese, bauxite, graphite

Green Statistics

Major Conservation Initiatives:
- ARCTIC TREATY AND CONVENTION: India
- CITES: Afghanistan, Bangladesh, India, Nepal, Pakistan, Sri Lanka
- LAW OF THE SEA: Afghanistan, Bangladesh, Bhutan, India, Pakistan, Sri Lanka
- MARPOL: India
- MIGRATORY SPECIES: India, Pakistan, Sri Lanka
- OCEAN DUMPING: Afghanistan, Nepal
- RAMSAR: India, Nepal, Pakistan, Sri Lanka
- WORLD HERITAGE: Afghanistan, Bangladesh, India, Nepal, Pakistan, Sri Lanka

Protected Areas: Afghanistan: less than 1%, Bangladesh: 1%, Bhutan: 19%, India: 4%, Maldives: figures unavailable, Nepal: 7%, Pakistan 10%, Sri Lanka: 11%

Green Rating

2•3 The area is threatened with population explosion and limited resources, yet local actions are beginning to empower local citizens.

Between 1973 and 1982, India's forest cover declined from 17 percent to 12 percent. Less than 15 percent of Bangladesh is forested. Most of this depletion is due to the timber industry, which produces much-sought rosewood and sandalwood. Some 144,750 square miles of forest remain (mostly tropical deciduous forests) in India.

Bagunda, India, is plagued by droughts due to deforestation. Residents say the real reason behind India's environmental degradation is people's alienation from their land. Indigenous people in the area claim superior knowledge of local ecology and are seeking returned control of their forests and water resources. In the last 100 years, indigenous people's access to and regulation of the forests has gradually decreased. Nearly one-third of the land in South Asia belongs to the governments. Estimates show that approximately half this land is controlled by government departments, further removing control from the people. Many groups in India and Pakistan are beginning to work with villagers on afforestation nurseries and anti-desertification programs.

Because South Asia's main economic activity is agriculture, its most vital resource is water. Much of the commercial energy is derived from hydroelectric power. The long dry season and variable rainfall have always presented problems in this region, and low rainfall is often supplemented by river water or aquifers. Drilling for water often reaches depths of 22 yards. Massive dams destabilize the earth, increasing the potential for earthquakes. Groundwater is used faster than it is replaced. One-third of India is drought-prone. Although land degradation is common in this region, few areas have reached the point of desertification. Much fertile soil still exists, particularly in Bangladesh.

Bangladesh has one of the highest annual rainfalls in the world. Frequent floods spread water that has been polluted by faulty sewage systems. Similarly, India's rivers and streams are extremely polluted, yet most of its water is used for irrigation. Freshwater sources in Sri Lanka are also polluted and depleted.

South Asia has over 2,000 species of birds, 500 species of mammals and several hundred reptiles and amphibians. Hunting has reduced the royal Bengal tiger, native to Bangladesh, from tens of thousands to hundreds. Domesticated cattle erode habitat on

Q: *How many gallons of water are used for every minute a person is in the shower?*

which wildlife depends.

To protect some species from extinction, 9,650 square miles in 15 different reserves have been set aside, where hunting is illegal. National parks exist for single species as well as entire ecosystems. There are 250 reserves across South Asia.

PROMISING INITIATIVES

IN SOUTH ASIA THERE IS A GROWING EFFORT to empower communities to make their own environmental decisions. SAARC—South Asia Association for Regional Cooperation—is developing poverty-alleviation strategies. Two members from each country represent their people. By the year 2000, SAARC hopes to establish an education program to ensure that all children ages 6 to 14 have a primary education. SAARC is a rare example of a region sharing common goals, yet funding poses an ongoing problem.

HOT SPOT

PAKISTAN AND INDIA ARE IN A STANDOFF regarding nuclear weapons. Pakistan wants to organize talks with the US about nuclear nonproliferation in the region, but India refuses. India claims the US wants South Asia to disarm without an international commitment to reduce nuclear weapons. There is no evidence that either country possesses a nuclear bomb.

Sources: Agence France Presse, *The Christian Science Monitor*, Inter Press Service, *New York Times, The Washington Quarterly, 1992 Information Please Environmental Almanac* (Houghton Mifflin).

"Civilization's problems would not all be solved if growth of the human population were halted humanely by limiting births. They would not necessarily be solved if the world population were ultimately reduced to a more or less permanently sustainable size. Society might still be plagued by racism, sexism, religious prejudice, gross economic inequity, threats of war, and serious environmental deterioration. But without population control, none of these problems can be solved; halting growth and then moving toward lower numbers simply would give humanity an opportunity to grapple with them. The old saying still holds: Whatever your cause, it's a lost cause without population control."

—*Paul R. Erlich,*
Greenhouse Glasnost

A: *5 to 7 gallons*
Source: The Student Environmental Action Guide, *Earth Works Press*

EARTH JOURNAL

East Asia

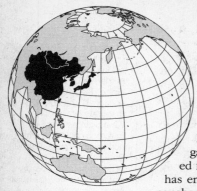

East Asian countries have grown to be economic giants since World War II, with their rate of Gross National Product (GNP) growth outpacing nearly every other region in the world. With that growth has come consequences: Acid rain formed by industrial pollutants has poisoned lakes; industrial waste, garbage and untreated sewage have contaminated freshwater and ocean coasts; and air pollution has enshrouded cities. In Benxi, China, the smoke coughed out by steel and iron mills is so thick for six months a year, that one can see only 50 yards. The city disappears from satellite view for weeks on end. In Taiwan, rivers are so thick with petrochemical company discharge they can be set aflame.

When Japan saw that the side effect of its economic miracle was the notorious air pollution of Tokyo and Osaka, it became the first East Asian nation to take steps to clean its environment. Japan has reduced the sulfur dioxide and nitrogen oxide emissions of its power plants, thanks in part to $2-million smokestacks that bring pollution levels down to one-sixth of levels 30 years ago. Because Japan relies heavily on expensive energy imports, it has also learned to become extremely energy efficient. For example, the country redirects the heat in its natural gas turbines to generate additional electricity.

Pollution levels continue to rise in Japan's major urban areas, however. The environment is the leading concern of Japanese citizens, a 1992 poll showed, making it an issue politicians and industry leaders must address. (The same poll found that Americans ranked the environment as a concern behind crime and prices.) In 1991, the Keidanren, a Japanese big-business group, adopted a charter calling for Japanese industry to become more ecology-conscious.

Global concern about the environment is also pressuring Japan. In November 1991, Japan announced that it would ban all driftnet fishing by January 1993. Driftnets are plastic nets 30 miles long and 30 feet deep that commercial fishing boats stretch across the ocean to catch tuna and squid. But the nets trap all creatures that swim in their path, including whales, sharks and dolphins. Media

China, Hong Kong, Japan, Mongolia, North Korea, South Korea, Taiwan

Regional Statistics

Population: 1,293,800,000

Land Area: 4,373,617 square miles

Geography: This region is widely varied geographically. The Himalayas dominate much of China, creating fertile valleys. Agricultural lands are found in northern Mongolia and parts of China, while the Gobi Desert dominates southern Mongolia. A series of mountain ranges runs through North and South Korea, with many coastal harbors in South Korea. Japan is an archipelago extending 1,744 miles in the East China Sea.

Climate: Ranges from humid to semitropical to montane to desert (in southern Mongolia).

Resources: Timber, coal, natural gas, limestone, metals, marble,

Q: *How long would it take to fill the Louisiana Superdome with garbage that Americans throw out?*

exposure moved many foreign consumers to stop buying tuna that was not "dolphin-safe," and this, combined with the Bush Administration's threat of trade sanctions, forced Japan's hand.

At the 1992 Earth Summit, Japan became a major environmental player through its many discreet back-door meetings with foreign leaders. Japan endorsed the strong global warming and biodiversity treaties that the US adamantly rejected. Japan also pledged $7 billion in environmental aid to developing countries over the next five years, almost twice as much as the US proposes to spend.

Yet Japan has fought nearly every concession to environmental concerns to the bitter end. Its complete ban of driftnet fishing will start six months after the deadline set by a 1989 UN moratorium. At the 1992 Convention on International Trade in Endangered Species (CITES) meeting held in Kyoto to discuss the trade of endangered species, Sweden had to withdraw a proposal to restrict the trade of Atlantic bluefin tuna because of the protests of Japan, which controls two-thirds of the trade. Japan continues to hunt whales in the Antarctic, claiming that they are taken for scientific research, though the whales' meat is simply sold. Japan also continues to be the leading importer of timber, shipping in large amounts from Southeast Asian countries where deforestation is at a crisis level. Even Japan's environmental aid has been criticized as merely opening markets for Japan's world-leading environmental technology.

Throughout East Asia, lofty goals for industrial output place pressure on natural resources. China's decision to cut back on its timber imports, previously the world's second-highest total, has caused its own forests to dwindle further. Almost half of China's highlands have serious erosion problems, with loose topsoil washing into the silt-laden Yellow River. South Korea, despite US trade sanctions, remains the last Asian country to use driftnets. Taiwan gave in to the sanctions and signed an agreement to end the practice by July 1992.

Trade in endangered species is fair game in East Asia, which leads the world in illegal wildlife trade. In response to apparent government indifference, Traffic, a watchdog organization affiliated with the World Wildlife Fund, has set up a branch in Taiwan.

fish, copper, iron ore, copper ore, tungsten, graphite, gold, silver

Green Statistics

Major Conservation Initiatives:
• ANTARCTIC TREATY AND CONVENTION: China, Japan, North Korea, South Korea
• CITES: China, Japan
• LAW OF THE SEA: China, Japan, Mongolia, North Korea, South Korea
• MARPOL: China, Japan, North Korea, South Korea
• MIGRATORY SPECIES: None
• OCEAN DUMPING: China, Japan, North Korea, South Korea
• RAMSAR: Japan
• WORLD HERITAGE: China, North Korea, South Korea, Mongolia

Protected areas:
China: 1%, Japan, South Korea: 6%, Mongolia, North Korea: less than 1%, Hong Kong, Taiwan: 0%

Green Rating

4 This area is beginning to pay more attention to environmental protection—time will tell if it will be lip service or genuine commitment.

: *12 hours*
Source: The Denver Post *1991 Colorado Recycling Guide*

"Some feel that the population explosion in some parts of the world may have pushed the search for answers beyond the framework of traditional liberal values. Several people I talked to mentioned China. Though I am sure these people view China's repressive society as intolerable, they still look at China's improving living standards and rapidly declining birth rate and find repression an acceptable alternative to the degradation and death of millions."
—*Wade Greene,* New York Times Magazine

"Japan's green movement is beginning to take a new direction. People now realize that opening one's own shell is opening the world, and that being able to open the world is dependent upon whether or not one can open oneself. Here, in particular, is where we find participation in the green movement by people involved in spiritualism and shamanism."
—*Oe Masanori,* Japan Environment Monitor

CITES also vowed at its Kyoto meeting to regulate trade in endangered species among nonmember nations, such as Taiwan and South Korea. This motion was apparently directed toward Taiwan, the leading trader in powdered rhinoceros horn, a traditional Chinese remedy for fever and other ills, including epilepsy. The medicine sells for $2,240 per gram, about eight times the price of gold, and Taiwan has stockpiled nearly ten tons of horn from 4,000 slaughtered rhinos. Because rhino horn is so expensive, rhinos continue to be poached, though only 11,000 are estimated to remain in the world.

A hot topic of discussion at the CITES conference was the East Asian demand for bear gall bladders, which are used in traditional medicines to treat liver, stomach and spleen ailments. With the population of Asiatic black and brown bears dwindling, the procurers have apparently shifted to North America, where bear carcasses are now found with only their gall bladders missing. Also, bears bought at auctions are being shipped to "bear farms" in East Asia. Penned in cages, the bears have their livers continually drained through catheter tubes. The bile sells for $70 per teaspoon.

South Korea is concerned over whether North Korea is stockpiling nuclear weapons. Under a peace plan between the two countries, North Korea allowed international inspectors to view nuclear facilities in September 1992. However, the North Korean government allowed inspection of only certain sites, which did not reassure South Korea and the US.

Even the remote country of Mongolia is not immune to the lure of economic growth. Mongolia plans to increase its industry and its now-small share of world markets. Yet it is also considering how to accomplish this without harming the environment. It's a tough balance, but the experience of its East Asian neighbors over the last 30 years may provide some important lessons.

Sources: *Boston Globe, Los Angeles Times, Washington Post,* Inter Press Service, *1992 Information Please Environmental Almanac* (Houghton Mifflin).

: *What percentage of American trash is incinerated?*

REGIONAL REPORTS

Southeast Asia

Over the past 20 years, Southeast Asia has become one of the fastest-growing economic regions in the world. To sustain this growth and to provide for their ever-increasing population, Southeast Asian countries have relied heavily on exports of timber, rice and petroleum. To pave the way, these countries have cut down a huge portion of their forests, as much as one-third in the Philippines. Reforestation has hardly kept pace, and Southeast Asian nations recently increased timber exports to counteract a drop in petroleum revenues caused by a drop in oil prices.

Each year, nearly 9,375 square miles of forest, an area the size of New Hampshire, are cut down. In total, 1.3 million square miles of this region's rainforests have been destroyed. If Malaysia and the Philippines continue to fell forests at their current pace, in ten years they will not have enough timber remaining to satisfy their own needs.

Western environmental groups have called for a boycott of Southeast Asian forest products, causing leaders there to scoff at what they consider first world self-righteousness. "The industrialized nations shouldn't pass their responsibility for the present environmental problems onto developing nations," said Indonesia's President Suharto in October 1991. Developing countries are as entitled as industrialized nations to improve their standard of living through the harvest of their natural resources, he added.

Southeast Asian leaders reaffirmed their favoring of GNP growth over old growth at a summit meeting in January 1992. Environmental issues were not on the agenda. Summit leaders dwelt primarily on the importance of competing economically with the rest of the world and agreed to work at forming their own economic trading bloc to compete against blocs like the European Community.

The small, elite industrial and landowning class of Southeast Asia benefits most from the exploitation of natural resources. So do foreign business interests, among them leading American lumber companies such as Georgia Pacific and International Paper. And foreign trade partners such as Japan and the United

Brunei, Cambodia, Indonesia, Laos, Malaysia, Myanmar (Burma), Philippines, Singapore, Thailand, Vietnam

Regional Statistics

Population:
458,300,000

Land Area: 1,730,106 square miles

Geography: Tropical rainforests (as much as 75 percent of the land in Brunei) cover much of this region. The several archipelagos of volcanic islands, including the Philippines, Indonesia and Singapore, consist of more than 20,000 separate islands. Dense mangrove swamps cover Malaysia, and fertile alluvial plains cover most of the Indochina Peninsula.

Climate: Tropical rainy and humid subtropical. There is no winter. Annual rainfall exceeds annual rates of evaporation.

Resources: Timber, precious stones, petroleum,

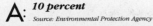

A: *10 percent*
Source: Environmental Protection Agency

rubber, coal, natural gas, nickel, cobalt, copper, gold, tin, bauxite, tungsten, fluorite

Green Statistics

Major Conservation Initiatives
• ANTARCTIC TREATY AND CONVENTION: None
• CITES: Cambodia, Indonesia, Malaysia, Singapore, Thailand, Philippines
• LAW OF THE SEA: Cambodia, Indonesia, Malaysia, Myanmar, Singapore, Thailand
• MARPOL: Indonesia, Myanmar, Singapore
• MIGRATORY SPECIES: None
• OCEAN DUMPING: Cambodia
• RAMSAR: None
• WORLD HERITAGE: Indonesia, Malaysia, Philippines

Protected Areas:
Indonesia: 8%, Laos, Myanmar: less than 1%, Malaysia, Vietnam: 3%, Philippines: 2%, Singapore: 4%, Thailand: 9%, Brunei, Cambodia: 0%

Green Rating

3 Southeast Asia has not yet put the environment on its political agenda. The population puts enormous pressure on forests and other fragile environments. Environmental groups there are just beginning to make an impact.

Kingdom benefit from low timber prices.

Foreign lumber companies do not share their riches with a large segment of the Southeast Asian population, the rural poor. Landowners who wish to cut down more forests for export or for farmland continually force out tenant farmers, who move elsewhere and cut more trees to make room for themselves. Even projects that are supposed to benefit all, such as hydroelectric dams, come at the expense of the poor.

To construct the Hoa Binh Dam in Vietnam, for example, the tenth-largest hydroelectric project in the world, builders flooded a fertile rice valley, displacing more than 50,000 inhabitants. The builders provided little compensation to these people for their land and did not prepare them for their new lives. Though the reservoir is supposedly rich in fish, they did not teach the farmers how to fish or provide any boats or nets. The farmers now cling to steep mountainside villages and struggle to produce small harvests. Many eat tree bark and roots to survive. None of the villagers has received any electricity.

Companies that want to do business in Southeast Asia are likely to find a friend in government. "We can survive here because the government isn't too strict on wastewater regulations," said the manager of an Indonesian Levi's acid-washed jeans factory. In Indonesia, bribery is part of politics, with the environmental ministry responding to hundreds of cases of logging violations by noting them, not prosecuting them. Philippine environmental enforcement agencies have also been notoriously corrupt.

Large tracts of cleared forest in tropical Thailand and Indonesia have already become desert-like because unprotected topsoil washes away in torrential rains. Cleared forests also probably provided the fuel for another ecological disaster: forest fires, such as those in Indonesia in 1991, when thick smoke billowed into Malaysia, Singapore and Thailand. The fires, which burned from August to October, were the worst in logged areas strewn with left-behind logs, branches and leaves.

The human-made environmental problems of Southeast Asia are being battled, however, by grassroots nongovernmental organizations (NGOs). One NGO attracted a strong enough public outcry to cause the planners of a $2.7 billion hydroelectric dam to scale down their plans. Originally, the Pa Mong

: How many people starve to death each year worldwide?

Dam along the Mekong River was supposed to displace 350,000 people; now it will force only 50,000 to resettle. Other NGOs question the state's right to parcel out forests that were once common to all. Others have called for the rural poor to be in charge of reforestation efforts, to ensure that they will always have enough trees to fulfill their basic needs. This plan has met with resistance from business interests.

HOT SPOT

ONE LARGE PROBLEM THAT REMAINS TO BE RESOLVED is the environmental damage inflicted on Vietnam, Laos and Cambodia during the Vietnam War. The Cambodian countryside is booby-trapped with hundreds of thousands of land mines that will require an estimated four to five years to remove. The worst damage, however, is in South Vietnam. To get rid of the dense jungle that gave North Vietnamese guerrillas cover against US troops, the US military used the herbicide Agent Orange to wipe out nearly 4.5 million acres of trees. This created vast spreads of wasteland dotted only by bare trees. It also left behind a highly toxic dioxin, TCDD.

Scientists have uncovered no definite connection, but in areas affected by Agent Orange large numbers of adults are now dying of cancer, particularly liver cancer. Birth defect rates are high, as are the rates for infant mortality and miscarriages. Some doctors have said that South Vietnam is a living laboratory in which to measure the effects of dioxins, but few Western doctors are aware of the situation there. The US has had little to do with Vietnam since the end of the Vietnam War.

Sources: *Los Angeles Times*, *The Gazette* (Montreal), Inter Press Service, Reuters Information Services, *Journal of Asian Studies*, *The Progressive*.

"Properly managed, forests can enrich human life in a variety of ways which are both material and psychological. Poorly managed, they can be a source for the disruption of the environment of the entire region. However, through the centuries we have seen a pattern repeated. The misuse of axe or saw, of fire or grazing, causes forest destruction. This leads to disruption of watersheds, to the erosion or loss of fertility of soils, to siltation and flooding in stream valleys, and to loss of the continued productivity of the land on which man must depend."
—Raymond F. Dasmann, Environmental Conservation

A: *11 million*
Source: The Colorado Daily, *Boulder, Colorado*

EARTH JOURNAL

South Pacific

The South Pacific is less-densely inhabited than most regions of the world, so superficially the environment is lush and appears untouched. Many of these countries have beautiful coral reefs and expansive rainforests. However, the region is plagued by the fear of islands being submerged under rising sea levels. Efforts are being made to simultaneously preserve wildlife and prevent overgrazing. And drought is a serious problem in some areas.

Experts claim that even a moderate change in sea level would threaten atoll islands. A comprehensive coastal management plan is necessary to help deal with both the rising sea level and other impacts of global climate change. Global warming could raise the sea level as much as eight inches by the end of this century. This seemingly small amount is enough to jeopardize life on the islands, possibly making them uninhabitable.

Australia plans to provide the region with $5 million for environmental protection and a biodiversity conservation management project. These programs are expected to enhance the region's capacity to monitor climate change by upgrading meteorological equipment and operational skills.

Drought has dried up lakes in New Zealand, which is particularly problematic since the country relies on hydroelectricity for 80 percent of its energy. New Zealand is suffering from its worst drought in 100 years. Drainage of water basins is monitored in New Zealand to protect animals and fish. Due to the drought, many advocate violating this regulation to increase available water. The power industry has urged people to reduce their power consumption by 10 percent to 20 percent, and energy experts criticize New Zealand for not doing enough to find other power sources, such as wind and geothermal energy. Scientists theorize that El Niño may be contributing to New Zealand's water crisis (see "El Niño," page 7).

Environmentalists in Australia blame land degradation on farmers and their millions of sheep, while farmers blame a population explosion of red and gray kangaroos. A three-year study is in the works to

Australia, Nauru, New Zealand, Micronesia, Papua New Guinea, Solomon Islands, Fiji, Kiribati, Tonga, Tuvalu, Vanuatu, Western Samoa

Regional Statistics

Population: 26,077,000

Land Area: 3,274,975 square miles

Geography: Consisting of over 585 islands with different geographic features, this region lies just south of the equator, in the South Pacific Ocean.

Climate: Ranges from rainy tropical to dry desert to humid subtropical areas, where seasons are more pronounced.

Resources: Iron ore, zinc, lead, tin, coal, oil, gas, copper, uranium, nickel, gold, timber, silver, fish, bauxite, copra

Green Statistics

Major Conservation Initiatives:
• ANTARCTIC TREATY AND CONVENTION:

Q: *According to independent scientific estimates, over the next 70 years, how many deaths could result from the 1986 Chernobyl nuclear disaster?*

measure animal impact. The goal is to develop strategies to sustain and restore the degraded land.

Only 400,000 koalas remain in Australia, contrasted with the millions that inhabited this country earlier in this century. Eucalyptus trees, on which the koala depends for food and habitat, have been cut to make way for development projects. Several city governments are working together to protect a 39-square-mile koala habitat.

In addition, the introduction of nonindigenous animals to Australia has had disastrous effects. Native species have been displaced by competition and predation. Six species of deer from Southeast Asia and Europe have caused severe erosion in New Zealand. The Indian water buffalo damages wetlands, and brumbies (wild horses) compete with livestock for food. The European rabbit is a pest to crops and trees.

In the past decade, the plight of the Aborigines has drawn to the forefront, revealing the conflict of culture since the Europeans arrived in the region. Land rights are of foremost concern, but spiritual and cultural survival are also concerns. Aborigines have lower life-expectancy rates than whites, less access to running water, plus higher rates of malnutrition and respiratory infections, and dramatically higher imprisonment rates. Most Australian Aborigines live in conditions equivalent to those in underdeveloped countries of the world.

HOT SPOT

CONTROVERSY IS ARISING AGAIN in Tasmania, Australia, an area renowned for ardent environmentalism. Environmentalists there acquired momentum a few years ago when 20,000 people flooded the town of Hobart to protest the damming of the Franklin River, which runs through a lush Tasmanian forest. At issue this year is a broader concern. Tree lovers are demanding a ban on all logging in Tasmanian forests. Presently, one-quarter of the forests are protected in reserves. Since the upheaval over the Franklin dam, legislation has passed that threatens $20,000 fines and year-long jail sentences to those who interfere with logging practices. Reports do not show the new law to be a deterrent.

Sources: Inter Press Service, *Los Angeles Times, Marketing News*, The Reuter Library Report, *Whole Earth Review*, Xinhau General News Service.

Australia, New Zealand, Papua New Guinea
- CITES: Australia, New Zealand, Papua New Guinea
- LAW OF THE SEA: Australia, Fiji, New Zealand, Papua New Guinea, Solomon Islands
- MARPOL: Australia, Tuvalu, Vanuatu
- MIGRATORY SPECIES: None
- OCEAN DUMPING: Australia, New Zealand, Papua New Guinea, Solomon Islands
- RAMSAR: Australia, New Zealand
- WORLD HERITAGE: Australia, Fiji, New Zealand

Protected Areas:
Australia: 5%, New Zealand: 11%, Fiji, Papua New Guinea, Solomon Islands: less then 1%, Kiribati, Tonga, Tuvalu, Vanuato, Western Somoa: 0%, Nauru, Micronesia: figures unavailable

Green Rating

1 Low populations and sustainable practices make this area a bright spot on the environmental map. Australia is one of the countries leading the way in environmental policy.

A: *As many as 500,000*
Source: Greenpeace Action

EARTH JOURNAL

South America

When José Lutzenberger was fired as Brazil's environmental minister in March 1992, a stormy tenure came to a sad end. In 1990 the appointment of Lutzenberger, a world-renowned crusader for the environment, was hailed as evidence of Brazil's clear intention to value a healthy environment as greatly as economic growth.

But Lutzenberger was powerless against economic interests. His ministry received little money from an administration shackled by debt and deficit.

The Brazilian government severely underfunded its environmental protection agency, IBAMA, in 1991. Often IBAMA could inspect a logger's operations only when the logger picked up the expenses of the visit, inviting corrupt officials to issue logging permits in ecologically sensitive areas and causing a frustrated Lutzenberger to declare before a World Bank gathering in New York: "IBAMA is a branch office of the timber industry."

It was not the first time Lutzenberger had accused the Brazilian government of corruption before an international audience. But with the administration feeling the heat of such charges, then-Brazilian President Fernando Collor de Mello fired Lutzenberger and IBAMA's president, Eduardo Martins. Collor de Mello was impeached by congress in late September 1992 on grounds of corruption.

Lutzenberger fought within an administration that did not wish to restrain economic development. His struggle typifies the struggle of environmental concerns against the economic goals of South American nations. The continent's deforestation problems, for example, began with the region's dependence on agriculture, as farmers and ranchers continually cut and burned trees to clear the way for short-lived farmland and pasture. Paraguay will have few forests left by 2010. Ecuador has lost 95 percent of its coastal forests, threatening 2,500 species with extinction. South America loses nearly 1 percent of its forests each year, the second-highest rate in the world.

In Brazil, home to one-quarter of the world's tropical rainforests, the destruction has slowed, thanks partly to the government's elimination of subsidies

Argentina, Bolivia, Brazil, Chile, Colombia, Ecuador, French Guiana, Grenada, Guyana, Paraguay, Peru, Suriname, Trinidad and Tobago, Uruguay, Venezuela

Regional Statistics

Population: 312,330,000

Land Area: 6,882,087 square miles

Geography:
A continent connected to Central America by a tiny piece of Panamanian land, it touches the Atlantic Ocean in the east and the Pacific Ocean in the west. Dense Amazonian rainforests, snow-covered peaks of the Andes, Paraguayan lagoons, and sandy, arid coastal strips make up this atypical region. A great diversity of extremes rule—from the fertile plains of the pampas in Argentina to the volcanic mountains in Ecuador.

Climate:
The Atacama Desert of

 What percentage of US energy consumption is from non-renewable sources?

for forest clearing. Some 3,600 square miles of Brazilian forest were cut in 1991, compared to 5,600 in 1990, and 36,000 in 1987. But deforestation is expected to increase in the next decade as Southeast Asia, the world's leading timber exporter, is rapidly being logged out. The potential for the Brazilian timber industry has been estimated at $600 billion. Although Brazil presently exports only 5 percent of the world's timber, it is expected to lead the world in exports by the end of this decade.

To fulfill industrial demands for electricity, Chile has proposed the construction of six hydroelectric dams along the Bío-Bío River. The Pehuenche natives, who would be displaced, plus those who wish to preserve the river's natural beauty, are battling the industrial development proponents, including Chile's national environment secretary, Rafael Asenjo. Chile, like Argentina, Colombia and Brazil, has again become a hot spot for foreign investment, so the pressure to develop its natural resources is sure to increase.

The new investment situation is a change from the 1980s, when South America's heavy foreign debt paralyzed its economies. Banks trying to recoup their losses on their failing loans began to pass off South American IOUs at 30 cents for every dollar owed. Several environmental agencies invested in this paper, thinking that they were buying out South American debt at bargain prices, freeing their economies, and obligating a country to repay them by investing in protected national parks and other environmental projects. Yet these debt-for-nature swaps have not worked as advertised.

First, the country remains just as strapped as before the debt purchase. Second, since the country couldn't pay back the bank, there's no guarantee it will be able to repay the environmental organization. Finally, simply buying land and calling it a national park does not prevent loggers and miners from exploiting it. The park service must be adequately budgeted to enforce the borders, which is not often the case. The real winners in debt-for-nature swaps have been the banks that unloaded the IOUs.

Companies in the US, where waste disposal regulations are strict, have found Argentina to be a friendly place in which to dump waste of questionable content. The US Environmental Protection Agency insists that the sewage being shipped to Argentina's Patagonia region is nontoxic, "Grade A" sludge, but Argentine

Chile is one of the driest places on Earth, with no measurable rainfall. Most of South America is tropical and humid, with cooler temperatures along the coast.

Resources:
Timber, minerals, oil, copper, iron ore, tin, zinc, gold, silver, bauxite, fish, uranium, antimony, bismuth, sulfur, tungsten

Green Statistics

Major Conservation Initiatives:
• ANTARCTIC TREATY AND CONVENTION: Argentina, Brazil, Chile, Colombia, Ecuador, Peru, Trinidad and Tobago, Uruguay
• CITES: Argentina, Bolivia, Chile, Colombia, Ecuador, Guyana, Paraguay, Peru, Suriname, Trinidad and Tobago, Uruguay, Venezuela
• LAW OF THE SEA: Argentina, Bolivia, Brazil, Chile, Colombia, Paraguay, Suriname, Trinidad and Tobago, Uruguay
• MARPOL: Brazil, Colombia, Ecuador, Guyana, Peru, Suriname, Uruguay
• MIGRATORY SPECIES: Chile, Suriname, Uruguay
• OCEAN DUMPING: Argentina, Bolivia, Brazil, Chile, Colombia, Suriname, Uruguay, Venezuela
• RAMSAR: Bolivia, Brazil, Chile, Suriname, Venezuela, Ecuador
• WORLD HERITAGE: Argentina, Bolivia, Brazil, Colombia, Ecuador, Guyana, Paraguay, Peru, Uruguay, Venezuela

A. *92.5 percent*
Source: Union of Concerned Scientists

Protected Areas:
Brazil: 2%, Chile: 16%, Ecuador: 38%, French Guiana: 0%, Paraguay: 3%, Trinidad and Tobago: 3%, Venezuela: 10%, Colombia, Suriname: 5%, Argentina, Bolivia, Peru: 4%, Guyana, Uruguay: less than 1%, Grenada: figures unavailable

Green Rating

3.4 This region has vast resources, yet they are in danger. It has some good initiatives, but it remains unclear if these will be carried out, or if they will be just for show. This area has perhaps the greatest opportunity for real improvement.

"People complained and pricked the consciences of others, the government acted, and, finally, companies themselves have learned it is better not to pollute.... Today no one is willing to propose a major industrial construction project in Brazil without including the costs of an environmental-impact study. Anyone who ignored it would be shunned by others."

—*Jesus Marden Dos Santos,* World Press Review

Greenpeace scientists are doubtful. "If these wastes are such good stuff, why don't they keep them in New Jersey?" asked Mario Epelman, a Greenpeace physician.

The lower regions of South America may also be affected by the ozone hole over Antarctica. In Chile's Ultima Esperanza ("Last Hope") province, rabbits are blinded by cataracts and sheep have inflamed corneas. Nature's sense of timing also seems to be off: Geese breed in autumn instead of spring, and flamingos arrive a season early. However, no hard scientific connection has been found between these recent events and the increased levels of ultraviolet light in areas close to Antarctica.

A major problem in South America's development is maintaining the health of its burgeoning population, particularly in its cities. As South America becomes more urbanized, the need for safe water grows. Only 10 percent of wastewater in South America is treated. A large amount of sewage and industrial waste is dumped into rivers and streams directly, or contaminates groundwater from landfill seepage. One result of this water pollution was a continent-wide cholera epidemic in September 1991 that killed 2,909 people in Peru alone. Many fear that such an epidemic could strike again.

PROMISING INITIATIVES

BRAZIL AND ECUADOR HAVE demarcated certain territories solely for indigenous people. In Brazil in 1992, after considerable pressure from international human rights and environmental organizations, President Collor set aside 36,000 square miles for the Yanomamis and ordered miners to vacate the mineral-rich region. Shortly thereafter, Collor designated 19,000 square miles for the Kaiapós. Ecuador's 20,000 Amazon Basin natives received full title to 3 million acres of homeland in May 1992.

Ecuador's agreement stipulates that non-Indians cannot be forced out of the area and that the state will continue oil exploration in the areas. In the Yanomami region, miners have been spotted flying food and fuel into the area, indicating they intend to return.

Sources: *Washington Post, New York Times, Los Angeles Times, Chicago Tribune, The Gazette* (Montreal), Greenwire, *Financial Times, Latin America Weekly Report, The Independent, The Ecologist,* Commission for the Creation of a Yanomami Park, *1992 Information Please Environmental Almanac* (Houghton Mifflin).

: *What percentage of original vegetation has Madagascar already lost?*

REGIONAL REPORTS

Central America

Central America's equatorial rainforests were ravaged and ignored during the violent civil wars that raged there during the 1980s. But with Nicaragua and El Salvador proclaiming peace in the 1990s, and Guatemalan armies conducting peace talks in 1991, governments can start to wage war on environmental problems instead of opposition forces.

Of the region's original rainforest, only 18 percent remained by the late 1980s. To stem this widespread deforestation, traditional livelihoods will need to change. Agriculture is the dominant source of income in this impoverished region. Subsistence farmers grow staple crops such as rice and beans to sustain families and villages. However, large landowners have forced these farmers to move to create more livestock pasture and more cropland for exports such as coffee, sugar cane, bananas and cotton. The subsistence farmers, whose numbers are growing, have moved to marginal areas where they fell wood for fuel and crop space.

Without trees to hold it down, topsoil has been swept away by wind and rain. The earth left exposed is nonarable, forcing farmers to quickly find new, unspoiled areas to farm for another short period of time. Loss of forest also means loss of habitat, threatening some species with extinction.

Environmental organizations find it hard to fight economics. Still, there have been victories, albeit partial ones. Some 12 percent of Costa Rica is set aside as national parks and reserves, in a policy of nature preservation that began in the 1980s. Another 15 percent is listed as protected national park buffer zones, but is being badly deforested because the financially strapped central government cannot adequately protect the zone borders. In Guatemala, a group of unarmed university students, farmers and chicle gatherers have defended the borders of the Mayan Biosphere Reserve, a 4-million-acre national park in the El Petén region. Loggers have tried to scare off the group by inciting colonists to set guard posts aflame.

In Honduras in 1992, after a public outcry, the government refused to grant the US-based Stone Container

Antigua and Barbuda, Bahamas, Barbados, Belize, Commonwealth of Dominica, Costa Rica, Cuba, Dominican Republic, El Salvador, Federation of St. Kitts and Nevis, Guatemala, Haiti, Honduras, Jamaica, Nicaragua, Panama, St. Lucia, St. Vincent and the Grenadines

Regional Statistics

Population: 56,944,000

Land Area: 279,503 square miles

Geography: This region has many island countries. Tropical forests average 30 percent of the total land coverage. Mountain chains tend to be volcanic creating fertile volcanic soil.

Climate: This region lies just north of the equator and is marked by heavy rains and tropical temperatures.

Resources: Salt, timber, aragonite, crude oil, nickel, gold, silver, lead, zinc, bauxite, gypsum, copper

A: *90 percent*
Source: Protect Our Planet Calendar, Running Press Book Publishers

Green Statistics

Major Conservation Initiatives:
- ANTARCTIC TREATY AND CONVENTION: Cuba
- CITES: Belize, Costa Rica, Cuba, Dominican Republic, El Salvador, Guatemala, Honduras, Nicaragua, Panama
- LAW OF THE SEA: Barbados, Belize, Costa Rica, Cuba, Dominican Republic, El Salvador, Guatemala, Haiti, Honduras, Nicaragua, Panama
- MARPOL: Panama
- MIGRATORY SPECIES: Jamaica, Panama
- OCEAN DUMPING: Costa Rica, Cuba, Dominican Republic, Guatemala, Haiti, Honduras, Panama
  RAMSAR: Guatemala, Panama
- WORLD HERITAGE: Belize, Costa Rica, Cuba, Dominican Republic, Guatemala, Haiti, Honduras, Jamaica, Nicaragua, Panama

Protected Areas: Costa Rica: 12%, Cuba: 8%, Dominican Republic: 11%, El Salvador: 1%, Guatemala: 1%, Honduras: 5%, Panama: 17%, Commonwealth of Dominica, Haiti, Nicaragua: less than 1%, Bahamas, Belize, Jamaica: 0%, Antigua and Barbuda, Federation of St. Kitts and Nevis, St. Lucia, St. Vincent and the Grenadines: figures unavailable

Green Rating

1•2 Good policies here, but the costs of war are high.

Corporation a 40-year concession to log more than a million acres of forest. A major factor in the decision may have been a failure to agree on a fair price. In Panama in 1992, Texaco dropped a proposal to drill for oil in a marine park along the Caribbean coast. Panama is now offering the region to smaller oil companies.

With peace, the region is awakening to the profitable possibilities of ecotravel. In Costa Rica, which boasts both a Caribbean and a Pacific coast, plus lush rainforests and cascading whitewater rivers, tourism is a $300-million-a-year industry, on the same level as banana and coffee exportation. Belize, with the world's only jaguar preserve and the world's second-longest barrier reef, has appointed a minister of tourism and the environment. Guatemala houses the archaeological ruins of Olmec and Maya cultures, while Honduras has ruins, cloud forests and Caribbean reefs.

Yet tourism is not always the best solution. In Belize, for example, living coral along the barrier reefs are turning gray or black from a disease likely caused by the silt churned up by divers. Officials have had to shoo off tourists who were trampling across a reef in tennis shoes.

PROMISING INITIATIVES

SOME COMPANIES ARE EXPLORING the profit possibilities in preserving Central American forests. Merck, one of the world's largest pharmaceutical firms, is investing $1 million in Costa Rica's National Institute for Biodiversity, to train and pay *campesinos* to collect plant species of interest. Merck hopes to find plants that can spawn profitable new drugs; the institute hopes to classify more of the estimated half million plant and animal species (5 percent of the world's total), that live in Costa Rica. Another aim is to encourage conservation of local forests. Central American environmentalists hope that Merck will properly compensate the Costa Ricans who share their knowledge of rainforest plants.

Sources: *New York Times, Los Angeles Times,* Newhouse News Service, Tropical Conservation Newsbureau (Rainforest Alliance), *The Nation, New Statesman and Society, 1992 Information Please Environmental Almanac* (Houghton Mifflin).

: *What is the total number of people living on Earth?*

REGIONAL REPORTS

North America

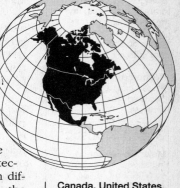

In this 500th anniversary of the "discovery" of America, some believe there are more reasons to lament than to rejoice. Plants and trees are disappearing, some species endemic to this region are extinct, and industrial and agricultural success has led to severe contamination of air and water. Yet the US and Canada maintain the highest environmental standards of any countries in the world, and are looked to as leaders in the protection of the environment. To unify a region with different needs and economic conditions, Canada, the US and Mexico are hammering out the terms of the North American Free Trade Agreement (NAFTA). These three countries have different regulations regarding the environment, and many fear NAFTA will weaken the environmental standards of the US and Canada.

Mexico complains that stringent environmental laws discriminate against trade. For instance, a California state initiative demanding the labeling of carcinogenic and toxic chemicals on packaging may be abolished because it restricts free trade—Mexico does not have such regulations. The US also bans tuna imports from any country whose fishing fleet kills dolphins at a significantly higher rate than the US. An international trade panel found this ban discriminates against Mexican trade. Proposed revision of US law leaves environmentalists increasingly skeptical about free trade.

Trade disputes are settled by NAFTA or the General Agreement on Tariffs and Trade (GATT), which supervises the world trading system. GATT is perceived as less sympathetic to issues that obstruct trade. Under NAFTA, disputes would be decided by trade officials, advised by environmental and technical experts. Environmentalists want disputes settled under GATT moved, but they are not convinced that NAFTA will listen to their concerns.

Many environmental groups believe that promoting industrial growth in Mexico will exacerbate an already severe pollution problem. Only a fraction of Mexico's cars burn unleaded gasoline, and the dangerous air quality led to a state of emergency in

Canada, United States, Mexico

Regional Statistics

Population:
364,343,000

Land Area: 8,154,348 square miles

Geography: There are major mountain chains in this region: the Appalachians in the east; the Sierra Nevada and the Cascades in the west; and the Rocky Mountains in the western part of the US, reaching up into Canada and including the Continental Divide. Much of the southwestern and western US is arid or semi-arid, rocky terrain. Vast fertile farmlands and prairies extend to the north and south in central North America. Mexico is predominantly arid, with volcanic mountains in the southern exterior. Canada consists of vast central plains, and islands to the north in the Arctic Ocean.

A: *More than 5 billion*
Source: *Children's Television Workshop*

Climate: Ranges are dramatic: from hot, dry desert, to humid continental on the coasts, to subarctic and tundra in the north.

Resources: Petroleum, timber, ores, silver, copper, gold, lead, zinc, minerals

Green Statistics

Major Conservation Initiatives:
- ANTARCTIC TREATY AND CONVENTION: Canada, US
- CITES: Canada, US, Mexico
- LAW OF THE SEA: Canada, Mexico
- MARPOL: US, Mexico
- MIGRATORY SPECIES: None
- OCEAN DUMPING: Canada, US, Mexico
- RAMSAR: Mexico, US
- WORLD HERITAGE: Canada, Mexico, US

Protected Areas: Canada: 4%, Mexico: 3%, US: 9%

Green Rating

1 This area has some of the strongest environmental regulations in the world. North Americans may be on their way to more environmental awareness.

"I have an 'Ecology Now' sticker on a car that drips oil everywhere it's parked."
—Mark Sagoff, Earth Ethics

March 1992 in Mexico City, when levels of pollution far exceeded acceptable standards (the pollution measured 398 points on a scale where 100 points verges on dangerous).

There is concern that under NAFTA, more Canadian and US companies will move to Mexico to take advantage of less-stringent environmental laws. US exporting companies in Mexico, called *maquiladoras,* are the major contributors to border pollution. Critics claim companies take advantage of the cheap labor provided by some 500,000 Mexican maquiladora workers, and exploit the country's lax environmental laws. The average wage of a maquiladora worker is $1.73 per hour, while most Mexican companies pay $2.17 per hour. Proponents of free trade say the maquiladoras provide jobs that would otherwise not exist.

The US and Mexico have agreed on an $840 million cleanup plan for the border area. SEDESOL, Mexico's version of the EPA, has fined 14 companies more than $10 million in the past year—money that will be returned if the companies improve their equipment. However, some environmentalists claim that most of SEDESOL's detection methods are faulty, and that their information is erroneous.

On January 1, 1989, Canada and the US instituted a free trade agreement. Although the Canadian minister for international trade claimed this was not an environmental agreement, it does involve energy and agricultural policies, forest management practices and pesticide regulation. Under the agreement, both countries forgo the right to regulate the development of energy resources for export markets, even for conservation purposes.

Since the trade deal between Canada and the US passed, licenses were granted for two of the largest energy projects in Canada's history—natural gas development in the Mackenzie Delta and James Bay II, a hydroelectric project. Although James Bay II will not provide power to the state of New York as previously planned, the province of Quebec remains its biggest proponent. Quebec has passed legislation to weaken environmental assessment and reduce public hearing requirements. However, the project is on hold because of New York's action.

Critics contend the US-Canada free trade agreement obscures environmental problems in the US. Although

---

Q: *Which US national park is home to the largest glacier in the continental US?*

access to Canada's natural resources seems beneficial initially, conservation of nonrenewable resources within the US is imperative. For instance, 48 states have less than 5 percent of their original forests remaining. At current logging rates, all unprotected forests in Washington and Oregon will be gone by 2023. Use of Canada's resources merely forestalls the inevitable depletion of reserves. The development of enormous energy projects in Canada also deters the advancement of energy conservation policies.

Potential conflict is evident when one compares pollution regulations in the US and Canada. Canada is working toward increasing energy efficiency and conservation, to save money and lower greenhouse gases, acid rain and toxic emissions. By comparison, on June 25, 1992, the Bush Administration issued a rule that will give manufacturers the right to increase the amount of hazardous pollutants they pour into the atmosphere by 245 tons a year, beginning in the mid-1990s. Under this rule, a company can raise the amount of toxins it emits by applying for an increase. It may legally go ahead with the emissions while the application is pending.

The US-Canada Free Trade Agreement was meant to harmonize environmental standards. NAFTA maintains the same objectives. Questions and fear remain.

HOT SPOT

THE AIR IS THICK in places across this region, as pollution continues to be a problem. Prevailing winds carry pollution from Canada and the US to the Arctic, and an air mass traps these contaminants in the lowest part of the atmosphere, creating the Arctic haze. This haze constricts visibility and increases solar radiation enclosed in the troposphere. The change in radiation could ultimately alter the Northern Hemisphere's climate, making efforts to reduce pollution the foremost concern for all countries in this region. In the past year, pollution has increased: A state of emergency concerning air quality was declared in Mexico City on March 17, 1992; and the US has gutted clean air regulations.

Sources: *Daily Report for Executives, The Financial Times, Investor, Inc., 1992 Information Please Environmental Almanac* (Houghton Mifflin), Levitt Communications, *New York Times, Wall Street Journal, The State of Canada's Environment* (Government of Canada), State News Service.

"The air is the ultimate global commons—mixed and moved around the globe by the winds; shared by all living things; used and reused for many different purposes. It not only sustains life but, in the ozone layer, shelters it from harsh ultraviolet rays of the sun and buffers the Earth from extremes of hot and cold. The air must increasingly be seen as a common resource, not a common sewer."
—Business Week

"Does the wanton subjugation of nature by our species have a causal connection with the wanton subjugation of women by men?"
—*David Quammen, Outside*

"One can exist for days without food or water or companionship or sex or mental stimulation, but life without air is measured in seconds. In seconds."
—*Caskie Stinnet, Countryside*

 Mt. Rainier National Park in Washington, home to Emmans Glacier
Source: The W.A.T.E.R. Foundation

EARTH JOURNAL

The Arctic

Although this region is remote, its climate cold, and its population sparse, there are many areas of environmental concern relating to community waste, oil and gas, global warming and Arctic haze. The native Inuit, who contribute to and suffer from these problems, are the major inhabitants of the Arctic.

Before 1972, no laws governed land in the Arctic. Industries started and finished projects, accumulating and leaving behind waste. Some of these wastes (chemicals, fuels, fuel drums) are hazardous and still remain.

Today at least nine federal agencies, two territorial governments, four provinces, native agencies and various special commissions prioritize conservation and regulation of this region. Some of the groups include the Arctic Marine Conservation Strategy, the Inuit Regional Conservation and Indigenous Survival International strategies, and the Task Force on Northern Conservation Strategy. Between 1991 and 1997, the Arctic intends to monitor water pollution, clean up waste disposal sites, and help communities achieve their environmental goals.

Waste disposal is a problem for all Arctic communities. Certain areas, like Resolute in the Northwest Territories, Canada, discharge untreated sewage directly into the surrounding waters. There are some lagoons (human-made pools for sanitary treatment), but they tend to overflow into surface drainage systems. These wastes take an inordinate amount of time to decompose, 10 to 25 times more slowly than in warmer areas. Although there is concern about waste disposal, the people lack funds to implement change.

The Arctic National Wildlife Refuge, sometimes referred to as the American Serengeti, is at the center of a territorial tug of war. An Anchorage-based oil company, Aleyska Pipeline Service Co., wants to dredge and transport surging hot oil from the refuge to Prince William Sound. This remains a controversial issue in the US, between those who wish to drill the oil, and conservation advocates who fear destruction of wildlife habitat—of animals hunted by native people for food.

Regional Statistics

Geography: The Arctic is really an ocean, bordered by the northern tips of three continents: North America, Europe and Asia. The borders of the Arctic Circle are difficult to draw, and some scientists set the boundary at the point where the taiga (subarctic evergreen forests in Siberia, Eurasia and North America—one of the Earth's largest forests) gives way to sparse tundra.

Land temperature can reach -40°F, yet Arctic Ocean waters never dip below 32°F.

The one-third of the Arctic Ocean that is underlaid by continental shelves is one of the world's most abundant fishing grounds, yielding one-tenth of the annual global catch. During the spring, three-quarters of a million marine mammals inhabit the rich waters of the area.

The land, rich in resources, ecologically sensitive and sparsely populated, is in a location that makes the

 What percentage of landfill garbage is plastic?

After the March 1989 oil spill in Prince William Sound, a settlement of $1.1 billion was forged between Alaska, federal prosecutors and Exxon. Overall, Exxon has spent $2.5 billion in its cleanup efforts. Yet, many environmentalists and residents remain critical of the company, since the damage to wildlife and the marine environment is worse than Exxon predicted. Joseph Hazelwood, captain of the *ExxonValdez*, was exonerated of all charges because of a law stating that those who report accidents within 20 minutes of a spill are absolved of personal responsibility. Prosecutors say this law was created for small spills from small tankers.

Some 25 villages and communities in Alaska still suffer from social and economic problems which the Exxon spill left in its wake. For example, on March 24, 1989, Valdez was inundated by divers, biologists, oil specialists, reporters and bureaucrats. In one week, the population expanded from 2,300 to over 5,000. Exxon hired cleaning crews for $17 an hour, leaving local businesses without employees. Crime and rents increased. By the summer of 1989, the population of Valdez had reached 12,000—five times its original size. Residents blame higher divorce and alchoholism rates on the change in their community brought on by the spill.

Global warming threatens to alter life in the Arctic. Scientists predict the biggest changes in temperature will occur in the high-latitude zones. Although a minor temperature increase is anticipated in summer months, a large increase is expected in winter. In the Arctic, the impact of even a one-degree temperature variance can be enormous. The harsh climate protects the wildlife and the treeline, as indigenous flora and fauna are products of these low temperatures. Any melting could also affect the foundations of roads and buildings that are constructed on permanently frozen subsoil (permafrost).

Sources: *The State of Canada's Environment* (Government of Canada), Organization for Economic Co-operation and Development (OECD), *World Resources: A Guide to the Global Environment, 1992-93* (Oxford University Press).

Arctic of immense geopolitical importance. Unlike Antarctica, the sovereignty of Arctic-rim states over the land is never in question.

The US obtains over 20 percent of its oil from its land in the Arctic.

Although home to many indigenous people, the Arctic is regarded as a comparatively safe and convenient environment for the deployment and operation of strategic weapons systems.

Climate: The mean temperatures are less than 50°F in the warmest month, less than -22°F in the coldest month, and below 14°F average for six or more months of the year.

Resources: Oil, natural gas

Green Rating

1•2 Genuine efforts are being made to improve the Arctic environment. This area will be a test case of modern indigenous people's abilities to protect and preserve their lands.

A: *13 percent*
Source: "The Pulse of the Planet" radio program

PART 3
ECOCULTURE

IV

ARTS & ENTERTAINMENT

"The real character of a man is found by his amusements," said Sir Joshua Reynolds, the 18th-century painter. Whether your favorite amusements are books, magazines, TV, movies, computer games, or music, this chapter will enlighten and amuse you.

EARTH JOURNAL

Green Stars

Celebrities are news makers and, lately, much of the news they are making is green news. The number of celebrities who give their name, time, effort and money to environmental causes is astounding. This past year, they were everywhere, all doing their part by participating in various organizations, conferences and star-studded fundraising galas. Earth Journal takes a look at some of this past year's eco-related events and the big names that gave their support.

The First Annual Environmental Media Association Awards program: This ceremony, honoring people of the creative community who have used their special talents to advance environmental concerns, attracted, among others, **Ted Turner, Jane Fonda, Sting, Steven Seagal, Sally Field, Barbra Streisand, Chevy Chase** and **Ed Begley, Jr.**, who was the highlight of the event when he rolled up to the festivities in his electric car.

Don Henley

Walden Woods Project fundraiser: **Don Henley's** pet project to

People

People are the news makers of the environment. And increasing numbers of people are making news—from celebrities to teenagers, people are changing their world. Earth Journal brings you the names and faces of the new eco-movers and shakers.

THE 10 MOST INTERESTING ENVIRONMENTALISTS OF THE YEAR

1. **Peter Matthiessen** of the US. A renowned naturalist and writer, Matthiessen has published three environmental books in the past 24 months: *African Silences* (Random House, New York, 1991), a compelling description of his journeys through the heart of Africa; *Lake Baikal* (Sierra Club Books, San Fransisco 1992), a journal of his trip to the largest freshwater lake in the world, now threatened by the former USSR's environmental devastation; and *Shadows of Africa* (Abrams, New York, 1992), an appreciation of African wildlife.

2. **Dr. Helen Caldicott** of Australia. A world-renowned environmental activist and physician, Dr. Caldicott toured the US this year to publicize her new book, *If You Love This Planet: A Plan to Heal the Earth* (W.W. Norton, New York, 1992). Audiences across the country were moved by her stirring lectures: "Over my dead body!" is how Dr. Caldicott describes her approach to the nuclear power industry.

3. **Richie Havens** of the US. Composer and singer Havens, who spoke for a generation with his classic performance of the song "Freedom" at Woodstock more than 20 years ago, has founded "The Natural Guard"—inner-city clubs that help children learn more about the environment. Havens believes children will transform the future: "These are kids that know they live on a *planet*, so they are really our first planetary generation," Havens says. "They want to *be* the change."

4. **Vice President Al Gore** of the US. Throughout the 1992 presidential elections, Gore brought the environmental debate to the forefront of American politics. Author of the best-selling *Earth in the Balance*

Q: *According to the US government, what percentage of the nation's 575,000 bridges are in serious need of repair?*

ARTS & ENTERTAINMENT

(Houghton Mifflin, Boston, 1992), Gore says "the new organizing principle" of all people on Earth "must be the effort to save the global environment."

5. **Helena Norberg-Hodge** of England. Founder of the International Society for Ecology and Culture, Norberg-Hodge is the author of *Ancient Futures: Learning from Ladakh* (Sierra Club Books, San Francisco, 1991), a riveting account of her experiences in that Himalayan countryside. A trained linguist and the first outsider to master the Ladakhi language, Norberg-Hodge describes the Ladakhi plunge from contentment to anxiety as they moved from a sustainable, cooperative, agricultural system to an increasingly Westernized, money-based culture.

English environmentalist Helena Norberg-Hodge.

6. **Maurice Strong** of Canada. As secretary-general of the 1992 United Nations Conference on Environment and Development in Rio de Janeiro, Strong led heads of state and environmentalists in heated debates on the future of our planet. Rich versus poor, North versus South, powerful versus weak—Strong called his job "a Herculean task," but his commitment never wavered. "Rio must be seen as the start of a process, not the end," he said.

7. **Wangari Maathai** of Kenya. The first black African woman to receive a doctoral degree in eastern Africa, Professor Maathai founded the Greenbelt Movement to plant millions of trees in her native Kenya. She was arrested, jailed and reportedly beaten in 1992, in the

save Walden Woods and its nearby pond from condominium development attracted **Jack Nicholson, Cher, Harry Hamlin, Jessica Lange** and **Ric Ocasek**, as well as many other guests who all shelled out $250 to get into the chic Tatou nightclub in Aspen, Colorado.

The Dalai Lama

The **Earth Summit** in Rio: The largest international conference on the environment attracted the usual amount of hangers-on as well as foreign dignitaries and concerned stars. Among the conference-goers were spiritual leader **the Dalai Lama, Shirley MacLaine, Roger Moore, Olivia Newton-John, Bianca Jagger, River Phoenix, Jeremy Irons, Edward James Olmos**, then-US senators **Al Gore, Tim Wirth** and **Larry Pressler**, as well as musical performances by **Placido Domingo, the Beach Boys** and **John Denver**.

Numerous other independent conferences and awards in 1992 attracted big names. Some stars took matters into their own hands. **Jimmy Buffett** staged a fundraising concert in Fort Lauderdale, Florida, to aid his effort to gain independent control of

A: *23 percent*
Source: *Campaign for New Transportation Priorities*

the Save the Manatee Club from its parent Florida Audubon Society. U2 lead singer **Bono** donned a radiation suit and joined demonstrators in Cumbria, England, to protest the proposed building of a second nuclear plant on the site. Proceeds from **Michael Jackson's** tour of Africa were to be used for children's groups and environmental matters.

midst of a hunger strike protesting government policies that threaten public lands. "I was able to see that if I had a contribution I wanted to make, I must do it, despite what others said. That it was all right to be strong," Maathai said in a 1991 interview.

8. **Fritjof Capra** of the US. "Mindwalk," a feature film based on Capra's book *The Turning Point* (Bantam, New York, 1987), introduced thousands of moviegoers to a holistic view of the Earth as a living, breathing organism. Dr. Capra, a theoretical physicist, is founder of the Elmwood Institute, a think tank in Berkeley, California. "Every single entity, every organism, every part—everything in nature!—owes its characteristics and identity to its relationship with other things," Capra says. "It's the very essence of the new ecological world view—to shift from thinking in terms of objects to thinking in terms of relationships."

9. **Stephan Schmidheiny** of Switzerland. One of the richest men in the world, Schmidheiny (who holds a major interest in Leica, the optical and scientific corporation), has spent two years organizing his fellow business leaders to increase their environmental responsibility. Schmidheiny has a 4,942-acre farm in Costa Rica, where he tests environmentally friendly agricultural techniques. He spoke at the Earth Summit in Rio de Janeiro and organized his fellow industrialists for a mini-summit in Geneva, Switzerland.

10. **Jessica Tuchman Matthews** of the US. Her syndicated column on environmental topics lights up the op-ed page of newspapers like *The Washington Post*. As vice president of the World Resources Institute, she edited *Preserving the Global Environment: The*

What's Green and What's Not

GREEN	BROWN
wild dolphins	Flipper
tagua nut buttons	plastic buttons
cloth diapers	disposable diapers
Native Americans	Columbus Day
vegetable gardens	lawns
organic wine	Chateau Rothschild
Grand Canyon	Euro Disneyland
electric cars	gas guzzlers
thinking ahead 7 generations	thinking ahead 7 seconds
bicycling	motoring
solar power	nuclear power
Greens	Reds
recycled paper	dioxin-bleached paper
fountain pens	disposable pens
trees	Bushes
Mindwalks	car chases
Earth First!	Earth last
herbs	'burbs
ecofeminism	animal-tested cosmetics
Nuclear Valdez	*Exxon Valdez*
quantum shamans	juice boxes
ivory on elephants	ivory on people
tribes	bribes
quail	Quayle
peace in action	piece of the action

Q: *How many hours per day on average do women in India spend looking for firewood?*

Challenge of Shared Leadership (W.W. Norton, New York, 1990), a collection of experts' essays on topics ranging from climate change to overpopulation to deforestation and species loss.

THE ENVIRONMENTAL YOUTH MOVEMENT

THE ENVIRONMENT concerns everyone, especially those who will inherit the Earth as we move toward the 21st century.

Increased environmental awareness in the past few years has had a definite impact on today's youth, as environmental issues rank among their top concerns. In an annual survey of college freshmen conducted by Alexander S. Austin of UCLA, Austin found "a resurgence of student interest in influencing social values and changing the political structure." In the same survey, the percentage of students who said it was "essential" or "very important" that they do their part in cleaning up the air and saving the rainforests in Brazil has doubled over the past four years. This attitude has spawned numerous examples of grassroots activism and initiatives among today's youth.

From coast to coast, and around the world, kids are pulling together to do their part. In Brandywine, Maryland, a group of 12-year-olds thwarted corporate dumping in nearby streams by circulating petitions and working with county officials.

In Massachusetts, an Ipswich middle-school group, Students Against Vandalizing the Earth, has forced vending machine operators to adopt more stringent recycling programs. The group's efforts led the city to place a referendum on the November ballot.

In Salt Lake City, a group of Jackson elementary-school students were disturbed upon finding over 100 barrels of abandoned toxic waste in close proximity to their school. After being rebuffed by local health officials, they took their case to the Denver branch of the Environmental Protection Agency, where their concerns no longer fell upon deaf ears. Barrel removal commenced within three months.

Other concerned students put out their own newsletters and information designed to educate and inform people about salient issues that face us today. A vegetarian magazine called *How on Earth*, for example, is produced and run entirely by teens, to publicize animal, vegetarian and environmental issues and to get other students interested and involved.

Earth Pledge of Allegiance

I pledge allegiance to Earth
For the sake of all its future plant,
Animal, and human forms of life.
One living planet
Revealing God;
Wholly lovable,
With clean air, water, and soil,
Liberty and justice for all.

Even celeb hang-outs are getting into the act. Propped on the roof of the **Hard Rock Cafe** in Los Angeles, is an "environmental scoreboard," flashing environmental statistics linking world population and remaining acreage of the world's rainforests, designed to heighten environmental awareness of local motorists and passersby.

Another example of the popularity of green stars occurred this year in France, where **Jacques-Yves Cousteau** was named the country's most popular person.

EcoPhilosopher Extraordinaire

British philosopher Iris Murdoch says philosophy means "looking at something you've always taken for granted, and suddenly seeing how very strange it really is."

Peter Russell, author of *A White Hole in Time* (HarperCollins, New York, 1992), is a Cambridge University-trained mathematician and physicist who looks at

5 hours
Source: Protect Our Planet Calendar, Running Press Book Publishers

modern industrial society and sees something "very strange" indeed. Russell believes we're on the brink of a White Hole in Time, "a moment of unimaginably rapid transformation" when "the end of our evolutionary process is a spiritual awakening into complete knowledge of the universe."

Einstein remarked that, after the atomic bomb, everything changed except our way of looking at things. *A White Hole in Time* will forever change your way of looking at things.

Chatwin Revisited

Bruce Chatwin, the British adventurer and writer who died in 1989 at the age of 49, was a secret wrapped inside an enigma. "What Bruce most admired, and wanted for himself, was the life of a nomad," said his friend and fellow writer, David Plante. Chatwin's famous travel books include *The Songlines*, set in Australia, and *In Patagonia*, set in the remote Patagonia region of Chile.

For fans who can never get enough Chatwin, Sierra Club Books has produced a posthumous volume, *Nowhere Is a Place: Travels in Patagonia* (Sierra Club, San Francisco, 1992), with text by Chatwin and his fellow travel writer, Paul Theroux. The words are a transcription of their command presentation to the Royal Geographical Society in London.

In May 1992, more than 2,000 children from over 30 countries met at the United Nations Assembly Hall in New York to take part in the eighth annual Global Youth Forum. This United Nations education program provides an opportunity for youths to share their interest in gaining a greater knowledge of the environment.

To what do we owe this recent surge of environmental responsibility among today's children? Many people believe that it begins with more activist-oriented ecology and conservation programs taught in schools. Kids are not only being trained to recognize potential environmental hazards in the home and their neighborhoods, but also are being taught to go out and do something about it. With this training comes a confident attitude that they *can* change old habits and long-standing public policy.

KID HEROES

Kid Heroes of the Environment (EarthWorks Press, Emeryville, CA, 1991) has true stories of kids who are saving the planet. As kid hero Ryan Eliason says: "The question isn't, 'Do I make a difference?' It's 'Do I want to make a positive difference or a negative one?'" Young environmentalists featured in the book include:

• Bryan Kaplan, age 13, from New York state, raised over $4,000 to protect wilderness areas and the animals living in them.
• Laura-Beth Moore, age 13, started a neighborhood recycling program in Houston.
• Kory Johnson, age 10, started a group called Children for a Safe Environment to stop an incinerator from burning hazardous waste near her town.
• Tanja Vogt, age 17, got her New Jersey school to ban Styrofoam lunch trays and use paper trays instead.
• Lyle Solla-Yates, age 10, from Florida, raised money to save the lives of manatees and other endangered animals.
• Melissa Poe, age 9, of Tennessee, started a club, Kids For A Clean Environment, which now has chapters all over the world.
• Brady Landon Mann, age 9, started a school recycling program in Vancouver, Washington.
• Matt Fischer, age 13, built a compost bin and teaches others about composting organic waste for gardening.

If you have a nominee for *Kid Heroes*, write to EarthWorks Press, 1400 Shattuck Ave., Ste. 25, Berkeley, CA 94709.

: *How many US states currently have laws requiring recycling?*

ARTS & ENTERTAINMENT

Spirituality and Ecology

- By Michael Dowd -

To look for a technological solution to the ecological crisis would be a lethal mistake. Scientific analysis points, curiously, toward the need for a quasi-religious transformation of contemporary cultures.

—*Paul Ehrlich*

We are in the midst of perhaps the greatest shift in human awareness in 3 million years, certainly the greatest shift since the dawning of the neolithic era 10,000 years ago. The magnitude of Copernicus and Galileo's announcement that Earth was not the center of the universe was nothing in comparison to the magnitude of the revolution of consciousness that has already begun during these last decades of the 20th century. History will one day clearly show that we are at a major turning point, not merely in the human story, but in the story of the universe itself. What follows is an attempt to outline the nature of this transformation, this planetary eco-spiritual awakening.

For the past several thousand years, we Westerners have generally thought of ourselves as separate from and superior to the rest of nature. If we believed God existed at all, we imagined that God resided off the planet, in the heavens. We also assumed that the more we could control the forces of nature, the more "progress" we were making. Until recently, these beliefs were rarely discussed; they were taken for granted. They are some of the assumptions about reality that we have inherited in the West. While it could be argued that these beliefs are directly or indirectly responsible for much of the ecological devastation taking place on the planet today, they have also made possible enormous scientific and technological breakthroughs. Ironically, some of these scientific breakthroughs are now the foundation for an eco-spiritual awakening that might be our only hope of planetary salvation.

Recent discoveries in biology, chemistry, physics and astronomy indicate that the universe is nothing

Resources for Ecological Spirituality

• *The Dream of the Earth,* by Thomas Berry, Sierra Club Books, San Francisco, 1990. An excellent introduction to one of the most prominent eco-theologians alive today. Berry explores the implications of our scientific creation story with regard to energy, technology, ecology, economics, education, spirituality, patriarchy and more. Contains a helpful annotated bibliography.

• *EarthLight Magazine of Spirituality and Ecology,* Unity with Nature Committee of Pacific Yearly Meeting, Religious Society of Friends (Quakers), 608 E. 11th St., Davis, CA 95616. The best periodical of its kind on the market.

• *Sacred Land, Sacred Sex—Rapture of the Deep: Concerning Deep Ecology and Celebrating Life,* by Dolores LaChapelle, Finn Hill Arts, Silverton, CO, 1988. An insightful and challenging look at how Western culture has brought the world to the brink of ecological suicide and the direction we must move to be saved from this fate. A most provocative and comprehensive treatment . . . both educational and empowering.

• By Miriam Therese MacGillis: "Earth Learning and Spirituality" (five-hour videotape); "New Earth Education" (three-hour

: **7, plus the District of Columbia**
Source: *National Wildlife* magazine

videotape); and "The Fate of the Earth" (90-minute audiotape), Global Perspectives, P.O. Box 925, Sonoma, CA 95476, (707) 996-4704. MacGillis is a leading popularizer of the new cosmology.

• *World as Lover, World as Self*, by Joanna Macy, Parallax Press, Berkeley, CA, 1991. A wonderful synthesis of experiential deep ecology, despair-work, general systems theory and engaged Buddhism, this book contains a wealth of wisdom and compassion that is essential for the healing of our world.

• *Ishmael*, by Daniel Quinn, Bantam Books, New York, 1992. Winner of the Ted Turner award for environmental fiction, this book is full of wisdom essential to planetary salvation.

• *Gaia: The Human Journey from Chaos to Cosmos*, by Elisabeth Sahtouris, Pocket Books, New York, 1989. An excellent introduction to the Gaia theory, the scientific understanding that Earth itself is alive, rather than merely a planet with life *on* it.

• *Thinking Like a Mountain: Toward a Council of All Beings*, by John Seed, et al., New Society Publishers, Philadelphia, 1988. This collection of essays, group exercises and poetry is designed to help facilitate personal deep ecology experience.

• *States of Grace: Spiritual Grounding in the Postmodern Age*, by

like the great machine we have supposed it was for the past 300 years. Rather, scientists are beginning to suggest that the universe is more like an evolving, maturing organism—a living system. The human can now be understood as a being in whom the universe, after some 15 billion years, having reached this degree of complexity, can consciously reflect on itself, its meaning, who it is, where it came from and of what it is made. "The human person is the sum total of 15 billion years of unbroken evolution now thinking about itself," as Teilhard de Chardin said over half a century ago. The scientist looking through a telescope is literally the universe looking at itself. The child entranced by the immensity of the ocean is Earth enraptured by itself. The student learning biology is the planet learning in consciousness, with awareness, about how it has functioned instinctively and unconsciously for billions of years. The worshipper singing praises is the universe celebrating the wonder of the divine Mystery from whence it came, and in which it exists. We humans are the means by which the universe can sense and perceive its beauty with conscious awareness. We are not separate beings *on* Earth, we are a mode of being or an expression *of* Earth. We did not come into this world, we grew out of it . . . in the same way that an apple grows out of an apple tree. Just as every cell of my body is part of a larger living being that I call "me," so are we each an integral part of a living planet that is our larger self, our larger body. As physicist Brian Swimme is fond of saying, "The planet Earth, once molten rock, now sings opera."

This shift, from seeing ourselves as separate beings placed on Earth to seeing ourselves as a conscious expression of Earth, is a major shift in our understanding of who and what we are. It is a shift at the most basic level: our identity or sense of self. It is the core of the revolution in eco-spirituality. We can now see that the Earth is not a planet with life *on* it; rather, it is a living planet. Indeed, scientists have discovered that over the last 4.5 billion years, Earth has maintained its average body temperature, while the sun's temperature has increased 25 percent to 30 percent. Earth has also maintained the salinity of its oceans and the chemical makeup of its atmosphere in a way that is possible only for a living system. This scientific understanding of Earth as a living being is

: *What substance has been found to have the same pH balance as battery acid?*

called the "Gaia theory" (named by James Lovelock after the ancient Greek goddess who symbolized Mother Earth).

Beyond the scientific evidence, however, the fact that Earth is a living system just makes sense. The physical structure of the planet—its core, mantle and mountain ranges—acts as the skeleton or frame of its existence. The soil that covers its grasslands and forests is like a mammoth digestive system in which all things are broken down, absorbed and recycled into new growth. The oceans, waterways and rain function as a circulatory system that provides life-giving blood, purifying and revitalizing the body. The algae, plants and trees provide the planet's lungs, constantly regenerating the entire atmosphere. The animal kingdom provides the basic functions of the nervous system, a finely tuned and diversified series of organisms sensitized to environmental change. Humanity itself is the capacity of the planet for conscious awareness, or reflexive thought. That is, the human enables Earth to reflect on itself and on the divine Mystery out of which it has come and in which it exists. We allow nature to begin to appreciate her own staggering beauty and to feel her own magnificent splendor.

It will take quite some time yet before this cosmology becomes universally accepted, and before this shift in perspective transforms our beliefs and institutions. It will, eventually, as surely as we all know that the Earth revolves around the Sun—although most people 500 years ago believed such a notion was absurd and heretical. As the philosopher William James observed about the history of ideas, "A new idea is first condemned as ridiculous, then dismissed as trivial, until finally it becomes what everybody knows."

Michael Dowd is an ecologist, pastor and author of EarthSpirit: A Handbook for Nurturing an Ecological Christianity *(Twenty-Third Publications, Mystic, CT, 1991) and the forthcoming* Living Sacred Story: Science, Deep Ecology, and the Destiny of Humanity.

Charlene Spretnak, HarperCollins, San Francisco, 1991. This book is about reclaiming the core teachings and practices of Buddhism, Native American spirituality, Goddess spirituality and the Semitic traditions (Judaism, Christianity and Islam) for the well-being of the Earth community as a whole.

• By Brian Swimme: *The Universe Is a Green Dragon: A Cosmic Creation Story*, Bear & Company, Santa Fe, NM, 1984; "Canticle to the Cosmos," Tides Foundation, NewStory Project, 134 Colleen St., Livermore, CA 94550. Swimme is a physicist who has studied extensively with Thomas Berry. *The Universe Is a Green Dragon* is an alluring introduction to the new cosmology. "Canticle to the Cosmos" is a 12-part video lecture series designed to be used for academic classes, small group study or personal enrichment.

• *The Universe Story: From the Primordial Flaring Forth to the Ecozoic Era*, by Brian Swimme and Thomas Berry, HarperCollins, San Francisco, 1992. A telling of the story of the universe with a feel for its sacred dimensions, this book is already being hailed by ecologists, scientists and theologians as one of the more significant works of the 20th century.

A: *Acid rain in West Virginia*
Source: Global Environmental Issues, *Routledge publishers*

Environmental Magazines

Following is a list of some of the many periodicals that cover the environment.

American Farmland
Published quarterly by the American Farmland Trust, 1920 N St. NW, Ste. 400, Washington, DC 20036.
News and features on farming practices that lead to a healthy environment.

American Forests
Published bimonthly by American Forests, P.O. Box 2000, Washington, DC 20013.
American Forests emphasizes the benefits of trees and forests. It presents the latest issues, lifestyles, how-to information, adventures and travels.

American Horticulturist
Published bimonthly by the American Horticultural Society, 7931 E. Boulevard Dr., Alexandria, VA 22308.
Gardening news and features, plus colorful photos.

The Amicus Journal
Published quarterly by Natural Resources Defense Council, 40 W. 20th St., New York, NY 10011.
The Amicus Journal is the quarterly, award-winning publication of the Natural Resources Defense Council. The journal covers national and international policy and has been called "the leading journal of environmental thought and opinion."

Publishing

Even in a time when most people get their information from television, print is still a lively area of information dissemination. There are still those who prefer to read, whether books, magazines or newspapers. Below are just a few of the books published in 1992 that address the environment. Earth Journal has highlighted the top ten books of the year, all of which are "must reads." Following them are titles organized by general topic.

TOP 10 ENVIRONMENTAL BOOKS OF THE YEAR

1. ***Beyond the Limits: Confronting Global Collapse, Envisioning a Sustainable Future***, Donnella Meadows, Dennis Meadows and Jorgen Randers, Chelsea Green, Post Mills, VT, $19.95.

Using World 3, a systems dynamic computer model, the authors project the necessary steps to "ease down" the global economy's enormous demands on the planet. This is a fascinating sequel to their groundbreaking 1972 book, *The Limits to Growth*.

2. ***The Threat at Home: Confronting the Toxic Legacy of the US Military***, Seth Shulman, Beacon Press, Boston, $23.

Shulman reports on our nation's biggest polluter: the US Military. From the Rocky Flats nuclear weapons plant, to the "national sacrifice zones," Shulman uncovers the Pentagon's "war" on American lands.

3. ***Earth in the Balance: Ecology and the Human Spirit***, Vice President Al Gore, Houghton Mifflin, Boston, $22.45.

Vice President Al Gore proposes a "Global Marshall Plan" to save us from "an environmental holocaust." Gore led the US Senate delegation to the Earth Summit in Rio de Janeiro, and his book is a well-researched overview of environmental problems and solutions.

4. ***Global Warming: Understanding the Forecast***, Andrew Revkin, Abbeville Press, New York, $29.95.

Revkin covers every aspect of the controversial global warming issue, in easy-to-follow terms. If

---

Q: How many people in the US and Canada die prematurely each year as a result of cardiac or respiratory problems attributed to air pollution levels?

you find newspaper reports on global warming more confusing than enlightening, read Revkin's compelling book.

5. ***The Rolling Stone Environmental Reader***, the editors of *Rolling Stone*, Island Press, Washington, DC, cloth $25, paper $15.

A fun-to-read—but serious—collection of environmental reporting, from Tom Horton's "Paradise Lost" (about the *Exxon Valdez* oil spill) to P. J. O'Rourke on "The Greenhouse Affect" and Bill McKibben on "Milken, Junk Bonds and Raping Redwoods."

6. ***State of the World 1992***, Lester Brown, et al., W.W. Norton, New York, cloth $19.95, paper $10.95.

The annual Worldwatch Institute report on our "progress toward a sustainable future" is always compulsively readable and thoroughly enlightening. Lester Brown is optimistic: "We know what we have to do, and we have the technologies needed for the Environmental Revolution to succeed."

7. ***Beyond Beef: The Rise and Fall of the Cattle Culture***, Jeremy Rifkin, Dutton, New York, $21.

Livestock consume one-third of all the grain produced in the world; one beefsteak takes 1,200 gallons of water to produce; livestock are a leading cause of topsoil depletion and pollution. There's a Chinese saying: "To know and not to act is not to know." If you're still eating beef, you need to know what's in this book.

8. ***The Elder Brothers***, Alan Ereira, Alfred A. Knopf, New York, $23.

BBC filmmaker Alan Ereira visited the Kogi tribe, direct descendants of a pre-Columbian civilization in the mountains of Colombia. The Kogi have broken their isolation to bring an urgent message of environmental stewardship to the world.

9. ***Ecocide in the USSR***, Murray Feshbach and Alfred Friendly, Jr., Basic Books, New York, $24.

Ecocide covers the environmental wreckage wrought by years of Soviet totalitarianism: the Chernobyl nuclear disaster, the disruption of ecosystems at Lake Baikal (the world's largest freshwater lake) and widespread health problems from massive pollution across the former USSR.

Animals
Published bimonthly by the Massachusetts Society for the Prevention of Cruelty to Animals, 350 S. Huntington Ave., Boston, MA 02130.
Animals appeals to a wide general audience with features on both pets and wildlife.

The Animals' Agenda
Published monthly by the Animal Rights Network, Inc., 458 Monroe Turnpike, Monroe, CT 06468.
The Animals' Agenda covers animals and environmental issues, and provides a forum for discussion of problems and ideas.

Animals International
Published quarterly by the World Society for the Protection of Animals, Park Place, 2 Langley Lane, London, SW8 ITJ, England.
International news on animals, and people working to protect them.

The Animals' Voice Magazine
Published quarterly by the Compassion for Animals Foundation, Inc., P.O. Box 34347, Los Angeles, CA 90034.
The Animals' Voice comprehensively covers animal defense issues worldwide.

Audubon
Published bimonthly by the National Audubon Society, 700 Broadway, New York, NY 10003.
Audubon reports on the state of the Earth. Coverage includes ecology, conservation,

A: *50,000 annually*
Source: Greenpeace Action

10. ***Millennium: Tribal Wisdom and the Modern World,*** David Maybury-Lewis, Viking, New York, $45.

Anthropologist Margaret Mead and others have pointed out that, for more than 99 percent of human history, people lived in small tribes. What can we learn from the tribes that remain on Earth today? How can we incorporate tribal wisdom into our urban lives? Anthropologist Maybury-Lewis takes the reader on a world tour to answer these questions.

Adventure

Battle for the Elephants, Iain and Oria Douglas-Hamilton, Viking, New York, $30.

Fighting for their lives and against the striking indifference of the international conservation establishment, the Douglas-Hamiltons struggled in their attempt to save elephants from the devastating ivory trade. A beautifully photographed adventure story of courage and dedication.

Bush for the Bushman: Need "The Gods Must Be Crazy" Kalahari People Die?, John Perrott, Beaver Pond Press, Greenville, PA, $14.95.

As civilization encroached on the African bush, the African Bushpeople's numbers dwindled from perhaps 300,000 before the Dutch arrived to less than 70,000 today. An expedition to Botswana led Perrott to his encounter with a small clan of Bushpeople that prompted him to defend the lives of the Bushpeople.

Economic Policy

On Common Ground: Managing Human-Planet Relationships, Ranjit Kumar and Barbara Murck, John Wiley & Sons, New York, $24.95.

Explores fundamental development, environmental concepts and sustainable management of Earth resources. Assists managers, policymakers and planners who want to incorporate environmental ethics into their business plans.

Cool Energy: Renewable Solutions to Environmental Problems, Michael Brower, MIT Press, Cambridge, MA, $12.95.

Renewable energy is the only viable alternative to

---

wildlife, policy, recreation and technology. It offers views of problems and proposed solutions, and it celebrates natural beauty that is to be preserved.

Backpacker
Published eight times a year by Rodale Press, Inc., 33 E. Minor St., Emmaus, PA 18098.

Features trips, products and outdoor news for backpackers.

Borealis
Published quarterly by the Canadian Parks and Wilderness Society, Ste. 1335, 160 Bloor St. E., Toronto, Ontario M4W 1B9, Canada.

This magazine has photos, features and news about Canada's environment.

Buzzworm
Published bimonthly by Buzzworm, Inc., 2305 Canyon Blvd., Ste. 206, Boulder, CO 80302.

As an independent magazine reporting on national and international environmental issues, *Buzzworm* strives to offer balanced and comprehensive coverage. Great photography and features, including eco-travel, green business, organica, ecohome and much more.

Calypso Log
Published bimonthly by the Cousteau Society, Inc., 8440 Santa Monica Blvd., Los Angeles, CA 90069.

Calypso Log's purpose is to protect and improve the quality of marine life for present and future generations.

Q: *How much does the wasted fuel and higher insurance premiums resulting from traffic congestion cost Americans in a given year?*

resource-depleting and polluting oil, gas and coal. A comprehensive review of progress in the field of renewable energy technologies—solar, wind, biomass, hydroelectric and geothermal—since the mid-1980s.

The Gaia Atlas of Green Economics, Paul Ekins, Mayer Hillman and Robert Hutchison, Doubleday, New York, $16.

What are the hidden costs of the growing international economy? The destructive obsession with economic growth is a threat to the planet's survival. A guide to what governments and people can do to build a sustainable society—to create prosperity while maintaining a healthy environment.

Saving Cities, Saving Money: Environmental Strategies that Work, John Hart, Resource Renewal Institute, Sausalito, CA, $15.95.

Ideas that will help city governments and concerned citizens direct the essential transition from disastrous urban sprawl to environmentally sensitive and economically beneficial urban management. How cities can promote efficiency through the products they buy, the contracts they sign and the facilities they maintain.

Sustainable Cities: Concepts and Strategies for Eco-City Development, edited by Bob Walter, Lois Arkin and Richard Crenshaw, Eco-Home Media, Los Angeles, $20.

Leading experts from their respective fields present concepts and strategies for ecological city designs. Includes topics like resource management, community relations and construction techniques.

Toward a Sustainable Society: An Economic, Social and Environmental Agenda for Our Children's Future, James Garbarino, Noble Press, Chicago, $19.95.

Criticizing current economic priorities and the world consumption of natural resources, Garbarino examines the effects an unrestrained consumer-oriented economic system will have in the future.

Environmental History

From Eros to Gaia, Freeman Dyson, Pantheon Books, New York, $25.

A: *About $25 billion annually*
Source: Campaign for New Transportation Priorities

The City Planet
Published monthly by The City Planet, 4988 Venice Blvd., Los Angeles, CA 90019.

The City Planet offers up-to-date information on ecological products and services, organic gardening, socially responsible investments, vegetarianism, water and energy conservation and more.

Common Sense on Energy and Our Environment
Published monthly by Common Sense on Energy and Our Environment, P.O. Box 215, Morrisville, PA 19067.

Common Sense explores all aspects of the scientific, technical, economic, and political components of energy and environmental issues.

Countryside
Published by the Hearst Corp., 959 Eighth Ave., New York, NY 10019.

A glossy magazine that celebrates country life. *Countryside* is filled with environmental news, organic gardening and cooking features, and stories about people who have farms or green businesses in the countryside.

The Consumer's Guide to Planet Earth
Published biannually by Schultz Communications, 9412 Admiral Nimitz NE, Albuquerque, NM 87111.

The Consumer's Guide to Planet Earth is an information-packed resource for Earth-friendly products and services including adventure/

ecotravel, organic foods, mail order catalogs, household and gardening products, personal care products, solar energy, natural pet products and much more.

Co-op America Quarterly
Published quarterly by Co-op America, 2100 M St. NW, Ste. 403, Washington, DC 20037.

Co-op America Quarterly covers positive alternatives and practical strategies for creating a more just and sustainable society. Each issue contains information on environmental investing, green consumers and the Boycott Action News.

Defenders
Published bimonthly by Defenders of Wildlife, 1244 19th St. NW, Washington, DC 20036.

Defenders provides provocative, in-depth coverage of major US and foreign wildlife conservation issues.

Design Spirit
Published quarterly by Design Spirit, 438 Third St., Brooklyn, NY 11215.

Design Spirit supports "architecture that harmonizes with nature" and includes updates on nontoxic building materials, feng shui, earthquake-proof building and traditional timber framing.

Ducks Unlimited
Published bimonthly by Ducks Unlimited, Inc., 1 Waterfowl Way, Memphis, TN 38120.

Ducks Unlimited includes stories on wetland and avian conservation.

An immensely broad range of ideas, people, contemporary history and discoveries of many sorts, with homages to Eros, the god of youthful passion, and Gaia, the fertile, life-giving, mother planet: Earth.

The Sacred Hoop: A Cycle of Earth Tales, Bill Broder, Sierra Club Books, San Francisco, $12.

A combination of rich illustrations, myth, history, archaeology and literature that convey an understanding of how Western cultures relate to Earth. A series of powerful stories from prehistory to the settling of the American West.

A White Hole in Time: Our Future Evolution and the Meaning of Now, Peter Russell, HarperSan Francisco, $20.

A challenging journey that begins with the Big Bang, progresses through physical-chemical-biological phenomena and cultural history, and ends with modern humans lost in the chaos of their own technological achievements. When the humans reach a spiritual "supernova," an inner awakening occurs, ending the dysfunctional attitudes and behavior.

Environmental Issues

Beyond the Beauty Strip: Saving What's Left of Our Forests, Mitch Lansky, Tilbury House Publishers, Gardiner, ME, cloth $26.95, paper $15.95.

Lansky's extensive research details the destruction of the forest by companies interested in profit, not in sustainable harvests. Challenges industrial forest management and reveals corrupt practices like the use of a "beauty strip," a thin strip of trees hiding deforested areas from public view.

The Great Thirst: Californians and Water, 1770s-1990s, Norris Hundley, Jr., University of California Press, Berkeley, CA, $25.

A 200-year history of Californians and their search for water is told through a colorful cast of characters with constantly shifting political alliances. Hundley confronts California's water problems and predicts the grim reality of California's dwindling resources and heavily consuming population.

 In an American Council on Education poll, what percentage of college freshmen said that the environment was their top social concern?

ARTS & ENTERTAINMENT

Living with the Land: Communities Restoring the Earth, edited by Christine Meyer and Faith Moosang, New Society Publishers, Philadelphia, cloth $34.95, paper $9.95.

A lively collection of inspiring accounts of communities from around the world that are taking back their lands and waters and using them in ecologically and economically sustainable ways. Eighteen success stories from rural and urban settings, from Nigeria to the Philippines.

No Risk Involved: The Ken McGinley Story, Survivor of a Nuclear Experiment, Ken McGinley and Eamonn P. O'Neill, Trafalgar Square, North Pomfret, VT, $29.95.

A brutally frank account of Ken McGinley's battle for justice for the nuclear "guinea pigs" in the face of high-level and influential opposition. Frightening new evidence that should force authorities into a re-examination of the whole nuclear issue.

The Primary Source: Tropical Forests and our Future, Norman Myers, W.W. Norton, New York, $10.95.

Tropical forests form the most diverse and complex ecosystem on Earth, containing 40 percent of all living species. The current status of the tropical forests, what is happening to them and what can be done are discussed in this informative account of the environmental crisis that will result if the exploitation is allowed to continue.

Rubbish!: The Archaeology of Garbage, William Rathje and Cullen Murphy, HarperCollins, New York, $23.

Garbage is a reflection of modern life and character. An analysis of one-quarter of a million pounds of garbage done at the University of Arizona has yielded insights into politics, economics and demographics, as well as the idiosyncrasies of human behavior.

The State of Native America: Genocide, Colonization and Resistance, edited by M. Annette Jaimes, South End Press, Boston, cloth $40, paper $16.

A collection of essays by noted Native American authors and activists exploring the circumstances confronted by native people in the contemporary US.

E: The Environmental Magazine
Published bimonthly by Earth Action Network, Inc., P.O. Box 5098, Westport, CT 06881.
E Magazine is a not-for-profit independent magazine covering a wide range of environmental issues. Each issue contains feature articles on key issues and campaigns; industry and consumer product trends; interviews with leading advocates and thinkers; reviews of books, films and videos.

Earth
Published bimonthly by Kalmbach Publishing Co., 21027 Crossroads Cir., Waukesha, WI 53187.
Earth magazine is designed to present what humans know about the planet we live on.

Earth Island Journal
Published quarterly by Earth Island Institute, 300 Broadway, Ste. 28, San Francisco, CA 94133-3312.
Earth Island Journal delivers "local news from around the world." Eyewitness accounts from the environmental front lines—from Malaysia to Moscow—with global eco-news and exposés.

Earth First!
Published eight times a year on the old pagan European nature holidays, P.O. Box 5176, Missoula, MT 59806.
Earth First! is an independently owned newspaper within the broad Earth First! movement.

Earthwatch
Published bimonthly by

: *89 percent*
Source: The Student Environmental Action Guide, *EarthWorks Press*

Earthwatch Expeditions, Inc., 680 Mount Auburn St., P.O. Box 403, Watertown, MA 02272.

The mission of *Earthwatch* is to improve human understanding of the planet, the diversity of its inhabitants and the processes that affect the quality of life on Earth.

Earth News
Published bimonthly by A Mother Earth Newsletter, 5126 Clareton Dr., Ste. 200, Agoura Hills, CA 91301.

Earth News covers environmental issues and environmentally correct products.

Earthword
Published quarterly by the Eos Institute, 1550 Bayside Dr., Corona del Mar, CA 92625.

Earthword is a journal of environmental and social responsibility.

The Ecologist
Published bimonthly by MIT Journal, 55 Hayward St., Cambridge, MA 02142.

For 20 years *The Ecologist* has provided a forum for social and environmental activists who are seeking to change current development policies.

Equinox
Published bimonthly by Telemedia Publishing, 7 Queen Victoria Rd., Camden East, Ontario K0K 1J0, Canada.

A Canadian journal of discovery exploring such topics as environmental issues, travel, history and heritage, profiles and scientific news.

Superpigs and Wondercorn: The Brave New World of Biotechnology and Where it All May Lead, Dr. Michael W. Fox, Lyons and Burford, New York, $21.95.

Biogenetic research is capable of producing new life forms whose effects may alter the intricate balance of nature in ways no one can predict. A provocative survey of a dramatic new technology and an impassioned plea to use these new tools in the long-term interests of the global ecosystem.

World of Waste: Dilemmas of Industrial Development, K.A. Gourlay, Zed Books Ltd., Atlantic Heights, NJ, cloth $55, paper $17.

An examination of the environmental costs of the escalating production and disposal of waste. A unique challenge to existing political assumptions.

Natural History

Baikal: Sacred Sea of Siberia, Peter Matthiessen, photographs by Boyd Norton, Sierra Club Books, San Francisco, $25.

Lake Baikal is more than a mile deep, contains one-fifth of the fresh water on Earth and is endangered by acid rain and pollution from industries on the lakeshore. Matthiessen describes his journey to this sacred landscape and interweaves the text of folklore, history and myth with powerful, color photographs.

The Eye of the Elephant: An Epic Adventure in the African Wilderness, Delia and Mark Owens, Houghton Mifflin, Boston, cloth $45, paper $22.95.

After being expelled from Botswana for writing the controversial *Cry of the Kalahari*, Delia and Mark Owens found a new African paradise in the North Luangwa Valley of Zambia, an area with no roads, buildings or people. Discovering that the indigenous wildlife was being killed by poachers at an alarming rate, the Owenses initiated a sustainable economic development plan that became a battle for their lives.

Global Warming and Biological Diversity, edited by Robert L. Peters and Thomas E. Lovejoy, Yale University Press, New Haven, CT, cloth $45, paper $22.95.

The first detailed discussion of the consequences of global warming for ecosystems. Experts present the

: *In an average day, how much electricity does a person in the US consume?*

ARTS & ENTERTAINMENT

responses of animals and plants to climate changes and human activities such as deforestation. The information is factual and of concern to all living things.

The Great Bear: Contemporary Writings on the Grizzly, edited by John A. Murray, Alaska Northwest Books, Bothell, WA, $14.95.

Collectively, these essays offer provocative insights into the status of the grizzly in North America. A gathering of some of the finest writings on the grizzly over the past 40 years, by authors such as Edward Abbey, John McPhee and William Kittredge.

On Nature's Terms, edited by Thomas J. Lyon and Peter Stine, Texas A&M University Press, College Station, TX, cloth $35, paper $14.95.

A collection of essays featuring writers such as Barry Lopez, Rick Bass and William Kittredge. These are vivid stories of understanding and coming to terms with nature as seen through the authors' experiences in places like the Grand Canyon and the Florida Keys.

Shadows of Africa, Peter Matthiessen, art by Mary Frank, Abrams, New York, $34.95.

Concentrating on Africa, where the indigenous species population has declined significantly, this lyrical and inspiring collection of words and images is a tribute to African wildlife. A portion of the authors' earnings will be donated to Wildlife Conservation International.

Where the Bluebird Sings to the Lemonade Springs: Living and Writing in the West, Wallace Stegner, Random House, New York, $21.

Reflective thoughts woven together with essays on the West and appreciations of other writers. An indication of Stegner's passion for the past, for the future, for truth and for the ranches of the Western US.

A World Between Waves: Writings on Hawaii's Rich Natural History, edited by Frank Stewart, Island Press, Washington, DC, cloth $24.95, paper $15.95.

Fourteen essays on the natural history of Hawaii by renowned American writers such as Peter Matthiessen and John McPhee. A passionate call to protect the fragile Hawaiian landscape.

Environmental Action Magazine
Published quarterly by Environmental Action, 6930 Carroll Ave., Ste. 600, Tacoma Park, MD 20912.
    Environmental Action founded in 1970 by the organizers of the first Earth Day, includes news, reviews and resources for activist and armchair readers alike.

Forest & Conservation History
Published quarterly by the Forest History Society in association with Duke University Press, 701 Vickers Ave., Durham, NC 27701.
    The Forest History Society is a nonprofit educational institution, founded in 1946. Their quarterly magazine includes stories on all aspects of our forests.

Garbage: The Practical Journal for Environment
Published bimonthly by Old House Journal, 2 Main St., Gloucester, MA 01930.
    Practical features on living an environmentally sound life.

Geographical
Published monthly by Hyde Park Publications Ltd., 2323 Randolph Ave., Avenel, NJ 07001.
    This magazine of the Royal Geographical Society features exploration, research and geographical knowledge.

The Green Consumer Letter/The Green Business Letter
Published monthly by Tilden Press, 1526

About 70 killowatt-hours
Source: Union of Concerned Scientists

Connecticut Ave. NW, Washington, DC 20036.

The Green Consumer Letter is the independent voice of environmentally safe shopping, written by Joel Makower, author of *The Green Consumer*. *The Green Business Letter* is the companion resource to *The Green Consumer Letter*. It is written for anyone involved or interested in business who is concerned about the environment.

Great Expeditions
Published quarterly by Sahara Publications Ltd, P.O. Box 8000-411, Abbotsford, BC V2S 6H1, Canada.
Great Expeditions provides information on exploration, research, adventure, outdoor recreation and travel.

Green
Printed by Severn Valley Press, COMAG, Tavistock Rd., West Drayton, Middlesex UB7 7QE, England.
Green is an independent magazine for the environment, based in England.

Green MarketAlert
Published monthly by MarketAlert Publications, 345 Wood Creek Rd., Bethlehem, CT 06751.
Green MarketAlert tracks the business impacts of green consumerism. Covers marketing/advertising, regulation, corporate strategies, green consumer surveys, products/packaging, international developments and more. Also provides resources for

Resources

Alternative Energy Sourcebook: A Comprehensive Guide to Energy Sensible Technologies, edited by John Schaeffer, Real Goods Trading Corporation, Ukiah, CA, $16.

More than a catalog of alternative energy supplies, this is a complete resource for energy independent living. Thirty-eight chapters organized into eight major sections, including the efficient home, mobility and just-for-fun.

The Directory of National Environmental Organizations, US Environmental Directories, edited by John C. Brainard, Roger N. McGrath, St. Paul, $54.

A directory to over 675 nongovernmental organizations, with brief descriptions of each organization. This expansive guide lists environmental organizations alphabetically and also includes a subject index, geographical index and federal agencies index.

The Encyclopedia of Environmental Studies, William Ashworth, Facts on File, New York, $60.

Over 3,000 entries covering a broad range of topics. A source of accurate and concise information on technical and general topics of environmental concern, supported by diagrams and tables.

EcoLinking—Everyone's Guide to Online Environmental Information, Don Rittner, Peachpit Press, Berkeley, CA, $18.95.

EcoLinking is the first guide to the use of computer technology by scientists, environmentalists and concerned citizens to share ideas and research. This book details how computer networks, bulletin boards and online services can be put to work to save the planet.

The Environmental Sourcebook, Edith C. Stein, Lyons and Burford, New York, $16.95.

A compilation of information pertinent to anyone interested in the environmental movement. Includes significant issues, active environmental groups, foundations that support environmental groups and ideas for obtaining more information from periodicals, books and magazines.

Q: *In an average day, how much electricity does a person from Latin America, Africa or Asia consume?*

ARTS & ENTERTAINMENT

Environmental Success Index 1992, Renew America, Washington, DC, $25.

A directory listing more than 1,600 environmental success stories in the US. Contact names and descriptions are given for each of the programs, which represent community, business, government and environmental organizations.

Environmental Telephone Directory, 1992-1993, Government Institutes, Inc., Rockville, MD, $79.

There's an old saying: "It's not what you know, but who you know." Anyone who has tried finding that "who" in a federal or state governmental office knows how challenging it is. Not any more. If you are an environmental activist, you can't afford not to have this 235-page publication next to your phone.

Global Assembly of Women and the Environment: Partners in Life, edited by Joan Martin-Brown and Waafas Ofosu-Amaah, United Nations Environment Programme and Worldwide Network, free, $5 postage.

The Global Assembly was the culmination of a series of four regional assemblies of women and environment, that were held to engage women in an assessment of the environmental conditions in their respective regions, advance the development of women's environmental networks and review regional blueprints for environmental action. This summary of the proceedings includes regional environmental concerns and international environmental contacts.

The Green Encyclopedia, Irene Franck and David Brownstone, Prentice Hall General Reference, New York, cloth $35, paper $20.

The complete A-to-Z sourcebook for students, activists and ordinary citizens who want to understand and solve environmental problems. Over 1,000 entries.

World Resources 1992-1993, The World Resources Institute, The United Nations Environment Programme and The United Nations Development Programme, Oxford University Press, New York, $19.95.

An indispensable guide to the global environment. A source of data and ideas addressing the natural resource and environmental problems of the developing world.

implementing green business strategies.

Harrowsmith Country Life
Published bimonthly by Camden House Publishing, Inc., a division of Telemedia Publishing Inc., Ferry Rd., Charlotte, VT 05445.
A magazine of country life that deals with topics such as animal life, forestry and gardening.

Health
Published ten times a year by Bill Kupper, 275 Madison Park, Ste. 1314, New York, NY 10016.
A magazine covering all aspects of health, including psychology, medicine, nutrition, fitness and apparel.

In Context
Published quarterly by Context Institute, P.O. Box 11470, Bainbridge Island, WA 98110.
In Context explores and clarifies what is needed for a humane, sustainable culture—and how we can get there.

International Wildlife
Published bimonthly by the National Wildlife Federation, 8925 Leesburg Pike, Vienna, VA 22184.
A magazine focusing on wildlife all over the globe. Stunning color photos in every issue.

Landscape Architecture
Published monthly by the American Society of Landscape Architects, 4401 Connecticut Ave. NW, Fifth Fl., Washington, DC 20008-2302.
A magazine that covers the art of

A: ***About 3 killowatt-hours***
Source: Union of Concerned Scientists

landscape architecture, including natural solutions to water problems, park preservation and landscaping.

Mother Earth News
Published bimonthly by Mother Earth News, P.O. Box 3015, Harlan, IA 51593-4255.

Mother Earth News covers sound environmental practices in nature and at home, land issues, travel and wildlife.

Mother Jones
Published bimonthly by the Foundation for National Progress, 1663 Mission St., Second Fl., San Francisco, CA 94102.

A magazine devoted to people, politics, environmental issues, cultural profiles and critiques of current literature.

National Geographic
Published monthly by National Geographic Society, 1145 17th St. NW, Washington, DC 20036.

National Geographic is a source of information about the world and its people dealing with the changing universe and human's involvement in it. Stunning photography and in-depth articles.

National Wildlife
Published bimonthly by the National Wildlife Federation, 8925 Leesburg Pike, Vienna, VA 22184.

National Wildlife magazine is published to create and encourage an awareness among the people of the world of the

Wildlife

Bears: Their Life and Behavior, William Ashworth, photographs by Art Wolfe, Crown Publishers, New York, $40.

Bears combine immense physical power with one of the keenest intelligences in the animal kingdom. A vivid photographic portrayal of the three North American bear species complemented by an enlightening text on bear behavior.

Dangerous Birds: A Naturalist's Aviary, Janet Lembke, Lyons and Burford, New York, $21.95.

Practical but humorous essays on birds and the passions of birders. Through reflections on birds ranging from sparrows to hummingbirds, Lembke explains how the interest birds inspire makes them "dangerous."

The Diversity of Life, Edward O. Wilson, Harvard University Press, Cambridge, MA, $29.95.

A comprehensive presentation of the processes that create new species and the reasons why global diversity has diminished over the past 600 million years. The world is threatened by the "sixth great spasm" of extinction, which has been created entirely by humans. The scientific evidence described makes a plea for specific actions that will enhance populations and preserve habitats.

Elephants: Majestic Creatures of the Wild, edited by Jeheskel Shoshani, illustrated by Frank Knight, Rodale Press, Emmaus, PA, $40.

Featuring over 300 color photographs and an authoritative account of elephants assembled by internationally renowned scientists. The informative text includes the evolutionary development of both African and Asian elephants, as well as details on efforts to save elephants from extinction.

Raptors: Birds of Prey, John Hendrickson, Chronicle Books, San Francisco, cloth $29.95, paper $18.95.

A comprehensive volume of the world's birds of prey—eagles, hawks, falcons, owls and vultures. In this beautiful showcase, Hendrickson presents captivating photographs and detailed information compiled during 30 years of research, tracking, photographing and writing.

 What is the ratio of plastic flamingos to real flamingos in the US?

ARTS & ENTERTAINMENT

CHILDREN'S BOOKS

In 1992, more environmental kids' books were published than ever before. In fact, many bookstores had to set up new sections to shelve them. Listed below are just a few of the titles that appeared in 1992. Prices are for clothbound books, unless otherwise listed.

An Adventure in New Zealand, The Cousteau Society, Simon and Schuster, New York, $14.

With brilliant photographs, the Cousteau Society takes the young reader on a journey through Maori customs, active volcanic zones, a glacier and the incredible creatures of New Zealand.

All About the Frog; All About the Turtle, William White, Jr., Ph.D., Sterling, New York, $14.95 each.

Informational books for older children that include facts about the history and development, habitat, lifestyle and survival of frogs and turtles.

Antarctica, The Last Unspoiled Continent, Laurence Pringle, Simon and Schuster, New York, $15.

Pringle describes the exploration, geology, climate, wildlife and ecology of a unique land. Antarctica is a frozen mass of environmental history and possible solutions to environmental threats of the future.

need for the proper management of those resources upon which life depends: the soil, the air, the water, the forests, the minerals, the planet life and the wildlife.

National Parks
Published bimonthly by National Parks and Conservation Association, 1776 Massachusetts Ave. NW, Ste. 200, Washington, DC 20036.

National Parks defends, promotes and improves our country's national parks system while educating the public about parks.

Natural History
Published monthly by the American Museum of Natural History, Central Park West at 79th St., New York, NY 10024.

A magazine covering wildlife, travel, cultural profiles, science and history.

Natural Health
Published bimonthly by East West Natural Health, 17 Station St., Box 1200, Brookline Village, MA 02147.

Natural Health covers natural health, environmental issues, psychology and other topics.

Ocean Realm
Published monthly by Ocean Realm, 342 W. Sunset Rd., San Antonio, TX 78209.

Beautiful color photography and stories about life in and on the world's oceans.

Orion
Published quarterly by The Myrin Institute, 136 E. 64th St., New York,

 700 to 1
Source: The Daily Camera, *Boulder, Colorado*

NY 10021.

Through science, art and the humanities, *Orion* points to the connections between nature and human culture, social policy and environmental health—both local and global.

Outside
Published monthly by Outside, 1165 N. Clark St., Chicago, IL 60610.

Outside covers the great outdoors, with features, photos and news.

Planetary Citizen
Published quarterly by Stillpoint Publishing, Meetinghouse Rd., Box 640, Walpole, NH 03608.

Planetary Citizen covers politics, the environment, animals and more.

Preserve
Published bimonthly by Publisher's Service Corp., P.O. Box 557, Smyrna, DE 19977.

Preserve's goal is to increase public awareness of environmental issues and support efforts to take positive action on these issues.

Rock & Ice
Published bimonthly by Eldorado Publishing, P.O. Box 3595, Boulder, CO 80307.

New products, climbing tips, color photos and feature stories for rock climbers.

Safe Home Digest
Published biannually by Safe Home Digest, 24 East Ave., Ste. 1300, New Canaan, CT 06840.

Safe Home Digest is dedicated to helping people improve their

The Bear Family, Dieter Betz, Tambourine, New York, $15.

Full of beautiful photographs taken by the author, this book tells the story of the lives of grizzly bears in Alaska, including one mother bear, Teddy, who takes care of her two cubs as she teaches them to survive.

The Coloring-Activities Book of Endangered Species, RuthEllen Stec, RMS Publishing, Birmingham, MI, paper $4.95.

A fun and educational coloring book that teaches children about the fragile populations of endangered species worldwide.

The Curiosity Club: Kids' Nature Activity Book, Allene Roberts, John Wiley & Sons, New York, paper $12.95.

An educational and hands-on book that "allows children to discover for themselves the wonder and importance of the natural world."

Dancers in the Garden, Joanne Ryder, Sierra Club Books, San Francisco, $15.95.

A lyrical and beautifully illustrated portrait of a hummingbird and his mate and their lives in the garden.

Dinosaurs to the Rescue!, Laurie Krasny Brown and Marc Brown, Little, Brown and Co., Boston, $14.95.

A guide that discusses the environmental problems confronting children. The animated story offers fun and attainable ways kids can help protect the planet.

Discover My World Mountain; Discover My World Desert, Ron Hirschi, Bantam Little Rooster, New York, $13 each.

Beautifully illustrated with the watercolors of Barbara Bash, this series takes the young reader on a journey through different environments and allows the child to discover the animals depicted in their indigenous habitats.

Eagles: Hunters of the Sky, Ann C. Cooper, Roberts Rinehart, Niwot, CO, paper $6.95.

An informational book that includes educational activities and describes the place of the eagle in the Native American culture as well as in its own natural habitat.

: *What is the estimated life span of a nuclear reactor?*

ARTS & ENTERTAINMENT

The Forgotten Forest, Laurence Anholt, Sierra Club Books, San Francisco, $14.95.

An inspirational tale about the fate of the last forest hidden within the busy walls of the city, and the children who relish its existence and fight for its survival.

The Gift of the Tree, Alvin Tresselt, Lothrop, Lee & Shepard, New York, $14.

Henri Sorensen's amazing paintings illustrate this story of the role of an oak tree in the cycle of nature.

The Girl Who Loved Caterpillars, Jean Merrill, Putnam & Grosset, New York, $14.95.

Based on an ancient Japanese scroll, Merrill's book is a vivid story of a free-spirited girl determined to be herself, even within the rigid confines of the 12th-century Japanese court.

It's My Earth, Too, Kathleen Krull, Doubleday, New York, $13.50.

A fancifully illustrated book about the fragile and often limited elements that make up our Earth. The young narrator describes his place in the world and ways in which children can help save our planet and improve its condition.

Junkyard Bandicoots and Other Tales of the World's Endangered Species, Joyce Rogers Wolkomir and Richard Wolkomir, John Wiley & Sons, New York, paper $9.95.

This illustrated book includes stories about more than 35 endangered species of mammals, birds, fish, reptiles, amphibians and invertebrates. The stories explain why the animals are disappearing and what is being done to protect them.

Likeable Recyclables, Creative Ideas for Reusing Bags, Boxes, Cans, and Cartons, Linda Schwartz, Learning Works, Santa Barbara, CA, paper $9.95.

A creative how-to guide for kids to teach them how to create toys and tools by using recyclable materials found at home.

Mother Earth, Nancy Luenn, Atheneum, New York, $13.95.

A book that celebrates the Earth through the

home environment. Reports on the latest safe and healthful building products, interior design, home testing kits, alternative paints and finishes, low-toxic cleaning products, "green" auto care and more.

Sierra
Published bimonthly by Sierra Club, 730 Polk St., San Francisco, CA 94109.

Sierra is the award-winning magazine of the Sierra Club. Beautiful color photography and lively responsible editorials inform you about today's environmental issues.

The Sciences
Published bimonthly by the New York Academy of Sciences, 2 E. 63rd St., New York, NY 10021.

The Sciences contains absorbing essays by leading scientists and reviews of new science books.

Sea Frontiers
Published monthly by Sea Frontiers, UM Knight Center, 400 SE Second Ave., Fourth Fl., Miami, FL 33131-2116.

Sea Frontiers is a joint publication of Nature America, Inc., and the International Oceanographic Foundation. Color photos and stories about the sea.

Sea Kayaker
Published quarterly by Sea Kayaker, Inc., 6327 Seaview Ave. NW, Seattle, WA 98107-2664.

Trips, tips, news and photos for sea kayakers.

A: *40 years*
Source: Nuclear Information and Resource Service

Summit: The Mountain Journal
Published quarterly by Summit Publications, 111 Schweitz Rd., Fleetwood, PA 19522.

Summit is a full-color magazine that covers mountains from every angle: adventure, travel, fiction, photos.

Utne Reader
Published bimonthly by LENS Publishing, 1624 Harmon Place, Ste. 330, Minneapolis, MN 55403.

Often called a *Reader's Digest* for the hip, every issue of *Utne Reader* contains stories from the best of the nation's alternative magazines and newspapers.

Vegetarian Times
Published monthly by Vegetarian Times, P.O. Box 570, Oak Park, IL 60303.

News, features, personality profiles and recipes from the vegetarian world.

Whole Earth Review
Published quarterly by POINT, 27 Gate Five Rd., Sausalito, CA 94965.

Whole Earth Review has book and product reviews as well as stories dealing with alternative, sustainable lifestyles and world issues.

Wild Earth
Published quarterly by Wild Earth, P.O. Box 492, Canton, NY 13617.

Dave Foreman, co-founder of Earth First!, is *Wild Earth*'s executive editor. *Wild Earth* is a wilderness and biodiversity magazine.

simple yet poetic language of the author and the beautiful paintings of illustrator Neil Waldman. The book teaches children to enjoy Mother Earth's gifts, and to return them to her "with respect and love."

The Moon of the Chickarees; The Moon of the Fox Pups; The Moon of the Salamanders, Jean Craighead George, HarperCollins, New York, $15 each.

Each book in the series, named for the lunar cycles, describes a unique animal in its habitat, and celebrates the natural world, and shows the reader how precious our environment really is.

Mr. Raccoon and his Friends, Eugene J. McCarthy, Academy Chicago Publishers, Chicago, cloth $16, paper $10.

Senator McCarthy's amusing book, illustrated by Julia Anderson-Miller, is an environmentally relevant tale about the lives of forest animals.

Our Yard Is Full of Birds, Anne Rockwell, Macmillan, New York, $13.95.

A beautifully illustrated book about a boy who learns about all the wonderful birds in his garden.

Projects for a Healthy Planet, Shar Levine and Allison Grafton, John Wiley & Sons, New York, paper $9.95.

Levine and Grafton present simple and fun activities that inspire understanding and respect for the environment. The book teaches children about pollution, and what they can do to limit pollution through conservation and recycling.

Recycle!: A Handbook for Kids, Gail Gibbons, Little, Brown and Co., Boston, $14.95.

A lively book that describes the process of recycling in five different areas—paper, glass, aluminum cans, plastic and polystyrene. The book also suggests what kids can do to recycle regularly and includes many "fantastic facts about garbage and recycling."

A River Ran Wild, Lynne Cherry, Harcourt Brace Jovanovich, New York, $14.95.

An illustrated book describing the history of the Nashua River in Massachusetts that encourages children to value and preserve natural resources.

 When aluminum is made with recycled material, by what percentage is pollution reduced during production?

ARTS & ENTERTAINMENT

Save My Rainforest, Monica Zak, Volcano Press, Volcano, CA, $14.95.

A picture book based on a true story of a boy who makes a pilgrimage to the dying rainforest and then, upon his return, finally wins an audience with the president of Mexico.

The Simple People, Tedd Arnold, Dial Books, New York, $14.

An insightful allegory about a community that temporarily loses sight of its values. The story illustrates how greed can hurt the environment.

Thirteen Moons on Turtle's Back, Joseph Bruchac and Jonathon London, Philomel, New York, $15.95.

A collection of poems based on native legend that give voice to the thirteen moons of the year and reveal the wonder of the seasons and the beauty of the sublime land.

Townsend's Warbler, Paul Fleischman, Harper-Collins, New York, $13.

Illustrated with paintings of the 1800s by artists such as Bierstadt and Audubon. Fleischman describes a journey of naturalist John Townsend and the bird that he discovers.

Where Does Our Garbage Go?, Joan Bowden, Doubleday, New York, $10.

Full of pop-up pictures and engaging wheels and flaps, this informative book describes the processes of recycling and includes many ways in which families can reduce, reuse and recycle.

Wildlife Art News
Published bimonthly by Pothole Publications, Inc., 3455 Dakota Ave. S., St. Louis Park, MN 55416-0246.

This magazine of wildlife art explores the world of animals, birds and habitat with nearly 200 pages of full-color photos in each issue.

Wildlife Conservation
Published bimonthly by New York Zoological Society, Bronx, NY 10460.

Through its unique blend of exploration, adventure and scientific excellence, *Wildlife Conservation* inspires appreciation for and understanding of wild creatures and wild habitats.

WorldWatch
Published bimonthly by Worldwatch Institute, 1776 Massachusetts Ave. NW, Washington, DC 20036.

WorldWatch offers in-depth coverage of international environmental issues.

Z Magazine
Published 11 times a year by The Institute for Social and Cultural Communications, 116 St. Botolph St., Boston, MA 02115.

Essays and news articles on political and environmental issues, by leading journalists.

A: *95 percent*
Source: The Student Environmental Action Guide, *EarthWorks Press*

Green Radio

Radio is one of this country's most accessible forms of mass communication. Whether you're at home, in your car or at work, listening to the radio is an easy and relaxing way to keep abreast of what's happening on the eco-news front. In just about every local market, there are environmental programs and even some all-eco-stations with a variety of programs for the environmentally concerned listener. *Earth Journal* has compiled a list of some of the best national environmental radio programs. To find out about programs in your area, tune in to your local public or community station, or contact the National Federation of Community Broadcasters, (202) 393-2355 or the Corporation for Public Broadcasting, (202) 879-9600.

E-Town

Now in its second season, "E-Town" has grown and is now carried on 100 stations nationwide. "E-Town" recently received a $25,000 grant from the EPA as part of a new national government environment education program. "What 'E-Town' is about is reaching a non-environmentalist audience through music," says producer and host Nick Forster. "E-Town" features a variety of musical acts, as well as special guests, and includes items like a debate on the merits of meat eating as opposed to vegetarianism.

Media

Media—television, radio, film and computer software and networks—is one of the most rapidly developing and responsive forms of information dissemination and entertainment. Earth Journal *brings you the hottest sources for new environmental information.*

Graham Greene, star of "Clearcut."

ENVIRONMENTAL FILM

A VIRTUAL EXPLOSION HAS OCCURRED in the number of environmentally themed movies available today. In addition to documentaries and nature/wildlife specials that have been plentiful in the past, we are now seeing the "greening of Hollywood," as producers try to capitalize on growing viewer interest in the environment. As a result, there are numerous feature-length, high-budget eco-films with plans for many more in the making. Celebrities are vying for roles and rights to movies with plots about saving the whales.

Q: *Who is the only person ever to have a US national park named after him?*

ARTS & ENTERTAINMENT

National movie theater chains have also signed on, and we have seen the advent of environmentally related public service announcements. As the environment continues to be recognized as an important issue by more and more people in the 1990s, we can expect more films featuring it in a leading role.

Earth Journal provides a closer look at some of the best, most recent environmentally themed films, including documentaries, children's features, hard-hitting exposés and instructional films. Many of these have won acclaim at environmental film festivals. Most of these titles are available on video cassette.

Amazon, 97 min., 1992, Finland. Director: Mika Kaurismaki. Producer: Mika Kaurismaki and Pentti Kouri. Distributor: Cabriolet Films, (310) 472-2062.

Modern drama in which Finnish banker and his two daughters have an epiphany in the Brazilian rainforest.

Antarctica, 43 min., 1991, US. Director: John Weiley. Co-producers: Australian Film Finance Corporation, and Museum of Science and Industry, Chicago. Distributor: Museum of Science and Industry, Chicago, (312) 684-1414.

This incredible IMAX film celebrates the adventurous and courageous explorers and scientists who traverse the awesome vastness and uncover the haunting mysteries of the world's southernmost continent.

Big Bang, 4 min., 1991, Italy. Writer/Director/Producer: Bruno Bozetto. Distributor: Italtoons, (212) 730-0280.

Famed Italian animator Bruno Bozetto presents a humorous but worrisome solution to the garbage disposal problem.

The Bitter Winds of Bitterfeld, 40 min., 1991, Germany. A film by Rainer Hallfritzch, Margit Miosga and Ulrich Neumann. Distributor: Margit Miosga, Berlin, (030) 312-4365.

Clandestine filming of the ecological disasters (resulting from a chemical plant's waste) in this former East German city, which has earned Bitterfeld the distinction of being the dirtiest city in Europe.

Black Harvest, 75 min., 1992, Australia. Directors: Robin Anderson and Bob Connolly. Producer: The

Earthwatch Radio
"Earthwatch" is a two-minute radio feature on the environment, produced and distributed free by the Institute of Environmental Studies and the Sea Grant Institute at the University of Wisconsin. Reaching approximately 100 radio stations nationally, it has prompted many additional spin-off projects such as an environmental newspaper feature series, and bound paperback collections of "Earthwatch" scripts for schools and libraries.

Pulse of the Planet
A daily two-minute program funded and provided free of charge to radio stations by the Du Pont Corporation, "Pulse of the Planet" addresses issues such as endangered species, waste management, global warming, rainforests, energy issues and recycling. The program combines eco-sounds with conversations with leading scientists on issues to give the listener a global perspective on environmental concerns.

Radio Expeditions: The Unheard World
A joint effort of the National Geographic Society and National Public Radio, "Radio Expeditions" are hour-long specials that visit different locations around the globe and examine the world of natural sound. Featured programs vary from hearing low-frequency alligator communications in southwest Florida to traveling to the

: *Teddy Roosevelt, former President of the US*
Source: The W.A.T.E.R. Foundation

sub-Antarctic basin, where underwater sound blasts are used to help scientists measure global warming while potentially harming whales and other marine life indigenous to the area.

USA Outdoors

A Sun Radio Network project, "USA Outdoors" is a weekly two-hour, live call-in program that covers all aspects of the outdoors and the environment. Hosted by Dr. Kris W. Thoemke, a knowledgeable scientist, avid outdoorsman and experienced radio broadcaster, "USA Outdoors" is a program for people who enjoy being outside, care about the environment and are thirsty for more knowledge on both subjects.

Green Television

The United Nations Conference on Environment and Development in Rio de Janeiro in June was a springboard for many networks' attempts to garner viewers interested in environmentally-themed programming. Network executives have begun to realize that viewers remain interested in the environment. MTV news anchor Tabitha Soren says, "The environment is one of the most important issues for our viewership."

Earth Journal has compiled a list of networks that are leading the commitment to eco-broadcasting.

Australian Film Commission, 8 West St., North Sydney, NSW, 2060 Australia, (61 02) 925-7333.

The third film from Connolly and Anderson studies the intrusion of modern culture on the Ganiga, an aboriginal tribe in the highlands of Papua New Guinea.

Black Robe, 100 min., 1992, US. Director: Bruce Beresford. Producers: Robert Lantos, Stephane Reichel and Sue Milliken. Distributor: The Samuel Goldwyn Company.

A look at the travels of a French Jesuit missionary in 17th-century Canada, and the confrontations and experiences he has with the indigenous people of the land.

Clearcut, 98 min., 1991, Canadian. Director: Richard Bugajski. Distributor: Northern Arts Entertainment.

"Clearcut" features Graham Greene (star of "Dances with Wolves" and "Thunderheart") as a Native American trickster spirit who leads a land rights attorney on a vision quest through the Canadian forest. Native American groups have praised the movie for its accurate depiction of the battle between loggers and Natives.

Chemical Valley, 58 min., 1991, US. A film by Mimi Pickering and Anne Johnson. Producers: Appalshop Inc., (606) 633-0180.

A profile of the town of Institute, West Virginia, site of a giant Union Carbide plant—the only plant still operating that produces the same gas responsible for killing 8,000 people and the permanent disabling of more than 200,000 people in Bhopal, India. The film serves as a testament for people struggling to make industries responsible for the poisons they produce.

Cuberoid, 10 min., 1991, USSR. Director: Konstantin Bronzit. Producer: Ada Staviskaya. A production of Panorama Studio, Leningrad. Distributor: Guild Television Ltd., London, (71) 323-521.

This abstract, animated short is a wonderful metaphor for the interconnectedness of life on this planet and the inherent greed of humans who have destroyed our environment.

Daughters of the Dust, 113 min., 1992, US. Director: Julie Dash. Producer: Julie Dash and Arthur Jafa.

 What percentage of water used in American homes is flushed down the toilet?

ARTS & ENTERTAINMENT

Distributor: Kino International, (212) 629-6880.

Focusing on a family of Gullahs, descendants of West Africans who lived off the Georgia/South Carolina coast around the turn of the century, this film looks at the spiritual conflicts among the generations as they struggle to maintain their ancestral identity, but find that task difficult when migration to the mainland becomes inevitable.

Deadly Deception: General Electric, Nuclear Weapons and Our Environment, 29 min., 1991, US. Oscar Winner for Best Documentary, Short Subject. Director/Producer: Debra Chasnoff. Distributor: INFACT, (617) 742-4583.

This shocking exposé of shady General Electric (GE) practices concerning the production of nuclear weapons made a big splash on Oscar night. The film charges GE with failing to clean up the Hanford Nuclear Reservation site in Washington state, as well as knowingly contaminating employees with cancer- and birth-defect-causing radiation and asbestos at a Schenectady, New York, site.

The Death that Creeps from the Earth, 33 min., 1991, Germany. Producer: Michael Wenning. A production of the World Uranium Hearing in cooperation with The Heinrich Boll Foundation.

A compelling look at greed and its devastating impact on indigenous people all over the world. Activities such as uranium mining, underground nuclear testing and overall injustice are shown to have negative impacts on indigenous people worldwide.

The Drain, 5 min., 1991, US. A film by Matthew F. Leonetti, Jr., (213) 459-5426.

Fifteen-year-old surfer Leonetti is upset about the water pollution at his favorite surfing spots after the first "big flush" through the sewers in Santa Monica Bay during rains the previous winter.

Ferngully: The Last Rainforest, 72 min., 1992, US. Director: Bill Kroyer. Producers: Wayne Young and Peter Faiman. Distributor: 20th Century Fox.

This conservation message to youngsters attracted an all-star cast, including Christian Slater, Cheech Marin, Tommy Chong, Tone-Loc and Robin Williams as the voices of Ferngully's characters. The animated

Public Broadcasting System (PBS)
PBS has had a slew of excellent eco-programming in the past, with Nova and "National Wildlife" specials winning awards. Its commitment continued with several new series that debuted in 1992, including: the "Nature" series; "Millennium: Tribal Wisdom and the Modern World," a ten-hour look at indigenous people and issues in their lives; and eight National Audubon specials, tackling such issues as water and land pollution, elephant endangerment in the Congo, and a look at ecotourism in Kenya.

PBS offers programming for the younger viewer as well, continuing its award-winning presentations of "3-2-1 Contact Extras," supplemental shows to its regular "3-2-1 Contact" science news magazine for kids.

Turner Broadcasting System (TBS)
TBS has continued its programming of "Network Earth," plus the popular children's eco-crime fighting show, "Captain Planet and the Planeteers." In addition, TBS aired the "National Geographic Explorer" series, "One Child, One Voice," chronicling seven children from around the world and their hopes and fears for their future, and many other specials.

The Discovery Channel
This network was named "the most environmentally committed cable network" in a 1991

: *33 percent*
Source: Protect Our Planet Calendar, Running Press Book Publishers

survey of American cablewatchers by *Green Action Monitor*. The Discovery Channel has broadcasted features such as "In the Company of Whales," a look at the intelligence and grace of one of Earth's most intriguing species. Discover also regularly airs "Rediscovering America," which features environmental episodes, "Those Incredible Animals," "Wildlife Journey" and more.

The Winners

LOS ANGELES—The Environmental Media Association announced the finalists in each of eight categories for the organization's Second Annual Environmental Media Awards. This year's awards were sponsored by Georges Marciano and Guess? Inc.

Listed are the winners (in italic) and runners-up for the 1992 Environmental Media Awards.

Television Episodic Drama
L.A. Law, "I'm Ready For My Close-Up, Mr. Markowitz" (20th Television—NBC)

The Trials of Rosie O'Neil, "Environmental Robin Hood" (Rosenzweig Productions—CBS)

The Young Indiana Jones Chronicles, "British East Africa, 1909" (Paramount Television/Lucasfilm Ltd.—ABC)

Television Episodic Comedy
Dinosaurs, "Endangered Species" (Disney

film gives children a look at the perils facing rainforests today.

Final Test for the Earth, 15 min., 1991, US. Producer/Director: David L. Brown. Distributor: Ennergone Films, (415) 468-7469.

Focusing on the Semipalatinsk movement in the former Soviet Union, this is a heartening story of the Nevada-Semipalatinsk movement to stop nuclear testing.

From the Heart of the World: An Elder Brothers Warning, 87 min., 1990, UK/Colombia. A film by Alan Ereira. A production of the BBC. Distributor: Mystic Fire Video, (212) 941-0999.

One of the Earth's last ancient tribes, the Kogi of Colombia protect themselves from the modern world by isolation, secrecy and suspicion of outsiders. Despite four centuries of isolation, they let the BBC's Alan Ereira film them for one reason: to warn the rest of us "who only understand machines, who rip and tear at the body of the Earth," to stop damaging the planet or we will all "perish."

Incident at Oglala, 86 min., 1992, US. Director: Michael Apted. Producer: Arthur Chobanian. Distributor: Spanish Fork Motion Picture Company.

This docu-drama examines the events leading up to the controversial 1977 conviction of Native American activist Leonard Peltier for the murder of two FBI agents on the South Dakota Pine Ridge Indian Reservation.

The Last of the Mohicans, 110 min., 1992, US. Director: Michael Mann. Distributor: 20th Century Fox.

A thrilling recreation of 18th-century America, when the Mohicans were fighting for their lives among the French and British. Native American activist Russell Means stars as Chingachgook, "the last of the Mohicans," and Daniel Day-Lewis portrays his adopted son, Hawkeye. Based on the James Fenimore Cooper novel of the same name.

The Lunatic, 90 min., 1992, Jamaica. Director: Lol Creme. 1880 Century Park E., Los Angeles, CA 90067, (310) 553-1707.

A Jamaican film that profiles the conflict a local

: *In the process of producing one beef steak, how many gallons of water are needed?*

village idiot has over whether to follow the comfortable life he knows of friendship with trees, or to venture into the vices and debauchery of the human world.

Magnetic Field, 26 min., 1991, US. Director: Mary Benjamin. Producer: E. Shain. Distributor: Mary Benjamin, (213) 578-2131.

An original drama about a young advertising executive who is both romantically involved with her boss and morally conflicted about her job. She is assigned to an electric company account, then learns about a potentially devastating ecological danger—electromagnetic waves. She must make a tough choice about where her values truly lie.

Medicine Man, 1992, US. Director: John McTiernan. Distributor: 20th Century Fox.

Shot in Mexico, this film casts Sean Connery and Lorraine Bracco as research scientists searching for the cure for cancer in the rainforest. One of the first major environmental films to draw well at the box office, "Medicine Man" combines beautiful scenery, an urgent environmental message and the romantic tension of Bracco and Connery.

Mindwalk, 110 min., 1992, France. Director: Bernt Capra. Producer: Andrianna Aj Cohen. Distributor: Triton Pictures, (310) 275-7779.

"Mindwalk" is a riveting, fictional conversation between a scientist, a poet and a politician that takes place as they roam the French isle of Mont St. Michel. The film brings modern physics to life.

Our Vanishing Forests, 60 min., 1992, US. Director/Producer: Arlen Slobodow. Distributor: Public Interest Video Network, (301) 656-7244.

This documentary examines what is happening to American forests, and finds that their worst enemy is none other than the government agency that is supposed to protect them, the US Forest Service.

At Play in the Fields of the Lord, 180 min., 1991, US. Director: Hector Babenco. Producer: Saul Zaentz. Distributor: Universal Pictures.

Peter Matthiessen's novel comes alive in Babenco's adaptation of the tragic effects of modern Americans

Television/Jim Henson Productions/Michael Jacobs Productions—ABC)

Dinosaurs, "Power Erupts" (Disney Television/Jim Henson Productions/Michael Jacobs Productions—ABC)

The Simpsons, "Mr. Lisa Goes to Washington" (Gracie Films/Fox—FBC)

Television Special
"Audubon: Battle for The Great Plains" (Turner Broadcasting—TBS)

"One Child, One Voice:" (Turner Broadcasting—TBS)

"Network Earth: Save the Earth Special: Personal Effects" (Turner Broadcasting—TBS)

Children's Programming/Animated
Captain Planet and the Planeteers, "Fare Thee Whale" (Turner Broadcasting/DIC Entertainment—TBS)

Captain Planet and the Planeteers, "Summit to Save Earth" (Turner Broadcasting/DIC Entertainment—TBS)

Widget, "Sort It Out" (Zodiac Entertainment—Syndication)

Children's Programming/Live Action
"50 Simple Things Kids Can Do to Save the Earth" (David Eagle Productions/Earthworks Group—CBS)

"Popular Little Planet/Get Busy: How Kids Can Save the Planet"

A: *2,607 gallons*
Source: 50 Simple Things You Can Do to Save the Earth, *Earth Works Press*

(Children's Television Workshop—PBS)

"The Day the Earth Threw Up" (Nickelodeon—PBS)

Daytime Programming
"The Bold and the Beautiful" episode #1103 (Bell-Phillips Television Productions—CBS)

"Days of Our Lives" episode #6742 (Corday Productions—NBC)

Feature Film
"FernGully: The Last Rainforest" (20th Century Fox)

"Medicine Man" (Cinergi Productions/Hollywood Pictures—Buena Vista)

"Mindwalk" (Triton Pictures)

"Thunderheart" (Tribeca Productions/Waterhorse Film Productions—Tri-Star)

Music Video
"Get Busy: Parts I & II" (Children's Television Workshop—PBS)

"Mercy, Mercy Me (The Ecology)," Marvin Gaye (Motown Record Company)

"Right Now," Van Halen (Warner Brothers Records)

"Saltwater," Julian Lennon (Atlantic Records)

on a fierce tribe of almost Stone-Age natives living deep in the Amazon jungle.

Raspad, 101 min., 1992, USSR. Director: Mikhail Belikov. Producer: Mikhail Kostiukovskii. Distributor: MK2 Productions, (212) 265-0453.

Director Belikov deftly weaves a dramatic film about the 1986 Chernobyl nuclear disaster. His ability to poignantly illustrate the true effects of the catastrophe through his fictional characters is superb. The film serves as a moving metaphor for the entire Soviet collapse.

A River Runs Through It, 123 min., 1992, US. Director: Robert Redford. Distributor: Columbia Pictures.

Based on the novel of the same name, by Norman Maclean, "A River Runs Through It" is an elegiac look at a turn-of-the-century Montana family in which "there was no clear line between religion and fly-fishing." Filmed on location in Montana, it celebrates the beauty of the rivers and mountains of the high plains.

Sun Bear: Vision of the Medicine Wheel, 90 min., 1992, US. Producer/Distributor: Four Directions Productions Company, P.O. Box 11C, Valley Head, AL 35989.

Profile of Sun Bear, of Ojibwa descent, and his quest for a spiritual ecology movement that involves walking in harmony with all forces of the land.

Thunderheart, 1992, US. Director: Michael Apted. Producers: Robert De Niro, Jane Rosenthal, John Fusco. Distributor: Tri-Star Pictures, (310) 280-1419.

This drama is based on Native American reservations' explosion into violence in the 1970s, after being subjected to countless environmental hazards by the government.

Tropical Rainforest, 39 min., 1992, US. Director: Ben Shedd. An IMAX film. Imax Corporation, (416) 960-8509.

This film, made in awesome 70-millimeter IMAX process, chronicles the rainforest's 400-million-year development and its state today.

Wallace Black Elk/Vision of the Sacred Pipe, 60 min., 1992, US. Producer/Distributor: Four Directions

: *For many American schools, what expenditure is at least equal to the amount of money spent on textbooks?*

Productions Company, P.O. Box 11C, Valley Head, AL 35989.

In this program, Wallace Black Elk, the Lakota Elder from South Dakota, takes the viewer into the most sacred of the ceremonies of the *Cannupa* (sacred pipe), and into the *Inipi* (sweat lodge). He shares insight into the Earth's current changes, as well as the plight of future generations.

What Your Doctor Won't Tell You About Cancer, 78 min., 1991, US. A film by Chuck and Lisa Wintner. Distributor: Malibu Video, (213) 456-5186.

A controversial exposé of the environmental causes of cancer. The film also offers prevention and alternative cancer care techniques.

ECOLINKING

- By Don Rittner -

DURING MOST OF THE DAY AND NIGHT when people are going about their normal routine, more than 40 million people are communicating in a special way—by using their personal computers. Writers send their copy to editors; environmental groups post calls to action; business plans are sent across the country in seconds; and literally millions of scientists, students and professionals around the world carry on dialogue and debate. What is amazing is that all this takes place without one person leaving his or her office or home.

You too can become part of this exciting online global community. All you need is access to a personal computer, a modem, communications software and a phone line. According to the latest US census figures, more than 75 million Americans own a personal computer, and almost one-quarter of them have a modem, an inexpensive device that lets your computer send information over phone lines. With personal computer prices constantly dropping, this figure is likely to rise rapidly. You may already be ready to EcoLink!

There is a tremendous amount of information residing on mainframe and personal computers, and, it is available to the worldwide environmental community through networks. Perhaps the most important asset is linking up with people who share your interests in preserving the environment no matter where they live, be it Albany, New York, or across the ocean in Zambia.

Computer Software

There is a lot of environmentally focused software available for different kinds of computers; we highlight some that may interest you.

Global Recall (Macintosh)
Available from the World Game Institute, University City Science Center, 3508 Market St., Philadelphia, PA 19104, (215) 387-0220. Price: $85.

America faces a major "awareness" problem as 1992 sees the 500th anniversary of Columbus's "discovery" of America, the birth of the European Economic Community and the International Space Year. Our educational failure when it comes to geography makes it no surprise that many students think Wyoming is another country.

The World Game Institute wants to change that and has produced an interactive geography program, using Hypercard, called Global Recall for the Apple Macintosh computer. Its purpose is to make learning geography and facts about the world easy and interesting. It succeeds at both.

Earthquest Explores Ecology (Macintosh or DOS)
Available from Earthquest, Inc., 125 University Ave., Palo Alto, CA 94301, (415) 321-5838. Price: $59.95.

A great teaching program. Features some 15

A: *Trash disposal*
Source: The Denver Post *1991 Colorado Recycling Guide*

interactive games, six biosphere simulations and dozens of ecology topics. Uses Hypercard, which is included. For ages 10 to adult.

Earthquest (Macintosh or DOS)
Available from Earthquest, Inc., 125 University Ave., Palo Alto, CA 94301, (415) 321-5838. Price: $59.95.

Another great teaching aid, Earthquest takes you on a journey of discovery of the Earth, its people and the environment. With a multimedia presentation of maps, charts, graphics, animation, music and more, users learn about the world.

World Resources 1992-93 Data Base Diskette (DOS)
Available from WRI Publications, P.O. Box 4852, Hampden Station, Baltimore, MD 21211, (800) 822-0504. Price: $119.95 plus $3 shipping.

This resource provides quick access to authoritative information on virtually any issue involving the world's natural resources. It contains figures and statistics and enables users to search, manipulate, chart and download data. A summary of the disk is published in *World Resources 1992-93*, which is provided at no extra cost when the disk is ordered.

SimEarth, SimAnt, SimLife (Macintosh or DOS)
Available from Maxis, 2 Theatre Square, Ste. 230, Orinda, CA 94563-3346, (510) 254-9700.

Let's look at a small sample of what you can get from some of these resources:

Global Networks

The Internet. Sponsored by the US government, the Internet is open in part to the public. The Internet is actually a few thousand separate computer networks that talk to each other. Thousands of the world's leading scientists are on the Internet, and it spans more than 120 countries. You can send private mail across the world in seconds. Discussion lists on a number of environmental and scientific topics are open for free membership. Here you can debate the issues with the best minds around the world. In addition to private mail and discussion lists, you have access to some of the best library card catalogs, databases and bulletin boards around the world. Once on the Internet, you can exchange mail and resources with people on Bitnet, Usenet and Fidonet (three other global networks), and many of the commercial online services.

Bitnet. Bitnet is mostly an academic network and is similar to the Internet. You can send private mail, participate in environmental discussions and retrieve documents from areas called servers.

Usenet. Usenet is a series of discussion groups, often carried by the other networks, and many environmental topics are discussed.

Fidonet. Fidonet is an amateur computer network, similar to the ham radio network, in which more than 10,000 private bulletin boards share mail and discussion groups, called echos, with each other. This system works remarkably well and now spans more than 40 countries, including the former Soviet Union. There probably is a Fido BBS (see "Bulletin Boards," page 211) in your city.

Online Services

Commercial online services usually charge an hourly fee to access their systems. There is a tremendous amount of resources you can get, and you'll be meeting up with people who share your interests.

American Online. This Virginia-based company is the most user-friendly of them all. American Online

Q: By recycling a daily newspaper every day, how many trees could just one person save every year?

ARTS & ENTERTAINMENT

provides its telecom software free for DOS, Macintosh and Apple II platforms. Moreover, it uses a graphical user interface. A series of icons and buttons makes finding information on American Online painless.

CompuServe. The granddaddy of online services, CompuServe has been around since 1979 and provides many services.

Network Earth Forum. If you are a fan of TBS' "Network Earth" TV show on Sunday nights, you will love the Network Earth Forum on CompuServe. You can discuss each episode with the producers of this great environmental show immediately after watching it. Also, you can suggest ideas for new shows, participate in debates on a variety of topics and download files.

Econet. This nonprofit network, part of the Institute for Global Communications, devotes itself solely to environmental issues. Hundreds of environmental organizations post announcements, calls to action, requests for letter writing, discussions and even their newsletters for all to read. Econet has gateways to other networks so you can send mail to people on the Internet, American Online, CompuServe and more than a dozen others.

Bulletin Boards

A computer Bulletin Board, commonly called a BBS, is found in almost every city in the US and throughout the world. Here, one individual creates an electronic storehouse of information on any subject imaginable. There are many bulletin boards devoted to environmental issues, and all carry hundreds of downloadable files, reports, articles, news and discussions. Most are free to access. Those that participate in the Fidonet link you with more than 10,000 bulletin boards around the world.

Online Libraries

You can prepare research bibliographies in minutes instead of having to thumb through hundreds of indexes at your local library. Bibliographical retrieval companies like Dialog and BRS offer the computer user an inexpensive set of databases that cover science, environment and technology. An average search takes 10 minutes (compared to weeks at the

Price: $59.95 to $69.95.

These games, from the makers of SimCity, offer users the chance to control each universe with an extensive set of manipulable variables. In SimEarth, you create a world and monitor climate, types of life forms, continental drift, biomes, energy usage, civilization and more. The game is based on the Gaia theory. In SimAnt, you, the black ants, battle the red ants and take over the house. Watch out for insecticides. SimAnt is based on *The Ants* by Harvard biologists E.O. Wilson and Bert Hölldobler. SimLife, Maxis' newest game, lets you control the genetics of plant and animal life. You monitor the world's soil, humidity, temperature, genetics and many other variables. The trick is to create sustainable life.

CD-ROM

Environmental Data Disc (Macintosh) Available from PEMD Education Group, P.O. Box 39, Cloverdale, CA 95425, (707) 894-3668; Fax (707) 894-5200. Price: $149.

Imagine having thousands of pages of raw environmental data at your fingertips and being able to analyze and study them at the click of a button! This CD-ROM contains over 125 megabytes of data, including temperature and precipitation data for over 800 US stations; worldwide food, agriculture and demographic data; energy, economic and trade data;

Source: The Denver Post *1991 Colorado Recycling Guide*

atmospheric ozone data for over 100 stations; general environmental data from the World Resources Institute (1990-91) *Guide to the Global Environment*; and more. The Environmental Data Disc is a valuable analysis tool that all scientists, researchers, teachers and students working in the environmental community should have.

The data is compiled from the World Resources Institute, Carbon Dioxide Information Analysis Center, Food and Agriculture Organization of the United Nations, Organization for Economic Cooperation and Development, World Health Organization and World Ozone Data Center.

GAIA Environmental Resource (Macintosh) Available from Wayzata Technology, Inc., P.O. Box 807, Grand Rapids, MN 55744, (800) 735-7321. Price $249.

Environmental organizations often need a good assortment of illustrations to use in their newsletters and other publications. Artist Josepha Haveman has compiled GAIA Environmental Resource, a collection of hundreds of full-color and black-and-white illustrations with environmental themes, and more.

Haveman has scanned in more than 500 photographs/illustrations and created a database of other useful environmental information, combining it all on one CD-ROM disc.

library). Online searching is available 24 hours a day, seven days a week.

This is just a small sample of the environmental resources available to you if you use a personal computer. There is no doubt that we live today in a global village. The tools are there to communicate with each other: to build bridges of understanding and cooperation regardless of race or culture. Joining this community is as simple as turning on your computer.

ARE YOU COMPUTER "GREEN"?

- By Don Rittner -

MORE THAN 75 MILLION AMERICANS own a personal computer, and this number is rapidly climbing as computers become more affordable. Many more use computers in their workplace (more than 40 million Intel-based PCs and 7 million laser printers use 18.2 billion kilowatt/hours of electricity per year).

Most people think of computers as relatively pollution free, but the *act* of computing is not. Here are a few tips to help make your computing more environmentally friendly.

Electronic Mail

If your office does not have your computers networked together, do it. The use of electronic mail for interoffice correspondence can save a tremendous amount of paper. American offices last year generated more than 775 billion pages of paper—that's 14 million tons of paper per year, or 238 million trees.

Printers

If you use a dot matrix or laser printer, there are a few things you can do. First, be sure to use recycled paper (and envelopes) and remember to use the back side of sheets that you print as drafts. There is nothing wrong with using the second side of the sheet. This cuts your consumption of paper in half. Proof your work before you print. Most wasted paper results from typos.

If you use cloth-type ribbons in your dot matrix printer, you can re-ink those ribbons. In fact, you can get up to 15 to 20 re-inks per ribbon, and the print is usually darker than that from newer ribbons. This

 Of the 47 chemical plants ranked highest in carcinogenic emissions by the EPA, how many of them are involved in plastic production?

also reduces the cost per ribbon. Many computer user groups have re-inkers and charge $1 to re-ink (versus $5 to $15 per new ribbon).

Toner Cartridge Recycling
For laser printer users, many toner cartridge manufacturers now recycle those cartridges and donate money to environmental organizations. Some pay you! They pay for the UPS shipping, too. Also, there are companies that will recharge your toner cartridge for considerably less than the cost of a new one ($40 compared to $90). Considering that more than 98 percent of the 15 million cartridges sold in 1991 ended up in landfills, you can see how important it is to recycle them. Below are some companies that recycle toner cartridges.

Apple Clean Earth Campaign, (800) 776-2333. Donates 50 cents to National Wildlife Federation and Nature Conservancy per cartridge. Call them, and they will send you a prepaid UPS shipping label.

Canon Clean Earth Campaign, (800) 962-2708. Canon has the same deal as Apple.

Dataproducts Imaging Supplies Division, (800) 423-5095. Dataproducts will pay you $10 for each Canon SX cartridge, plus pay for shipping if you send 28 or more cartridges at a time.

Lexmark Operation Resource, (800) 848-9894. Recycles the six IBM laser printer models in the 4019 and 4029 series. Will send you a postage-paid container. They give the returned cartridges to a workshop for the handicapped that makes money by selling the parts to recycling companies.

Computer Magazines
Don't throw away those already-read computer magazines. You can recycle those, too. Donate them to your local public library, user groups, doctor's office, health clubs, even laundromats.

Don Rittner is the author of EcoLinking—Everyone's Guide to Online Environmental Information, *Peachpit Press, Berkeley, CA, 1992, (800) 283-9444. You can reach Rittner via Internet mail at AFLDONR @AOL.COM.*

A: *35*
Source: Greenpeace

ART

Relief Globe Slides
(Macintosh or DOS) Available from NOAA/National Geophysical Data Center, Code #/GC4, Dept. 880, 325 Broadway, Boulder, CO 80303, (303) 497-6338. Price $45.

Need a good "solid" picture of the Earth? The National Geophysical Data Center has published a set of 20 slides, using digital data generated from a database of land and sea-floor elevations based on a five-minute latitude/longitude grid. The images generated from this data comprise a stunning series of 14 global views of the Earth, as well as a world view with earthquake epicenters, crustal plate boundaries, the famous "Pacific Ring of Fire" earthquake zone and a close-up view of plates and epicenters in the Americas. The images are in full color, shaded relief, and show both land and undersea topography (no water).
—*Don Rittner*

Theater

Robert Schenkkan's **"The Kentucky Cycle"** was one of the major theatrical productions of the year, winning the 1992 Pulitzer Prize for best play. It was a box office success in Seattle and Los Angeles. The play dramatizes the destitute lives of eastern Kentucky hill people between 1775 and 1975, and specifically focuses on "the devastation of both inhabitants and landscape, poisoned and ravaged by generations of greed." "The Kentucky Cycle" consists of nine one-act plays, each of which explores important "turning-point episodes" in the lives of seven generations of an Irish immigrant family living on a small plot of "dark and bloody" Kentucky land.

Running during the Earth Summit in Rio, **"A Midsummer-Night's Dream in the Amazon Forest"** took William Shakespeare's work and set it in the contemporary Brazilian rainforest. The play opens with a documentary film, directed by German filmmaker Werner Herzog, also the play's director, which focuses on the ecological devastation of Kuwait. Also running during the Rio Summit, was **"Forest of the Amazon"** a ballet that tells the tale of a goddess who falls in love with a rubber tapper. This, in turn, angers both the tribe that worships the goddess and the forest itself. This "poetic

Arts

Visual and performing arts often reflect the society in which they develop. The emergence of environmentally themed art is a natural step in the evolution of art. Earth Journal highlights some of the most exciting environmental artists and their work.

PAINTING

Joan Giordano is an artist living in New York City who focuses most of her creative efforts on environmental projects. Giordano was born and raised on Staten Island and studied at the Museum of Modern Art. She received her MFA from Pratt Institute and has since exhibited at museums and galleries in New York and Paris. Giordano's present undertaking is the formation of "Urban Artists for the Earth" (UAE), a group of urban artists who hope to raise the consciousness of New Yorkers to the importance of urban environmental issues. UAE has recently created a mixed-media collage of environmental art that was donated to the New York chapter of the Audubon Society. It will be auctioned with the proceeds going to protect the rainforest. UAE plans to sponsor bimonthly projects to address urban environmental issues.

Renowned artist **Robert Rauschenberg** has painted "Last Turn, Your Turn," the official poster to promote the Earth Pledge. The Earth Pledge—"I pledge to make the earth a secure and hospitable home for present and future generations"—was launched by the United Nations Conference on Environment and Development and is intended to raise environmental consciousness worldwide.

A tremendous and brilliant **mural on the corner of Pico and Union streets** in Los Angeles strives to warn passersby of the horrific and immediate threats to our ecological balance. The mural combines modern graffiti style with colonial Mexican church decoration, and is further influenced by traditional Mexican *pulque* art. The project celebrates Earth Day and was created by an ecumenical consortium of crews.

Carolyn Lankford paints to capture the soul of a place, to communicate its spirit so that the viewer

 How many batteries are thrown away each year by Americans?

ARTS & ENTERTAINMENT

A recent painting by Raleigh Kinney.

feels as if he or she knows the place as intimately as Lankford does. Although Southwestern landscapes are her first love, all of her works reflect her love of color and texture. Lankford graduated summa cum laude from the University of New Mexico and was heading for a career in research and writing until she realized her intense desire to communicate through painting.

Raleigh Kinney, formerly a public school teacher for 15 years, is a watercolor artist of national recognition. His experiences as a painter span more than 20 years, and his works can be found in collections across the country. The founding member of the Midwest Watercolor Society, Kinney focuses much of his work on the mystical aura of the past generations of the American Southwest. His earthy hues and soft, flowing shapes lend a sense of magnificence to his paintings and their subjects.

ECO-PHOTOGRAPHY

There is a saying: "The only requirement for a good photo is that it be memorable." The photographers listed below have taken memorable photos of our environment—and if we cannot forget the photos, we cannot forget the endangered places, people and animals in the photos.

James Balog of Boulder, Colorado, has gained international attention for his striking photos of wild animals posed against stark studio backdrops, collected in his book, *Survivors: A New Vision of Endangered Wildlife*. Balog says: "I have no desire to perpetuate the

fantasy" of a work, as described by choreographer Dalal Achar, is set to music by Brazilian composer Heitor Villa-Lobos.

"The Christopher Columbus Follies: An Eco-Cabaret," performed by the Underground Railway Theater of Arlington, Massachusetts, presents a musical tour of song and dance routines, satirical skits and puppetry. It focuses on the legacy of Christopher Columbus and his impact on Native Americans. The touring company, which has been performing for more than 15 years, engages audiences worldwide with its dramatic representation of Columbus's impact on the Americas.

Sculpture

Martin Puryear, at age 50, is considered a master sculptor. His works of cedar, oak, poplar, ash and hickory have "restored an unfamiliar warmth to the look of current art," according to *The Washington Post*. The winner of a MacArthur Foundation "genius" prize, a Guggenheim Foundation grant and a sojourn at the American Academy in Rome, Puryear occupies a secure and prominent position at the forefront of modern sculpture. He brings an exciting new vision to the minimalist movement. His sculptures are contemporary in that they draw from simple forms yet are crafted with perfection and expertise. His

A: *2.5 billion*
Source: The Denver Post *1991 Colorado Recycling Guide*

intellectual depth, as expressed through his sculptures, has placed Puryear in limited company. If you have the chance to see it, his work is not to be missed.

William Wareham, the first Artist in Residence sponsored by Norcal Waste Systems' "Make Art, Not Landfill" program, uses discarded car bumpers, steel springs, sheet metal, appliance parts and a host of other so-called "garbage" items in his sculptures. For his Norcal residency, Wareham created over 40 works of art, ranging in size from just a few inches to 50 feet. Wareham is primarily interested in geometric abstractions. He maintains a studio in San Francisco.

Mel Chin is an eco-sculptor with a unique vision. Chin is a traditional sculptor by trade but has recently begun to bring his efforts down to Earth, literally. Chin, backed by a $10,000 grant from the National Endowment for the Arts, has worked on projects such as "Revival Field" in St. Paul's Pig's Eye Landfill. "Revival Field" consists of an enclosed garden that cleans contaminated land and groundwater sources by using hyper-accumulators, plants that extract heavy metals from the earth. Following the Pig's Eye project, Chin traveled to the Netherlands to demonstrate the power of hyper-accumulators to the Dutch.

An image from an upcoming book of photography by James Balog.

romantic mirages of traditional wildlife photography. Instead, I have created images of animals in exile from a lost Eden." His work has been featured in *National Geographic*, *Buzzworm* and many other magazines.

Peter Beard is an American who lives in Nairobi, Kenya, and is known for his incredible photos of Africa. His many books include *Eyelids of Morning*, *End of the Game* and *Longing for Darkness*.

Jim Brandenburg is a nature and wildlife photographer perhaps best known for his book of Arctic wolf photos, *White Wolf: Living with an Arctic Legend*. His early intent was to be a wildlife biologist, but he says he is glad he became a photographer instead, because "I think I can actually make much more of a difference in this world by reacting to something emotionally and sharing that emotion with others. I really believe you have to come in through the back door if you are going to change peoples' attitudes toward the land and toward nature; you have to do it in a spiritual way rather than a factual way. We certainly have all the facts we need."

Gary Braasch, based in Portland, Oregon, has published photographs in *National Geographic*, *Life*, *Audubon*, *Orion*, *Geo*, *Oceans*, *Nature*, *Omni*,

Q: *Which developing country has initiated a plan to derive 5,000 megawatts of energy from the wind by the year 2000?*

ARTS & ENTERTAINMENT

Outside and many more magazines. He specializes in landscapes, aerial and science photos.

Marilyn Bridges received her MFA in photography from the Rochester Institute of Photography and is the winner of a Guggenheim Fellowship. Her most recent book, *Planet Peru*, is an aerial photographic journey through Peru.

Carr Clifton is a native Californian who currently lives in the northeastern Sierra Nevada. He was the primary photographer for the Sierra Club book *Wild By Law*, and his work is featured in *California: Magnificent Wilderness*, *The Hudson: The River That Flows Two Ways* and *New York: Images of the Landscape*.

W. Perry Conway, based in Louisville, Colorado, became interested in nature photography as a boy growing up on the Kansas prairie. His work has been featured in *Audubon, Buzzworm, National Geographic, National Wildlife* and several books.

Stephen J. Krasemann is a wildlife and environmental photgrapher based in Sedona, Arizona, who has worked for *National Geographic* and the Nature Conservancy. He was the still photographer on the Disney feature film *Never Cry Wolf*.

Frans Lanting's work has appeared in many magazines, including *Audubon, Buzzworm, Equinox, Natural History, Smithsonian* and *Travel & Leisure*. He says, "In a way, photographers are responsible for painting too rosy a picture. I try to make up for that by on the one hand photographing the beauty, and on the other hand documenting what really goes on."

David Muench is a well-known nature photographer who believes that "words are not as eloquent as seeing through images." Muench says: "There is, it seems, hardly any place where a human hand has not touched—or photographed. The idea of nature may become extinct!"

Galen Rowell is a leading nature photographer whose books include *Mountain Light, The Yosemite* (text by John Muir) and *My Tibet* (text by the Dalai Lama).

Seattle artist **Pam Beyette** is a modern alchemist, turning industrial junk (discarded wiring, trashed circuit boards, rusting machine parts) into powerful art. She mines copper, aluminum, bronze and brass from trash heaps and combines them into gleaming structures. Microsoft Corp. recently bought her "Industrial Medallion" (fashioned from computer circuit boards and electronic components), and AT&T Gateway Tower in Seattle commissioned her "Industrial Amulets" (made with cable and foil). Beyette says she enjoys sifting through "industrial relics" and "contemporary artifacts" discarded by factories and corporations.

Chester Fields is a painter and sculptor living in Veradale, Washington. His works vary from extraordinary and lifelike portraits of Native Americans to exquisite, enormous bronze casts. Fields has gained wide recognition and praise as he has successfully diversified his media. With many of his subjects remanded to the endangered species list, Fields's sculptures and paintings may be among the only forms by which future generations will know the glorious creatures of our planet.

Choctaw artist **Randall Chitto** speaks Chatta, the language of his native tribe, through his art. His inventive and popular clay turtles speak the legends that

A: *India*
Source: Union of Concerned Scientists

have been passed down to him through the generations, and Chitto strives to ensure that these tales do not vanish as many of his people have. His turtles embody the spirit, the traditions and the tribal beliefs of his people. Chitto loves to work with clay, and he speaks of the supernatural experiences he often has as he molds legends of the past into modern artwork. "These creatures can tell you many things," he says.

Cloud Eagle, a.k.a. Ernest Mirabal, is a Nambe Pueblo stone carver who devotes much of his creative energy to discovering and furthering his sculptural heritage. Cloud Eagle is a Tewa; his people came to the Pajarito Plateau during the 1200s as part of the massive Anasazi migration. The petroglyphs etched into the black basalt by his ancestors captured Cloud Eagle's imagination and inspired him. After graduating from the Institute of American Indian Art in Santa Fe, New Mexico, Cloud Eagle decided to combine his knowledge of stone carving with the spiritual heritage that inspired him so deeply within the pueblo. His sculptures strive to capture the traditional spiritual energy so vital to his people and to ensure that these traditions continue to inspire others.

Kevin Schafer of Seattle specializes in tropical rainforest, Central America, deforestation and wildlife photos. He has published in *National Geographic, Natural History, Intrepid* and *Outdoor Photographer*, as well as other magazines.

Sebastiao Salgado is famous for his striking black-and-white photos of poor children, families, miners, workers—and their natural and unnatural surroundings—in third world countries.

Wendy Shattil and Bob Rozinski are a photography team specializing in nature. They are based in Denver, and their photos of birds, flowers, geology, glaciers, animals, weather and more can be found in environmental and nature magazines.

Tom Till, based in Moab, Utah, is a landscape photographer whose work has been featured in *Arizona Highways, Audubon, Outdoor Photography, Outside* and many other magazines.

Stephen Trimble lives in Salt Lake City, Utah. He specializes in photos of Western Native Americans and their lands. His credits include *Audubon, Buzzworm, Natural History, Outside, Sierra* and *Wilderness* magazines. His latest book is *The People: Indians of the American Southwest.*

Larry Ulrich, based in Trinidad, California, photographs wild American animals and landscapes. His credits include *Arizona Highways, Audubon, Natural History, Outside* and *Sierra*. He is the author/photographer of *Oregon* and *Arizona: Magnificent Wilderness.*

Art Wolfe of Seattle has published photographs in *Backpacker, Esquire, National Geographic, Natural History, Outside, Smithsonian* and many other magazines. He photographs birds, wildlife and landscapes all over the world.

Hiroaki Yamashita is a leading Japanese nature photographer who has published a book of photos of Yakushima Island called *Ancient Grace: Inside the Cedar Sanctuary of Yakushima Island.*

: *How much hazardous waste is produced in the US every day?*

ARTS & ENTERTAINMENT

Fashion

Being fashionable in the 1990s requires more than keeping up with the latest trends—it means using less, buying consciously and making do with what you have. It also means supporting companies that strive to make a difference.

FASHION GROWS GREEN

STYLE IS NO LONGER THE ONLY FACTOR in deciding what to pull on in the morning. The chic, sophisticated,

This newspaper advertisement by Chanel stimulates environmental awareness through irony.

Beauty: Is It Up in the Air?

It may seem ironic to some and logical to others that Los Angeles, a city long obsessed with beauty, has a terrible smog problem. Do more people in LA use hair spray than in other cities? That may be the case. Every time a starlet gives her hair a spray, she adds volatile organic compounds (VOCs) to the air. When VOCs and other pollutants like car fumes mix, smog results.

Hair spray companies have rushed to produce hair sprays with reduced VOCs, responding to new clean-air regulations in California and the probability of legislation in other states. Redken's Clean Air line includes a super-holding spray, and Vidal Sassoon now makes Air Spray, which is propelled by pumped-in air.

To make matters worse, you may have to give up (or at least feel guilty about) wearing panty hose. Researchers at the University of California at San Diego say that panty hose are indirectly to blame for depleting the ozone layer. They say that adipic acid, used in making nylon, emits nitrous oxide, a chemical harmful to the ozone layer. Panty hose producers are trying to eliminate adipic acid emissions during the manufacturing process.

Environmental Armor

It's an environmental war zone out there. Experts tell us that there is a new

A: *About 700,000 tons*
Source: Protect Our Planet Calendar, Running Press Book Publishers

environmental threat lurking in dark movie theaters, electronics stores, even in our own living rooms and kitchens. While lead- and PCB-contaminated seafood, pesticide-poisoned vegetables and damaging ultraviolet rays are still a problem, we are also being bombarded with harmful electromagnetic fields (EMFs).

Yet there is hope. Some of the latest fashions double as environmental "chain mail." An Italian textile manufacturer, Lineapiu, has created a fiber that protects the body against EMFs. The yarn, called Relax, is a blend of rayon and carbon fiber that partially reflects and absorbs EMFs. Apparently, the material reduces exposure to EMFs emitted by televisions and other household appliances between 0.1 megahertz and 300 megahertz but is ineffective against EMFs produced by high-frequency microwave ovens and power lines. First used in space suits and industrial protective clothing, Relax is now being exploited by French designer Azzedine Alaia, who says that it shields against environmental attack while bringing psychological peace to the wearer. Other designers are expected to jump on the bandwagon and use the fiber in their 1993 collections.

Save the Earth, One Button at a Time

Just think, if everyone bought two tagua nut

eco-conscious consumer of the 1990s ponders how her or his choice of clothing will affect overflowing landfills, poisoned water supplies and rainforests. Growing environmental awareness among consumers has forced changes in almost all segments of business. Many companies have jumped on the bandwagon, making dubious claims about their products' "environmental friendliness." The fashion industry has not hesitated to exploit the new eco-awareness. According to *Green MarketAlert,* a bulletin that analyzes the impacts of "green" consumerism on business, green products, which include health and beauty aids, brought in $26.1 billion in 1991, and that figure is expected to grow by 30 percent each year. Everything from hair-care products to fragrances to sportswear reflect environmental concerns, some more genuinely than others.

Eau de Ocean

Perfume makers around the world are creating fragrances based on ocean elements, and some of the major perfume houses comb marine environments (such as the sea fields off the northern coast of Brittany or Acadia National Park in Maine) for the essence of hundreds of species of algae and sea plants. The plants are brought back to labs, where they are analyzed and chemically synthesized into fragrances. Paloma Picasso describes her new cologne for men, Minotaure, as something that "might have washed up on a beach." Safari by Ralph Lauren contains notes of the sequoia tree as well as seaside wildflowers. Environmental concerns have also encouraged major perfumeries to leave alcohol out of their fragrances. Ten years ago, Giorgio Beverly Hills, an immediate success with its high concentration of bold scents like jasmine, rose and gardenia, was the epitome of the excess of the 1980s. Ten years later, in 1992, fragrances have turned to a more subtle, natural, fresh, "larger than life outdoors" spirit.

Frugality, Practicality, Versatality!

Many new styles reflect a turn away from the excess and glitz of the 1980s. Frugality, practicality and versatility have become new words in the fashion vocabulary. A New York marketing research company, Yankelovich Monitor, explains that people are embarrassed and even ashamed of the amount they spent in the late 1980s. The firm explains the new tendency as

 Q: *Wood means what to over 2 billion people worldwide?*

a transformation from the "shop till you drop" mentality to a "drop shopping" mentality. As a result, there is a return to classic, less expensive and more durable clothes that offer the consumer more versatility and flexibility. This "anti-status" trend, combined with the effects of the environmental movement, has had a major impact on the fashion industry. Fashion houses such as Emanuel Ungaro and Giorgio Armani are creating "secondary" lines that boast price tags 30 percent to 40 percent lower than their "signature" lines. Businessmen will be wearing down parkas and field jackets, called "rough wear" by Ralph Lauren, over business clothes for work. While designers like Claude Montana and Nino Cerruti have mixed expedition gear with sport coats and slacks for the last few years, outdoor companies like Patagonia, Eddie Bauer and Timberland have found a new market niche in the cities of the Northeast and expect to outnumber designer labels in sales. The new look expresses a person who is equally comfortable riding the elevator to the 18th floor of a Manhattan office building and climbing Mt. McKinley—someone who will not bow to Seventh Avenue's every fashion whim.

Fashion's Green Thumb

Perhaps the new cultivation of "garden style" carries the emphasis on practicality and durability (though maybe not frugality) to its logical end. With the environmental movement has come a new appreciation of simple "country" living. With increasing urban pollution and social problems, the rural life offers a vision of environmental hope and social harmony. And the fashion world, which has long focused on urban centers for inspiration, sees the country as the next "look" to use. One might ask, why not garden in old clothes? The answer might be: What if the Joneses showed up when I was in the garden, wearing old clothes? That just would not do.

Gardening is an "in" activity and, as with any trend, requires appropriate dress. Several magazines and catalogs, like Smith and Hawken, L.L. Bean and the *New York Times* magazine, show gardening clothes that offer the gardener everything from environmentally sensitive hats and organic cotton T-shirts in "garden colors," to French gardening gloves and shatterproof gardener's sunglasses. One waterproof field jacket sold by Timberland costs over $1,000, and a gardening buttons, how much we could do to save the South American rainforests! Saving the environment one small step at a time has found its logical conclusion in buttons made with nuts from Ecuadorian rainforests. The latest "hook" in eco-consumerism, tagua nut buttons can be found on Smith & Hawken's farmer's canvas jacket, which sells for $89. The company's Spring 1992 catalog claimed that wearing the buttons represents a more dedicated effort to protect the Earth's resources than wearing "Save the Earth" pins because the tagua palm nut harvest contributes to Ecuador's social and ecological development. The nut is gathered by indigenous people in Ecuador without stressing or degrading the tropical rainforest. The groups promoting the buttons, the Tagua Initiative and Conservation International, claim that income from the sale of the buttons encourages local populations to continue harvesting the nut and thereby preserve the forest. Without financial incentive, these people might find other sources of income that destroy the rainforest.

Go Organic

Here are some companies offering organic cotton clothes of all styles.

Earthlings, 205 S. Lomita Ave., Ojai, CA 93023, (805) 646-7770.

Environmentally smart fashion for kids! This

A: *A primary energy source for living*
Source: Union of Concerned Scientists

company produces an adorable line of children's clothing made of plain organically grown cotton and Fox Fibre Colorganic, a fabric fashioned from cotton grown naturally in various colors. Earthlings also offers buttons made from tagua nuts and natural jute accessories ranging from flower-shaped pins to barrettes to belts.

EcoSport, 28 James St., South Hackensack, NJ 07606, (800) 486-4326.

A complete line of good, basic wear made entirely from organic cotton. Everything from T-shirts, sweatshirts and tank tops to tunics, cardigans and leggings. This company also makes products for babies.

Green Cotton, imported by Ardent Intimates, Ltd., 1685-B Scenic Ave., Costa Mesa, CA 92626, (714) 434-0212.

Manufactured by Novotex in Denmark, Green Cotton is a line of sportswear and sleepwear for women, men and children made with organic and long-staple combed cotton.

Esprit, 900 Minnesota St., San Francisco, CA 94107, (415) 550-3725.

Ecollection, Esprit's newest line, incorporates Fox Fibre Colorganic, unbleached, undyed linens, and cotton poplins, denim and knits.

Levi Strauss, 1155 Battery St., San Francisco, CA 94111, (800) USA-LEVI.

Levi's Naturals is a

watch from Smith & Hawken will put you out $72. Yet doing the right "eco-friendly" thing probably means sticking with that old wool overcoat you've been wearing for years or gardening in secondhand overalls and boots. Unfortunately, no matter how "green," fashion continues to be defined by change and consumption.

The Demise of the Business Suit

To go along with the trend toward more versatile, practical clothing, many offices are allowing employees to wear casual attire. When a company in Montvale, New Jersey, lost its air-conditioning for three weeks, employees were allowed to come to work in cooler, more casual clothes like chinos and polo shirts. Employee attitudes and productivity improved greatly. Other companies have adopted "business casual" policies, and many report that informal clothes enhance the work atmosphere by blurring the lines between employee and boss. Sales of casual twill trousers at Levi Strauss and Company brought in $850 million in 1991, while profits at Brooks Brothers dropped by 50 percent. The new casual ethic means that men do not need to fork out a minimum of $450 for a suit, and women need not don panty hose in the heat of summer. But in addition to financial benefits, better morale and work efficiency, casual dressing could mean benefits for the environment. With employees shedding the suit during the summer, companies can cut down on air-conditioning, which burns electricity. In addition, casual clothes do not require costly dry-cleaning, which means that fewer dry-cleaning chemicals are used. And unfettered by the need to change into a suit at the end of a commute, many more people will use nontraditional transportation, such as walking or riding a bicycle.

Cotton *Sans* Chemicals

The environmental movement has brought a growing emphasis on natural fibers such as cotton. Cotton futures jumped to the highest they have been since February 1967 after the US Census Bureau announced that domestic cotton consumption is on the rise. Yet cotton growers rely heavily on pesticides. Two times as many pesticides per acre, or 5.4 pounds, are applied to cotton than to all other US agricultural crops combined. Dangerous defoliants are used before cotton can be harvested mechanically, and formaldehyde is often

: *According to recent NASA data, how much is the ozone layer in the Northern Hemisphere decreasing each year?*

used during the spinning process. Cotton is certainly "natural," but is by no means environmentally benign.

While sportswear companies and mail-order catalogs have emphasized natural fibers like cotton in their collections over the past few years, "green" consumers have become even more subtle and are demanding organically grown cotton. Major clothing manufacturers like Patagonia, Levi Strauss and Company, Esprit International and European manufacturers like Novatex A/S and Bo Weevil are using organic cotton. European producers buy organically grown cotton from Turkey, where farmers have developed nonchemical methods to control insects and fertilize the soil. Patagonia, Inc., donated $1 million to environmental organizations in 1990 and is undergoing an "environmental review process" to analyze all materials and processes used during manufacturing. Esprit International has introduced a new line of clothing called Ecollection that uses only organic cotton. Currently, 80 percent of Esprit's clothes are made with cotton and, ultimately, the company hopes to use organic cotton for all its garments. The company also uses recycled packaging and does not use plastic to tag merchandise.

Some companies are concerned with more than just the chemicals involved with growing cotton. Smith & Hawken not only uses cotton organically grown in Texas, it is also developing a nontoxic dyeing process with a Mexican firm that uses iron oxides instead of heavy metals. Sally Fox, a yarn spinner and entomologist, has come up with an alternative to dyed cottons. Over the past nine years, Fox has bred insect resistance in cotton plants. During that time, she has developed plants that naturally produce brown and green cotton. Fox has patented her colored fibers, called Fox Fibre Colorganic and hopes to eventually produce pinks and blues. Levi Strauss and Company signed a contract in 1991 to buy a thousand acres of her colored cotton. Fox says her cottons are attractive because they do not require chemically costly dyeing, but also because they grow darker with age instead of fading like dyed cottons. Although there is growing interest in organic and colored cottons, supply comes nowhere close to meeting demand. Currently, interested companies must settle for cotton produced with fewer chemicals than usual, but cotton growers in California and Texas have heeded the public's call and are finding new ways to grow the fiber organically.

line of naturally colored brown jeans and jean jackets fashioned from strains of colored cotton.

O Wear, P.O. Box 77699, Greensboro, NC 27417-7699, (919) 547-7886.

O Wear is a casual line of mix-and-matchable knit tops and bottoms, for both men and women, made of 100 percent certified organic cotton.

EarthCare, Inc., P.O. Box 326, Hickory, NC 28603, (704) 327-3633.

Created by EarthCare in 1990, Valley Care is a line of socks, T-shirts, camisoles and underwear made with only natural fibers.

Wearables Marketing, 401 Ocean Front Walk, Venice, CA 90291, (310) 399-5608.

This company produces a line of purely organic cotton products called Go Organik—socks for men, women and youths, infant wear and even clothing for prematurely born babies (preemies).

Wearable Integrity, 1725 Berkeley, Santa Monica, CA 90404, (310) 449-8606.

The company claims this is the first complete natural cotton sportswear line for women. Dresses, shirts, pants and skirts are fashioned from 100 percent organic cotton and Fox Fibre Colorganic and adorned with tagua nut buttons.

An average of 1.7 percent to 5 percent annually
Source: The Campaign for Safe Alternatives to Protect the Ozone Layer

Music

Songs Like These

Many recording artists have become aware that the environment is a worthy topic. Here is a varied selection of artists and songs that deal with the environment.

One of the only soundtrack albums that deals directly with the theme of the environment came from director Wim Wenders's 1992 film, **"Until the End of the World."** The movie is a tale of love and technology set in 1999, under the looming threat of nuclear destruction. Wenders handpicked artists either to write songs for the movie or to perform cover tunes that prevailed with that theme. The musical guests who appear on the soundtrack and wrote the tunes are **U2, Elvis Costello, Lou Reed, Neneh Cherry,** U2 producer **Daniel Lanois, REM, Patti Smith** and **Depeche Mode.**

Julian Lennon has shown a great deal of interest in the environmental movement. Lennon's album "Help Yourself" has a couple of tunes that express his concern for our planet, and Lennon made a special guest appearance at "E-Town," an environmental radio program, in spring 1992.

In fall 1991, **Robbie Robertson** (formerly of The Band) released a new project called "Storyville" on Geffen Records. Two of the most rocking tunes on the CD, "Shake This

THE GREENING OF THE POP MUSIC WORLD

- By Wendy Kale -

Music has long been a movement for change. While some music echoes the way things are, other music offers a taste of the fringe, which eventually becomes an echo of the way things are. As the environment becomes a more popular issue, it begins to appear as a topic of songs. Earth Journal offers some of the newest and some of the oldest environmental pop music.

THIS WAS AN IMPORTANT YEAR for the music world and for the environmental movement, as both joined forces to enlighten the record-listening and concert-going public about the environmental needs of our planet.

This year, not only did major artists contribute their musical talents to fostering environmental excitement, but also new acts joined to make their audiences aware of the global situation.

Midnight Oil is one of the music industry's biggest proponents for cleaning up the environment and saving the land of indigenous people. In 1992, this Australian band released a live album, "Scream in Blue," on Sony Records. The material on the CD was recorded between 1982 and 1990 at various environmental gatherings, including the Dantree Rainforest Benefit in Sydney, Australia, and a concert recorded in front of the Exxon Building in New York City.

In late 1991, Don Henley, formerly of the Eagles and well known as a singer/songwriter in his own right, started work on his Walden Woods Project. This was a wide-ranging collection of musicians, authors and environmental activists trying to raise $8 million to purchase 68 acres of land from developers. This is the same Massachusetts land where philosopher Henry David Thoreau lived in the 1850s. The Walden Woods Project's active members include Bonnie Raitt, Meryl Streep, author Alex Haley and leaders from the National Audubon Society, the Sierra Club and the National Parks and Conservation Association.

Henley toured in late 1991, not only promoting his

: *What is the average time it takes for a recycled can to make it back to the store shelves?*

ARTS & ENTERTAINMENT

Richie Havens, founder of the Natural Guard.

music and the Walden project, but also calling attention to his new book, *Heaven Is Under Our Feet*. The work contains personal environmental pieces written by famous musicians Paula Abdul, Sting, Bette Midler, Jimmy Buffett, Arlo Guthrie and Janet Jackson, as well as other celebrated people from the film and entertainment arenas.

The Walden group was able to purchase a parcel of the land after a May 1992 two-day benefit show held at Universal Amphitheater in Los Angeles. The first night, country superstars like Clint Black performed, and the rockers played the second night of music. That bill included Henley, John Fogerty from Creedence Clearwater Revival, Roger Waters of Pink Floyd fame and Big Head Todd and the Monsters.

Richie Havens, the folksinger and activist whose Town" and "Go Back to Your Woods," both take people's relationship to the natural world as an index of their own well-being.

Little Village was a group that came out in early 1992. It comprises some of the best singer/songwriters of the decade, **John Hiatt, Nick Lowe, Ry Cooder** and **Jim Keltner**. The band's debut album, "Little Village," on Reprise Records, had a calypso-flavored "Do You Want My Job?" The tune is about an impoverished fisherman who's forced to work on a tanker that dumps nuclear waste into the ocean.

The very popular and environmentally aware band **Deee-Lite** has a new ode to ozone on its 1992 release, "Infinity." The tune is titled, "I Had a Dream I Was Falling Through a Hole in the Ozone," and one of the main lines proclaims, "Convenience is the enemy."

Folkie singer/songwriter **Tracy Chapman** put out a new album in 1992 called "Dreaming in a World." Chapman ponders a lot of society's problems—including the environment—on this record. "Short Supply" is Chapman's official environmental protest song.

MC 900 Foot Jesus is the name of a thrash-rap act that relies heavily on lead trumpet. The group's debut album came out on IRS Records in 1991 and features an instrumental avant-garde piece called "Ozone."

The Samples released their third CD in

A: *Approximately six weeks*
Source: 50 Simple Things You Can Do to Save the Earth, *EarthWorks Press*

1992, on W.A.R. Records. The Samples have featured environmental tunes on all their records, since the environment is a major concern to the various players in the band. "Another Disaster" is on the new release. The Samples also have a tune called "Giants," about the ruination of the rainforests. The guest artist on this song is the famed **Branford Marsalis**.

The Scottish group **Del Amitri** came out with a new recording, "Change Everything," on A&M Records in summer 1992. The band has one tune on the CD that deals with leveling towns and nature, all in the name of progress.

"Kiko" is the 1992 release from the Los Angeles-based band **Los Lobos** on Warner Bros. Records. "Arizona Skies" on the CD pays homage to the wonders of nature. "Wicked Rain" is about acid rain and nuclear fallout.

Australian **Paul Kelly** and his band, **the Messengers**, put out a new release in 1992 on Dr. Dream Records. It's called "Comedy." One tune on the album "Buffalo Ballet," is about how the buffalo were almost hunted to extinction.

In 1991 **Merl Saunders** and his band put out "Blues from the Rainforest," a musical suite. The record, on Summertone Records, is raising money to support the Rainforest Action Group. Songs on the CD include "Blues from the Rainforest," "Dance of

rendition of "Freedom" at Woodstock is an American classic, has entered the 1990s with a new mission: saving the Earth. Havens grew up in New York City, and he has founded The Natural Guard to help city kids learn about the environment.

CD packaging hit the music and environmental news this year. The CD packaging debate heated up over the use of the cardboard "longbox" and the plastic jewel-box in which CDs were often wrapped. The wasteful CD packaging may be on its way to becoming history, since the longbox wastes over 21 million pounds of paper annually. The new look of 1992 is the Eco-Pak. This is a longbox-sized device made of plastic and paperboard that easily folds into the size of the now out-dated jewel-box. Unfortunately, although this was supposed to be used by many record labels by the spring of 1992, few are using the new design. One band that was brave enough to blaze the way for the Eco-Pak is the dance band Deee-Lite. This popular act uses Eco-Pak for their summer 1992 release, "Infinity."

A Boulder, Colorado-based band, the Samples, has taken this packaging concept one step further. The Samples have totally done without plastic in packaging: Everything in their package is recycled paperboard, and the printing is all done in environmentalist-approved soy ink. There's more— the Samples' new release, "No Room," contains an outer cardboard sleeve that can be cut out to create an instant postcard to be sent to the band's record company, W.A.R.

In November 1991, music business types got together to protest the plans to build a hydroelectric dam in James Bay, Quebec. Not only would the dam hurt the area's environment and wildlife, but also it would threaten the Cree tribe with extinction. The "Ban the Dam Jam" was held at the Beacon Theater in New York City, and folks like John Doe of X, Jackson Browne, Mike Scott of the Waterboys and ex-Talking Heads leader David Byrne all pitched in their musical talents to aid the cause.

In April 1992, the major rainforest benefit of the year was held at Carnegie Hall in New York City. The bash raised over $500,000 and environmental awareness with the guest appearances of Sting, Don Henley, Elton John, Natalie Cole and James Taylor.

Taylor decided to do some environmental good

: *How much does every US nuclear test cost?*

deeds of his own on his 1992 summer tour. On at least two dates, in Denver, and Albuquerque, New Mexico, some proceeds of his concerts went to the National Resources Defense Council.

The "Don't Bungle the Jungle" benefit was held on May 19, 1992, in New York City. Top designers donated fashions to be modeled by members of Deee-Lite and the B-52s. The fashion show was staged to increase awareness of the rainforest situation; it raised over $150,000.

This year a new wave of musicians emerged on the music scene, a wave of aboriginal pop musicians from Australia. One of the primary bands is Yothu Yindi, whose members are from Arntenland in the tropical area of Australia. The members of Yothu Yindi have become the first aboriginal rock superstars of the Outback. The band won an Australian grammy for its hit single "Treaty," from the CD "Tribal Voice" (on Hollywood Records). The band's big ecological news: They've found a new way to recycle tree trunks. The band's new album is filled with the deep drones of Australia's most famous instrument, the didgeridoo. The instrument is made out of a hollowed tree trunk and is played by breathing into it, creating a variety of sounds.

Archie Roach is another native Australian musician who's been hitting the news. His most recent album, "Charcoal Lane" (Hightone Records), is about the aborigines' struggle to hold onto and maintain their environment in the Outback. Roach also makes extensive use of the didgeridoo in his work.

Earth Day 1992 was celebrated all over the country. The event that garnered the most attention was the celebration that took place in Foxboro, Massachusetts, for over 30,000 fans. The event featured performances by Midnight Oil, Steve Miller Band, the Kinks, Fishbone and the Violent Femmes.

Raleigh, North Carolina, was another major spot for musical-related Earth Day celebrations. Some 10,000 fans showed up to see a variety of alternative music acts and to help environmental organizations. The music was provided by Buffalo Tom, Flat Duo Jets, Firehose, the Connells and Dillon Fence.

The Seattle music scene and local musicians have long been supporters of preserving the environment. On June 6, 1992, a "Rock and the Environment" concert was held at Music in the Gorge, southeast of the Fireflies" and "Blue Hill Ocean Dance."

Luka Bloom, a folk artist native to Ireland, has been taking America by storm. In 1992, Bloom released his second CD on Warner Bros. called "Acoustic Motorbike." Naturally, the title song is about the pleasure of pedaling on a bicycle. Bloom also pays tribute to the loss of the buffalo and the land belonging to Native Americans.

The Levellers are a popular British band that finally hit the US in summer 1992. Besides writing Earth-conscious tunes, the band also tours with groups like Earth First! (which set up info booths at the band's concerts). "One Way" is a Levellers tune about polluting the planet.

Toad the Wet Sprocket became a popular club band in the US in 1991, and by 1992 they were one of the most popular college and alternative acts in the country. The group's 1991 release, "Fear" (on Sony), pays homage to the environment. The first cut on the CD is "Walk on the Ocean," about leaving the polluted city to find peace by the shore.

Bruce Cockburn put out several environmentally aware albums in 1991. A remix of some of Cockburn's earlier material, "The Bruce Cockburn Primer," was released on Sony and contains the classic tune "If a Tree Falls." And Cockburn released an album of brand new material in 1992, called

A: *From $15 to $80 million per test*
Source: Greenpeace

"Nothing but a Burning Light." Ever environmentally minded, Cockburn wrote the first tune on the album, "A Dream Like Mine," about balancing nature.

The **Merchants of Venus** were former members of **Lone Justice** and **David Johansen's** band, as well as folks that played on albums for artists like David Wilcox. The trio joined forces in 1991 to release a debut album on Elektra Records. The Merchants have one tune on the album, "Kiowa," about bringing balance and harmony to Mother Earth and the Native American lands.

Australian rockers **Midnight Oil** had several tunes centered on fighting the system to keep the environment safe on their live 1992 CD, "Scream in Blue."

Food for Feet is the alter-ego trio of the Los Angeles band **Oingo Boingo**. Food for Feet embarks on a very different musical journey, singing tunes about the environment and racism. On Food for Feet's first release on Dr. Dream Records in 1991, the group sings an ode to the Earth "Paradise."

The Rain is a new acoustic-rock band from England that enjoys singing songs about preserving the planet and nature. The band's first release, "Rain," came out in 1992 on Sony Records and featured the tune "Mother Earth."

Tribe After Tribe is a group of displaced South Africans who now live in Los Angeles but still sing

Seattle, near Vantage, Washington. Headliners for the event included Seattle's own local-turned-national metalheads, Queensryche, an "unplugged" version of Heart (who also call Seattle home), the Walkabout, Bananafish, Metal Church, Runners of the Big Wave and the War Babies. In addition to these musicians, several bands had taped environmental messages for the audience. There were information booths providing tips on recycling and conservation, and members of environmental groups spoke between musical sets.

On June 13 and 14, 1992, the "Health and Harmony Music and Arts Festival" was held in Santa Rosa, California. The annual event brings together musical artists, an arts and crafts fair and an environmental expo. This year Merl Saunders and the Rainforest Band, Charlie Musselwhite and a variety of Brazilian and Caribbean bands performed. The Environmental Exposition showcased products, services and organizations that support the environment. The displays taught people how to conserve, recycle and treat the planet with respect.

In MTV news relating to the environment, prior to the Earth Summit in Rio de Janeiro in June 1992 artists like Madonna, Seal and REM made taped video pleas to President Bush to change his anti-environmental views before the conference.

MTV wasn't the only source of video pleas to Bush. In May, singer and environmental activist Olivia Newton-John sent a video message to the president featuring Cher, John Forsythe, Tony Danza, Pierce Brosnan, Jane Seymour, the cast of "Cheers" and an entire class of kindergarteners. The video urged Bush to attend the summit in Rio. Newton-John served as host for a concert there, and she has been the goodwill ambassador for the United Nations Environmental Programme since 1990. Besides being an animal rights activist, she is concerned about the packaging of her CDs. Her record label, Geffen, used recycled paper for Newton-John's last two album covers. Her new release in July 1992, "Back to Basics," has a built-in punch-out card to send to President Bush asking him to take steps to help the environment.

The Wetlands Preserve club in New York City has been doing its best to live up to its name. Not only does the club bring in top national and regional bands, but also, several nights a week, it holds

 Which country has an average birth rate of 7.5 children for every woman?

ARTS & ENTERTAINMENT

educational programs on the environment, including speakers from the Rainforest Action Group, the Federal Land Action Group and the Nature Forest Council. The club also holds musical benefits for organizations like the Rainforest Action Group and the Sierra Club. The folks who work at the Wetlands call it "maintaining an Earth station." The club supplies free literature and periodicals on the environment, and there's always an environmental petition to sign. The club's staff says that everything at the Wetlands is environmentally friendly. They use only products that don't harm the environment for cleaning, they use energy-efficient lights and they recycle everything. In addition, in fall 1992 the Wetlands sponsored a music tour called "Green the Music Industry."

Michael Jackson launched his "Heal the World" tour in June 1992, and was quoted in the media as saying, "The only reason I am going on tour is to raise funds for the newly formed Heal the World, an international children's charity that I'm spearheading to assist children and the ecology."

Jimmy Buffett is no longer wasting away in Margaritaville. The singer/songwriter worked in 1992 to aid the manatees in Florida. Buffett is the spokesman for the Save the Manatee Club. Manatees are mild-mannered, slow-moving sea mammals in danger "because idiots in speedboats are running them down," he says. In May 1992, Buffett broke away from the Florida Audubon Society because of its alleged lack of interest in this cause. Buffett gave a concert in Fort Lauderdale, Florida, that same month to raise funds for Save the Manatee Club, which has over 30,000 members.

Groups from a local to an international level around the world are performing and bringing attention to the issues most important to them: a horrible breakup, being in love, dwindling natural resources and other environmental ills. Tune in.

Wendy Kale is music columnist for the University of Colorado's Colorado Daily *newspaper, based in Boulder, Colorado.*

about the situations on their native continent. The band's debut release came out in 1991 on Arista and contains two songs about the environment and the dangers to indigenous people.

Other bands that have songs paying tribute to the Earth include: **Col. Bruce Hampton and the Aquarium Rescue Unit**, who sing a tune called "Planet Earth" on their 1992 Capricorn Records release; **Poi Dog Pondering**, who pay tribute to a Hawaiian rainforest area in "Jack Ass Ginger" on their 1991 CD "Volo, Volo"; and comedy rockers **Spinal Tap**, who included an ode to nature "Stinking Up the Great Outdoors," on their 1992 album, "Break Like the Wind." The **Chickasaw Mud Puppies** have a tune dedicated to locusts called "Chicada" on their 1992 "8-Track Stomp" record, and Australian rockers **Yothu Yindi** have a song called "Treaty" on their album "Tribal Voice" that sarcastically protests imperialist land plunder. And last, but not least, the **B-52s** released a new album in summer 1992 called "Good Stuff," featuring the let's-change-the-world environmental tune "Revolution Earth."
—*Wendy Kale*

A: *Ethiopia*
Source: Newsweek *magazine*

ECOHOME

I*f a home is a "machine for living," make sure your "machine" is environmentally friendly. Whether you're choosing food, clothes, cleaning products or a car, the decisions you make affect all of us.*

Heating and Cooling

According to the *Green Lifestyles Handbook* (Jeremy Rifkin, Henry Holt, NY), the average American home expends 50 percent to 70 percent of its energy on heating and cooling. The equivalent of 380,000 barrels of oil would be saved daily if every household lowered its regular temperature by four degrees, Fahrenheit. If each household raised its air-conditioning temperature four degrees, an energy equivalent of 130,000 barrels of oil per day would be saved. In addition, any temperature reduction below 68.6 degrees saves 3 percent per degree on heating costs. When air-conditioning temperatures are raised, every degree above 78 saves 5 percent in cooling costs. Air-conditioning reduction also lowers CFC emissions.

When heating with wood, use wood that gives the most heat: black birch, hickory, live oak, locust, northern red oak, rock elm, sugar maple and white oak. Woods that give the least heat are: alder, aspen, balsam fir, basswood, cedar, cottonwood, hemlock, northern white cedar, red fir, spruce and sugar pine.

Nuke it

Microwave ovens use about the same amount of energy as conventional ovens, but cook most foods in less than half the time—saving half the energy.

Home

Home means many things to many people—a place to put your feet up, a place where things are as you want them, a place to really relax—but most people probably don't think of their homes as being dangerous to their health or the environment. But many homes are inefficient and may cause discomfort. Earth Journal *takes a look at what you can do to have a greener home.*

THE FUTURE OF HOUSING: RE-DESIGNING THE AMERICAN DREAM

IN ALL CULTURES, a basic level of existence demands that we provide a set of universal daily routines dealing with sustenance, energy, water and waste management (nutrient cycling). Our houses are graphic examples of the interaction these cycles of biology, ecology and agriculture have with architectural and engineering methods. An array of behaviors, values and attitudes create our lifestyle patterns and reflect subtle philosophical orientation to the Earth, ourselves and our sense of community. We vote daily through our consumer behavior. This lifestyle individually and collectively influences the well-being of ourselves and the planet.

It is curious that economics and ecology have been divorced, for they stem from the same word, *eikos,* meaning "house." Ecology is the science of our house, and economics is the management of our house. The Law of the Circle reflects the ecological reality that to be sustainable a system must be circular and interdependent. Circulation, reuse and metamorphosis are the underlying processes that fulfill the Law of the Circle in our biological world.

Presently, our polluted environment is a by-product of the way we have organized our economy. A highly efficient, nonpolluting lifestyle stands in direct contrast to our current economic traditions of accumulating short-term material wealth in a consumer-oriented society driven by technological obsolescence. Our current lifestyles use nature's resources as capital in accumulating money, regardless of the long-term environmental degradation. Progress decrees that the collective knowledge of science and

 In an average school cafeteria, what percentage of the total garbage generated is food waste?

technology is best utilized when producing wealth. Our society defines wealth in terms of ability to acquire and collect things. So it seems that the roots of the problems of environmental degradation lie in Western culture's philosophy about nature. The design of our spaces reflects this lifestyle pattern and consumption philosophy.

For most of us, our homes provide a sense of permanence, an insurance policy against the environment. They isolate us from our natural surroundings instead of inviting the environment in to improve our lives. We create nice big houses and forget where our water comes from in our coastal desert. The energy that powers, heats and cools our homes is creating pollution at the generation site. We pay a great deal, and lose nutritionally, from eating imported foods that are out of season in our region. Our landfills are filling up, and the diversity in our waterways is a skeleton of what once existed. The health effects from this lifestyle are seeping in everywhere. The Consumer Product Safety Commission identified 150 different chemicals regularly found in our houses and linked to cancer, allergies, psychological abnormalities and birth defects. There are simple nontoxic alternatives to just about everything we are accustomed to wearing, walking on and applying to our homes.

Our housing designs are out of control economically and socially because they carry large unacknowledged operation costs. What are the basic design principles of our modern civilization? The American Dream house uses more energy than any other structure in the history of humankind and has not changed in 50 years. Most builders, until recently, assumed energy was cheaper than materials and labor and designed accordingly. Our emphasis has been on the construction of houses that reflect the cost of development, the economic and energy costs to build the structure, with little regard for its long-term environmental costs. The long-term success of housing will be achieved when we fully determine the extensive network of resources and impacts needed to maintain what we call home.

Presently our homes are energy sinks, creating toxicity and impacting space with limited consideration of their surroundings. Few houses use climatically sensitive designs to maximize the productive

Recycling

One of the easiest and most direct things people can do in their homes is practice the three R's—reduce, reuse and recycle. Reduce: Only purchase products that are durable and necessary. Reuse: Discard or recycle products only as a last resort—try reusing things in art projects or home improvements. Recycle: Whenever possible, buy products that are recyclable, and made from and packaged in recycled materials. Implementing these easy steps can be a creative and rewarding way to reduce your impact on the environment.

Household Energy Efficiency

About 15 percent of the electricity we use around our homes goes into lighting. Using light intelligently and taking advantage of efficient lighting technologies can save 50 percent to 90 percent of that energy. Those savings translate into lower electric bills, less acid rain and nuclear waste, and fewer greenhouse gases emitted.

Fluorescent: More Light for Less
Ninety percent of the energy consumed by an ordinary incandescent bulb is given off as heat rather than visible light. Fluorescent lights use one-quarter of the energy regular bulbs use to produce the same light.

  A: *Nearly 50 percent*
Source: The Student Environmental Action Guide, *Earth Works Press*

Replace ordinary incandescent bulbs with compact fluorescent lamps (CFLs). Compact fluorescents fit into your regular sockets and last nine to 13 times longer. Although they cost considerably more initially, compact fluorescents actually save you money by lowering your electric bill. Typically, an 18-watt compact fluorescent will save $40 in electricity and $5 to $10 in replacement bulbs. In addition, that one bulb will spare the Earth up to 2,000 pounds of carbon dioxide and 20 pounds of sulfur dioxide from a coal-fired plant.

Compact Fluorescent Caveats
• CFLs are not a wise investment in places they won't be used much: The payback period is too long. Keep regular bulbs in your closets and use CFLs in the kitchen, office and reading area.

• CFL bulbs are bigger than regular bulbs and may require the use of a lamp harp extender to fit in some lamps.

• Buy CFLs with electronic ballasts. They give better light.

• Not all CFLs work well outdoors in cold temperatures. Make sure you get the right model for outdoor use.

Indoor Lighting Tips
• Use natural daylight whenever you can. A three-by-five-foot window in direct sunlight lets in more light than 100 standard 60-watt bulbs.

potential of the site, minimize its overhead and automate the maintenance. Climatically sensitive architecture, like adobes in the southwestern US, the under-ground sod house of the plains and the New England saltboxes, reflects bioregional authenticity based on necessity. These houses, by design, have minimized maintenance and maximized productive yields. If we are building houses for the future, we need to ask how much they will cost to run, as well as how much they will cost to build.

What are the incoming support systems necessary to drive our homes? What are the outgoing products and pollutions our houses create? A simple input-output analysis using the basic universal human needs of energy, food, water and nutrient cycling can be a beginning. These four are interrelated and can potentially create a resource bank that lessens the demand on the utility, sewer and water companies, as well as the supermarket. Future housing needs to be "interest producing." We need to create a full circle interchange with: 1) the producers of our existence, plants, which are also our primary solar collectors; 2) the consumers and creators of pollution, humans and animals; and 3) the full-cycle recyclers, our soil neighbors, microorganisms, which begin the process of regeneration. Social and community issues also need to be added to begin a geomorphological model of sustainability.

—*Bill Roley*

Reprinted by permission from Earthword: The Journal of Environmental and Social Responsibility, *"The Future of Housing," by Bill Roley, Volume II, Special Eco-Cities Edition, 1991.*

ENERGY-EFFICIENT HOUSE PLANS

HOUSE PLAN BOOKS AND STOCK PLAN PACKAGES can provide builders (as well as people looking to contractors to construct a new home) with invaluable sources of information, ranging from simple design and construction ideas to full sets, working drawings and construction documents. Numerous organizations produce or distribute these, including magazines, owner-builder schools, architecture firms, utility companies and state energy offices, to name a few. One excellent way to track down many such sources is to contact the National Appropriate Technology Assistance

: *How many pounds of pesticides are used worldwide every year?*

ECOHOME

Service, (800) 428-2525, or the Conservation and Renewable Energy Inquiry and Referral Service, (800) 523-2929, and request that they send you their most up-to-date source lists for energy-efficient house plans. Make the most of any such inquiries by specifying the type of plans you're looking for: passive solar, superinsulated, earth-sheltered, log construction, attached greenhouse, etc. The following house plan sources offer guidance for energy-efficient building.

American Ingenuity, Inc., 3500 Harlock Rd., Melbourne, FL 32934-8407, (407) 254-4220.

Offers energy-efficient, superinsulated dome kits using non-CFC foam and a patented prefabricated, panelized building system. For those homeowners who do not select one of the standard house plans, they can design a custom floor plan.

 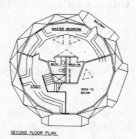

Plans like these from American Ingenuity offer unique, energy-efficient, environmentally responsible homes.

Architectural Designs, Davis Publications, Inc., 380 Lexington Ave., New York, NY 10017.

Five issues per year, available at newsstands. Each issue contains approximately 200 plans for traditional and contemporary homes, including energy-efficient, solar and vacation home designs.

Best-Selling Home Plans, Hachette Magazines, 1633 Broadway, New York, NY 10019, (800) 526-4667.

Bimonthly. Features over 150 plans by various architects and designers. Each issue includes plans for about 30 solar homes incorporating passive solar features such as greenhouses, clerestories and masonry storage walls. Some include earthberming, super-insulation and active solar systems.

- Turn off lights in any room not being used.

- Although halogen bulbs aren't as efficient as fluorescent, they are often two to three times as efficient as incandescent, last four times longer and give better light.

- Light-zone your home: Concentrate lighting in reading and working areas and where it is needed for safety.

- To reduce overall lighting in non-work spaces, remove one bulb out of three in multiple light fixtures and replace it with a burned-out bulb (for safety). Replace other bulbs in non-work spaces throughout the house with the next-lower wattage.

- Use one large bulb instead of several small ones in areas where bright light is needed.

- If you're buying a new lamp, consider the advantages of three-way switches. Use the brightest setting for reading and the lowest for watching TV.

- Reflector floodlights in directional lamps can provide the same light with half the wattage; i.e., substitute a 50-watt reflector bulb for the standard 100-watt one.

- Try 25-watt reflector flood bulbs in high-intensity portable lamps. They provide about the same amount of light, but use less energy than the 40-watt bulbs that normally come with these lamps.

A: *Approximately 4 billion pounds annually*
Source: Greenpeace Action

- Keep all lamp fixtures clean.

- Light-colored walls, rugs, draperies and upholstery reflect light and therefore reduce the amount of artificial light required.

Outdoor Lighting

- Turn off decorative gas lamps unless they are essential for safety. Just eight gas lamps burning year-round use as much natural gas as it takes to heat an average-size home for a winter heating season. By turning off one gas lamp, you can save $40 to $50 a year in natural gas costs.

- Use outside lights only when they are needed. One way to make sure they're off during daylight is to put them on a timer or photocell unit that automatically turns them off when it gets light.

- Solar-powered outdoor path lights are an ideal way to light your night walkway.
—*Marshall Glickman*

Reprinted by permission from Green Living, "Energy Efficiency Around the House," by Marshall Glickman, Summer 1992.

Better Homes and Gardens Home Plan Ideas, Special Interest Publications, 1716 Locust St., Des Moines, IA 50309-3023.

Quarterly, available at newsstands. Includes plans for a wide variety of passive, earth-sheltered and superinsulated homes.

The Bloodgood Plan Service, 3001 Grand Ave., Des Moines, IA 50312, (800) 752-6728.

Provides plans for energy-conserving houses of traditional and contemporary styling, ranging in size from 800 to 5,000 square feet.

Energy Saving Homes, Vol. 1, No. 1, Home Building Plan Services, Inc., 2235 NE Sandy Blvd., Portland, OR 97232, (503) 234-9337.

Contains plans for 57 energy-efficient homes.

Integrated Energy House Design Book, Passive Solar Environments (PSE), 821 W. Main St., Kent, OH 44240, (216) 673-7449.

Lists 23 solar house plans available from PSE.

NU Solar Homes Planbook, Northeast Utilities, P.O. Box 270, Hartford, CT 06141, (203) 665-5000.

Discusses considerations in energy-conserving house construction and lists eight plans for traditional-in-appearance New England-style homes that are passive solar by design.

Passive Home Designs, Florida Power & Light, P.O. Box 029100, Miami, FL 33102, (305) 442-8770.

Brochure presents two homes designed for each of the three climatic regions of Florida.

Solplan 5: Energy Conserving Passive Solar Houses, The Drawing-Room Graphic Services, Ltd., Box 88627, North Vancouver, BC V7L 4L2, Canada, (604) 689-1841.

Presents plans for 21 energy-efficient homes.
—*Robert Sardinsky and the Rocky Mountain Institute Staff*

Reprinted by permission from The Efficient House Sourcebook, *by Robert Sardinsky and the Rocky Mountain Staff, Rocky Mountain Institute, Old Snowmass, CO, 1992.*

: *What country has 16 percent of the world's cattle, 54 percent of its buffalo, and 21 percent of its goats, yet is only 2 percent of the world's land area?*

ECOHOME

Home Care

Whether your home is a model ecohome or just the opposite, you can make a difference without changing your home. You can clean with nontoxic cleaners, use nontoxic pest control methods, and protect yourself from other household dangers.

HOUSEHOLD CLEANING

REPLACING YOUR HOUSEHOLD CLEANING KIT with nontoxic materials is easy and inexpensive. In fact, the items that most people already own—ammonia, oven cleaners, furniture polish, scouring powder, disinfectant, glass cleaner—are probably more costly than the concoctions that can be made with basic ingredients. Simple, natural mixtures can clean many areas of the house.

Other than for monetary reasons, why switch? The Federal Hazardous Substances Act regulated by the Consumer Safety Commission finds cleaning products to be some of the most dangerous substances in the home. The product labels should be enough to deter us: toxic, corrosive, flammable, combustible under pressure, radioactive. Why do we use them? Hopefully the following suggestions will make us all realize how easy it is to change our ways of cleaning with little extra effort. What could be more important than protecting ourselves in the places we call home?

To begin cleaning, you must have the basic tools. The fundamental implements include a spray bottle, cellulose sponge cloth, latex gloves, rags (old T-shirts), sponges, cotton mops, buckets and pails.

To make your own all-purpose spray cleaner, use 1 teaspoon of borax, 1/2 teaspoon washing soda, 2 tablespoons vinegar or lemon juice, 1/4 teaspoon vegetable-oil-based liquid soap and 2 cups of very hot tap water. Pour this combination into your spray bottle and cleanse away! This mixture also works as a disinfectant. Increase the ingredients for cleaning floors, and simply use a bucket instead of a spray bottle. For fragrance, a fragrant tea or essential oil can be added. The cost of all these ingredients may sound more expensive than a bottle of Fantastic, but remember: A little bit goes a very long way.

Need to clear the air? All that is necessary is water and the flower of your choice. Pulverize the petals in a blender, place the mush in the bottom of a tea pot,

Lead In Your China

In March 1992, the *Washington Post* reported that California, which has the strictest anti-toxics law in the country, released a list of 657 china patterns made by 21 manufacturers that meet the state's standards for lead.

The California state attorney general's office and the Environmental Defense Fund sued 16 china companies on grounds that the level of lead in certain china patterns is too high.

Lead, which is used in glazes of china and other ceramics, as well as in lead crystal, can leach into food or drink. The heavy metal is known to cause neurological and other problems and is especially dangerous to children and pregnant women.

A list of safe china patterns was released by the china companies to help consumers who want to buy new china. David Roe, a senior attorney for the Environmental Defense Fund, said the list was released to demonstrate that it is possible to manufacture china without high levels of lead. "There is no reason to have this chemical in our plates," he said.

The names of patterns that have levels of lead above the standard were not released, because the state attorney general's office did not want to discourage companies that cooperated during the lead study.

The Food and Drug Administration

A: *India*
Source: International Herald Tribune

recommends that food, especially acidic foods like tomato sauce and orange juice, which leach lead at a faster rate, not be stored in ceramic dishes that might contain lead.

To obtain a copy of the list of lead-safe china, send a self-addressed envelope with 52 cents postage to: Office of the Attorney General, Press Office, 1515 K St., Sacramento, CA 95814; or Environmental Defense Fund, 5655 College Ave., Ste. 304, Oakland, CA 94618.

Renovating Uncovers Lead Paint

The Conservation Law Foundation, an environmental group in Boston, recently warned that home renovations often spread lead contamination, especially if a contractor sands down old paint, since children and adults can be hurt by breathing lead-filled dust.

Until a few years ago, lead-based paint was commonly used on windows, doors, stairs, columns and trim. According to the *New York Times*, a study done by the federal Department of Housing and Urban Development found that three-quarters of all housing units built before 1980 have lead paint somewhere, and single-family houses typically have more lead-based paint than apartments. Metal surfaces, including radiators and railings, are especially likely to be covered with lead-based paint.

and cover with a cup or two of boiling water. Let cool, pour into a jar, and leave it to seep for four or five days. Strain and place the fragrant water in various parts of the house. The flower scent can also be supplanted by lemon, grapefruit, orange, cloves or cinnamon sticks.

White bread eats up odors if a loose piece or two is placed in the back of the refrigerator. Vinegar and water is the best mixture for cleaning the fridge, as well as the cutting board.

The fireplace full of dusty, black soot is best attacked with 1/4 cup of washing soda in 2 gallons of hot tap water. The more washing soda, the more rinsing needed, so use sparingly.

The best way to clean a greasy, grungy oven is with a box of baking soda and 1/4 cup of washing soda. Sprinkle water on the bottom of the oven, then cover with baking soda and washing soda. Add more water on top of the soda mixture and let it sit over night. The grime will come up effortlessly with a rag.

What it comes down to is this: Baking soda absorbs odors and deodorizes. Borax disinfects, deodorizes and inhibits mold. Washing soda cuts through grease. A little here and a little there, keeping in mind the mixtures above while adding flowers, oranges or whatever appeals to the senses, should keep your house in order, your wallet full and your health up to par.

If you prefer, you may buy natural cleaners from natural food stores and some grocery stores. See the "Directory to New Green Products," page 292.

Resources:

Nontoxic, Natural and Earthwise, Debra Lynn Dadd, Jeremy P. Tarcher, Los Angeles, 1990.

Clean and Green, the Complete Guide to Nontoxic and Environmentally Safe Housekeeping, Annie Berthold-Bond, Ceres Press, Woodstock, NY, 1990.

PEST CONTROL METHODS AND MATERIALS

THE FOLLOWING LIST CONTAINS some nontoxic (or almost nontoxic) methods and materials for use in our pest control efforts. Note that this list includes both concepts and products. While corporations not surprisingly prefer to promote products over concepts

 Q: *How many pounds of compostable leaves and grass clippings are currently dumped in US landfills each year?*

ECOHOME

(easier to sell), these concepts are essential to the process of "least toxic" pest control and need to be recognized and promoted.

Sanitation: Nothing is more important in "least toxic" pest control than keeping your home clean. A clean house means no free lunch for the bugs. Train yourself not to leave food or dirty dishes out overnight. Try to clean the areas around and under stoves and refrigerators at least once every few years.

Moisture Control: Many insects are as attracted to moisture as they are to food. Fix leaks and drips promptly. Try to keep the area around sinks wiped dry.

Caulking: Insects need homes too, and because they are small, they prefer small homes. Caulking cracks eliminates insect hiding places and is always worth the effort. Concentrate caulking efforts in the kitchen, bathroom, and around doors and windows.

Screening: Screening is a very important component of "least toxic" pest control. It is especially useful in keeping out houseflies, mosquitoes and night-flying moths. Also make sure your crawl-space vents are tightly screened to keep out rodents and opossums.

Insecticidal Soap: Insecticidal soap is a specially formulated soap product used mostly against soft-bodied insects. Indoors, it can be used for flea control. Outdoors, it is useful against many ornamental and garden pests.

A: *24 million pounds*
Source: Environmental Defense Fund

The lead content of paint has been restricted nationally since 1978, but significant amounts of lead still exist in buildings and in soil around buildings where lead paint has been scraped off.

Lead in paint is most hazardous when it is disturbed. Experts recommend handling lead-based paint as carefully as asbestos. Lead poisoning effects can range from extreme irritability to anemia to brain damage. Children are particularly sensitive.

The state of Maryland offers a free booklet on the prevention of lead poisoning, which can be requested by telephone at (410) 631-3857.

Cockroaches: How to Control Them

Cockroaches have been crawling all over the Earth for more than 300 million years. That's more than 297 million years before our human ancestors evolved.

Sanitation: While miracles do happen, no one gets rid of roaches without cleaning up. While completely eliminating sources of food and water is often impossible, the cleaner your home, the easier it will be to get rid of the roaches.

Caulking: Roaches are able to move about in cracks as small as one-sixth inch and feel right at home in cracks just three-sixteenths inch tall. Not very big. Caulking eliminates these roach cracks.

Sticky Traps: Used one per room, sticky traps help you determine where your roach problem is centered, so that you can concentrate your attack where it will count. Used together in groups, they can catch enough roaches to put a significant dent in a population.

Boric Acid Dust: For serious infestations, boric acid dust can be applied very lightly in baseboard cracks, wall voids, false ceilings, under refrigerators . . . every place a cockroach might hide. Boric acid dust has been used for generations and still works great. It is of fairly low toxicity, but never apply boric acid dust in areas where children or pets could get into it.

Cockroach Growth Regulators: Especially in areas where cockroach levels are high or chronic, growth regulators can be very useful. Like their counterparts, the flea growth regulators, they are, in effect birth control for bugs. Growth regulators are of low toxicity and last for about four months.

—*Dan Stein*

Reprinted by permission from Least Toxic Home Pest Control, *by Dan Stein, Hulogosi Communications, Inc., Eugene, OR, 1991.*

IGRs: Insect Growth Regulators are synthetic versions of the natural growth regulators found in insects. Not nonchemical, they are chemicals of low toxicity to humans and pets that prevent insects from reaching maturity so they can't reproduce.

Pheromones: Pheromones are insect scent hormones that are often used as bait in traps to trick unsuspecting males into thinking they have found a female. Each insect has a specific pheromone that attracts only its own kind. Many pests can be monitored or controlled with pheromone traps.

Repellents: Repellents, not surprisingly, repel pests. There are not many commercially available besides mosquito repellents, but some herbs have short-term repellent properties (citronella, eucalyptus, wormwood).

Biological Control: More commonly used in agriculture, biological control is the intentional use of "good" organisms against "bad" ones. It is not all that practical in the home at this point, but gecko lizards are used in the South to control roaches (their use has a couple of drawbacks—they bark and bite.) Minute, harmless Trichogramma wasps can be used to control meal moths. One common example of biological control is the use of cats to catch mice.

Home Remedies: All sorts of home remedies exist. Some have been passed down for generations and still don't work (i.e., cucumber peels to discourage roaches). Others work, but have not been scientifically documented. As long as your home remedy does not involve the use of something poisonous, by all means, give it a try. You may stumble onto something great.

—*Dan Stein*

Reprinted by permission from Least Toxic Home Pest Control, *by Dan Stein, Hulogosi Communications, Inc., Eugene, OR, 1991.*

Q: *How much oil is spilled into the Earth's oceans every year?*

ECOHOME

Food

Food is one of the most enjoyable necessities. And it has impacts. By being conscious of the effects of what you eat and how you eat, you can make a difference in the environment. Earth Journal *offers* information to help you make healthy choices for you and the planet.

FOOD ADDITIVES

FOOD ADDITIVES ARE USED TO LENGTHEN SHELF LIVES and allow food manufacturers to use inexpensive ingredients. Most additives are considered safe; several have been proven harmful to your health. However, many additives are used despite inadequate testing. Avoiding processed foods and eating a variety of foods can help minimize the potential health hazards of additives.

Acesulfame-K
Acesulfame-K, which is distributed under the brand names Sunette or Sweet One, was approved by the FDA in 1988 as a sugar substitute.

The tests on which the FDA based its approval show that acesulfame-K causes cancer in animals, and therefore is a potential cancer threat to humans. Yet the FDA approved the additive.

Artificial Colors
Many foods and drinks in the US are artificially colored. Artificial colors can be naturally derived or synthetic. Synthetic dyes like Blue No. 2, Green No. 3, Red No. 40 and Yellow No. 5 are commonly used to color food.

Several synthetic dyes have been banned by the FDA, and many scientists question the potential health effects of synthetics. The FDA has banned Red No. 3 from many cosmetics and some foods because it causes thyroid tumors in rats. However, 180,000 pounds of Red No. 3 were approved for use in 1990 on foods such as maraschino cherries, gelatin desserts and pistachio nuts.

Aspartame
Approved for use in soft drinks in 1983, aspartame quickly became a popular artificial sweetener known as NutraSweet or Equal. Because some people have an intolerance to aspartame known as phenylketonuria

Shopping for Safe Foods

According to Patricia Hynes, author of *Earth-Right* (Prima Publishing, Rocklin, CA), diets with minimal exposure to toxic substances in foods are filled with fresh, whole foods, including: organically grown fresh fruits and vegetables; organically grown whole grains and legumes; eggs from free-range hens; lean meat or poultry from free-range animals; nonfat milk and low-fat cheeses; ocean fish; and honey, date sugar, barley malt syrup, fig syrup and rice syrup (in small quantities). These foods are available at large supermarkets nationwide (usually in specialty sections) and at natural food stores.

When you are buying vegetables and fruits you should look for these organic labels:

Certified Organically Grown means that no pesticides, fungicides, ripening agents (such as Alar) or chemical fertilizers are used during the growing or shipping process. The farm has been certified by a state authorized agency.

Transitional Organic or **Transorganic** foods have been grown on a farm that has not completed the required certification process, which is three years in most states. Pesticides, ripening agents and other chemicals are not sprayed on the plants, but residues may remain in the soil.

 At least 8 million tons
Source: Protect Our Planet Calendar, Running Press Book Publisher

IPM (Integrated Pest Management) farming techniques use natural insect repellents and pest control methods. Small amounts of insecticide sprays may be applied in an emergency, especially on some fruit species. However, no ripening agents are applied to the produce.

Drinking Water Safety

According to *EarthRight*, you should test your drinking water if these conditions exist:

1. A private well or an unregulated water system is your drinking water source and you have a septic system.

2. Lead is present from either lead pipes, lead service connections or lead solder on copper pipes.

3. The color, taste or odor of your water changes.

4. The local water company has fewer than 3,300 customers. Small public utilities have fewer monitoring requirements than large water facilities.

5. A landfill or waste dump, a farm that uses pesticides, an industrial park, a military base, a chemical manufacturing plant or an underground tank storing hazardous materials is nearby.

Water Resources:
EPA Emergency Planning and Community Right-to-Know Hotline, (800) 535-0202.

(PKU), the FDA requires that all packaged foods containing aspartame be labeled as such.

Scientists speculate whether aspartame can affect brain functions and behavioral changes. Despite studies that found an increased risk of brain tumors in rats that were fed aspartame, the FDA has not pursued mandatory testing of it.

BHA and BHT

BHA and BHT are two similar chemicals that prevent oxidation and retard rancidity in foods that contain

Vegetable Curry

1 1/2 cups unsweetened desiccated coconut
2 cups hot water
3 tablespoons unsalted butter
2 medium onions, diced
2 teaspoons minced ginger
4 garlic cloves, minced
1 1/2 teaspoons turmeric
2 teaspoons ground cumin
2 tablespoons ground coriander
1/4 teaspoon cayenne pepper, or more to taste
2 carrots, very thinly sliced
1 small (1 1/2 pound) cauliflower, broken into bite-size florets (about 4 cups)
1 cup diced green beans, fresh or frozen
1 cup canned chick-peas, rinsed and drained
1/2 teaspoon salt
Hot cooked white or brown rice

To make the coconut milk, combine the coconut and hot water in a blender or food processor and blend 2 minutes. Strain the coconut milk in batches through a sieve, pressing out all the liquid from the pulp with the back of a large spoon. Discard the coconut. You should have about 1 1/2 cups coconut milk.

Melt the butter in a large skillet over medium heat. Add the onions, ginger and garlic and sauté, tossing often, 10 minutes, or until the onions begin to brown. Sprinkle on the spices and stir to mix thoroughly. Cook this mixture 2 minutes to blend the flavors.

Stir in the coconut milk and bring to a boil. Add the carrots, cauliflower, green bean, chick-peas and salt, and toss to coat the vegetables with the sauce. Cover the pan and cook 5 minutes. Remove the cover and continue to cook the curry, tossing often, until the vegetable are tender and the sauce has thickened—about 10 minutes more.

Serve with some rice on the side, and drizzle a spoonful of sauce over the rice.

Reprinted by permission from Quick Vegetarian Pleasures, *by Jeanne Lemlin, HarperCollins, New York, 1992.*

 In America, how many aluminum cans are recycled every second?

oil. BHA is listed as a carcinogen in California and is considered a possible carcinogen by the International Agency for Research on Cancer. The FDA is currently testing the preservative, now used in hundreds of processed foods.

BHT has been researched inconclusively. Some studies show that it causes cancer, and others indicate that it prevents cancer.

Carrageenan
Carrageenan is a substance obtained from seaweed and used to stabilize foods like ice cream, hot cocoa and infant formulas to prevent fats and proteins from separating. Recently, it has been used by McDonald's to lower the fat content of its McLean Deluxe Burger. Research has not disproved its safe standing as a food additive.

Guar Gum
This additive is used as a thickening agent in ice cream, salad dressing and pudding. It is found in guar plants (similar to soybean plants), which are then used to feed cattle. Guar gum is considered safe, although it has caused stomach and esophagus blockages in people who ingested large quantities. When used as a food additive, guar gum is present in very small quantities.

Lecithin
Derived from soybeans and eggs, lecithin is a food additive used in mayonnaise, margarine, chocolate, and vegetable oil sprays to prevent blended ingredients from separating and fats from becoming rancid. Some health food stores sell lecithin as a cholesterol-lowering supplement, even though its effects on blood cholesterol levels have never been scientifically proven. The FDA considers lecithin a safe food ingredient.

Mono- and Diglycerides
These additives are types of fats used as emulsifiers in baked goods and frozen entrees to create a smoother texture. Mono- and diglycerides are used in small quantities and are considered safe.

Monosodium Glutamate (MSG)
MSG can cause headaches, tightness in the chest and a burning sensation in the forearms and the back of the neck. The MSG industry claims that reactions to

EPA Safe Drinking Water Hotline, (800) 426-4791.

Home Water Treatment Devices

Aeration is an effective way to remove foul smells, radon and volatile pollutants from your water.

A **carbon filter** removes chlorine, radon, and general taste and odor problems. It is effective on organic chemicals like pesticides, herbicides and industrial waste. It will remove a small amount of microorganisms or toxic minerals. This filter, however, does not remove nitrates, bacteria or metals.

Reverse osmosis removes 80 percent to 98 percent of organic chemicals, including most pesticides, heavy metals and lead. This process will not remove 100 percent of most chemicals and uses large amounts of water.

Distillation removes minerals like nitrate and sodium, heavy metals including lead, and many organic chemicals. Distillers boil water and produce steam. Distillation removes bacteria but results in bland-tasting water. The filter must be cleaned regularly.

Chlorination eliminates or reduces bacteria and viruses to safer levels. This process does not remove nitrates or other chemicals, and may produce excess chlorine or

A: *1,500*
Source: The Recycler's Handbook, *Earth Works Press*

chlorinated compounds in your drinking water.

A **neutralizer** treats acidic water but may increase its sodium content.

Ultraviolet systems use UV light to kill microorganisms. They disinfect the water, but do not remove radon, toxic minerals or organic chemicals.

Deionization removes minerals with charged ions but leaves behind other pollutants. It is an effective method for removing dangerous minerals like lead.

Food Resources

There are a variety of books offering enlightened opinions on the environmental and health impacts of the products you ingest. Here are a few of the options:

Beyond Beef, Jeremy Rifkin, Dutton Books, New York, 1992.

Diet for a New America, John Robbins, Stillpoint Publishing, Walpole, NH, 1987.

Diet for a Poisoned Planet, David Steinman, Harmony Books, New York, 1990.

The Green Consumer Supermarket Guide, Joel Makower, Penguin Books, New York, 1991.

May All Be Fed, John Robbins, William Morrow, New York, 1992.

Safe Food: Eating Wisely in a Risky World, Michael

Barbecued Tempeh

1 (8 oz) package tempeh

BBQ sauce:
1 1/2 Tablespoons natural ketchup
1 Tablespoon each honey and onion powder
1 teaspoon each tamari and lemon juice
1/4 teaspoon cayenne
1 garlic clove, pressed
Few drops toasted sesame oil or smoke oil (brushed on last after cooking)

1) Halve and steam tempeh for 10 minutes.
2) Mix sauce together.
3) Brush sauce over both sides of tempeh fillets and barbecue or broil for 10 minutes on each side.

Add buns and trimmings. Serves 2 people.

Reprinted by permission from A Vegetarian Ecstasty, *by Natalie Cederquist and James Levin, GLO, Inc., San Diego, 1990.*

MSG are exaggerated. MSG must be listed on all packaged foods that contain the additive.

Nitrate
Sodium nitrate and sodium nitrite are meat preservatives that have been used for centuries. They maintain the red coloring, inhibit the growth of botulism-causing bacteria and add to the flavor.

Nitrate is not harmful. However, bacteria in foods and the human body can easily convert it to nitrite. When nitrite combines with compounds known as secondary amines, it forms powerful cancer-causing nitrosamines. This threatening combination occurs in digestion or at high temperatures such as occur in frying. Bacon is especially susceptible to this reaction.

Pressure from the CSPI and Ralph Nader's Center for the Study of Responsive Law led to a ruling that makes it illegal for nitrate to be used in processed meats.

Saccharin
This sugar substitute was used for nearly a century before scientists questioned anything about the sweetener except its bitter aftertaste. In the 1970s, several studies linked it with cancer in laboratory animals. The FDA banned saccharin in 1977. However, public opposition to the ban was so persuasive that Congress was convinced to override the FDA's decision and exempt saccharin from normal food-safety laws.

: *What is the average annual energy bill for America's hot tubs?*

WHERE THE BUFFALO ROAM

WHAT MEAT HAS FEWER CALORIES and less fat and cholesterol than chicken? The answer, surprisingly, is *buffalo*.

"The popularity of buffalo meat is going up intensely, especially in the American West," says Charlie Maas, executive vice president of Denver Buffalo Company. "We sell 15,000 pounds each month to Safeway supermarkets in the Rocky Mountain region, and we sell to restaurants across the US and in Japan. People try it in a restaurant, it tastes great, and they start buying it."

"Demand for buffalo meat has outstripped supply," says Lorie Liddicoat, executive secretary of the American Bison Association. "People want it; it's a great alternative to beef."

The Denver Buffalo Company runs about 750 buffalo on their 14,000-acre Sweet Ranch, near Ramah, Colorado. No growth hormones, stimulants or

Buffalo is being promoted as the meat of the 1990s.

antibiotics are used on the animals. Their buffalo meat commands high prices: as much as $23 per pound for steak, and $3 or $4 per pound for ground meat.

Breeders believe only 100,000 buffalo exist in the US. Ninety percent of those animals are privately owned, and the rest are in Yellowstone and other parks. Buffalo are basically nomadic animals and wander too far, too quickly, to be let loose to graze on public lands, so they are most-often raised on ranches. Sweet Ranch cowboys use motorcycles and snowmobiles to manage their herd, Maas explains,

: *Around $200 million each year*
Source: 50 Simple Things You Can Do to Save the Earth, *Earth Works Press*

F. Jacobson, Ph.D., Lisa Y. Lefferts and Anne Witte Garland, Living Planet Press, Venice, CA, 1991.

Save Three Lives: A Plan for Famine Prevention, Robert Rodale, Sierra Club Books, San Francisco, 1991.

Green Wine

For almost every food item there is a less environmentally damaging alternative. And here's one for wine: Callaway Vineyard and Winery, Southern California's largest premium winery, uses a unique environmental program to produce some of California's finest white and red wines. They use an Integrated Pest Management system employing parasitic wasps, praying mantises and ladybugs to control crop-damaging insects; natural grasses grow between vineyard rows to distract leaf-eating insects. Platforms and nesting boxes were built to encourage red-tailed hawks and owls to provide 24-hour rodent control. Callaway also uses a drip irrigation system that saves enough water annually for 460 households of four people. Callaway wines are available in liquor stores nationwide.

"because buffalo wear out a horse. Buffalo can run just as fast, or faster, than a horse."

Billionaire Ted Turner, who recently converted his Montana ranch from cattle to buffalo, has called buffalo meat "the wave of the future."

For more information contact Denver Buffalo Company, 1109 Lincoln St., Denver, CO 80203, (800) BUY-BUFF; or American Bison Association, P.O. Box 16660, Denver, CO 80216, (303) 292-BUFF.

A few arguments for minimizing or eliminating meat from your diet:

The Hunger Argument
- Number of people who will die as a result of malnutrition this year: 20 million.
- How frequently a child dies as a result of malnutrition: every 2.3 seconds.
- Number of people who could be adequately fed using the land freed if Americans reduced their intake of meat by 10 percent: 60 million.
- Percentage of corn grown in the US eaten by people: 20.
- Percentage of corn grown in the US eaten by livestock: 80.
- Percentage of oats grown in the US eaten by livestock: 95.
- Percentage of protein wasted by cycling grain through livestock: 90.
- Pounds of potatoes that can be grown on an acre: 20,000.
- Pounds of beef produced on an acre: 165.
- Percentage of US farmland devoted to beef production: 56.
- Pounds of grain and soybeans needed to produce a pound of edible flesh from feedlot beef: 16.

The Environmental Argument
- Cause of global warming: greenhouse effect.
- Primary cause of greenhouse effect: carbon dioxide emissions from fossil fuels.
- Fossil fuels needed to produce a meat-centered diet vs. a meat-free diet: 3 times as much.
- Percentage of US topsoil loss directly related to livestock raising: 85.
- Number of acres of US forest cleared for cropland to produce meat-centered diet: 260 million.
- Amount of meat imported to US annually from Central and South America: 300 million pounds.
- Percentage of Central American children under the age of five who are undernourished: 75.
- Current rate of species extinction due to destruction of tropical rainforests for meat and other uses: 1,000 per year.

The Natural Resources Argument
- User of more than half of all water used in the US: livestock production.
- Gallons of water to produce a pound of wheat: 25.
- Gallons of water to produce a pound of meat: 2,500.
- Years the world's known oil reserves would last if every human ate a meat-centered diet: 13.
- Years they would last if human beings no longer ate meat: 260.
- Calories of fossil fuel expended to get 1 calorie of protein from beef: 78.
- Calories of fossil fuel to get 1 calorie of protein from soybeans: 2.
- Percentage of all raw materials (base products of farming, forestry and mining, including fossil fuels) consumed by the US that is devoted to the production of livestock: 33.
- Percentage of all raw materials consumed by the US needed to produce a complete vegetarian diet: 2.

For more information contact Earthsave, Box 949, Felton, CA 95018-0949.

 According to current population growth estimates, how long will it take for the Earth's population to double to ten billion people?

ECOHOME

Health

Your health is increasingly becoming an environmental issue—no surprise, since the majority of illnesses can now be attributed to environmental causes. Earth Journal *explains the news and offers tips to keep you healthy.*

WIRED FOR CANCER?

UNLESS IT'S BUZZING AT 7 A.M. ON A MONDAY MORNING, it's hard to think of your clock radio as dangerous. But according to a growing number of researchers, if you sleep with the electric clock by your head, it was bad news all night long.

In a 1990 draft report, the EPA took a major, if timid, step toward joining the growing number of scientists waving red flags about studies showing an increased cancer risk from the electromagnetic (EM) fields that surround electric devices. The EPA cited a "possible" cancer link. But EM expert Robert O. Becker, MD—author of *Cross Currents*, a provocative investigation of what's starting to be called "electropollution"—says EPA seriously underestimated the EM hazard. "It's not just 'possible,'" insists the

This Teslatronics meter measures magnetic fields generated by AC power lines, computers, and other electrical equipment.

Toxic Medications

Any kind of drug or medication in your home carries the potential risk of accidental poisoning from an overdose. Most home medicine cabinets are filled with over-the-counter, nonprescription drugs—painkillers, antacids, allergy medicines, cough syrups, laxatives and sleeping pills. The pharmaceutical industry is a multibillion dollar industry offering more than 200,000 products. However, if you want to try less hazardous and often less expensive remedies, there are several that have been used for centuries and can be bought at health food stores or made at home.

Natural Remedies

Headaches
According to Debra Lynn Dadd, author of *Nontoxic, Natural & Earthwise* (Jeremy P. Tarcher, Los Angeles), nine out of ten headaches are the result of apprehension, anxiety, depression, worry and other emotional states. Over-the-counter pain relievers provide little if any relief for this type of headache, says the FDA. Other common types of headaches are migraine headaches and hypertensive headaches, caused by a sudden rise in blood pressure. The FDA advises that these types of headaches not be treated with over-the-counter analgesics either. Other headaches are caused by

A: *43 years*
Source: Children's Television Workshop

inflammation—of the sinuses, of membranes surrounding the brain or even from a developing brain tumor. For these conditions, medical care is recommended, not over-the-counter drugs. So when is it appropriate to take an analgesic for a headache? The FDA recommends that analgesic be taken if you have a fever or a hangover, but otherwise analgesics should be avoided.

Dadd recommends a cup of strong peppermint tea and taking a short nap for temporary relief from occasional headaches, regardless of the type. Or try teas made from rosemary, catnip or sage.

Feeling Nauseous?
Try Nausea Relief Ginger Ale:
6 ounces fresh ginger root
2 cups water
1 1/4 cups honey

• Peel and finely chop the ginger (you should have about 1 cup).
• In an enamel or stainless steel saucepan, bring the ginger and water to a boil, then simmer for five minutes. Remove from heat and let stand for 24 hours, covered with a cloth (a kitchen towel works fine).
• Strain through two layers of cheesecloth (buy at a hardware store), and squeeze the pulp in the cloth to extract all possible juice.
• Return juice to saucepan, add honey, and bring to a boil over moderate heat, stirring to dissolve the honey. Simmer for five minutes.

professor of orthopedic surgery at the State University of New York's Upstate Medical Center in Syracuse. "It's quite real."

What exactly *is* an EM field? It's an invisible force that surrounds every electric current and appliance. You can't see or feel it, but you can see its effects. Pass a live wire over a compass, and the needle moves because the field generated by the current exerts an electromagnetic force on the compass' magnetized needle.

Since 1979, many studies have shown a disturbing association between long-term exposure to power-line fields—including neighborhood electric lines—and several types of cancer in children and utility workers.

To minimize EM risks, authorities recommend:
• In general, keep your distance from appliances and limit your exposure to EM fields.
• Don't place beds or cribs against walls with major appliances nearby or on the other side.
• Move electric clocks, clock radios and telephone answering machines at least four feet away from the head of your bed.
• Fluorescent bulbs generate much stronger fields than incandescent lamps. A foot or so is a safe distance from most incandescent lamps; keep at least three feet from fluorescent.
• Don't use electric blankets. Use quilts or regular blankets instead.
• TVs and video display terminals (VDTs) generate fields all around them, not just out from the screen. Sit at least arm's length from VDTs and six feet from 19-inch television screens. TVs with "instant-on" features have currents running through them even when they're off. It's prudent to keep a safe distance at all times.

It's also prudent to check the background field in your home. Fortunately, even if it's high, major rewiring is rarely necessary, says John Banta of Safe Environments, a California company specializing in home EM inspection. High background fields can often be eliminated by plugging appliances into different outlets.

How can you check your home? Several companies now market home EM meters, which cost $30 to $150, depending on their sophistication. I'm no electrician, so I called Safe Environments, which charges $200 for a complete inspection. It was money well spent. I felt

 How many nuclear weapons and reactors are deployed at sea by the five nuclear navies?

nervous about the tangle of utility lines a mere 10 feet from where my five-year-old son sleeps, but the meter, bless it, read less than one milligauss. The discovery of the health effects of EM fields proves that back in the 1950s, my parents were right when they admonished me to sit farther back from the TV.

—*Michael Castleman*

Reprinted by permission from Green Living, "Wired for Cancer?," by Michael Castleman, Summer 1992.

Resources:
The Body Electric, Robert Becker, MD, and Gary Selden, William Morrow, New York, 1985.

Cross Currents, Robert Becker, MD, Philomel Books, New York, 1990.

Currents of Death, Paul Brodeur, Simon and Schuster, New York, 1990.

The Electric Wilderness, Joel Ray and Andrew Marino, San Francisco Press, San Franciso, 1986.

Electromagnetic Man, Cyril W. Smith and Simon Best, St. Martin's, New York, 1989.

Electromagnetism and Life, Robert Becker, MD, and Andrew Marino, State University of New York Press, New York, 1982.

The Microwave Debate, Nicholas Steneck, MIT, Cambridge, MA, 1984.

Microwave News, P.O. Box 1799, Grand Central Station, New York, NY 10163, (212) 517-2800.

Power Over People, Louise B. Young, Oxford University Press, New York, revised, 1992.

The Zapping of America, Paul Brodeur, Norton, New York, 1978.

- Cool, pour into a bottle and refrigerate.

To use: Mix a soupspoon of syrup in a glass of carbonated mineral water or club soda, more or less to taste.

The Story of Cosmetics

Throughout history cosmetics have played a magical and enhancing role for both women and men, psychologically and physically. They promise consumers alluring images of youth and beauty, enhanced attractiveness—flawless skin, full heads of hair, renewed soft hands, longer and stronger nails. More sexuality. Cosmetics often deliver on their promises—for a price.

Cosmetics in the US represent a $20 billion industry. Some 20,000 different products are sold annually. Yet this industry is regulated by a division of the Food and Drug Administration that employs only 27 people and spends less than half of 1 percent of its budget on cosmetic safety surveillance, according to Ruth Winter, author of *The Consumer's Dictionary of Cosmetic Ingredients* (Crown, New York). The 156-page Federal Food, Drug and Cosmetic Act of 1938, which regulates this product group, contains one page on cosmetics. The industry is largely self-regulated. Most of the information consumers receive comes from advertising—not the most objective source.

A: *16,000 weapons and 500 reactors*
Source: Greenpeace

Yet many questions have arisen, particularly in the last 15 years, about the short-term and long-term safety of the cosmetics. The answers, thus far, aren't reassuring.

Scientific evidence is mounting about the toxicity of cosmetics. We now know cosmetics can cause cancer, chromosomal damage and other chronic diseases.

Beauty Without the Poison

In 1991, after nearly seven years of research, Estee Lauder introduced its Origins cosmetics line. This line of cosmetics contains safer ingredients not derived from or tested on animals; wherever possible, the packaging is recyclable. William Lauder, the 31-year-old general manager of Origins, notes the Origins target market is the customer with a conscience: "an intelligent, issues-oriented woman." Says *Newsweek*, "There are enough cosmetics consumers to justify Estee Lauder's investment of some $20 million." There are currently 100 Origins stores nationwide.

Bath and Body Works, a division of The Limited Inc., a major national retailer based in Columbus, Ohio, is one of the nation's fastest-growing cosmetic chains. Bath and Body Works assures customers its products are made wholly with natural ingredients and are cruelty-free. It plans 500 stores nationwide by the mid-1990s.

SIMPLE WAYS TO REDUCE FORMALDEHYDE IN YOUR HOME

ACCORDING TO THE FORMALDEHYDE INSTITUTE, the production of formaldehyde and formaldehyde-containing products accounts for 8 percent of the US gross national product. It is used in a wide variety of products. It is used in clothing to make it wrinkle resistant. Formaldehyde may also be in the material your kitchen cabinets are constructed of, and it may well be an ingredient in your shampoo.

What kinds of health effects can formaldehyde cause, and is it a problem for everyone? The answers are many and no. There is plenty of scientific proof that cigarettes cause lung cancer, but we have all heard of somebody who smoked his or her whole life, only to die at 90 of some other cause. Human beings vary not only in height, weight and hair color, but also in how much pollution their particular metabolisms can tolerate. Substances like formaldehyde and cigarettes affect each one of us differently. Some people aren't bothered at all, while the health of others is threatened. Yet we should avoid as much formaldehyde exposure as we can, just as most of us concerned about our health don't tempt fate by smoking.

Formaldehyde is carcinogenic in animals, and is suspected to have the same effect on humans. However, there are many other health effects of formaldehyde that should also be of concern. The most common complaints are mucous membrane problems such as eye, nose or sinus irritation, sore throats, runny noses, sinus congestion or coughs. Formaldehyde can also cause a wide range of other symptoms such as breathing difficulties, chest pain, wheezing, headaches, fatigue, nausea, difficulty sleeping, diarrhea or vomiting. Women may experience menstrual irregularities, and formaldehyde can occasionally trigger asthma attacks. It may even be a contributing factor in Sudden Infant Death Syndrome (SIDS).

One of the most insidious properties of formaldehyde is its ability to sensitize people to other pollutants, such as petrochemicals. Once someone has been sensitized, he or she will begin to react to extremely small exposures to formaldehyde, levels that were previously not a problem. The person then may require a house and furnishings that are completely free of

: *What percentage of tropical plants have been screened for potential medicinal drug use?*

formaldehyde. This can be very expensive because formaldehyde is used in so many products.

The people most at risk are young children, the elderly, those who are already ill, and pregnant women. This explains why construction workers aren't often affected—they are usually healthy adult males in the prime of life.

Over the last 20 years, manufacturers have drastically reduced the formaldehyde emission levels of products such as particle board, which is held together with "interior glue" containing formaldehyde. But there are still people who have complaints, sometimes very severe ones. A few manufactured wood products are being stamped with a "low emissions" label, but they still give off more formaldehyde than they would if they were made with exterior glue.

Kitchen cabinets are almost always loaded with formaldehyde because of the particle board shelving and veneered end panels. Many ready-made cabinets are also sprayed with a very potent formaldehyde-based finish. Fortunately, there are some solutions that can be used if you desire a low-formaldehyde kitchen. A standard countertop can be covered with laminate on top, bottom and all edges, to seal in a good deal of the formaldehyde emissions from the particle board core. Stainless steel, Corian and ceramic tile are less polluting but more costly choices. If cabinets are being custom made, most shops (at an added expense) can build the cases out of solid wood. There are even quality designer-style steel cabinets being manufactured (by St. Charles Mfg. Co., 1611 E. Main St., St. Charles, IL 60174, (708) 584-3800), and, for those on a tight budget, Sears also has a line of steel cabinets.

Unless you are chemically sensitive, and assuming you are in reasonably good health, your system can probably tolerate small to moderate exposures. Unfortunately, moving into a typical new house or one that has been recently remodeled may very easily result in large continuous exposures. This can be minimized by choosing materials carefully.

—*John Bower*

Reprinted by permission from Green Alternatives, *formerly* Greenkeeping, *"Healthy Building," by John Bower, March/April 1992, P.O. Box 28, Annandale-on-Hudson, NY 12504.*

Cosmetics industry consultant Allen Mottus recently told the *New York Times* that the trend of safer, socially conscious cosmetics is "a fad that's gone mainstream."

What was formerly a cottage industry has become a major growth segment within the $20 billion cosmetic industry: Clinique cosmetics are fragrance-free; Mennen baby oils are hypo-allergenic; Intelligent Skincare products are "allergy-tested, dermatologist-tested and fragrance-free"; Vidal Sassoon aerosols use air propellants instead of older, potentially more toxic chemicals; the Body Shop is a Europe-based sensation—offering customers natural cosmetics with minimal allergens and no chemical carcinogens. The chain has 66 stores in the US and plans 500 by the mid-1990s.

But let's not be fooled by the hype. Masqueraders on the shelves may look good, but a closer look at the ingredients reveals some charades. Many of these products use the same petrochemical-based materials that are part of the mega-toxification of our planet. They can hardly be called "Earth friendly." We need to begin making educated choices.

—David Steinman

Reprinted by permission from Green Alternatives, *formerly* Greenkeeping, *"Beauty and Safe Cosmetics," by David Steinman, March/April 1992, P.O. Box 28, Annandale-on-Hudson, NY 12504.*

A: **Less than 1 percent**
Source: National Wildlife *magazine*

Vegetable Growing Tips

1. Rotate crops to allow different parcels of land to rest.

2. Plant green manure like clovers, winter rye, vetch, alfalfa, fava beans, barley, millet and oats to enrich soil with nitrogen, attract earthworms and retard erosion. Plow under before flowering.

3. Use natural fertilizers such as manure, fish emulsion, seaweed and leaf mulch. Always compost.

4. Nurture beneficial insects. Lacewings, ladybugs, praying mantises (all available commercially), parasitic wasps, spiders, and tiger and whirligig beetles devour insect pests.

5. Encourage larger predators: Attract bats with ponds and bat houses; insect-eating birds with various berries, houses, food and birdbaths; frogs and toads with ponds; and snakes and lizards with cool cover areas.

6. Apply natural pesticides like the Bt (Bacillus thuringiensis) type of bacteria (available in stores as spore dust or liquid), diatomaceous earth, and commercial insecticidal soaps or mild soaps, such as Ivory.

7. Rely on homemade traps for insects—jars or buckets of water with sugar or molasses and yeast.

Gardening

Gardening is one of the best and most direct ways to be in touch with the Earth. You can produce your own food, eat healthy and feel good about your impact on the planet. Earth Journal *offers tips, ideas and information to help you plant a better world.*

GARDENING AT HOME

THE BEST WAY to learn about and care for the Earth is to get your hands dirty. Gardening can be an art form regardless of how much space you have in your yard to plant flowers, sprout rows of vegetables or cultivate luscious fruits. But aesthetics are not the only benefit from organic gardening—economics and health might just be a catalyst for a different kind of growth.

In comparison to food produced by commercial agriculture, per acre, a home garden can reap two to four times more food and save in produce costs. When we educate ourselves about various bio-intensive organic methods, this form of gardening can feasibly double our yields. Rewards also include a healthy diet free of polluting chemicals and increased nutritional food value.

Pesticides and fertilizers create residue in our food and contaminate our yards. Studies have shown that children who lived in homes where pesticides were used were four to six times more likely to develop leukemia. Pesticides are dangerous not only because of their side effects; they are also unsafe to keep around a house where children or pets could accidentally ingest them.

The principle of "ecogardening" lies in the discovery of how nature grows healthy plants. This involves biological, chemical and electromagnetic steps and undoubtedly many more processes that we do not yet understand. Naturally, the goal is to garden without pesticides. Organic gardening entails teaching your soil to be self-contained and learning what it needs to foster growth.

The study of natural ecosystems has led to a process called permaculture. With this model, gardens and farms can be developed that maintain diversity as well as stability. Included are methods of forest agriculture, edible landscaping, biological pest control, organic waste recycling and water/energy

: *How many trees does one ton of recycled paper save?*

efficiency. Integral also to permaculture are rainfall storage methods and increased availability of land for crop use.

Biodynamics, meaning "forces of life," is the study of the life processes of plants and animals, stating that soil is the fundamental component to plant, animal and human growth. Vegetables grown using biodynamic techniques taste better and last longer. Humans are not the only ones who benefit. When animals are given a choice, they prefer biodynamically grown grain.

Like many other environmentally sensitive methods, organic gardening is not more widely used because of habit. With a little time and thought devoted to organic gardening, it will help you relax, offer rewards, cost less and produce more.

CREATING AN EFFICIENT ORGANIC GARDEN

Sustainable Methods
- Landscape with edible plants, fruit and nut trees, flowers and herbs.
- Use native plants, which require fewer resources to maintain, and which are often threatened with extinction by plants imported from other regions.
- Create windbreaks with trees, bushes and other plants. This will protect your garden and attract beneficial birds and insects.
- Build a living fence out of trees and bushes, creating a habitat for wildlife and reducing your use of wood and metals.
- Use companion planting techniques. Organize your garden by lining up plants that support each other by providing benefits like shade, insect deterrence and growing support. For example, chives will chase aphids from roses.
- Minimize your use of plastic. Save toilet paper tubes to make biodegradable seedling pots. Cut tubes into two-inch lengths, place vertically in an old baking pan and fill with potting soil. When the seedlings are ready, they can be transplanted with minimal disturbance. Use paper mulch instead of plastic mulch.

Water-efficient Methods
- Xeriscape by using plants that do not require a lot of water. Use native plants and plants that can grow

8. Catch infestations early: Inspect daily. Cold blasts of water applied directly can help battle undesirable insects.

9. Handpick insects from plants in the early morning, when pests are usually sluggish.

10. Erect row covers of netting, or install commercially available spun fiber or floating row covers.
—*Elizabeth Berry*

Reprinted by permission from Countryside, *by Elizabeth Berry, July 1992.*

Climate and Lawn Grasses

You can maintain a healthy lawn without using an exorbitant amount of water if you choose the best grasses for your climate. According to *Caring for Lawns* by Mike Bowker, (Avon Books, New York) grass is divided into two types—a cool-season grass that grows in the cooler, northern climates, but withers in hot southern summers, and a warm-season species that flourishes in summer heat, but becomes dormant—turns brown—in colder weather. A combination of grasses is often used in areas where summers are hot and winters cold.

What to Plant:
South: Bermuda grass, St. Augustine grass, centipede grass.

Southwest: Bermuda grass, St. Augustine grass, zoysia.

 17 trees, and three cubic yards of landfill space
Source: The Student Environmental Action Guide, *EarthWorks Press*

Tropical: carpetgrass, Bermuda grass, centipede grass, bahiagrass and St. Augustine grass.

Mountain States, Western Canada: native grasses like buffalo grass, crested wheatgrass, blue grama.

Northeast, Eastern Canada: Kentucky bluegrass, fine fescues and perennial ryegrass.

Mid-Central US: fescues.

Midwest US, Central Canada: bluegrass, perennial ryegrass and fescue.

Pacific Northwest: Kentucky bluegrass and fine fescues.

Power Lawn Tools

Very few outdoor maintenance machines have pollution control devices incorporated into their technology. According to the Environmental Protection Agency, off-road gas machinery is a large contributor to suburban smog. A lawn mower can release as many smog-causing hydrocarbons into the air in an hour as a modern car.

Research conducted by the California Air Resources Board found that a chain saw operated for two hours emits as many hydrocarbons as a new car driven 3,000 miles. Modern machinery like cars are equipped with mechanisms that control engine variables to reduce pollution. Gardening tools have a few decades of

in your area with little additional water from you. Plants from open-pollinated seed are usually less demanding of water that those from hybrid seed.
• Install a drip irrigation system, which can save 60 percent to 80 percent of the water used in your garden.
• Use a water conditioner to reduce scale deposits on plant roots and assist plant intake of water and nutrients.
• Use squeeze nozzles on hoses.
• Save gray water, household wastewater that can also be a source of moisture for needy plants. The most practical place to start collecting water is in the kitchen. Keep a large pot or a gallon watering can near your sink, ready to catch leftover water. This includes water run from the tap before it gets hot, water used to rinse fresh vegetables or fruit, water used for steam-cooking, water used to boil eggs (which is also a good source of minerals) and water used to boil frozen food packets. Just remember to let hot water cool before tossing it on plants. Water used to rinse the dishes is also useful, provided it is free of nonbiodegradable soap, grease, fat or oil. Water from the last rinse cycle of clothes washers, if it is free of bleach, is also a good source of gray water. To collect this water, disconnect the drain hose just before the last rinse cycles and direct it into a large bucket. Use biodegradable soap. A drainpipe from a roof is another good spot to collect water with a large bucket.
• Water in the evening to minimize evaporation; run water slowly to allow the soil to absorb thoroughly.

Integrated Pest Management
• Identify the types of plants in your land and garden. Certain types of pests are attracted to certain types of plants, so you can get a good idea of the pests to look for.
• Begin setting traps to catch and identify pests. The most common type of snares are yellow sticky traps. Sticky traps are made of yellow cards or plastic covered with sticky material, which can be purchased or made with equal parts petroleum jelly and household detergent.
• Pinpoint the damage the pests inflict on the plants. Most pests have recognizable chewing or other destructive patterns.
• Keep a record of the date, number, type of bugs and pest control steps taken throughout the year.

 What is the oil energy equivalent consumed by an average American household each year?

- Clean your garden by picking up nearby fallen leaves, fruit, old boards and rocks.
- Apply integrated pest management control techniques: the release of predatory beneficial insects; the installation of lightweight insect barriers that keep pests from alighting on plants; the application of botanical sprays or dusts that are made from plants, not chemicals.
- Greenpeace recommends using pests to control less desirable garden pests. Ladybugs, bees, fly larvae, aphids, praying mantises, dragonflies, mites, spiders, toads, garter snakes and birds are all helpful in the garden.
- Handpick all visible offending pests.
- Plant collars will stop hatching larvae from burrowing into the soil around plants. Cut a piece of stiff paper, plastic or tar paper a foot square, shape it like a cone, and place it on top of the soil surrounding the plant. Use a paper clip to hold it together.
- Netting, such as cheesecloth, placed over a garden can help protect seedlings from chewing and egg-laying insects and will keep cats and birds away.

Natural Pesticide Recipes
- To kill snails and slugs, place a shallow pan filled with stale beer at infested areas or sprinkle salt on the ground.
- Place a large handful of tobacco in four quarts of warm water. After the mixture stands for 24 hours, dilute and apply with a spray bottle. (This solution is also poisonous to humans, so be careful.)
- Mix four quarts water, two tablespoons garlic juice (not garlic powder, it will burn the plant), 32 grams of diatomaceous earth (pulverized silica algae) and one teaspoon rubbing alcohol. This spray works for indoor and outdoor plants and can be frozen for future use.
- Mix two tablespoons of soap per quart of water. Or, mix 50 grams of dry soap per quart of water. Use only pure soap, as detergents will damage plants.

pollution-control catching up to do.

Lawn Pesticides

According to Jay Feldman, executive director of the Washington, DC-based National Coalition Against the Misuse of Pesticides, of the 36 most commonly used lawn pesticides, 13 can cause cancer, 14 can cause birth defects, 11 can have reproductive effects, 21 can damage the nervous system, 15 can injure the liver or kidneys, and 30 are sensitizers or irritants. Despite the potential hazards, American homeowners apply 67 million pounds of active lawn chemicals each year.

Bugs Adapt, Too

Pesticides may be useless against many pests: According to the Worldwide Insecticide Resistance Database, at the University of California at Riverside, over the last 40 years, nearly 600 species of insects have become resistant to major classes of insecticides. Insects can become resistant to particular pesticides in just a few growing seasons. This keeps farmers and manufacturers in a cycle of buying and creating new chemicals. Maybe it would be easier if *we* adapted and went organic.

A: *Over 1,200 gallons annually*
Source: Protect Our Planet Calendar, Running Press Book Publishers

Cars

Auto-Free Cities

• Since 1975, Singapore has charged a fee for each vehicle entering the city center—with the exception of buses, commercial trucks and cars holding more than four people. According to *What Works: Air Pollution Solutions*, published by the Environmental Exchange, Washington, DC, Singapore has successfully reduced congestion, traffic accidents and further road building through an expanded bus service, increased parking charges and high car taxes.

• In Geneva, Switzerland, car parking is prohibited outside offices in the central city. Commuters have no choice but to use the city's public transportation system.

Tire Waste

The best way to minimize personal tire waste is to buy long-life tires and keep them properly inflated, which will increase your gas mileage. Rotate and balance your tires every 6,000 miles. If possible, buy tires from dealers who guarantee they'll recycle your old tires. For more information contact: Directory of Scrap Tire Processors by State, Rubber Manufacturers Association, 1400 K St. NW, Washington, DC 20005, (202) 682-4800. They offer a free listing of companies that recycle tires.

Cars—a necessary evil? Perhaps. But necessity or luxury, they are a mainstay of America. If you must drive, there are some things you can do to reduce the impact of your car on the environment. Earth Journal *offers some tips for greener driving.*

AUTO-MANIA: WHAT IT'S DOING TO US

SINCE WORLD WAR II, our love affair with the automobile has changed the way we live. Unfortunately, the sprawling development that the car made possible has turned this love affair into an addiction. Government policies have generously subsidized car use and have been stingy about supporting mass transit for so many years that many people are now forced to rely on cars to get to work, shop or go to the doctor. In short, most of us need cars to survive.

To tackle air pollution from automobiles, we have enacted tailpipe emission standards; over the past two decades, individual motor vehicles have become significantly cleaner. But even though each vehicle belches fewer fumes, increases in total miles traveled have all but overwhelmed technological improvement. As we drive more, we pollute more. Americans now own more than twice as many cars as in 1960 and drive nearly three times as many miles. And our car use is growing faster than our population: The number of cars per capita has increased by about 60 percent in the past 30 years.

Cars have become such a sacrosanct symbol of our modern era and even our national identity that we seldom take a look at how they harm us. Here is some of what cars are accountable for:

Pollution. Cars are a triple threat: They are major contributors to toxic pollution, smog and ozone depletion.

• The Environmental Protection Agency estimates that motor vehicles (cars, trucks and buses) are responsible for more than half the incidence of cancer resulting from toxic pollution. Cars account for half of these emissions.

• Cars belch out more than two-thirds of the country's smog-forming chemicals.

• Automobile air-conditioners are considered the

 How many plant species face the possibility of extinction in the US in the next five years?

ECOHOME

largest single source of chlorofluorocarbons, the principal chemicals depleting our ozone layer.

Congestion. Americans now spend 1.5 billion hours each year stuck in traffic, and that figure is expected to triple by the year 2005. For the average person, this translates into six months of his or her life waiting at red lights and two years sitting in traffic. Counting time lost, gasoline burned, freight not sold and higher insurance premiums resulting from traffic tie-ups, the overall cost of congestion in our 39 largest metropolitan areas is estimated at $34 billion a year.

Inefficient Commuting. Two-thirds of all commuting cars in the US carry only their driver to work. If we added just one person to the average vehicle occupancy rate during rush hour, we would save 30 million to 40 million gallons of gasoline every day.

Land Use. In several of our major cities, 40 percent of the land is devoted to streets and parking lots. Ten percent of our arable land is paved, and more of our land is covered with pavement for cars than it is with housing for people; every year 100,000 Americans are uprooted to make way for more highways.

Junkyard Waste. Cars cost us after they die, as well. In a land already choking with waste, we junk 9 million cars a year, along with 250 million tires and 80 million lead batteries.

—*Mark Malaspina, Kristin Schafer and Richard Wiles*

Reprinted by permission from What Works: Air Pollution Solutions by Mark Malaspina, Kristin Schafer and Richard Wiles, The Environmental Exchange, Washington, DC, 1992.

BICYCLES:
THE ELEGANT POLLUTION SOLUTION

EACH PERSON WHO CHOOSES TO CYCLE instead of drive plays a significant role in reducing air pollution. One month of commuting five miles a day by bike instead of in a car achieves the following reductions:
- 3/4 pound of hydrocarbon emissions
- 6 pounds of carbon monoxide
- 1/2 pound of nitrogen oxides

And bicycles are cheaper than cars by any measure. Providing a parking spot in an urban center

Public Rail

Moving people by rail reduces air pollution. A trip via electronically powered subway or light rail produces 99 percent less hydrocarbons and carbon monoxide than the average single-occupant commute. Not only that, rail is cheaper to build, moves more people, and is safer.

• The average mile of light rail costs $10 million to $20 million; the average mile of urban interstate highway costs $100 million.

• One rail track can move as much traffic as an 18-lane highway.

• Since Amtrak was founded in 1971, 750,000 people have died in car crashes; 63 have died in Amtrak accidents.

Of the transit options, light rail is gaining ground. Heavy rail (underground, tube, metro, subway) operates on exclusive rights-of-way in tunnels or on elevated tracks, and each mile costs $70 million to $100 million to build. Light rail (trolley) is a much cheaper alternative that can run either on exclusive rights-of-way or with other traffic.
—*Mark Malaspina, Kristin Schafer and Richard Wiles*

Reprinted by permission from What Works: Air Pollution Solutions, by Mark Malaspina, Kristin Schafer and Richard Wiles, The Environmental Exchange, Washington, DC, 1992.

A: *More than 250 species*
Source: The Colorado Daily, Boulder, Colorado

Batteries

According to the Earth-Works Group, 60 percent of the world's lead supply comes from recycled car batteries. One car battery contains 18 pounds of toxic lead and a gallon of sulfuric acid. Some 330 million pounds of lead are polluting US landfills and, in turn, the groundwater, as a result of the 20 percent of car batteries not recycled. Lead is a toxin that can cause liver, kidney and brain damage. It is already responsible for the poisoning of 250,000 children a year.

Precautions:
• Check to see if your service station recycles batteries before you allow them to remove your car battery.

• Trade in your old battery for a new one at a store that guarantees it will be recycled.

• If you change your own battery, recycle it immediately. The EPA found that one-quarter of the people who change their own batteries have at least six of them in their garage. Battery acid leaks sometimes. The fumes and liquids are hazardous to eyes and skin.

Antifreeze

Most antifreeze is made with a petroleum product called ethylene glycol, that is hazardous to the environment and especially to pets (which it attracts because of its sweet taste). The most effective process of recycling antifreeze is

can cost as much as $1,750 per year. A secure bike locker costs as little as $200 and takes up much less space. A mile of urban interstate costs about $100 million; a mile of urban bike path costs about a thousand times less.

Not surprisingly, we have some serious catching up to do in the commuter's velodrome. In the Netherlands, about 30 percent of work trips and 60 percent of school trips are made by bicycle; in Denmark, about one-third of the country gets to work on two unmotorized wheels. Here we check in with a paltry 0.5 percent—but the numbers are growing. Cities that put time and energy into promoting and enabling bicycle use find that many people will at least give it a try, and some will make it a habit.

In Madison, Wisconsin; Eugene, Oregon; Gainesville, Florida; and Boulder, Colorado, 10 percent of all journeys are made by bicycle. Davis and Palo Alto, California, have even higher levels; about 25 percent of all trips are made by bicycle, resulting in enormous air-quality benefits.

What are some of the measures that pull potential pedalers out of their cars? Comprehensive, bike-friendly planning is key, and the cities that do it well build paths, set aside travel lanes for cyclists, provide amenities like parking and storage, and prepare bike route maps.

In Seattle, you can pedal 25 miles from the suburbs to downtown and almost never ride with cars. Seattle police patrol downtown areas on mountain bikes. The bike-on-transit program in San Diego, California, allows people to put bikes on buses. Several cities around the country sponsor Bike to Work days: in Boulder, Colorado, more than 500 people tried biking to work for the first time during a city-organized event. Many admitted that their respect for bicyclists increased after the experience, and a substantial number said they would try it again.

—*Mark Malaspina, Kristin Schafer and Richard Wiles*

Reprinted by permission from What Works: Air Pollution Solutions, *by Mark Malaspina, Kristin Schafer and Richard Wiles, The Environmental Exchange, Washington, DC, 1992.*

 If you let the faucet run while you shave or brush your teeth, how much water do you consume?

Ten Worst mpg Cars

Make/Model	City	Highway
Vector Acromotive W8	7	11
Lamborghini DB132/Diablo	9	14
Rolls-Royce		
Silver Spirit IV/Silver Spur II	10	14
Touring Limousine	10	14
Corniche IV	10	14
Bentley Brooklands/LWB	10	14
Bentley Continental	10	14
Ferrari 512 TR	11	16
Rolls-Royce		
Bentley Turbo R/LWB	11	16
Bentley Continental R	11	16

Ten Best mpg Cars

Make/Model	City	Highway
Geo Metro XFi*	53	58
Honda Civic HB VX*	48	55
Geo		
Metro*	46	50
Metro LSi*	46	50
Suzuki Swift (1.0-liter engine)*	46	50
Honda		
Civic HB VX	44	51
Civic*	42	46
Geo Metro LSi Convertible*	41	46
Honda Civic	40	46
Suzuki Swift (1.3-liter engine)*	39	43

*(with shift indicator light)

The Geo Metro XFi is the most fuel-efficient car on the market today.

called redistilling. Call a local service station or a hazardous waste facility to facilitate the process.

Oil

One quart of motor oil can pollute 250,000 gallons of water, according to the Earth-Works Group. About 62 percent of all oil-related pollution in the US is the result of improper disposal of used motor oil, which seeps out of containers and leaches into groundwater. Always make sure your used oil is recycled either at a local garage or a hazardous waste facility, and replace your oil with re-refined oil (recycled oil).

For oil recycling information:
Recycling Used Oil: 10 Easy Steps, EPA, Office of Solid Waste, 401 M St. SW, Washington, DC 20460, (202) 260-2080.

American Petroleum Institute, 1220 L St. NW, Washington, DC 20005, (202) 682-8000.

A: *3 to 5 gallons every minute*
Source: 50 Simple Things You Can Do to Save the Earth, *Earth Works Press*

Are Cities Inherently Unhealthy?

Over the years, people have migrated to the big cities, even though polls show people would rather live in small country towns. The post-World War II compromise was to live in suburbia and work in the city, but with commuting times up to three hours a day, the widespread separation of home from the workplace has become a serious problem that has increased stress—a negative for health.

In some ways, cities are healthier than rural areas. For example, the quality of drinking water is high in most cities, while a significant fraction of rural dwellers drink from polluted wells. City dwellers have prompt access to nearby emergency health services; rural dwellers must resort to distant emergency rooms. While the air is polluted in many cities—ozone particularly continues to be a disturbing problem—country dwellers are more threatened by insecticides and ticks. Congestion is much higher in the cities—an annoyance, but one that contributes to a lower rate of traffic fatalities than in the faster-moving vehicles on country roads and highways.

Above all, country dwellers lack the range and challenge offered by the city's best jobs, as well as the companionship and constant stimulation of urban social and cultural life. From the

Urban Ecology

Our urban environment affects us in ways we often overlook. Where we live impacts our social outlook, community concerns and quality of life. Earth Journal brings you insight into and information about our urban ecology.

- By Carl Anthony -

THE PARKER ELEMENTARY SCHOOL IN EAST OAKLAND, California, is a remarkable place. Its students are predominantly Latino and African-American, and the entire curriculum of the school is organized to give these inner-city kids an environmental education. According to a survey undertaken by Earth Island Institute's Urban Habitat Program, Parker Elementary is one of 50 projects in the San Francisco Bay area aimed at building ecological awareness and environmental justice for the region's African, Asian, Native American and Latino communities.

Kids from Parker take field trips to the surrounding mountains and coastal regions, and undertake projects to clean up their neighborhood. Yet the school is full of paradoxes.

"To be effective teaching environmental issues in inner city schools," says Running Grass, executive director of the Three Circles Center for Environmental Education, which is responsible for the innovative curriculum, "you must be a part of the struggle for educational justice. Overcrowded classrooms, lack of teaching materials and the stresses of the surrounding neighborhoods have an effect on the ability of teachers to teach and the students to learn."

Recently a teacher, showcasing the school to a group of funders, was explaining to an attentive fourth grade class the difference between needs and wants. "Needs are basic," she explained, "what you must have to survive. Wants, on the other hand, are what you would like to have, but don't necessarily need." She was trying to get the students to understand the difference between resources for survival and frills, to help them visualize waste in our consumer society. "Can anyone give me an example of a need? Something you absolutely have to have?" One student's hand was up and she called on him.

Q: *What is saved by recycling one ton of paper?*

"Guns," he replied. "You need guns."

The teacher, a little embarrassed, tried to gloss over the unexpected reply, paused, and then asked, "What do the rest of you think? Is this an example of a need?"

"Yes," the students chimed in, "you need guns."

This story illustrates the psychological and emotional denial at the heart of our urban crisis, frequently revealed within the environmental movement. To many environmentalists, the social environment is invisible. For them, the central problem is the relationship between human beings and nature. They ignore racial and class antipathies that are often at the bottom of ambivalent feelings toward cities. Young students at Parker Elementary School have not yet learned to make such arbitrary distinctions. They are aware of their surroundings and know what it takes to survive in a community under siege. Within a one-mile radius of the Parker school, there were over a hundred homicides last year.

Urban ecologists propose livable cities, often creatively addressing energy and natural resource conservation. But many proposals for sustainable cities reflect the bias of middle- and upper-middle-class entrepreneurs and visionaries, with little awareness of racial and class dynamics, a part of everyday urban experience. They criticize suburban single family zoning and over-reliance on automobiles, resulting in waste of precious fossil fuels, traffic congestion and air pollution. They dream of cities with bicycle paths and pedestrian pockets, solar neighborhoods, mixed-use buildings, European "mini communities" with work, home, public life and shops within walking distance. But they are at a loss to address racial discrimination, inequity, fear, violence and other fiscal and social ills at the root of our urban crisis. Yet these too must be addressed if we are to have an effective social movement for sustainable cities.

Paradoxes and contradictions around meeting basic needs are evident in many areas of inner-city environmental management, such as open space, transportation and housing. Community gardens sprout up on sites slated for affordable housing, often pitting community activists against one another. Recently, a developer of housing for the homeless was very upset with new lead-abatement regulations in San Francisco's Tenderloin District. Compliance

perspective of mental health, cities offer diversion that potentially makes it easier to break dependence on cigarettes or alcohol. (Alcoholic over-indulgence is a common hazard in rural areas, especially during inactive winters.)
—John Tepper Marlin

Excerpted from The Livable Cities Almanac, by John Tepper Marlin. Copyright © 1992 by JTM reports, Inc. Reprinted by permission of HarperCollins Publishers.

Environmental Stress and Population

• An overpopulated area can strain a community's ability to meet everyone's needs—leading to decayed urban conditions such as crime and homelessness. According to The Livable Cities Almanac, these are the least and most stressed metropolitan areas, based on rates of alcoholism, suicide, divorce and crime.

Best Ten:
1. Akron, OH
2. Paterson, NJ
3. Bismarck, ND
4. Sioux Falls, SD
5. Grand Rapids, MI
6. Syracuse, NY
7. Lincoln, NE
8. Buffalo, NY
9. Fort Wayne, IN
10. Madison, WI

Worst Ten:
1. Oklahoma City, OK
2. Phoenix, AZ
3. Fort Lauderdale, FL
4. Los Angeles, CA
5. San Francisco/Oakland, CA
6. Jacksonville, FL
7. Little Rock, AR
8. Miami, FL

: *17 trees, 7,000 gallons of water, and energy to heat a home for six months*
Source: Greenpeace Action

9. Las Vegas, NV
10. Reno, NV

City Drinking Water Quality
• Using four measures of quality: turbidity, chlorine, nitrates and pH.

Top ten out of 72 cities:
1. Savannah, GA
2. Hartford, CT
3. Oakland, CA
4. Sacramento, CA
5. Newport News, VA
6. Phoenix, AZ
7. Richmond, VA
8. Billings, MT
9. Columbia, SC
10. Raleigh, NC

Idaho

• Idaho has the cleanest air in the US, according to the *Green Index* (Island Press, Washington, DC).

• The Greener Pastures Institute found that Idaho has the third-greatest amount of land area in Bureau of Land Management and US Forest Service hands. Some 64 percent of Idaho is preserved land.

• According to the Environmental Protection Agency's Toxic Release Inventory, Idaho is second-best in the West behind Nevada in the total annual amount of toxic wastes released into the environment (15.1 million pounds, compared to California's 201.5 million).

The Best Place To Live

According to *Money* magazine, Sioux Falls,

with the new regulations would add $800 to the cost of each unit, enough to kill the deal. "My clients will die of overexposure from sleeping outside," he said, "long before they will be poisoned by lead."

A recent requirement that Bay Area public transit buses convert to alternative fuels is so expensive that routes already poorly serving the inner city must be cut back. Without relief, transportation control measures such as raising the cost of gasoline to reflect its true cost, or raising highway and bridge tolls, will penalize the poor.

Despite these paradoxes, the movement for urban sustainability has enormous potential for communities engaged in struggles for racial and social justice. This belief is the basis for Earth Island Institute's Urban Habitat Program in the San Francisco Bay area. The program is organized to promote multicultural, urban, environmental leadership. Its long-term goal is to develop a regional model of multi-ethnic, multi-class and multi-issue organizing helping to make the transition from destructive urban environmental practices to more sustainable ones.

If the San Francisco Bay area is any indication, interest in environmental issues is rapidly growing in the working class and communities of color. Since Earth Day 1990, at least a dozen groups in Bay area African-American, Asian-American and Latino communities emerged to fight toxics in the air, soil, water and built environment. Four new multicultural environmental networks have formed: the Coalition for Environmental Justice at the University of California at Berkeley, the Asian American Environmental Justice Network, the San Francisco People of Color Greening Network, and EDGE: An Alliance of Ethnic and Environmental Organizations, a coalition of established statewide organizations working on environmental justice. Complementing these projects, the Urban Habitat Program has developed new environmental justice programs in energy and transportation, urban agriculture, youth education and economic conversion.

Forging ties between environmental groups and advocates of racial and social justice could be extremely valuable in efforts to transform patterns of urban consumption and waste. Such alliances can help environmentalists deepen their understanding of connections between the environment and the

: *Which country was the first to incorporate ecotourism into a national policy?*

economy, protecting hard-won environmental gains and working to build a new economy based on diversity and greater local and regional self-reliance.

Communities of color are political majorities in 51 of the nation's 200 largest cities. The experiences, conditions and perceptions of these communities and are central to any successful movement for urban sustainability in the US. Advocates of urban ecology must join forces with these communities and their movements for public safety, educational reform and racial justice if they hope to realize the full potential of the sustainable-city idea in the nation's largest and most troubled urban areas.

What you can do:
1. Start an inventory of social justice organizations in your community. Learn the priorities of each community group and make new connections between environmental issues and community objectives. Remember that ecological design springs from the recognition that there is more than one right answer.

2. Acknowledge that organizations in the social justice community may have useful experiences related to urban ecology. Community development corporations, for example, know a great deal about how to put together land trusts and build affordable housing.

3. Support community groups on the front lines of the fight for environmental justice.

4. Use the arts as an organizing tool. For example, in the Bay Area, the Earth Drama Lab put together a successful series of "EcoRap" concerts, giving young people a chance to express their own views of environment and justice.

5. Remember, organizations in working-class communities, poor communities and communities of color may have a lot to learn about sustainable development. They also have a lot to teach.

Carl Anthony is president of Earth Island Institute and Director of its Urban Habitat Program. He teaches a course on Race, Poverty and the Environment at the University of California at Berkeley.

South Dakota, is the best place to live. It lacks pollution, traffic congestion, violent crime and state and city income taxes. In *Money*'s 1992 ranking of 300 metropolitan areas, Sioux Falls surpassed last year's top-ranked cities: Provo and Orem, Utah.

Environmental Research Spending

• In an environmental survey of the US, *Biocycle* magazine (April 1992) reported that New Jersey, Florida, Minnesota, Michigan and Pennsylvania spent the largest amount on recycling programs during the fiscal year of 1991. New Jersey spent $37.7 million, Florida—$30 million, Minnesota—$25 million, Michigan—$22 million and Pennsylvania—$15.4 million.

• The same study found that the five states that spent the most on air quality programs are California—$91 million, Florida—$12.8 million, New York—$11.4 million, Texas—$10 million and New Jersey—$8 million.

• The *Biocycle* survey found that energy research was given significantly less fiscal priority with the exception of Oregon, which spent $40 million on energy research in 1991. The four states following Oregon in energy research spending are California—$35 million, Louisiana—$4.4 million, Hawaii—$4 million and Nebraska—$2 million.

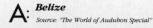
A: *Belize*
Source: "The World of Audubon Special"

VI
GREEN BUSINESS

"*G*etting and spending, we lay waste our powers"—and often we lay waste the Earth, as well. Green Business looks at investors, industrialists and working men and women who believe in sustainable growth: renewable energies; recyclable products; and goods and services that heal, rather than harm, the Earth.

Green Taxes

Friends of the Earth International, a worldwide environmental network, is lobbying for tax changes that would yield environmental and public health benefits, bolster sustainable economic vitality, increase employment in the US and eliminate environmentally harmful subsidies. Such an environmental tax reform package would tax products that pollute, deplete natural resources or otherwise degrade the environment, and perhaps lower other taxes that drain hard-earned income and savings or make the costs of labor or capital high. In addition to taxing pollution, small tax incentives could be provided to promote conservation and efficient use of natural resources.

Encouraged by the proposal, more environmentally sound methods of production, consumption and disposal would provide more jobs and reduce the long-term and immediate effect industry has on environmental health. Energy- and resource-intensive industries pollute the environment and use over 20 percent of total energy, but provide only 3 percent of jobs.

According to Friends of the Earth, the tax policy could be a powerful tool for environmental protection and sustainable resource use, by encouraging environmentally beneficial rather than harmful activities. For more information about environmental

The Greening of Business

Companies around the world are cleaning up their acts—or starting out green. Earth Journal brings you information on companies doing well by doing good.

SMALL COMPANIES PREVENT ENVIRONMENTAL PROBLEMS

SOME OF THE HOTTEST COMPANIES in the environmental industry hope to clean up in ways other than cleaning up.

The pollution-fighting business has long been the environmental industry's mainstay, accounting for the lion's share of revenues. But now some rising stars belong to a new breed of businesses more concerned with preventing future pollution sins than in correcting past ones, says Jeffrey Leonard, president of Global Environment Fund L.P., a Washington, DC, investment partnership.

As much as half of US environmental spending by the year 2000 will spring from "nonregulatory" factors rather than from antipollution laws, Leonard says. Examples include efforts to make manufacturing processes more efficient, to conserve energy and to meet consumer demand for nonpolluting, "green" products.

Besides trimming manufacturing waste, the ounces of prevention offered by the new breed will save companies pounds of cure for environmental problems—a particularly attractive combination of pluses when corporations are emphasizing cost containment. Sums up Michael Silverstein, a Philadelphia author who writes about the environmental industry: "The real action is going to be in avoiding environmental expenditures rather than making them."

The new field should be especially fertile for entrepreneurs. Rather than deep pockets or high-tech patents, it will require the ability to identify emerging social trends regarding the environment and to quickly harness existing technology to meet demand created by the trends, Leonard says. But the field may have pitfalls, such as consumer reluctance to pay more for green products. Here are snapshots of three of the new breed.

Q: *How much will the EPA spend this year to research the role of livestock flatulence in global warming?*

In Search of Green Consumers

Mo Siegel, a founder of tea maker Celestial Seasonings, calls himself "your basic environmentalist forever." So it's only natural he'd let his predilections prevail in a new business venture, Earth Wise, Inc.

The Boulder, Colorado, company is one of many commercial sprouts planted in the past few years to capitalize on consumers' stated desire to buy environmentally sound products. Interest in such products was near a peak in 1990 when the company launched its first contenders: detergents that biodegrade readily and trash bags made from recycled plastic. Last year it became a unit of Celestial Seasonings when Siegel—who had left Celestial Seasonings in the mid-1980s to travel the world—rejoined the closely held tea company as chief executive and sold his new environmental business to it.

Entering the green arena is no walk in the park, Siegel says. For one thing, there's "a major gap between what people say they want" in the way of green products and what they're willing to spend more for, he says.

Green products often cost more than traditional alternatives, making them especially vulnerable during a recession. One reason for Earth Wise's higher prices is that detergents that both biodegrade rapidly and clean as well as traditional alternatives require a "complicated blend" of relatively costly materials, Siegel says.

Skepticism about the benefits of purported green products represents both a problem and an opportunity, says Siegel. Hype by some makers of green products erodes consumer trust. But companies whose products stand out as environmentally beneficial when scrutinized by government regulators or consumer groups will gain a major competitive advantage. "I'm totally in favor of a crackdown on exaggerated environmental claims," he says.

Perhaps the biggest problem for emerging green-products companies will be getting shelf space. Grocery chains typically charge companies "slotting fees" of $10,000 or more to carry their products in a few stores in one town, Siegel says. Such fees can be prohibitive for small companies. "We calculated our product line would cost $1.5 million just to get into Boston," where a number of chains operate, he says.

Earth Wise now sells mostly through natural-food

taxes, contact Dawn Erlandson, Director of Tax Policy, Friends of the Earth, 218 D St. SE, Washington, DC 20003, (202) 544-2600.

Earth Share

Fundraising in the workplace, which (according to *Common Ground*) accounts for $3.2 billion in American charitable giving, until recently excluded most environmental causes. In 1988, Earth Share was founded to include 18 environmental foundations in payroll deduction fundraising in the US. Currently, Earth Share represents 40 groups and raises $7.5 million annually from federal, state and private sector employees. More than 50 corporations donated to Earth Share in 1992, demonstrating a concern for the preservation of the environment.

To add your employer to the list, contact Woody Sellers, The Conservation Fund, 1800 N. Kent St., Ste. 1120, Arlington, VA 22209, (703) 525-6300, or Earth Share, (800) 875-3863.

Earth Share Members and Beneficiaries
African Wildlife
  Federation
American Farmland
  Trust
American Forests
American Rivers
Americans for the
  Environment
Center for Marine
  Conservation
Citizen's Clearinghouse
  for Hazardous
  Wastes
Clean Water Fund

A: *$130,000*
Source: Harper's *magazine*

The Conservation Fund
Conservation International
Defenders of Wildlife
Environmental Action Foundation
Environmental Defense Fund
Environmental and Energy Study Institute
Environmental Law Institute
Friends of the Earth
INFORM
International Institute for Energy Conservation
The Izaak Walton League of America
Land Trust Alliance
National Audubon Society
National Coalition Against the Misuse of Pesticides
National Parks and Conservation Association
National Toxics Campaign Fund
National Wildlife Federation
Natural Resources Defense Council
The Nature Conservancy
Pesticide Action Network
Rails-to-Trails Conservancy
Rainforest Alliance
Safe Energy Communication Council
Scenic America
Sierra Club Foundation
Sierra Club Legal Defense Fund
Trust for Public Land
Union of Concerned Scientists
US PIRG Education Fund
The Wilderness Society
World Resources Institute
World Wildlife Fund

stores, which don't charge slotting fees. To give the company a second wind, Siegel is considering folding it into a joint venture with a large consumer-products company. But he remains optimistic about the business's long-term outlook.

Helping Utilities Conserve
One of the fastest-growing of the new environmental segments is the energy-efficiency business.

In the past few years, US electric utilities have committed billions of dollars to "demand-side management" programs aimed at getting their customers to use less electricity. Among other things, they are helping to pay for their commercial customers' purchases of compact fluorescent light bulbs, high-efficiency motors and other energy-saving equipment.

Utilities are farming out much of the consulting and installation work associated with demand-side management. Xenergy, Inc., of Burlington, Massachusetts, is one of a number of companies formed to take advantage of this growing market for "energy services."

Formed in 1975, Xenergy was one of the first companies to enter the energy-conservation business. Now it's growing explosively: Revenue jumped some 80 percent in 1991 to more than $20 million; the company's work force almost doubled to about 250. "I can't get people fast enough," says Stanley Kolodkin, co-founder and president.

Xenergy's role in demand-side programs typically begins with an "energy audit" of a utility customer's facility to identify the most effective ways to save electricity; everything from elevator motors to ventilation fans might be taken into account. Xenergy's analysis becomes a blueprint for installing energy-efficient equipment and helps determine how much of the customer's cost the utility will cover under its demand-side program. Xenergy then arranges the installation of the equipment and tracks the energy savings.

With their niche's electrifying growth rate, energy-services businesses are experiencing growing pains. Perhaps the biggest problem is their heavy reliance on utility programs to foster energy efficiency. Many of the programs are still in the pilot phase and hence are subject to sudden cutbacks. Still, "we don't see the push for demand-side management fading," says Steven Tadler, an energy-services expert at Advent International, a Boston venture-capital concern that

: *How do the relative annual budgets of General Motors (GM) and the environmental group Greenpeace compare?*

invested $4 million in Xenergy last year. "It's going to be very hard for utilities to site new generating plants" due to environmental and community concerns, prompting them to focus more on limiting demand for power than on expanding its supply. That, of course, will mean more business for companies like Xenergy.

One of Xenergy's strong suits is high-efficiency lighting, says Andrew Schon, vice president. The company has installed 28 lighting technologies in its headquarters, including novel energy savers such as high-efficiency fluorescent lights that automatically dim when the sun comes out and brighten when it sets or is obscured by clouds.

Designing "Interior Packaging"
Some of the fastest-growing components of trash—such as plastic packaging—are among the hardest to recycle. To environmentalists trying to foster more recycling, that's a headache. To Moulded Fibre Technology, Inc., it's a business opportunity.

Formed about a year ago, the Westbrook, Maine, company makes "interior packaging," the material inserted in product boxes to surround and protect what's inside. Such protective packaging is typically made of polystyrene, a plastic seldom recycled. Moulded Fibre's alternative is Reflexx, a material made from recycled newspapers and water using a proprietary process invented by co-founder Roger Baker. "It's akin to the process used to make egg cartons," says Peter Troast, the company's managing director.

Moulded Fibre's packaging is made entirely of recycled material and is itself easily recycled, claims Troast. "Our molded inserts can be recycled in any newspaper recycling network," he says. By contrast, to recycle polystyrene packaging samples that Moulded Fibre's customers send in as models, "our only choice at this point is to ship it to a center in Glens Falls, New York, where they charge us $7 a pound," he adds.

Moulded Fibre's future looks bright partly because of growing public pressure to recycle packages, which account for about 30 percent of trash. In Massachusetts, legislators are considering a bill that would require packages to meet recyclability standards, such as the including of certain percentages of recycled materials.

"Since our products are made of 100 percent recycled material, we can help a lot with the batting average" on recycled-content standards that

Green Relations—US and Mexico

An information clearinghouse has been designed to initiate environmental programs between US and Mexican companies. The Environmental and Energy Efficient Technology Transfer Clearinghouse in Mexico City, was set up in January 1992 by the World Environment Center to complement the North American Free Trade Agreement (NAFTA). The clearinghouse is a computer-based information system designed to give Mexican environmental and energy professionals access to US energy, environmental and trade information. It will access data from over 500 private sector and US government databases. The clearinghouse will also provide information on environmental pollution protection and control technologies, providers of environmental products and services, and US government hotlines. The clearinghouse is funded by the US Environmental Protection Agency and the US Department of Energy.

Eco-Photography

According to the *Washington Post*, every year Americans use almost 700 million rolls of 35mm film, nearly 100 million rolls of 110 and 126 film, and another 100 million or so packages of Polaroid-type film, disc film and other types.

: *Greenpeace's annual budget equals about four hours worth of GM operation*
Source: Greenpeace *magazine*

To alleviate the associated waste problems, Enco Photo, a mail-order film processor, evaporates silver-bearing solutions and stockpiles the metal residue for future sales. The company packages its own film brand in canisters recycled by customers. A nonprofit clearinghouse for environmentally responsible photography, PhotoGreen, accepts the materials Enco Photo cannot use.

Ask your local film processors if they recover the maximum amount of silver, use the least amount of water and discharge a minimum level of chemical waste. Do they recycle byproducts of processing like film canisters and paperboard packaging?

To have your film processed ecologically, get film mailers from Enco Photo Lab, P.O. Box 1151, Old Forge, NY 13420, (800) 659-3190. For information about environmentally responsible photography, send a SASE to: PhotoGreen, P.O. Box 124, Hampton, NJ 08827, (908) 537-7694.

GEMI

The Global Environmental Management Initiative (GEMI), a coalition of Fortune 500 companies, was formed in 1990 to foster environmental excellence by international business. GEMI was founded to protect the environment through improved management practices using Total Quality Environmental Management practices.

companies may soon have to meet, Troast says.

Reflexx's biodegradability also appeals to customers. A spokesman for Shiva Corp., a Cambridge, Massachusetts, seller of computer peripherals, says that's largely why his company packages modems in the material.

Troast says his company's packaging is comparable in price and performance to polystyrene. In some cases it damps vibration and protects products better than polystyrene does, he claims. But packaging made of a spongier plastic called polyurethane is more resilient than Reflexx, and hence may work better for protecting very fragile items.

Reflexx also is "gray-looking stuff," which may limit its use with "slickly packaged" products that are put on display in stores, says Troast. On the other hand, some environmentally correct customers think Reflexx's "look is aesthetically superior. They even ask us not to grind up the newspapers so much so that the letters will show" after the papers are recycled into packaging inserts.

—*David Stipp*

Reprinted by permission from the Wall Street Journal, *"Small Companies See Growth Potential in Preventing Environmental Problems," by David Stipp, June 1, 1992.*

WHAT SOME BUSINESSES ARE DOING TO SAVE THE EARTH

The Body Shop is committed to supporting and promoting social and environmental change for the better. Its packaging is minimal and most of its plastic containers can be refilled or returned to its stores for recycling. None of its products have ever been or will be tested on animals. It produces naturally based products and supports both Amnesty International and Happiness Is Camping.

Browning-Ferris Industries, Inc., one of the world's largest waste service companies, has received a preliminary certification from the Wildlife Habitat Enhancement Council to convert one of its North Carolina landfills into a permanent habitat for native and migratory wildlife species. The proposed 390-acre resting spot is expected to support owls, larks, frogs, toads, turtles, rabbits and other species. Completion of the project will take up to 20 years.

 How much water do Americans use as compared to Europeans?

GREEN BUSINESS

Anita Roddick, founder and managing director of The Body Shop, researching product ingredients in Brazil.

Automakers **Chrysler Canada, Ford Motor Company of Canada** and **General Motors of Canada** announced an agreement with the Canadian government and Ontario's provincial government that commits them to the voluntary reduction of the use, generation and release from their manufacturing facilities of toxic substances. The Automotive Manufacturing Pollution Prevention Project will emphasize source reduction and will target priority chemicals like chlorinated solvents, metals, CFCs, PCBs and volatile organic compounds. The project is part of Ontario's strategy, announced in September 1991, to encourage all stakeholders to make pollution prevention the primary means of achieving their environmental priorities.

Chrysler Corporation, Ford Motor Company and **General Motors Corporation** formed a partnership to conduct research on recycling and disposing of

Current GEMI Members
Allied Signal Inc.
Amoco Corporation
Apple Computer Inc.
AT&T
The Boeing Company
Browning-Ferris Industries
Digital Equipment Corporation
The Dow Chemical Company
Duke Power Company
Eastman Kodak Company
E.I. du Pont de Nemours Co.
Florida Power & Light
Merck & Company, Inc.
Occidental Petroleum Corporation
The Procter & Gamble Company
The Southern Company
Union Carbide Corporation
W.R. Grace & Company
Weyerhaeuser Company

Database Waste

The EPA has announced a free database available to companies that need the latest information on innovative hazardous-waste treatment technologies, such as bioremediation, thermal desorption, chemical treatment and soil washing. The database, VISITT, is available on diskette for IBM and compatible PCs. EPA plans to update the database annually. Call the VISITT Hotline at (800) 245-4505 or (703) 883-8448.

CERES Principles

Several US companies have offered to issue

 Americans use two to four times more water per person
Source: The Student Environmental Action Guide, EarthWorks Press

environmental reports to shareholders in exchange for a withdrawal of shareholder ultimatums. For the past three years, two shareholder groups have been presenting environmental-responsibility shareholder resolutions at annual meetings, demanding change in company environmental policies. The Coalition for Environmentally Responsible Economies (CERES) and the Interfaith Center for Corporate Responsibility (ICCR) placed resolutions on the ballots at the annual meetings of 67 companies in 1992. Among the demands were that the firms endorse the CERES Principles (previously known as the Valdez Principles), which (in addition to other things) urge companies to release documented information about environmental performance.

Although none of the 67 companies passed the shareholder resolutions, their administrators were forced to consider the CERES Principles. Amoco, AT&T, Chevron, Browning-Ferris Industries, Gannett, Monsanto, Occidental Petroleum, Polaroid, and Waste Management Inc. are among the firms that have already agreed to provide environmental performance reports in exchange for CERES and ICCR withdrawing shareholder resolutions. Presently, 48 companies have signed the CERES Principles. However, none of the Fortune 50

materials from used automobiles. The Vehicle Recycling Partnership's goals are to increase the amount of recyclable and recycled materials in vehicles and to develop guidelines on material selection and compatibility, bonding methods and materials, painting and design for disassembly.

CIBA-GEIGY Corporation built a reverse osmosis and ultra-filtration system to recycle water used in dye blending, recycling 99 percent of the water used in some operations.

Consumers Power Company has begun a program to recycle used refrigerators and freezers. The $65 million recycling program "Reduce the Use" allows customers to exchange a used appliance in working order for a $50 US Savings Bond. The utility has hired Appliance Recycling Centers of America to manage the pickup and processing of the used appliances.

Co-operative Bank, Manchester, England, has become the first UK bank to establish a set of social criteria used for determining the types of businesses it will accept as customers. A subsidiary of the Co-operative Wholesale Society, it has begun screening all of its corporate customers based on a set of 12 ethical criteria. The bank will refuse to do business with companies found to be involved in illegal activities, dealings with oppressive regimes, arms manufacturing, tobacco production, animal testing and exploitation, including destructive environmental practices.

Dow Chemical Company restored and enhanced surrounding wetlands through an employee program at a Joliet, Illinois, plant. Volunteers planted 255 acres of grasslands to increase nesting success of game and songbirds.

Ecover, a Belgian manufacturer of laundry detergents and other cleaning products with minimal environmental impact, has allocated about $4 million on a new addition to its factory, with about 30 percent of that amount going specifically for ecological features. Grass and plants will be planted on the roof of the factory to provide insulation. A reed pond will be used to treat the factory's wastewater, which can then be reused to water grass or clean machines in the

Q: *According to a recent CNN/Time Magazine poll, what percentage of Americans consider the environment to be a "very important issue"?*

factory. Ecover is considering the installation of small wind energy collectors to operate the closed-cycle wastewater treatment system.

Esprit designed Ecollection, clothes manufactured with minimal environmental impact, for the spring of 1992. The line, priced from $30 to $150, was created using few dyes or synthetic fibers and no harmful processes (no acid or stone-washing, no resin or formaldehyde finishing). Where needed, the clothes incorporate recycled metal or plastic (as in zippers), and the fasteners and other hardware are designed without electroplating or oxidation, which have by-products. Esprit also supports economically depressed areas and vanishing cultures. For example, Esprit's reconstituted glass buttons are made from old bottles by handicraft workers in Ghana.

Federal Express is running a test fleet of 113 vans built by Chrysler, Ford, General Motors and VEHMA (which is making the vehicles supplied by Southern California Edison). The vehicles will run on methanol, electricity, propane, reformulated gasoline and compressed natural gas. The project is a means of gathering information to influence public policy on alternative fuels.

Florida Power Corporation in partnership with beach communities agreed to recessed streetlight use to restore nesting habitat for endangered baby loggerhead sea turtles.

Georgia Gulf Corporation installed one of the nation's first large-scale, industrial, jet-aerated bioremediation projects to reduce air emissions and volatile organics in process water.

H.B. Fuller set aside 95 acres on its property as a nature preserve, underwritten by a $1 million grant from the company.

Henry Ford Hospital in Detroit is one of five hospitals working with Baxter Healthcare Corporation in a pilot program to recycle plastic materials from intravenous systems. Hospital workers can deposit plastic IV solution bags, pour bottles and overpouches into one container. Baxter will collect the materials

firms with which CERES has been in negotiation have agreed to the CERES Principles.

ICCR and CERES are working with companies to standardize environmental disclosure, using the CERES Principles report form as a guide. The form is a detailed environmental checklist. In the process of completing the form, a company is forced to acknowledge its behavior and practices.

In addition to the name change—which was made because of the negative connotations "Valdez" implies—some other program changes have been made. Companies are now allowed to endorse the principles rather than "sign" them, as a result of company lawyer disagreements with the word "sign." The report form and the principles have been altered in response to input from several companies and individuals. The changes include the removal of repetitive questions and other stylistic changes. The strategy and meaning of the principles remain the same.

Scrap and Reclaim

Los Angeles—Air quality agencies and private industry in the Los Angeles Basin are looking for innovative, alternative mechanisms to cut through the brown cloud of inefficiency that too often stalls pollution control programs.

California's South Coast Air Quality

94 percent
Source: National Wildlife *magazine*

Management District (SCAQMD), through its innovative RECLAIM program, is seeking to alleviate problems of control and implementation between agencies and industries. RECLAIM, which stands for Regional Clean Air Incentives Market, would limit air quality agencies' efforts to determine emission-reduction schedules for each industry, leaving the specifications for meeting those schedules to the discretion of the company. This would allow the company to determine the most efficient program for reducing its emissions.

Each polluting facility would be given an initial cap on air emissions, which it would then be required to reduce by a specified percentage each year to meet state and federal statutory requirements. SCAQMD expects these percentages to be between 5 percent and 6 percent for hydrocarbons (HC), and between 8 percent and 9 percent for nitrogen oxides (NOx) and sulfur oxides (SOx).

Trading Credits
What sets RECLAIM apart from previous approaches to air quality improvement is the leeway that it lends companies in meeting proscribed emissions caps; it creates a market in which they can earn and trade "pollution credits" among themselves. A pollution credit is earned for achieving greater emission reductions than required for a

from the participating hospitals and send them to a processor for sorting, cleaning and use in new products such as vinyl floor runners and mats. In addition, the hospital incinerates its waste, using a steam cogeneration system for heat.

Imperial Chemical Industries set up an Ecogrant program in which company employees and retirees actively support specific environmental projects. More than 30,000 people in the US and Canada qualified in 1992 to apply for funding for specific environmental projects benefiting their communities.

Kodak implemented a recycling program to help retailers and photoprocessing labs manage their solid photographic waste in an economical and environmentally responsible way. The recyclable film waste does not have to come from Kodak. All labs in the US may participate. Three independent recyclers and a transportation firm will collect and recycle specified solid waste from photofinishers. Kodak will provide special containers and materials for collection and shipping.

Lasertek has attempted to attack the production of over 54 million pounds of nonbiodegradable plastic toner cartridges used annually in more than 9 million laser printers. Lasertek has improved the remanufacturing process through inventions like an internal seal to prevent toner leakage in cartridges. Lasertek will pick up empty cartridges free of charge anywhere in the US and pay an incentive rebate or make a charitable donation to the National Wildlife Federation or the National Center for Missing and Exploited Children in your name. Call (800) 252-7374 for service.

McDonald's Corporation developed a list of 62 solid waste reduction initiatives, independently and in conjunction with the Environmental Defense Fund. The list includes an "Optimal Packaging Team" of interdepartmental managers who regularly review opportunities to reduce and/or eliminate secondary and shipping packaging. McDonald's established a new independent company called TriAce to work with restaurant operators to optimize solid waste management practices and develop recycling and composting infrastructures. The company also expanded

 What is the passenger carrying capacity of railroad tracks as compared with highway lanes?

its internal McRecycle Team to explore opportunities to use recycled materials in every area of its business.

The Mennen Company orchestrated a land-swap deal between local government, environmentalists and industry. The deal resulted in the preservation of 11 acres on Pyramid Mountain in New Jersey. Some 445 acres have been put under contract to be acquired for the park.

Nissan Motor Company, Ltd., of Japan started a recycling program for plastic automobile bumpers in Japan and Germany. Both programs are possible because of Nissan technology that enables the company to remove paint from the bumpers. Nissan will recycle the plastic for internal reuse in making air ducts, rear bumper parts, foot rests and transport pallets and will also recycle urethane chips into products such as floor mats.

Royal Bank of Canada helped fund the construction of the building that houses the Center for Biological Diversity in Georgetown, Guyana. The University of Guyana, the World Wildlife Fund and the Smithsonian Institution opened the center in June 1992 with a gift from the bank of $881,000. Royal Bank is among the first banks in North America to develop a comprehensive environmental policy.

Seven companies in the San Francisco area have formed the **Recycled Paper Coalition** to encourage corporate paper recycling and stimulate the demand for paper products made from recycled materials. BankAmerica and Pacific Gas & Electric initiated the coalition. Other members include Chevron, Pacific Bell, Safeway, George Lithograph and Wallace Computer Services.

The **Union of Wood and Plastics Manufacturers**, the **Central Federation of the German Wood and Plastics Processing Industry** and the **Association of German Timber Importers** pledged in February 1992 to import only tropical timber certified to have been harvested in a sustainable manner. The three groups called for the creation of a labeling program that would certify the origin of tropical timber and vowed to cooperate with environmental organizations working to protect forests.

certain time frame. This is the keystone of the RECLAIM program. A company that has the resources to modify its polluting systems ahead of its reduction schedule will receive pollution credits, which it can then trade or sell to other firms that may require additional time to meet their own schedules. Firms will also be able to use pollution credits to balance modification schedules for separate facilities within their companies. This would allow industry to achieve equal net benefits at considerably lower costs. As long as overall air quality improves, and the set quantitative objectives are met, why should it matter where the reductions begin?

Clunkers
In mid-1991, Union Oil Company (UNOCAL) took this line of thinking one step further, implementing a creative, highly successful experiment called SCRAP, or South Coast Recycled Auto Project. Driven by "enlightened self-interest," UNOCAL offered the public $700 for any pre-1971 vehicle that could be driven to a designated site, where it was then scrapped and recycled. In order to participate in SCRAP, owners had to prove that their vehicles had been registered and in use for at least six months, and thus were a real factor in the region's pollution woes. Older cars have been found to emit a disproportionately high share of the total US vehicle pollution.

  Trains can carry about eight times as many passengers per hour
Source: Campaign for New Transportation Priorities

California's Department of Motor Vehicles estimates that about 400,000 pre-1971 vehicles travel Southern California's freeways and side roads. Though they represent only 6 percent of the total vehicle population in the Los Angeles Basin and travel a mere 3 percent of the total vehicle miles, a US Office of Technology Assessment study determined that these aging vehicles are responsible for 22 percent of total emissions of hydrocarbons, 15 percent of carbon monoxide and 13 percent of nitrogen oxides.

SCRAP not only accomplished its objectives, but vastly exceeded UNOCAL's expectations. UNOCAL allotted a $5 million budget for the program, enough to purchase and scrap 7,000 vehicles. Contributions from individuals, a variety of corporations and private firms, and various regulatory agencies, such as the South Coast Air Quality Board, raised an additional $1 million, enough to scrap another 1,400 cars.

Good News
To gain a precise measure of SCRAP's effect on air quality, UNOCAL tested the tail-pipe emissions from each vehicle before relegating it to the jaws of the crusher. The average level of hydrocarbons exceeded UNOCAL's expectations by three times, while the carbon monoxide average was two times greater than predicted.

Westinghouse Electric is working with seven Polish power stations to implement a modernization and maintenance program for Poland's fossil-fueled power plants. Over the next ten years, Westinghouse will work with the power stations to improve the efficiency, output and environmental performance of 45 coal-fired units throughout the country.

GREEN CAREERS

WITH ESCALATING PRESSURE ON BUSINESSES to become more environmentally conscious, green professions are developing into a viable option for those who wish to make a career out of protecting the environment. Environmental management is not only wise, but also mandatory if a business wishes to remain competitive and lucrative.

What are some of the areas that need consultation and attention? Broadly, manufacturing, production and management strategies must be revamped. Companies are now forced to examine every aspect of their business from the extraction of raw materials to the disposal of the finished product, and each step in-between. This life-cycle assessment provides a framework for a company to examine product quality as well as cost, and subsequently creates careers for environmental specialists.

The general public now demands that a company be versed in the environmental components of its businesses. The environmental factor is a facet of decision-making for managers in areas such as purchasing, public relations, marketing, financial management, research and development, accounting, sales, personnel, training, strategic management and human resources.

Specifically, some of the jobs generated by the environmental crisis are listed below.
1. Environmental consultants are regularly called upon when companies are making transitions and managers need to become better informed about the necessary changes.
2. Banks need environmental investors and researchers, since their assets are lowered when they loan money to companies with contaminated land.
3. Nonprofit organizations provide careers in public interest foundations, think tanks, labor unions and trade associations. Analysts within these groups are hired to assess environmental issues relating to

On average, how much oil does it take for the world's farmers to produce one ton of grain?

education, safety and allocation of funds and resources.
4. Environmental services are needed to promote pollution control and waste management. Businesses are needed to create new technology for these clean-ups and to promote ongoing sensitivity toward the environment.
5. The petroleum industry needs environmental engineers, biologists and consultants to perform studies on the environment. Other natural resource industries must follow suit.
6. Chemical firms also need environmental engineers, as well as compliance administrators and product and marketing managers.
7. Organic foods have created a niche in industry, as people are concerned not only about the ingredients in their foods, but also about the pesticides used to grow them. Experts are needed for pest management, organic gardening, retailing of organic food and mail-order sales.
8. Environmental lawyers are consulted regularly by industry. Understanding of environmental law is an important component of decision-making to those in accounting, marketing, finance and management.
9. Insurance companies have had to acquire the cost of cleaning up waste sites left by firms carrying their policies. Environmentally aware underwriters are necessary, as is awareness by the writers of real estate and health care policies.

The above examples reveal the edge given to those in the job market who are environmentally educated. Environmental literacy can be part of a job's main responsibilities. If you are competing with someone who lacks environmental skills, you are more than likely to win.

Resources:
Environmental Careers: A Practical Guide to Opportunities in the 90s, David J. Warner, Lewis Publishers, Boca Raton, FL, 1992.

Environmental Jobs for Scientists and Engineers, Nicholas Basta, John Wiley & Sons, Inc., New York, 1992.

Green at Work: Finding a Business Career that Works for the Environment, Susan Cohn, Island Press, Washington, DC, 1992.

Only levels of nitrogen oxides conformed to the clunker emission forecast (still 11 times greater than from a 1990 vehicle). Some 20 percent of the vehicles exceeded the measuring capacity of the testing equipment altogether. Emissions from these vehicles, then, are more than 90 times dirtier than those of new models. With all tallies in, SCRAP was credited with removing 12.8 million pounds of pollutants from the air in the Los Angeles Basin.

Though UNOCAL was awarded a small delay for completing modifications at one of its facilities as a result of SCRAP, its goal was to prove a point, and this point was well taken. The US Senate and House of Representatives showed interest in the UNOCAL experiment, and the Subcommittee on Energy and Power of the House Committee on Energy and Commerce ordered the Office of Technology Assessment to examine SCRAP in further detail. The South Coast Air Quality Management District is also developing plans to include mobile sources in RECLAIM. Representatives of SCAQMD expect large-scale trading of pollution credits to begin by the end of 1993, pending its approval by the agency's governing board. It might just be high time to put that old car to rest.

A: *A little more than one barrel*
Source: Worldwatch *magazine*

Green Funds

In 1992, the number of socially responsible mutual funds available nearly doubled, according to *Investing for a Better World*. Here are a few ways you can make an investment with the environment in mind.

Calvert Social Investment Fund
4550 Montgomery Ave., Ste. 1000 North, Bethesda, MD 20814, (800) 368-2748; (301) 951-4800 in Maryland.

Calvert Group offers the nation's first and largest family of socially and environmentally responsive mutual funds, made up of many market, bond, equity and managed growth "balanced" portfolios.

Domini Social Index Trust
6 St. James Ave., Boston, MA 02116, (800) 762-6814.

A passively managed fund that invests in an index of 400 companies that have passed broad social screens. It focuses on firms with positive records in providing useful products, employee relations, community activities and the environment. It excludes firms deriving revenue from weapons sales, tobacco, alcohol, gambling or nuclear power.

Dreyfus Funds
200 Park Ave., New York, NY 10166, (800) 645-6561; (718) 895-1206.

A growth and income fund that invests in the stocks of

Green Investing

Although money may look green, it doesn't necessarily act green. As consumer consciousness grows, people are seeking ways to support businesses with an ecological and social conscience, and are refusing to support companies that degrade the Earth. Earth Journal helps you decipher the many options for your dollars.

PROFILE: GREEN CENTURY FUNDS

IN THE PAST SEVERAL YEARS, consumers have witnessed the emergence of several new options for investing their funds. The choices vary by level and quality of risk, rate of projected return and the manner in which the capital is invested.

According to the Social Investing Forum, the fastest-growing niche in money management is socially responsible investing. The socially responsible investing movement has grown continually over the last 20 years, and presently more than $650 billion is invested based on some degree of social screening criteria. In addition to the growth of this sector, there is increasing segmentation of investment by social issue. For many years, social investing meant primarily not putting money in "sin" stock companies that manufactured alcohol or tobacco. The effort was largely initiated by religious institutions. Then, in the 1970s, the movement expanded to avoid war industries—not investing in corporations that manufactured weapons.

In the 1980s, conditions in South Africa caught the attention of investors. Hundreds of billions of dollars were directed to companies that had severed their ties to South Africa. Over the past few years, screened investments have expanded to include environmental protection, equal opportunity for employees and justice in Northern Ireland.

Environmental Screens
One of the newest entries in the social investing arena is a screen for environmental management. Green Century Funds, based in Boston, is the "first family of no-load, environmentally responsible mutual funds," says Mindy S. Lubber, president of Green Century Capital Management, Inc., who, prior to

: *In 1945, US corn growers lost 3 percent of their crop to insects. Today, while pesticides are widely used, what is the crop loss?*

launching the fund, was with the Massachusetts Public Interest Research Group.

There are several environmental-sector funds that invest in companies in waste management or cleanup and remediation. The oldest among them is Fidelity Select Environmental Assets. Others include Kemper Environmental Services, Financial Strategic Environmental Services, Freedom Environmental and Oppenheimer Global Environmental. Those funds typically do not screen their investments.

"It is important to distinguish the Green Century Funds from environmental-sector funds, most of which purchase securities in companies whose business derives from solving environmental problems, even if those companies have consistently negative environmental records," explains Lubber.

Her group believes that companies ignoring environmental laws and paying fines will not be profitable in the long run. "At some point, they will have environmental problems [for their investors] due to their actions," she says. "Corporations should remain environmentally responsible because it affects their bottom line. It is in their own long-term self-interest."

Green Century seeks to enable investors to earn competitive returns through investments in companies with strong environmental records, as well as in companies with limited negative impacts on the environment. "As we enter a new century, what is good for business will increasingly be good for our environmental future," Lubber says.

Investment Criteria

Green Century's stated criteria for investing in companies clearly reflects a more aggressive approach to environmental protection. Criteria on the "will invest" list are companies that meet or exceed environmental regulatory requirements, strive to improve environmental performance in a proactive manner, and are committed to a healthier environmental future. This includes investments in renewable energy, recycling, sustainable agriculture and effective remedies for existing pollution problems.

Conversely, Green Century states it will not invest in companies that have a pattern of noncompliance with state or federal environmental laws, use excessive packaging, or produce harmful toxic chemicals or other products that are particularly harmful to the

companies with positive records in the areas of consumer protection, occupational safety and health, and environmental issues.

Freedom Environmental Fund
John Hancock Mutual Funds, P.O. Box 9116, Boston, MA 02205, (800) 225-5291.

This fund's prospectus states that it "will invest primarily in companies which the fund advisor believes contribute to a cleaner and healthier environment." The fund may invest in companies engaged in pollution control, waste management or pollution-waste remediation.

Global Environment Fund L.P.
1250 24th St. NW, Ste. 300, Washington, DC 20037, (202) 466-0529.

The partnership's primary objective is to realize long-term capital appreciation through investments that promote environmental improvement.

Green Century Funds
29 Temple Place, Ste. 200, Boston, MA 02111, (800) 93-GREEN; (617) 482-0800.

These funds are the first and only family of "no-load environmentally responsible mutual funds." The funds invest only in companies that meet or exceed environmental regulatory requirements and that also strive to improve their environmental performance with clearly stated policies and programs.

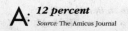
A: *12 percent*
Source: The Amicus Journal

Kemper Environmental Services Fund
811 Main St., Kansas City, MO 64105-2005, (800) 621-5027.

This fund states in its prospectus that it "will invest at least 65 percent of its total assets in the equity securities of issuers engaged in environmentally related activities."

New Alternatives Fund
295 Northern Blvd., Great Neck, NY 11021, (516) 466-0808.

Invests in common stocks of companies in the solar and alternative energy industries. It does not invest in petroleum- and atomic-based energy sources. However, the fund may invest in petroleum companies that are actively developing in areas such as solar electric generation, etc.

Oppenheimer Global Environmental Fund
Individual Investors, P.O. Box 300, Denver, CO 80201-0300, (800) 525-7048.

This fund invests in companies offering products, services or processes that contribute to a cleaner and healthier environment. It regularly invests at 65 percent of its assets in common stocks of environmental companies located in the US and in at least three foreign countries.

Parnassus Fund
244 California St., San Francisco, CA 94111, (800) 999-3505; (415) 362-3505 in California.

The fund does not invest in firms that environment. Like many other social investment funds, it also stays away from companies that manufacture tobacco products, produce nuclear energy or manufacture equipment for nuclear power or nuclear weapons, or that are primarily engaged in business in South Africa.

In addition, Lubber says, Green Century "offers more diversity in our investments than environmental-sector funds. Our criteria allow us to spread out the risk across the market. There are many Fortune 500 companies that are not doing anything to violate environmental laws."

Green Century Capital Management, Inc., is the fund's administrator and advisor. One hundred percent of the firm's net earnings are dedicated to non-profit environmental protection programs.

Because of the fund's no-load status, it is unlikely that many investors will be referred to it by stockbrokers. Investments are made directly to the fund through an agreement with a distributor. Green Century is concentrating on reaching environmental activists. "Our market is any person who wants to invest and also wants to make a statement about the environment," says Adrienne M. Shishko, Green Century's general counsel and vice president of marketing.

Prospective investors are reached through magazine advertisements, publicity and literature distributed through environmental organizations. "Because our proceeds are given to environmental groups, we work closely with them to promote the fund," says Shishko.

For more information contact Green Century Funds, 29 Temple Place, Ste. 200, Boston, MA 02111, (800) 93-GREEN.

—*Steven M. Rothstein*

Reprinted by permission from In Business, *"Investing in the Earth," by Steven M. Rothstein, August 1992.*

THE FIVE PRINCIPLES OF USING YOUR PRINCIPLES

Principle 1: Start by determining your financial needs and goals. In other words, start where any investor does. Do you want security, high income or the opportunity to make large capital gains? Since you won't be able to do all these things at once, identify the most important.

 Since 1959, how many dolphins have been accidentally killed by "encirclement," a wasteful net-casting procedure used in tuna fishing?

Principle 2: Identify your most important social concerns. Is nuclear power the most important issue to you? Perhaps women's rights or animal testing? Don't try to solve all the world's problems at once. Make priority judgments about what you consider the most important social issues at this time. Weigh the clear social positives against any potential social negatives in an investment decision.

Principle 3: Keep your ethically absolute social screens to a minimum. There is no such thing as complete ethical purity in the social world. A long list of absolutes covering every imaginable sin will simply make prudent investing impossible. However, it is possible and relatively simple to screen for some types of absolutes.

Principle 4: Decide what your investment can accept in terms of financial risk. And contrast this with potential social benefit. Would you be willing to increase some economic risk because the potential social benefit is so great? Would you be willing to sacrifice some economic return because the potential social return is so great? An example might be community development loan funds that provide an excellent service for the community but offer a return somewhat lower than market rates.

Principle 5: Decide what social risks are involved in your investment. Remember that the risk factor doesn't always go against social investments. People who got out of weapons-related stocks for ethical reasons, for example, may have been making very good decisions on financial grounds as well. Socially undesirable factors can prove financial liabilities.
—*Ritchie P. Lowry*

Reprinted by permission from Business Ethics *magazine*, "The Five Principles of Using Your Principles," by Ritchie P. Lowry, May/June 1992, 52 Tenth St., Ste. 110, Minneapolis, MN 55403.

generate electricity from nuclear power. It also looks for companies with a good environmental protection policy.

Pax World Fund
224 State St., Portsmouth, NH 03801, (800) 767-1729; (603) 431-8022 in New Hampshire.
A balanced fund that invests in life-supportive goods and services such as health care, education, pollution control and renewable energy. It excludes firms that are engaged in or that contribute to military activities, as well as firms in the liquor, gambling or tobacco industries.

Working Assets Money Fund
111 Pine St., San Francisco, CA 94111, (800) 223-7010; (415) 989-3200 in California.
Working Assets does not knowingly invest in companies that generate nuclear power or manufacture nuclear equipment or materials. Nor does it knowingly invest in corporations that consistently violate regulations of the Environmental Protection Agency or the Occupational Safety and Health Administration.

A: *More than 7 million*
Source: Greenpeace *magazine*

Green Catalogs

The market for environmentally sensitive products is expanding to include everything from eco-clothing and eco-toys to chocolate and buttons from the rainforest. Green catalogs benefit the environment by creating sustainable markets in impoverished areas of the world, promoting environmental awareness and donating a percentage of their profits to conservation projects. Below is a sampling of green catalogs.

Biobottoms, P.O. Box 6009, Petaluma, CA 94953, (800) 766-1254.

A catalog chock-full of fun clothing for your child. Biobottoms "fresh air wear" incorporates 100 percent cotton, bright colors and simple construction techniques to create comfortable, quality clothing for kids.

The Body Shop, 45 Horsehill Rd., Cedar Knolls, NJ 07927, (800) 541-2535.

This "value over profit" minded catalog features magnificent oils, cremes, shampoos and other natural body products, as well as T-shirts and infant supplies. The Body Shop's "Trade Not Aid" agreement encourages rainforest sustainability by employing indigenous people to gather product ingredients.

CARE, 660 First Ave., New York, NY 10016, (800) 422-7385.

The world's largest private relief and development organization

Green Consumerism

If you are a conscientious consumer, you can change the world with your buying power. Earth Journal *gives you information to help you make good choices for the Earth.*

WHEN PRODUCTS ARE TIED TO CAUSES

IT USED TO BE THAT THE LAST THING companies wanted to do was to take a stance on a controversial issue that could offend a single customer. Aiming to please everyone, they chose to help mainstream, do-good causes to which only a Scrooge would object: Toys for Tots, Easter Seals, the March of Dimes.

But a new generation of corporate executives, by linking their products to abortion rights, AIDS prevention, gun control and other issues that roil the nation, is gradually beginning to change the way many goods and services are sold. Point-of-purchase politics, as it is called, aims to appeal to some consumers but definitely not to others, winnowing them through a striking strategy of speaking out.

Companies are finding that their politics can be good for business. Leading adherents of point-of-purchase politics, like the Benetton clothing chain and Ben & Jerry's ice cream, are enjoying record sales and profits. Their success has emboldened more traditionally minded marketing giants to link their own products with causes.

Time Warner, Hearst, Sara Lee and Heublein are cautiously supporting causes that combine charity and politics. Sara Lee's Hanes apparel unit has donated hundreds of thousands of dollars in money and clothing to a foundation that helps the homeless. Heublein introduced a vodka by sponsoring the AIDS Danceathon in New York, organized by the Gay Men's Health Crisis.

Nigel Carr, director of brand planning for the Kirshenbaum & Bond advertising agency in New York, said that for many products, provocation can pay. "The meaning it has to people who don't like it is as important as the meaning it has to people who do," Carr said.

Many of the advertisements for these products read

: *In 1820, 71.8 percent of Americans were farmers. By 1990, what had the number fallen to?*

GREEN BUSINESS

like protest signs. The Esprit de Corp. clothing maker runs magazine advertisements in which young people assert what they would do about gun control, AIDS and abortion rights. "Keep a woman's right to choose . . . unless George Bush is free to baby-sit," one advertisement says.

Approach by Working Assets

The Working Assets Funding Service seeks customers for a long-distance telephone service that donates 1 percent of phone charges to groups including Amnesty International, Greenpeace and the American Civil Liberties Union. A Working Assets print advertisement shows a protester making an obscene gesture, under the headline: "Twenty years later, we've given people a better way to put this finger to use."

Many nontraditional marketers were 1960s activists, and most of their causes are left-wing. Point-of-sale politics is the latest example of how liberals, exiled from the White House for more than a decade, now pursue their goals through commerce—while conservatives, who have long dominated the world of business, have been achieving many of their goals through politics. You will probably not see pints of ice cream anytime soon tied to causes dear to Pat & Jesse (Buchanan and Helms)."

Longer-standing strategies for welding commerce to causes include credit cards that help benefit groups like the Sierra Club and mutual funds that invest only in companies deemed socially responsible. "If the political system doesn't express what we want, fine," said Peter Barnes, the president of Working Assets. "People will express themselves through the marketplace."

Some Political in Word Only

While companies often present themselves as altruistic crusaders for righteous causes, philanthropic groups and business executives are loudly clearing their throats. They question whether some businesses are merely basking in an aura of caring, without the commitment.

"Some companies that are political in their advertising have absolutely no political involvement," said Neil Kraft, Esprit's former image director, now at Calvin Klein in New York. "It's already becoming a gimmick, and that worries me."

To traditional marketers, activism has been a

publishes a catalog featuring toys, notebooks, backpacks and T-shirts. All proceeds from catalog sales go to CARE.

Coop America, 2100 M St. NW, Ste. 403, Washington, DC 20037, (800) 424-2667.

Pleasant, fun products from groups that are socially and environmentally responsible. An array of environmental foods, decorations and clothing.

Cultural Survival, 215 First St., Cambridge, MA 02142, (617) 621-3818.

Founded 20 years ago, this nonprofit organization provides money, technical support and markets for indigenous people's products all over the world. The catalog features Rainforest Crunch foods, as well as an in-depth selection of books about indigenous cultures.

The Daily Planet, P.O. Box 1313, New York, NY 10013, (212) 334-0006.

This catalog features such products as tin-can briefcases from Kenya and hand-pressed bark paper from India. The Daily Planet is a showcase of goods from around the world, and it donates 10 percent of its profits to environmental and social action groups.

Earth Care Paper, Inc., P.O. Box 770, Madison, WI 53707-7070, (608) 223-4000.

Offering a full line of fine paper products made from recycled paper, Earth Care's

A: *2.4 percent*
Source: The Daily Camera, *Boulder, Colorado*

catalog will fulfill all of your stationery and greeting card needs. The company was an original signatory of the CERES Principles for corporate responsibility, was a founding member of Business for Social Responsibility, works with Business Partnership for Peace to affect global economic policy in the third world and donates 10 percent of its profits to environmental and social programs.

Earth Tools, 9754 Johanna Place, Shadow Hills, CA 91040, (800) 825-6460.

A unique catalog featuring jewelry, clothing and furnishings from artisans around the world. The beautiful selection includes silk scarves woven by Hmong refugees and rag rugs braided by White Earth and Leech Lake Indian Reservation women.

Enviro Clean, 30 Walnut Ave., Floral Park, NY 11001, (516) 775-1425.

A necessary addition to the ecologically sensitive household, this catalog features a variety of safe alternatives to conventional cleaning and hygiene products.

My Favorite Planet, 1740 Broadway, New York, NY 10019, (212) 645-4641.

Recently upgraded from a brochure to a catalog, My Favorite Planet offers child-oriented products to the conservation-minded. The products are made of natural fibers, have no excess packaging, save

prescription for marketing suicide. It runs the risk of generating resentment among consumers who either disagree with the company's opinions or resent corporate proselytizing. Working Assets, for example, is on boycott lists of right-wing and fundamentalist religious organizations like the Christian Action Council, Barnes said.

Sea Change in Marketing

This new approach has become popular and possible, because of a sea change in marketing: primarily a shift from mass marketing, or being all things to all people, to market segmentation, or slicing the mass market into smaller pieces.

"We are trying to serve a niche, of people who believe in 'controversial' causes," said Barnes, whose company also lets consumers donate a nickel from each credit-card purchase to what its advertising calls "hard-hitting advocacy groups." "Sure, we're not trying to please everybody, like AT&T, MCI and Sprint do. It's fine with us that people who oppose a woman's right to choose not become customers."

That is the strategy being followed by a growing list of companies, primarily smaller concerns that portray themselves as socially committed Davids battling uninvolved Goliaths. Among them: the Kenar clothing company; the Ringer Corporation, which makes lawn-care products; Stonyfield Farm, a dairy-products company; and Kenneth Cole Productions, a shoe company.

Products Matched to Causes

"We're not a mass marketer," said Holly Alves, marketing director of Ben & Jerry's Homemade Inc. in Waterbury, Vermont. For every cause, the company has a product or flavor: a percentage of Peace Pops sales goes to pacifist groups, Rainforest Crunch benefits rainforest preservation and Wild Maine Blueberry aids that state's blueberry-growing Passamaquoddy tribe.

Companies often find that their messages suit their markets. "People feel good supporting people who believe in the same things they do," said Kenneth D. Cole, president of Cole Productions. He has been deeply involved in point-of-purchase politics since he ran an advertisement in 1985 in support of the American Foundation for AIDS Research.

By relating product pitches to real-life issues more

 Q: *According to the EPA, for every 1 percent of ozone depletion, how many new cases of skin cancer deaths occur in the US each year?*

important than the choice of tissue brands, point-of-purchase politics also appeals to those who disdain the hype and hoopla of conventional advertising.

Advertising Opinions
Companies involved in the phenomenon make their opinions known in their advertising, on their packages, on aisle displays and on signs at cash registers and other points of purchase (hence the term *point-of-purchase politics*).

Sebastian International of Woodland Hills, California, has established Club Unite (for Unity Now Is a Tomorrow for Everyone), which customers at about 6,000 hair salons nationwide can join. Members make donations to any of seven organizations, including the Design Industries Foundation for AIDS, and the Rainforest Foundation. In return, the salons hand out samples of Sebastian hair-care products.

"It's a little strange to be happening in a salon," said John Sebastian, the company's chief executive. "But where is it going to happen? A bank? A lawyer's office?"

20 Million Socially Conscious
For the 20 million Americans who, according to surveys, consider themselves socially conscious, point-of-purchase politics can distinguish a company as directly as a logo or a slogan.

Anita Roddick, founder and managing director of Body Shop International, a chain of body-care product stores well-known for backing controversial causes, said that many of these customers were "activists in the 1960s who didn't know we were not allowed to bring our values or ideals to work with us."

Stepping into a Body Shop can be like attending a protest rally. Product labels plead against testing cosmetics on animals. Signs at the cash register promote recycling and conservation. Bags ask shoppers to join Amnesty International. Volunteers offer voter registration on the spot.

Complaint About Benetton
But even among point-of-sale philanthropists, there can be hard feelings. Some executives accuse others of cynical opportunism. Critics have complained that a recent Benetton magazine advertisement, featuring a photograph of an AIDS patient at the moment of his death, cruelly exploited the issue, because the ad

energy and are made of recycled materials.

From the Rainforest, 133 Broadway, Ste. 1129, New York, NY 10010, (800) EARTH 96.
David Sadke, From the Rainforest's founder, defines his vision as "finding creative economic alternatives to save the rainforests." The company buys products from indigenous people and donates 5 percent of the profits to Cultural Survival.

Marketplace, 1461 Ashland Ave., Evanston, IL 60201, (708) 328-4011.
This nonprofit organization hails from Bombay, India, where women and men in the city's impoverished Golibar section have formed a cooperative that produces cotton dresses, jackets, shirts and batiks.

Montana Coffee Traders, 5810 Highway 93 S., Whitefish, MT 59937, (800) 345-JAVA.
Cafe Monteverde is a coffee "grown in harmony with the cloud forest of Santa Elena, Costa Rica," says R.C. Beall of Montana Coffee Traders. For every pound of Cafe Monteverde he sells, Beall sends $1 back to the Coope Santa Elena for its Special Projects Fund (education, habitat protection, etc.).

National Wildlife Holiday Cards and Gifts, 1400 16th. St. NW, Washington, DC 20036, (800) 432-6564.
The year 1992 marks

A: *20,000 annually*
Source: *The Campaign for Safe Alternatives to Protect the Ozone Layer*

the 40th anniversary of this catalog of nature-related cards and gifts published annually by the National Wildlife Federation. The catalog's proceeds fund efforts to preserve wildlife habitats.

The Natural Child, 611 Ash Ave., Ames, IA 50010, (515) 292-4471.

This catalog features children's clothing and toys made by craftspeople and small-scale manufacturers across the country. The clothes are made of all-natural materials that minimize waste and are intended to last for generations.

The Natural Choice, 1365 Rufina Circle, Santa Fe, NM 87501, (800) 621-2591; (505) 438-3448.

Features products ranging from natural wood finishes and thinners to hammocks made by indigenous people of El Salvador using handspun, locally grown cotton in a vast array of brilliant colors. A must for the environmentally conscious consumer.

The Nature Company, P.O. Box 188, Florence, KY 41022, (800) 227-1114.

Beautiful products for the ecologically sensitive shopper. The Nature Company's catalog features a host of products that celebrate nature. The company donates a portion of its profits to the Nature Conservancy.

Paraclete, 1132 SW 13th Ave., Portland, OR 97205, (800) 274-9649; (503) 274-5134.

did not include any information about AIDS charities or prevention.

The Philanthropic Advisory Service in Washington, DC, which collects data on charities, warns that some companies imply they will increase donations to causes if consumers buy more of their products, when in fact the companies have already agreed to donate fixed amounts.

Marketers are now striving for sincerity, stressing that their activism grows out of corporate philosophies— "mission statements," as they are known—and true commitment. While Alves assures investors that Ben & Jerry's "is a very marketing-driven company," she says it was founded on 1960s values that are embodied by the company's leaders. "It was not that marketing said, 'This would look good as an image for the company.'"

"Obviously Risks Involved"

"There are obviously risks involved," said Dan Osheyack, promotion director of *Entertainment Weekly*, a Time Warner magazine in New York that pegs all its promotional events to fighting AIDS. But considering the concerns of the entertainment industry and the magazine's readers, he added, "It's the right thing for us to do."

Larger mass-marketers, however, can be particularly vulnerable to backlashes. Sears, Roebuck & Company promised the Humane Society of the US 8 percent of the proceeds from sales of special stuffed animals in its "Great American Wishbook 1991" catalog. The National Rifle Association complained. Sears stopped selling the animals.

—*Stuart Elliott*

Reprinted by permission from the New York Times, *"When Products Are Tied to Causes," by Stuart Elliott, April 18, 1992.*

FEDERAL TRADE COMMISSION GUIDELINES

THE FEDERAL TRADE COMMISSION (FTC) instituted guidelines on environmental advertising and marketing claims in July 1992. The FTC guidelines apply to environmental claims included in labeling, advertising, promotional materials and all other forms of marketing, whether asserted through words, symbols, emblems, logos, depictions or product brand names. They regulate any claim about the environmental attributes

 Q: *By recycling waste, how many more jobs are created as compared with landfilling?*

GREEN BUSINESS

of a product or package in connection with the sale, offering for sale, or marketing of any such product or package. Below is a summary of the regulations.

Advertising

1. Qualifications or disclosures about a product's environmental qualities should be written clearly and prominently.

2. Environmental attributes or benefits of a product should be labeled in reference to the product, the product's packaging or a component of the product or packaging. For example, a glass soda bottle labeled "recycled" is made entirely from recycled materials, but the plastic cap is not, and therefore the entire product should not be referred to as recycled.

3. Environmental marketing claims should not be presented in a way that exaggerates the environmental attribute or benefit. An example of false advertising in this case would be when a manufacturer labels a package as having 50 percent more recycled material than before, though the manufacturer only increased the recycled material from 2 percent to 3 percent.

4. Comparative marketing statements should make the comparison sufficiently clear to avoid consumer deception. For example, an advertiser claims that their plastic diaper liner has the most recycled content. The advertised diaper does have more recycled content, calculated as a percentage of weight, than any other on the market, although it is still well under 100 percent recycled. This is intentional ambiguity used to deceive the customer.

Marketing Claims

1. It is deceptive to misrepresent that a product or package offers a general environmental benefit. For example, a product wrapper is printed with the claim "Environmentally Friendly." Textual comments on the wrapper explain that the wrapper is environmentally friendly because it is not chlorine-bleached. However, the production of the wrapper still creates and releases significant quantities of other harmful substances.

2. An unqualified claim that a product or package is degradable, biodegradable or photodegradable should be substantiated by competent and reliable scientific evidence that the entire product or package will completely break down and decompose into elements found in nature within a reasonably short

A nonprofit organization sponsoring "small-scale economic projects" in Mexico and Guatemala, the catalog features jackets, vests, belts, pants and other clothing sewn from colorful handwoven cloth.

Pueblo to People, P.O. Box 2545, Houston, TX 77252-2545, (800) 843-5257.

A catalog featuring toys, clothing, furniture, pottery and more, handcrafted in nine cooperatives in Central and South America. Pueblo to People is a nonprofit organization.

Real Goods Trading Corporation, 966 Mazzoni St., Ukiah, CA 95482-3471, (800) 762-7325.

The world's largest selection of alternative energy products. The company's full product line is depicted in the *Alternative Energy Sourcebook,* while the more frequent catalogs highlight merchandise of more seasonal interest.

Save the Children, P.O. Box 166, Peru, IN 46970, (800) 833-3154.

A catalog published by a nonprofit, international organization committed to helping children and their families with community development projects. A portion of every dollar earned is donated to programs in food production, skills training, maternal and child health, and emergency response to natural disasters around the world.

A: *Six times as many*
Source: The Denver Post *1991 Colorado Recycling Guide*

Seva, 8 N. San Pedro Rd., San Rafael, CA 94903, (800) 223-7382.

The Seva Foundation has provided eye care in India and Nepal for over a decade. It also sponsors literacy programs, shelters and clinics. The catalog features jewelry, T-shirts, wind chimes, books and other items handcrafted by Native Americans. All proceeds go to the foundation.

Seventh Generation, 49 Hercules Dr., Colchester, VT 05446, (800) 456-1177.

Offers easy-compromise items like water-conserving toilets and organic linens. The catalog features Rainforest Essentials, soaps, shampoos, conditioners and lotions made with ingredients hand-harvested from African and South American rainforests. Rainforest Essentials donates 40 percent of its profits to Cultural Survival.

Sounds True Audio, 735 Walnut St., Boulder, CO, 80302, (800) 333-9185, (303) 449-6229.

This "spoken word audio company" publishes over 100 audiotapes focusing on self-discovery, social responsibility and spiritual traditions, as well as other varied topics. With titles such as "The Fate of the Amazon Rainforest" and "Socially Responsible Investing," this fascinating catalog is sure to have something for everybody.

Toucan Chocolates, P.O. Box 72, Waban,

period of time after customary disposal. For example, a trash bag is marketed as "degradable," with no qualifications or other disclosure. The marketer relies on soil burial tests to show that the product will decompose in the presence of water and oxygen. The trash bags are customarily disposed of in incineration facilities or at sanitary landfills that are managed in a way that inhibits degradation by minimizing moisture and oxygen. Degradation will be irrelevant for those trash bags that are incinerated and, for those disposed of in landfills, the marketer does not possess adequate substantiation that the bags will degrade in a reasonably short period of time.

3. A claim that a product or package is compostable should be substantiated by competent and reliable scientific evidence that all the materials in the product or package will break down into useable compost in a safe and timely manner in a municipal compost or a home compost system.

4. A product or packaging should not be marketed as recyclable unless it can be collected, separated or otherwise recovered from the solid waste stream for use in the form of raw materials in the manufacture or assembly of a new package or product. Products that are made of both recyclable and nonrecyclable components should be adequately qualified to avoid consumer deception about which portions or components of the product or package are recyclable. A product or package that is made from recyclable material, but, because of its shape, size or some other attribute, is not accepted in recycling programs for such material, should not be marketed as recyclable.

5. A recycled content claim may be made only for materials that have been recovered or otherwise diverted from the solid waste stream, either during the manufacturing process or after consumer use. The manufacturer or advertiser must have substantiation for concluding that the pre-consumer material would otherwise have entered the solid waste stream. Distinctions must be made between pre-consumer and post-consumer materials when asserting a recycled content claim. An example of deception: A manufacturer routinely collects spilled raw material and scraps from trimming finished products. After a minimal amount of reprocessing, the manufacturer combines the spills and scraps with virgin material for use in further production of the same product.

: *What percentage of American trash is landfilled?*

6. It is deceptive to misrepresent that a product or package has been reduced or is lower by weight, volume or toxicity. Source reduction claims should be qualified to the extent necessary to avoid consumer deception about the amount of the source reduction and about the basis for any comparison asserted.

7. An unqualified refillable claim should not be asserted unless a system is provided for the collection and return of the package for refill or the later refill of the package by consumers with product subsequently sold in another package. A package should not be marketed with an unqualified refillable claim, if it is up to consumers to find new ways to refill the package.

8. A claim that a product does not harm the ozone layer is deceptive if the product contains an ozone-depleting substance.

A copy of the FTC guidelines can be obtained from the Federal Trade Commission in Washington, DC, (202) 326-2222.

CRUELTY FREE

IN RESPONSE TO THE INFLUENCES OF SOCIAL INVESTING and consumer pressure, several companies are making an effort to green their actions. However, many companies that have launched environmental campaigns continue to test their products on animals.

Ironically, there are several companies that score highly on ratings for corporate social and environmental responsibility, despite their continued practices of animal testing. Procter and Gamble, for example, is one of the Council on Economic Priorities' best-ranked firms. They claim to have reduced animal testing by 90 percent over the past five years.

People for the Ethical Treatment of Animals (PETA) keeps a list of firms that test on animals and those that do not. Its *Shopping Guide for Caring Consumers* lists the companies that have signed PETA's statement swearing that they do not and will not in the future conduct any animal tests or contract out any animal testing. Contact: PETA, Box 42516, Washington, DC 20015.

JUNK MAIL KIT

ACCORDING TO THE CONSUMER RESOURCE INSTITUTE, between one and two trees per person are sacrificed annually to create junk mail. Less than 10 percent of

MA 02168, (617) 964-8696.

With the ever-powerful American addiction to chocolate in its back pocket, Toucan Chocolates produces chocolates with the rainforest in mind. The company uses forest-gathered Brazil nuts and cashews in its products and donates a portion of the proceeds to Cultural Survival.

Trade Wind, P.O. Box 380, Summertown, TN 38483, (615) 964-2334.

Edward Sierra founded this nonprofit organization to meet "the thirst in North America for the deeply rooted spiritual values common to traditional native cultures, and a corresponding need in Indian country to sell goods at fair prices and become self-reliant." As may be expected, this catalog features beautiful Native American crafts, jewelry and quilts.

The Woman's Bean Project, 2347 Curtis St., Denver, CO 80205, (303) 292-1919.

This nonprofit group located in Denver's impoverished inner city prepares bean, chili, pasta and cornbread dry mixes. The project provides a safe place for women to gain self-sufficiency; they earn $4.25/hour, and receive free day care and free transportation to and from work.

A. *80 percent*
Source: The W.A.T.E.R. Foundation

A Shopping Checklist

1. Is this plastic or paper packaging necessary? Is a similar product available with less packaging?
2. Is this packaging recyclable? Does it contain plastic or mixed materials? Is there a metal or glass option?
3. Can I reuse this container for something else?
4. Can this beverage container be returned or recycled?
5. Can I buy this product (detergent, cereal, flour, shampoo) in a larger container or in bulk quantity?
6. Can I buy this product (razor, cups, utensils, napkins, diapers) in a nondisposable, reuseable form?
7. Is there a nonhazardous substitute for this cleaner, pesticide or solvent?
8. Can I use fabric shopping bags instead of paper bags or plastic bags?
9. Can I buy this fast food in paper containers instead of plastic or Styrofoam? Should I support the fast food industry at all?
10. Is this product worth a higher garbage bill or a more polluted environment?
11. Are all the components of this product biodegradable?
12. How much energy, land and water are used to produce this product?
13. Can this product be repaired rather than discarded?
14. How long will this product last?
15. If this product is something I seldom use can I borrow it or rent it?

this mail is recycled. Most of it is not read and ends up in landfills or incinerators.

To confront this wasteful situation, the Consumer Resource Institute has published a kit that allows the recipient to eliminate junk mail from his or her home, business or postal box. The consumer decides which type of unwanted mail to eliminate. The kit includes information, instructions and special adhesive labels and pre-addressed postcards. To order, send a check or money order for $6.75 to Consumer Resource Institute, Dept. JM-17, P.O. Box 2180, Mill Valley, CA 94942.

SHOPPING FOR A BETTER WORLD

ALICE TEPPER MARLIN'S MOTTO IS: "Turn your shopping cart into a vehicle for social change." Tepper Marlin has become the Diderot of corporate America, cataloging and describing its attitudes and behavior. But it wasn't until the late 1980s that she began to reach a wide national audience, when the Council on Economic Priorities (CEP), of which she is now director, began publishing its pocket-sized paperback, *Shopping for a Better World*. It has sold more than 800,000 copies.

With easy-to-read graphics and an encyclopedic analysis of corporate America—the latest edition rates 2,400 brand-name items—*Shopping for a Better World* has been wildly successful. CEP grades corporations on ten criteria, including the company's attitude toward the environment, treatment of women, family benefits, etc. According to a poll, 78 percent of the *Better World* readers said they had "switched brands" because of the ratings, and 64 percent said they referred to the shopping guide whenever they shopped. "Everybody is a consumer," says Tepper Marlin gleefully. "And it's easier to affect a consumer decision than it is to influence an investor decision. People make dozens of purchasing decisions a month, often between narrowly competing brand names, and a shift in half a percent of market share is a big deal for these companies."

Corporate Conscience

For the last six years, CEP has also sponsored America's Corporate Conscience awards. The nine winners of the 1992 awards were: Church and Dwight; US West; General Mills; Prudential Insurance; Supermarkets General; Donnelly Corp.;

: **What percentage of aircraft-sprayed pesticides reach their intended target?**

Tom's of Maine; Lotus Development Corp.; and Conservatree Paper Co.

Three "dishonorable mentions" were also given: to RJR Nabisco for its deceptive "Joe Camel" advertising; to Du Pont for the honor of being the "largest domestic emitter of toxic chemicals"; and to MAXXAM, Inc., for environmental destruction.

A new project launched by CEP is the Corporate and Environmental Data Clearinghouse, which has begun to issue in-depth reports on the environmental behavior of all 500 publicly held companies listed in *Standard and Poor's*. The first report, released in November 1991, examined the worst polluters in the petrochemical industry. The clearinghouse has also compiled a report on 20 of the worst polluters in the forestry products/paper industry.

The idea of providing consumers with boiled-down information about corporations is spreading. Recently, an editor at Japan's *Asahi News Journal* became interested in CEP's work and decided to run a special issue devoted to rating Japan's corporations. Tepper Marlin says that, since such a thing had never been done before, the editors decided to modify the grading system by designing new symbols. "Instead of our checks and X's, which they thought might be too offensive, they used a cracked egg as their lowest evaluation, which was to represent a company whose social responsibility is not yet born. I think it's a beautiful symbol. Their top evaluation was a bird in flight."

Tepper Marlin's plans for the future involve teenagers. With fervor, she says, "Teenagers spend $79 billion every year of their own money, and most of that is on consumer goods. The number one thing girls spend money on is clothes, and the second is fast food. Boys spend money first on fast food, and second on clothes." So, sometime soon, CEP will produce a shopping guide for teenagers. As for her own kids, though they are not yet college age, she says that she is looking forward to having more free time when they get older. "I'd like to return to being more of a workaholic," says Tepper Marlin.

Shopping for a Better World is available for $6.45 from the Council on Economic Priorities, 30 Irving Place, New York, NY 10003.

—*Micah Fink*

Reprinted by permission from In These Times, *"Tepper Marlin: Shopping Till They Stop," by Micah Fink, April 15, 1992.*

Cartons of Eggs and Milk

Nestling brown eggs, instead of white, into your shopping cart is a simple way to nurture the planet's genetic resources. Once you get past the shell, brown eggs do not differ significantly from white. But white eggs represent a genetic disaster in the making: more than 90 percent of the 3 billion eggs sold in the US each year come from one kind of chicken. In fact, virtually all commercial egg-laying chickens in the US come from just nine hatchery sources. What would happen to our egg supply if some pestilence wiped out that breed?

One thing is for sure. The white Leghorns that lay our white eggs cannot lay brown eggs. If even a moderate percentage of shoppers switched to brown eggs, the market would respond with a shift in our national egg-producing gene pool from one breed to two. Two breeds is still a very narrow genetic pool, especially given the large number of poultry breeds available, but it is a good start.

Milk presents a similar opportunity. Since 95 percent of our milk comes from just one kind of cow, the Holstein-Freisian, we can strike another small blow for gene-pool conservation by buying milk produced by other cows, such as Jerseys or Guernseys.

—*Martin Teitel*

Reprinted by permission from Rain Forest in Your Kitchen, *by Martin Teitel, Island Press, Washington, DC, 1992.*

A: **0.1 percent**
Source: The Amicus Journal

Directory to New Green Products

In Little House on the Prairie, Laura Ingalls Wilder and her sisters learn a little song to get them through the week's chores: "Monday, wash; Tuesday, iron; Wednesday, mend; Thursday, churn; Friday, bake; Saturday, clean; Sunday, rest!" And buy "green" products to use each day!

ALLENS NATURALLY FRUIT & VEGGIE WASH
Description: A ready-to-use, fast and convenient spray to aid in removing wax, soil, oily pesticides and chemicals from fruit and vegetables. **Green Qualifications:** Biodegradable, no animal testing. **Availability:** Cooperative and natural food stores or mail order. **Company:** Allens Naturally, P.O. Box 339, Farmington, MI 48332-0339, (313) 453-5410. **Corporate Environmental Policy:** 5% of mail-order sales donated to People for the Ethical Treatment of Animals (PETA).

ALLENS NATURALLY ULTRA LAUNDRY POWDER
Description: Concentrated to reduce packaging and increase value: one five-pound box does 75 loads of laundry. Just one scoop leaves clothes bright, clean and fresh smelling. **Green Qualifications:** Biodegradable, no animal testing. **Availability:** Cooperative and natural food stores or mail order. **Company:** Allens Naturally, P.O. Box 339, Farmington, MI 48332-0339, (313) 453-5410. **Corporate Environmental Policy:** 5% of mail-order sales donated to PETA.

AROMATHERAPY ROMANTIC RENDEZVOUS
Description: A warm floral blend of rose, armoise, passionflower, cognac, orris, patchouli, carrot seed, spearmint, black pepper, juniper berry and jasmine. **Green Qualifications:** Biodegradable, recyclable packaging, no animal testing. **Availability:** Health food stores. **Company:** Jason Natural Products, 8468 Warner Dr., Culver City, CA 90232-2484, (310) 838-7543. **Corporate Environmental Policy:** Support of the following groups: Beauty Without Cruelty, The National Anti-Vivisection Society, Amnesty for Animals and PETA.

BEARITOS LITE CHEDDAR PUFFS
Description: Made with real cheddar cheese; low-fat. **Green Qualifications:** Organic whole cornmeal, organic cornmeal, expeller-pressed canola oil. **Availability:** Natural food stores nationwide and some grocery stores. **Company:** Little Bear Organic Foods, 1065 E. Walnut, Carson, CA 90746, (310) 886-8200. **Corporate Environmental Policy:** Supports a wide range of environmental causes.

BRILLO STEEL WOOL PADS
Description: Recycled steel cleaning pad. **Green Qualifications:** Recycled carton, biodegradable detergent. **Availability:** Grocery stores. **Company:** The Dial Corp., Dial Tower, Phoenix, AZ 85077-1613, (602) 207-2869. **Corporate Environmental Policy:** To maximize the utilization of recycled materials in products, packages and manufacturing processes.

BUMKINS ALL-IN-ONE CLOTH DIAPER
Description: Diaper has a waterproof outer shell attached to soft 100% cotton flannel padding, elastic at legs and waist, Velcro-like closures. Also available: waterproof covers, cotton contoured diapers, cotton blankets, cotton towels, waterproof bibs. **Green Qualifications:** Recyclable, minimal packaging. **Availability:** Juvenile stores nationwide, or mail catalogs: Seventh Generation, Walnut Acres, Right Start Catalog. **Company:** Bumkins Family Products, 1945 E. Watkins, Phoenix, AZ 85034, (800) 553-9302. **Corporate Environmental Policy:** Internal recycling, recyclable packaging and reuseable products.

CITRA-SOLV
Description: Super-concentrated, all-purpose cleaner made from powerful yet natural citrus extracts. **Green Qualifications:** Made from the by-products of citrus peels. Contains only substances that have been judged biodegradable and environmentally safe. **Availability:** Grocery stores. **Company:** Chempoint Products Company, 188 Shadow Lake Rd., Ridgefield, CT 06877-1032. **Corporate Environmental Policy:** None.

Q: *How many pounds of red meat and poultry does an average American eat each year?*

GREEN BUSINESS

EARTHLINGS
Description: Environmentally friendly children's and infant's wear. **Green Qualifications**: Organically grown cotton, Foxfibre naturally colored cotton, Colorganics, unbleached, no formaldehyde, no resin, low-impact dyes, tagua nut buttons, recycling. **Availability**: Environmental specialty stores, children's wear stores and catalogs. **Company**: Earthlings, 205 S. Lomita Ave., Ojai, CA 93023, (805) 646-7770. **Corporate Environmental Policy**: Donates profits to environmental organizations.

ECOVER NON-CHLORINE BLEACH
Description: Super-concentrated hydrogen-peroxide-based bleach. Packaged in post-consumer (at least 50%) recycled, recyclable, high-density polyethylene bottle. **Green Qualifications**: Fully biodegradable, no synthetic perfumes or colors. No enzymes or optical brighteners, no chlorine bleach, not tested on animals. **Availability**: Natural food stores, environmental stores, mail-order catalogs, (PETA, Walnut Acres, Eco logicals, Eco-choice), select supermarkets. **Company**: ECOVER, Inc., 4 Old Mill Rd., Georgetown, CT 06829, (203) 853-4166. **Corporate Environmental Policy**: To develop, produce and bring to market cleaning products that are highly effective, yet have minimal impact on the environment.

ECOVER ULTRA CONCENTRATED LAUNDRY POWDER
Description: Highly concentrated, ecologically sound laundry detergent. **Green Qualifications**: Fully biodegradable, phosphate-free, no synthetic perfumes or colors, no enzymes or optical brighteners, no chlorine bleach, not tested on animals. **Availability**: Natural food stores, environmental stores, mail-order catalogs (PETA, Walnut Acres, Eco Logicals, Eco-choice), select supermarkets. **Company**: ECOVER, Inc., 4 Old Mill Rd. Georgetown, CT 06829, (203) 853-4166. **Corporate Environmental Policy**: To develop, produce and bring to market cleaning products that are highly effective, yet have minimal impact on the environment.

ECOWORKS PHOTOELECTRIC SMOKE DETECTORS
Description: Smoke detector. **Green Qualifications**: Contains no radioactive material, detects smoldering fires up to an hour faster than radioactive ionizing detectors and gives off fewer false alarms from cooking smoke. **Availability**: Wholesale and retail from EcoWorks. **Company**: Ecoworks, 2326 Pickwick, Baltimore, MD 21207, (410) 448-3319. **Corporate Environmental Policy**: Ecoworks is a supporter of 1% for Peace and contributes 10% of its profits to organizations working for a nuclear-free world.

ENVIROCARE PAPER PRODUCTS
Description: 100% recycled paper. **Green Qualifications**: No dioxin-producing chlorine bleaching agents in processing, no added inks, dyes or fragrances. **Availability**: Supermarkets throughout the US. **Company**: Ashdun Industries, Inc., 400 Sylvan Ave., Englewood Cliffs, NJ 07632, (201) 569-3600. **Corporate Environmental Policy**: Ashdun's CARE brand has a partnership with the American Forestry Association's Global Releaf program. The CARE products have a 900-number that consumers can call, and the charges for these phone calls are put toward planting trees.

ENVIROCARE HOUSEHOLD CLEANING AND DETERGENT
Description: Biodegradable cleaning and detergent products. **Green Qualifications**: Not tested on animals, no optical brighteners, no benzene, no petroleum solvents, no synthetic dyes or fragrances, no ammonia, no chlorine bleaching agents in processing. **Availability**: Supermarkets throughout the US. **Company**: Ashdun Industries, Inc., 400 Sylvan Ave., Englewood Cliffs, NJ 07632, (201) 569-3600. **Corporate Environmental Policy**: Ashdun's CARE brand has a partnership with the American Forestry Association's Global ReLeaf program. The CARE products have a 900-number that consumers can call, and the charges for these phone calls are put toward planting trees.

ESPRIT ECOLLECTION
Description: Clothing line. **Green Qualifications**: Clothing designed and produced using materials and processes with the least negative environmental impact or most positive social benefit. Chemicals are reduced or eliminated as much as possible in dyeing, washing and finishing. Fabrics include organically grown and naturally colored cotton, recycled and naturally colored wools, and rayon from sustainably harvested tree farms. **Availability**: Esprit outlets nationwide. **Company**: Esprit International, 900 Minnesota St., San Francisco, CA 94107, (415) 550-3612. **Corporate Environmental Policy**: Supports sustainable agriculture and forestry. Hangtags, brochures and packaging use recycled papers and plastics and are printed with soy-based inks.

A: *178 pounds annually*
Source: Worldwatch magazine

GRANNY'S OLD FASHIONED "E.Z. MAID"
Description: E.Z. Maid dish and multi-purpose cleaner. **Green Qualifications**: Coconut-based raw materials, recyclable. **Availability**: Health food stores and mail-order nationwide. **Company**: Granny's Old Fashioned Products, P.O. Box 256, Arcadia, CA 91066, (800) 366-1762. **Corporate Environmental Policy**: None.

GREEN COTTON
Description: Grown and handpicked cotton without the use of any synthetic chemicals. **Green Qualifications**: The life-cycle approach to "Green Cotton" insures its biodegradability. No use of CFCs, and utilization of post-consumer, recycled packaging throughout. **Availability**: Order through Ardent Intimates. **Company**: Ardent Intimates, 1685 B Scenic Ave., Costa Mesa, CA 92625, (714) 434-0212. **Corporate Environmental Policy**: Supports environmental groups and studies on a case-by-case basis.

KISS COLORS
Description: Naturally colored lipsticks. **Green Qualifications:** 100% natural, no animal ingredients, no preservatives, no animal testing. **Availability:** Natural food or health stores. **Company:** Kiss My Face, P.O. Box 224, Gardiner, NY 12525, (914) 255-0884. **Corporate Environmental Policy:** Extensive recycling program.

KITCHEN SAFE PRODUCT
Description: All-purpose cleaner with lemon and baking soda. **Green Qualifications:** Recyclable package, biodegradable product. **Availability:** Walmart stores nationwide, most grocery stores. **Company:** Miles, Inc., Consumer Household Products, Chicago, IL 60638, (800) 767-9927. **Corporate Environmental Policy:** None.

LIFE TREE FRESH AND NATURAL BATHROOM CLEANER
Description: Cleaner that helps prevent bacterial growth. **Green Qualifications:** Concentrated, biodegradable, cruelty-free, packaged in 60% recycled plastic bottles. **Availability:** Natural products and environmental specialty stores nationwide. **Company:** Life Tree, P.O. Box 1203, Sebastopol, CA 95472, (707) 577-0205. **Corporate Environmental Policy:** Committed to producing environmentally sound products with no harmful by-products or wasteful practices.

LIFE TREE FRESH MINT LIQUID SOAP
Description: A pH-balanced, extra-gentle, moisturizing liquid cleanser for washing hands, face and body. **Green Qualifications:** Contains fresh herbal extracts of aloe vera and calendula. **Availability:** Natural products and environmental specialty stores. **Company:** Life Tree, P.O. Box 1203, Sebastopol, CA 95472 (707) 577-0205. **Corporate Environmental Policy:** Committed to producing environmentally sound products with no harmful by-products or wasteful practices.

NATURAL CHEMISTRY'S ALL-PURPOSE CLEANER
Description: Exceptional new cleaning formulas that out-performs products that contain harsh toxic chemicals. These unique natural-enzyme formulas are completely safe, non-irritating and hypo-allergenic. **Green Qualifications:** Nontoxic, biodegradable, recyclable, minimally packaged, no animal products or phosphates. **Availability:** Natural retail stores throughout the country. **Company:** Natural Chemistry Inc., 244 Elm St., New Canaan, CT 06840, (203) 966-8761. **Corporate Environmental Policy:** Donates to various environmental and humane organizations. Produces and sells only products with no negative environmental impact.

NATURAL WORLD LEMON & BAKING SODA GENTLE CLEANSING SCRUB
Description: Nonabrasive, no chlorine bleach, concentrate, biodegradable, cruelty-free, safe and effective. **Green Qualifications:** Based on natural ingredients of lemon and contains no chlorine bleach, which is toxic. All products are biodegradable; no animal testing and minimal packaging. **Availability:** Natural retail stores throughout the country. **Company:** Natural World, Inc., Glenbrook Industrial Park, 652 Glenbrook Rd., Stamford, CT 06906, (203) 356-0000. **Corporate Environmental Policy:** None.

OASIS BIODEGRADABLE LAUNDRY DETERGENT
Description: The first laundry detergent designed and tested to be safe for plants and soil. **Green Qualifications:** Biodegrades entirely into plant nutrients. Ideal for gray water and septic systems, safe for sewer disposal. **Availability:** Health food stores nationwide. By mail from the Real Goods Trading Company, (800) 762-7325. **Company:** Oasis Biocompatible Products, 1020 Veronica Springs Rd., Santa Barbara, CA 93105, (805)

: *What is the number of gallons of water flushed in the US daily?*

682-3449. **Corporate Environmental Policy:** No animal testing, proceeds support research on designs for sustainable living, office is a laboratory for minimizing consumption with measured flows of all inputs and state-of-the-art hardware and policies.

ORANGECLEAN
Description: Household cleaner that removes the toughest dirt and stains from most surfaces. **Green Qualifications:** Totally biodegradable, no CFCs. **Availability:** Order by mail. **Company:** Appel Company, P.O. Box 2146, Littleton, CO 80161, (303) 781-3722. **Corporate Environmental Policy:** None.

O WEAR
Description: Men's and women's knitted apparel. **Green Qualifications:** Made from 100% certified organic cotton. **Availability:** Department and specialty stores nationwide. **Company:** O Wear, P.O. Box 77699, Greensboro, NC 27417, (919) 547-7886. **Corporate Environmental Policy:** The O Wear Alliance is a co-op effort between O Wear, local retailers and local environmental groups to benefit local environmental issues.

PURELY GROWN
Description: Hosiery for men, women and youth made from organically grown cotton. **Green Qualifications:** Certified organically grown cotton, recycled packaging printed with soy-based ink. **Availability:** Health food stores, environmental stores, department and specialty stores, mail-order catalogs. **Company:** The Wearables Marketing Group, Inc., 401 Ocean Front Walk, Venice, CA 90291, (310) 399-5608. **Corporate Environmental Policy:** To join other eco-warriors in the fight to continue to expand consumer awareness and to bring ecologically sound products to the market.

RENEW
Description: Renew brand trash bags are the first nationally distributed trash bags made of 100% recycled plastics, including 30% post-consumer waste. **Green Qualifications:** Contain the maximum percentage of recycled plastics possible, as well as the highest percentage of post-consumer material of any plastic trash bag on the market. **Availability:** Grocery, hardware, home center, discount, mass merchandisers, warehouse, office club and drug stores. **Company:** Webster Industries, 58 Pulaski St., Peabody, MA 01960, (508) 532-2000.

Corporate Environmental Policy: Renew recycles over 100 million pounds of plastics each year.

RING-OUT
Description: Concentrated liquid laundry detergent. **Green Qualifications:** Biodegradable and has a neutral pH. No phosphates. **Availability:** Order by mail. **Company:** Phoenix Dynamics Corp., 131 Finucane Place, Woodmere, NY 11598, (516) 374-1632. **Corporate Environmental Policy:** None.

ROSA MOSQUETA SUN PROTECTION HERBAL BUTTER
Description: A natural blend of herbal butters to moisturize skin and protect it from damage by UV rays. **Green Qualifications:** No synthetics, not tested on animals. **Availability:** Health food stores nationwide. **Company:** Aubrey Organics, 4419 N. Manhattan Ave., Tampa, FL 33614, (800) 282-7394. **Corporate Environmental Policy:** None.

SEVENTH GENERATION PAPER TOWELS AND NAPKINS
Description: 100% post-consumer paper towels and napkins. **Green Qualifications:** 100% post-consumer waste. **Availability:** Natural food stores on the East Coast or catalog orders. **Company:** Seventh Generation, Colchester, VT 05446-1672, (802) 655-6777. **Corporate Environmental Policy:** Donated over $100,000 to various environmental groups in the past two years.

SHARON'S FINEST TOFURELLA AND HEART'S D'LITE
Description: Low-fat tofu products. **Green Qualifications:** No CFCs; recyclable packaging. **Availability:** Health food stores, delicatessens, supermarkets, hospitals, health clubs, university cafeterias, and restaurants, by request. **Company:** Sharon's Finest, P.O. Box 5020, Santa Rosa, CA 95402, (707) 576-7050. **Corporate Environmental Policy:** 5% annual donation to rainforest preservation projects.

SKANDIA OIL FINISH
Description: Water-resistant oil finish for exterior and interior use. **Green Qualifications:** Natural, drying plant oils, citrus terpenes and balsamic turpentine. Free of petroleum distillates, mineral spirits and other potentially harmful substances. **Availability:** Catalog. **Company:** Eco Design Co.,1365 Rufina Cir., Santa Fe, NM 87501, (505) 438-3448. **Corporate Environmental Policy:** None.

A: *4,800,000,000 per day*
Source: Garbage magazine

STUART HALL GREEN LEAF PAPER PRODUCTS
Description: Paper products, including themebooks, memo pads and notebook paper, which are all 100% recyclable and 100% recycled post-consumer waste. **Green Qualifications:** Nontoxic, biodegradable, recyclable, minimally packaged. **Availability:** Discount and drugstore chains nationwide. **Company:** Stuart Hall Co., Inc., P.O. Box 419381, Kansas City, MO 61414, (816) 221-8480. **Corporate Environmental Policy:** Company commitment to recycling and source reduction in the office.

TEA TREE OIL ANTISEPTIC TOOTHPASTE
Description: Deodorizes the mouth by killing bacteria, dissolves plaque and stimulates circulation in the gums. **Green Qualifications:** Biodegradable, no CFCs, recyclable packaging. **Availability:** Health food stores and pharmacies. **Company:** Thursday Plantation, 118 Nopalitos Way, Santa Barbara, CA 93103, (805) 963-2297. **Corporate Environmental Policy:** None.

TWIGZIL
Description: Pencil handmade from gathered twigs. **Green Qualifications:** No trees are destroyed to make Twigzil pencils because the twigs are gathered from the ground. **Availability:** The Nature Company stores nationwide, Marshall Field's, Dayton & Hudson department stores. **Company:** The Green Consultancy, P.O. Box 91109, Santa Barbara, CA 93190, (805) 568-0017. **Corporate Environmental Policy:** Supports Global Releaf's tree planting program with a percentage of the sales. Packaging conveys information to consumers about Global Releaf.

VALLEYCARE SOCKS
Description: Cotton and wool natural socks. **Green Qualifications:** 100% natural fibers, all packaging uses recycled paper, no plastics. **Availability:** Outdoor and sporting goods stores, health food stores. **Company:** Earthcare, Inc., 1451 14th Ave. NE, Hickory, NC 28601, (800) 452-2532. **Corporate Environmental Policy:** Plant manufacturing and policies to promote Earth-friendly products.

VERMONT ORGANIC FERTILIZERS
Description: A full line of natural organic fertilizers for both retail and commercial use. **Green Qualifications:** Recycled from milk by-products, approved for certified organic crop production, packed in recycled and recyclable containers. **Availability:** Garden centers, hardware stores, mass merchants or mail-order through Seventh Generation. **Company:** Vermont Organic Fertilizer Company, 26 State St., Montpelier, VT 05602, (802) 229-1440. **Corporate Environmental Policy:** To provide gardeners with an all-natural organic fertilizer that reflects a deep commitment to the health and ecological balance of the environment

WESTSOY LITE NON DAIRY CREAMER
Description: All-natural creamer. **Green Qualifications:** Non-dairy, cholesterol-free, made with organic soybeans. **Availability:** Grocery stores nationwide. **Company:** Westbrae Natural, 1065 E. Walnut St., Carson, CA 90746, (310) 886-8200. **Corporate Environmental Policy:** None.

ZOO DOO
Description: Organic fertilizer. **Green Qualifications:** Biodegradable, no CFCs, packaging made from recycled plastic and is recyclable. **Availability:** Garden centers, catalogs and magazines nationwide. **Company:** Zoo Doo Compost Co., Inc, 5851 Ridge Bend Rd., Memphis, TN 38120, (901) 681-0775. **Corporate Environmental Policy:** 51% of profits go to zoos across the nation.

 What is the number of gallons of water that would be flushed in the US each day if toilets were replaced by ultra-low flush models?

Green Business Resources

Books

Beyond Compliance: A New Industry View of the Environment, edited by Bruce Smart, World Resources Institute, Washington, DC, 1992.

Companies With a Conscience, Mary Scott and Howard Rothman, Carol Publishing Group, New York, 1992.

A Consumer's Dictionary of Household Yard and Office Chemicals, Ruth Winter, Crown Publishers, New York, 1992.

The Corporate Environmental Profiles Directory, the Investor Responsibility Resource Center, Washington, DC, 1992.

The Environmental Entrepreneur: Where to Find the Profit in Saving the Earth, John Thompson, Longstreet Press, Atlanta, GA, 1992.

Environment Resource Guide, The American Institute of Architects, Washington, DC, 1992.

Environmental Packaging US Guide to Green Labeling, Packaging and Recycling, Thompson Publishing, Salisbury, MD, 1992.

Environmental Success Index 1992, Renew America, Washington, DC, 1992.

The Gaia Atlas of Green Economics, Paul Ekins, Mayer Hillman and Robert Hutchinson, Anchor Books, Doubleday, New York, 1992.

How Much Is Enough?, Alan Durning, W.W. Norton, New York, 1992.

Management for a Small Planet, W. Edward Stead and Jean Garner Stead, Sage Publications, Newbury Park, CA, 1992.

Recycler's Manual for Business, Government, and the Environmental Community, David R. Powelson and Melinda A. Powelson, Van Nostrand Reinhold, New York, 1992.

Magazines

Business Ethics Magazine, 52 Tenth St., Ste. 110, Minneapolis, MN 55403.

Buzzworm Magazine, 2305 Canyon Blvd., Ste. 206, Boulder, CO 80302.

Ecological Economics, The International Society for Ecological Economics, P.O. Box 1589, Solomons, MD 20688.

In Business Magazine, Box 323, 419 State Ave., Emmaus, PA 18049.

Newsletters

Business and the Environment, Cutter Information Corporation, 37 Broadway, Arlington, MA 02174-5539.

Capital Horizons, Lee Schwartz Associates, Charter House, 177 Angel Rd., London N18 3BW, England.

Common Ground, The Conservation Fund, 1800 N. Kent St., Ste. 1120, Arlington, VA 22209.

The Concerned Investor, Sacha Millstone Ferris, Baker, Watts, 1720 I St. NW, Washington, DC 20006.

Conscious Consumer and Company, The New Consumer Institute, P.O. Box 51, Wauconda, IL 60084.

EnviroLink, The Management Institute for Environment and Business, 1220 16th St. NW, Washington, DC 20036.

Good Money, Good Money Publications, Box 363, Worcester, VT 05682.

Green Alternatives, 72 Old Farm Rd., Rhinebeck, NY 12572.

The Green Business Letter, Tilden Press Inc., 1526 Connecticut Ave. NW, Washington, DC 20036.

The Green Consumer Letter, Tilden Press Inc., 1526 Connecticut Ave. NW, Washington, DC 20036.

Green MarketAlert, 345 Wood Creek Rd., Bethlehem, CT 06751.

The Environmental Investor's Newsletter, Chasen and Luck, 410 N. Bronson Ave., Los Angeles, CA 90004-1504.

Investing for a Better World, 711 Atlantic Ave., Fourth Fl., Boston, MA 02111.

Netback, Good Money Publications, Box 363, Worcester, VT 05682.

Responsive Investing News, 1 Liberty Square, 12th Fl., Boston, MA 02109.

Organizations

The Social Investment Organization, 366 Adelaide St. E., Ste. 447, Toronto, Ontario M5A 3X9, Canada, (416) 360-6047.

 1,536,000,000, a savings of 68 percent
Source: Garbage magazine

ECOTRAVEL

Huck Finn "lit out for the territory," and the rest of us are aching to go. Trying to escape the mal de mall of American monoculture, we dream of Costa Rican rainforests, Himalayan peaks, whale watching on the Alaskan coast. The question is: Will we do to those places exactly what we've done to our own once-pristine forests and prairies? Or will we be ecotourists?

EcoTravel News

To the Rescue?

Vital and unique ecosystems cannot be completely saved by ecotourism. According to the *Los Angeles Times*, the amount of rainforest saved by setting land aside for private nature preserves is a mere fraction of the 40 million to 50 million acres that are ravaged each year by logging, ranching and agriculture. Many of the most ecologically valuable areas are remote and inaccessible to tourists. Other ecosystems may have high scientific value but are only marginally interesting to tourists.

Growing Travel

The *Travel Industry World Yearbook* finds that nature travel is increasing. Ecotourism accounts for 10 percent to 20 percent of all tourism and is growing at a rate of 25 percent to 30 percent a year.

Native Help

A study by the World Bank, the US Agency for International Development and the World Wildlife Fund has found that tourism, including ecotourism, provides little benefit to most native populations.

Native Offering

Karen Ziffer of Conservation International explains one of the problems encountered when introducing ecotravel to indigenous people. "What we see as a limited, precious resource, they see as an undeveloped asset."

Nature's beauty provides an opportunity to reflect on the impact of our visits.

EcoTravel

- By Lisa Jones -

On one side of me, a Guatemalan conservationist was talking about the possibility of linking a series of small lodges in the Central American jungle by small plane or hot air balloon. "You know what happens to the jungle when they build roads," he said passionately.

On the other side of me, a Canadian tourism official revealed what was closest to his heart: "Children never finish what's on their plate when they're on holiday."

So went lunchtime conversation during the Second Annual Ecotourism Conference, held in fall 1992 in Whistler, British Columbia. The conference, like ecotourism itself, encompassed subjects from the mundane to the sublime.

What is ecotourism? The Ecotourism Society, a nonprofit group based in Alexandria, Virginia, defines it as "responsible travel which conserves environments and sustains the well-being of local people." This includes everything from energy-efficient travel to ensuring that money spent by tourists enriches the area in which it was spent, instead of being transferred to a corporate office in a distant city. Ecotourism can aid environmental conservation of precious places in many ways, including generating park fees and creating employment for local people as guides and hoteliers. Such financial benefits carry with them an important message: Nature is valuable.

Q: How much of all trash that is thrown away could be recycled into new products?

Because many of the planet's most biologically diverse areas are also economically impoverished, such a message is crucial if habitat and species are to be saved.

These may sound like high aspirations for an industry that has been justly blamed for environmental and cultural degradation over much of the planet. Hotels have blasted coral reefs and dredged mangrove swamps. Mass tourism has moved in and placed local people on the lowest rung of the imported economic ladder. Trekkers have left their garbage atop some of the planet's most spectacular mountains.

One native Hawaiian called tourism "the highest and biggest disease. It's the insult to the injury. Cellophane skirts and coconut bras for our women; that's what tourism gives us."

Ecotourism holds out the promise of relief from such exploitive tourism. But without proper planning and regulation, even the best-intentioned nature tourists can cause great damage. In the late 1980s, Kenya's national parks were marred by uncontrolled tourism as well as by poaching. More recently, Costa Rica's internationally lauded parks system has suffered from a rapid increase in visitors and a concurrent drop in parks funding.

"Visitor management plans should be prepared and funding for implementation designated long before mobs of tourists show up," says Megan Epler Wood, executive director of The Ecotourism Society. She adds that more tourist-generated funds need to be injected into the coffers of the agencies managing protected areas, as well as local communities.

"The whole vision of tourism benefiting parks and people often completely breaks down," Wood says, "because there are inadequate entrance fees to parks, or those entrance fees are never recycled into park protection funds."

While the challenges ahead are enormous, ecotourism projects are flourishing both on the drawing board and in the field. In Fiji, where luxury hotels rise out of the dilapidated neighborhoods of the islanders, Australian tourism consultant Marc Aussie-Stone and the people of the village of Vatukarasa are working on an antidote to destructive tourism.

"They wanted a project that didn't have tourists visiting their village," explains Aussie-Stone. "They didn't like being shopkeepers. They favored any project that They see their forest as a resource or asset and they want a return just the same as anybody."

EcoHotels

Around the world, hotels are participating in a movement to become more ecologically efficient. Several hotel chains have initiated environmental programs. The Stouffer Stanford Court Hotel in San Francisco recycles steam heat and uses the condensation to reheat water, as well as to replace water evaporated from the air-conditioning cooling towers. At a Days Inn in Fort Myers, Florida, shredded wastepaper is piled in a compost heap with grass clippings for use as mulch. The Radisson South in Minneapolis contracted with a company to use its food waste as livestock feed.

Many luxury hotels have also participated in the eco-hotel movement without sacrificing elegance or comfort. These hotels include: Hotel Bel-Air in Los Angeles; Mansion on Turtle Creek in Dallas; Hotel Hana-Maui in Hawaii; the Willard in Washington, DC; the Grand Traverse Resort in Traverse City, Michigan; and the Park Plaza Hotel and Towers in Boston. The Boston Park Plaza Hotel and Towers has a comprehensive environmental program, which includes thermal windows and water reducing devices on showerheads and in toilets. Table linens burnt by cigarettes are made into chef's aprons.

A: *Half of it*
Source: 50 Simple Things Kids Can Do to Save the Earth, *EarthWorks Press*

Environmental pamphlets are left on bedside tables, and water conservation suggestions are etched on brass plates in the bathrooms. Although many of the eco-hotel programs were not initiated to save money, hotels are learning that ecological business practices often increase profit.

Types of Ecotravel

Ecotravel does not automatically imply deprivation, sacrifice, struggle and pain. Some programs may restrict your amenities and encourage a back-to-nature philosophy, but other tours require little physical exertion and allow for comfort. According to the *Green Travel Sourcebook* (John Wiley & Sons, New York), there are three types of environmental travel.

Rugged travel involves a lack of comforts and amenities. Travelers may find themselves camping in the woods and trekking for hours up steep mountain trails. Volunteer programs might house the traveler in a communal living situation such as in a dormitory. Some trips may take travelers to live with natives, stay in their huts, bathe in streams and eat whatever food is provided on the communal fire. Travelers are often expected to share in the responsibilities of cooking, cleaning and setting up camp.

Soft adventure travel is less strenuous. Rather

included their traditional activities of feasting, talking, farming, singing, fishing and community gatherings. And it had to be run by the whole village."

The people of Vatukarasa are planning a resort that will be built as a "sister village" to their own Fijian village. The resort, which is planned to be complete in about two years, will be owned and controlled by the villagers. They will improve their farming and fishing techniques to supply the hotel and village with fresh produce and fish. A "cultural respect center" between the Fijian village and the hotel will explain local customs and culture and keep the tourists out of the village.

Perhaps the most intriguing part of the project is the financing: Guests will pay on a sliding scale, according to the strength of their own currency. "In this way, all peoples of the world will be able to afford to visit and share hospitality with other indigenous people," says Aussie-Stone. "The Japanese may pay $600 to stay; the Nepalese, $20." Some 70 percent of the hotel profits will be dedicated to other villages developing similar projects, with the understanding that those villages will do the same with their profits. "We're trying to set up a rolling system," Aussie-Stone explains.

On the other side of the Pacific, Rara Avis, a private forest reserve in Costa Rica, is dedicated to tourism, scientific research and the sustainable use of forest products. Money is generated by extraction of ornamental plants and wicker; tourists and scientists stay there, examining the forest canopy, the waterfalls and everything in-between. The place has generated employment in this remote corner of Costa Rica, and is testament to the belief that standing forest can be profitable. This message is especially important in Costa Rica, which is being deforested as fast as any country in the hemisphere, often at the hands of subsistence farmers.

Meanwhile, in Zimbabwe, much of the revenue generated by trophy hunting is being funneled into the coffers of local communities through a program called CAMPFIRE (Communal Areas Management Program for Indigenous Resources). This encourages local people to get involved in professional wildlife management and guiding, and is instrumental in changing negative attitudes about wildlife among rural farmers. Wildlife becomes an investment

 How many children worldwide died in 1991 as a direct result of waterborne diseases stemming from unsanitary drinking and bathing water?

instead of an impediment to making a living.

It is clear that travelers are increasingly interested in benefiting the places they visit. Some 85 percent of travelers interviewed by the Travel Industry Association of America in 1991 claimed that they are likely to patronize travel companies that help preserve the environment—and they are willing to pay more for the privilege.

Ecotourism isn't going to save the world, but it could help save part of it. Properly planned ecotourism is a tool with a wonderful ability to create incentives to protect our planet's natural and cultural diversity.

ECOTOURISM IN COSTA RICA: THE TOUGH ROAD FOR REMAINING NUMBER ONE

IN AUGUST 1992, Costa Rica received the Golden Compass Award from the Travel News Network Cable TV. With this award, 35 of the leading travel writers in the US branded this small Central American country the number one ecotourism destination in the world. As new investments are being made in the industry at meteoric rates, and as politicians place their highest hopes on tourism becoming the country's first source of foreign income in the years to come, Costa Ricans are clearly counting on things to stay that way.

But, if Costa Rica wishes to retain its position, it will have to invest substantially to strengthen its system of conservation areas, which is the base where this specialized brand of tourism rests. The country may soon reach its limit as a tourism destination, and the short-term benefits will not be properly distributed, while the social and ecological costs will have to be covered by the local populations and the government.

Even worse, shortsighted tourism development could exhaust resources. In this area, Costa Rica is under double jeopardy, since natural resources are not only the raw material for ecotourism development, but also the base for the country's development in general.

In only five years, tourism has grown to become Costa Rica's second-largest source of foreign income. In 1991, it generated $336 million, well beyond the authorities' expectations. The number of tourists has also grown beyond projections: This tiny country of 3 million inhabitants received half a million tourists in

than trekking the entire distance by foot, raft, horse or skis, participants are bused, flown or ferried to remote sites. Accommodations, meals and amenities are usually on board a ship or in a lodge or hotel. The staff sets up and breaks camp, cooks, serves the meals and cleans up if camping is required.

Sybaritic travel is often considered ecotravel, but would be scoffed at by ecotravelers of the rugged category. On sybaritic tours, participants can expect moderate to luxurious accommodations and amenities. Only if the accommodations are sensitive to the area and if the indigenous people benefit from the profits is such travel truly green.

"Ecotourists are, by definition, Green Travelers. However, Green Travelers are not necessarily ecotourists. Green Travelers include, in addition to those interested primarily in nature and the environment, individuals concerned with humanitarian matters, cultural interchange and intellectual enrichment. Often, such activities have little or nothing to do with flora or fauna."
—*Green Travel Sourcebook*

Travel Magazines

***BUZZWORM*: The Environmental Journal**
Published bimonthly by Buzzworm, Inc., 2305 Canyon Blvd., Ste. 206, Boulder, CO 80302.

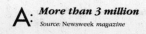
A: *More than 3 million*
Source: Newsweek *magazine*

Each issue of *Buzzworm* has a large eco-travel section focusing on a different area of the world. Recent issues have covered: Southern Africa, Australia, Central America, the American Southwest, Alaska, Hawaii and Scotland. Each issue includes photos, in-depth stories, environmental news and travel information on the featured area.

Cultural Survival
Published quarterly by Cultural Survival, 215 First St., Cambridge, MA 02142.

Cultural Survival is a 20-year-old nonprofit organization dedicated to preserving indigenous cultures. Its quarterly magazine is a good resource for travelers who want to understand the indigenous culture of the countries they visit.

Just Go!
Published quarterly by Lisa Tabb, 284 Connecticut St., San Francisco, CA 94107.

Aiming to fill the gap between luxury-travel magazines and backpacking magazines, *Just Go!* features trips for people with "middle range" incomes. Trips in recent issues included Maho Bay on St. John Island, Georgia O'Keeffe's "Ghost Ranch" in New Mexico and Tree Tops Guest House in Thailand.

National Geographic Traveler
Published bimonthly by the National Geographic Society, 17th and M Sts.

1991, and early projections for 1992 indicate the figure will increase substantially.

Inevitably, the rapid growth of the nature tourism industry has had an impact on the country's protected areas, which are the main attractions for this special brand of tourists. Already, the demand for use of some of the country's protected areas has grown significantly. If the availability of management, conservation budgets and human resources does not grow accordingly, the situation in those areas will become critical.

The rapid increase in visitation is well illustrated by Carara Biological Reserve, an attractive transition forest near the Pacific Coast that hosts a healthy population of the beautiful scarlet macaw. In 1989, 5,603 people visited Carara. The figure grew to 11,700 in 1991. By July 1992, Carara had already received 19,500 visitors, almost a 100 percent increase from the whole of 1991.

Perhaps more critical is the situation of Manuel Antonio National Park, a coastal forest with lush vegetation and white sandy beaches, which until recently was most popular with nationals. The number of visitors to Manuel Antonio climbed from 36,462 in 1982 to 152,543 in 1991. Today, more than 65 percent of the park's visitors are foreigners. The budget for managing the park and servicing these visitors is only $130,372 for 1992, which does not even amount to $1 per visitor.

The situation in these areas reflects what is happening at the National Parks Service: With the government's structural adjustments and its efforts to cut expenses in order to meet the demands of international financing agencies, the budget of the National Parks System has been cut in half, and the total personnel amounts to barely 250 people, including office, management and field personnel. Of the total $12 million National Parks budget for 1992, the Costa Rican government only provides $2.8 million, and only 0.5 percent of this comes from entrance fees to the parks. The rest comes from donations, making the National Parks System extremely fragile and dependent. At this time, the former owners of 20 percent of the country's protected areas have yet to be paid.

Considering this difficult situation, the National Parks System is turning to the tourism industry as a likely contributor to the country's conservation efforts. It makes sense, say the conservation

: According to Global ReLeaf, how much can a home's air-conditioning costs be reduced by one well-placed shade tree?

authorities, for the tourism industry to invest in what has become the industry's raw material: Costa Rica's protected areas.

As yet, the situation does not reflect this desire, as documented at Tortuguero, one of the most popular tourism destinations in the country, with its lush canals, its abundant wildlife and its world-famous green turtles. In 1991, Tortuguero received 47,376 visitors, neary 90 percent of them foreigners. Yet during that time, only 0.18 percent of the income generated by tourism in Tortuguero reached the park, which is the area's most important attraction, and only 5.91 percent of the income generated stayed in the Tortuguero community.

In order to guarantee a real contribution of tourism to Costa Rican conservation, it is not enough that well-intentioned private entrepreneurs support isolated efforts. The National Parks System needs to properly control the income generated by tourism visitation, and experience shows this has rarely been the case. Legal adjustments also need to be made, to allow the park service to manage and make the best use of this growing opportunity for income generation. These would allow, for instance, an increase in admission fees and the development of a basic service infrastructure for visitors, such as interpretive trails, visitors' centers and other educational and management services.

Costa Rica has earned an international reputation for its peaceful, democratic ways, and for its farsighted conservation models. For visitors, it holds a few other major advantages: its small size and an expansive and thorough road and transportation system, together with other important services.

Conservation authorities in Costa Rica have clearly expressed their willingness to open the country's natural heritage to the tourism industry. Yet worries remain among ecologists, because visitors are concentrating in a few areas, putting these under growing pressure. They hope to lessen the visitation impact by providing tourists with guidance and better services.

"Tourism can become the first really green industry," wrote members of the board of directors of CANATUR (the National Chamber of Tourism Entrepreneurs) in a recent publication.

A number of cooperative actions are already under way: The National Tourism Board is financing

NW, Washington, DC 20036.

Covering the US and the rest of the world, *National Geographic Traveler* is the travel-oriented progeny of the famous *National Geographic* magazine. *National Geographic Traveler* has a larger page size than *National Geographic* and is filled with travel stories and photos.

South American Explorers Club
Published quarterly by the South American Explorers Club, 126 Indian Creek Rd., Ithaca, NY 14850.

The nonprofit South American Explorers Club publishes a magazine devoted to adventure travel on the South American continent.

Guide Books

As computers, faxes and telephones put us in instant touch with the rest of the world, we are experiencing the greatest information revolution since the invention of the printing press. This new technology makes it quicker and easier than ever to update travel guides. Some of the many outstanding travel books of the past year are listed below. All books were published in 1992.

AFRICA

West Africa: A Travel Survival Kit, by Alex Newton, Lonely Planet Publications, Oakland, CA, $19.95.

This 750-page guide

A: *By half*

Source: The Student Environmental Action Guide, *Earth Works Press*

covers Benin, Burkina Faso, Cape Verde, Côte D'Ivoire, Gambia, Ghana, Guinea, Guinea-Bissau, Liberia, Mali, Mauritania, Niger, Nigeria, Senegal, Sierra Leone and Togo. It is packed with Lonely Planet's incomparable travel information, plus maps and photos.

EUROPE

Cycling France, by Jerry H. Simpson, Bicycle Books, Inc., Mill Valley, CA, $12.95.

Plan your own tour de France. Descriptions and maps of "the best bike tours in all of Gaul!" A jug of wine, a loaf of bread and thou, beside me pedaling in the wilderness. . . .

IRAN

Iran: A Travel Survival Kit, by David St. Vincent, Lonely Planet Publications, Oakland, CA, $14.95.

This 350-page guide to Iran opens with the author's assertion: "Iran is a strange sort of country." This is a fascinating, fact-filled guide, featuring color photographs, maps and loads of travel information.

MEXICO

West Mexico: From Sea to Sierra, by Charles Kulander, La Paz Publishing, Ramona, CA, $16.95.

An introduction to the vast and extreme geography of Sonora and the Baja Peninsula. The guide describes the endless beaches,

the preparation of carrying-capacity and environmental-impact-assessment studies for some parks, and a number of interdisciplinary planning commissions have been created to harness the development of ecotourism in Costa Rica. The National Parks System will invest in training personnel and adjusting its present structure to respond to the new demands and opportunities created by tourism. Support from the international community is necessary to meet the objectives of sustainable ecotourism development.

Betting on Costa Rica's traditions, CANATUR leaders say that tourism can be, at the turn of the century, the democratic option for development—by being a high generator of foreign income distributed in numerous small enterprises, and by investing in rural, economically depressed areas. Therefore, entrepreneurs say, tourism can be, at the end of the 20th century, what coffee was at the century's beginning: a democratic economic activity, leading to an equitable distribution of wealth and helping to maintain Costa Rica's tradition of social stability.

—*Ana Báez and Yanina Rovinski*

Excerpted from the 1992 World Congress on Adventure Travel and Ecotourism held by the Adventure Travel Society.

MONITORING ECOTOURISM

ECOTOURISM HAS BECOME a conservation and sustainable-development opportunity worldwide. Until recently, there has not been any discussion of a monitoring program to prevent abuse of the term "ecotravel." In May 1992, The Ecotourism Society received a grant from the Liz Claiborne/Art Ortenberg Foundation to research and develop a green evaluation program for international nature tourism services. The society will research the feasibility of a monitoring program that would be performed by consumers during the next year.

A consumer evaluation program would gather and distribute the first objective data on the conservation and sustainable development performance of international tourism services. Although a consumer evaluation system could not monitor all impacts of ecotourism on ecosystems and cultures, the program would provide a valuable source of information for tour operators, travel agents, lodge owners,

: *In sub-Saharan Africa, how many cattle have died as a result of overgrazing and desertification of land?*

governments, activists, consumers and the media.

In July 1992, a survey was mailed to 250 US-based adventure travel tour operators, collecting data on conservation and sustainable development performance of tourism services. Forty-three operators responded, most of them from the larger corporations. According to the MacKenzie and Company research for the Pew Foundation, just 46 nature tour operators in the US account for 93 percent of the profits in the industry. The results were compiled and

An Alaskan moose appears to be oblivious to dayhikers.

discussed by representatives from the Sierra Club, Earth Island Institute, Wilderness Travel, Mountain Travel, InnerAsia Expeditions, University of Washington, University of California at Oakland and Berkeley, and Colorado State University.

Some 500 consumers who have contacted the Ecotourism Society for information in the last two years, and a list of 1,000 consumers expressing interest in the environment, will be mailed another survey. These surveys will ask how consumers select tours and how ecotourism standards influence the selection process. In addition, a phone survey will be conducted of nonprofit organizations that contract wilderness tours to determine what role these organizations will have in the evaluation process. An inbound operator and lodge owner survey will be done in Costa Rica to examine operators and their views on the criteria for consumer evaluation. The compiled survey results will be reviewed by representatives of World Wildlife Fund, National Audubon and the Nature Conservancy.

uninhabited islands and mountains, as well as the history of the Mexican coastlines.
A Naturalist's Mexico, by Roland H. Wauer, Texas A & M Univ. Press, College Station, TX, $14.95.

A unique introduction to the majestic biological diversity of Mexico.

SOUTH AMERICA

Argentina, Uruguay, and Paraguay: A Travel and Survival Kit, by Wayne Bernhardson and Maria Massolo, Lonely Planet Publications, Oakland, CA, $16.95.

This inclusive guide invites travelers to discover the friendly people and magnificient natural attractions of southern South America, including the Falkland Islands.

SOUTH PACIFIC

Australian Bed and Breakfast Book, by J. and J. Thomas, Pelican Publishing, Gretna, LA, $9.95.

A fascinating guide to bed and breakfast places on the Australian continent.

Australia: A Travel Survival Kit, by Hugh Finlay et al., Lonely Planet Publications, Oakland, CA, $21.95.

This 835-page guide to Down Under is packed with facts, maps and a continent's worth of information.

New Zealand Bed & Breakfast Book, by J. and J. Thomas, Pelican Publishing, Gretna, LA, $9.95.

A thorough guide to bed and breakfast spots in New Zealand.

A: *Over 5 million*
Source: Global Environmental Issues, *Routledge Publishers*

UNITED STATES

The Big Outside, by Dave Foreman and Howie Wolke, Harmony Books, New York, $16.

Subtitled "A Descriptive Inventory of the Big Wilderness Areas of the United States." Facts, figures and information on the biggest wilderness areas in the US, from Yosemite in California to Moose River in Maine, with plenty in-between.

Colorado BLM Wildlands, by Mark Pearson and John Fielder, Westcliffe Publishers, Englewood, CO, $17.95.

An-all inclusive guide to Colorado's BLM lands. Includes over 70 color photos and 51 color maps.

Fifty Hikes in Massachusetts, by John Brady and Brian White, Countryman Press, Woodstock, VT, $12.95.

Features 50 hikes, "from the top of the Berkshires to the tip of Cape Cod."

The Hiker's Guide to Alaska, by Evan and Margret Swensen, Falcon Press, Billings, MT, $9.95.

A perfect companion for the Alaskan traveler, this guide introduces the reader to hiking in Alaska's national parks, wildlife refuges, national forests, wilderness areas and state parks. This eco-sensitive guide also aids the traveler by citing low-impact camping methods.

Maverick Guide to Hawaii, by Robert Bone,

In May 1993, recommendations will be presented to the board of directors of the Ecotourism Society.

PEOPLE-TO-PEOPLE OUTREACH

TRAVEL IS THE BEST FORM OF EDUCATION. For several decades, tourism has brought us closer to global understanding and acceptance of various cultures. Not only does tourism stimulate economies and distribute wealth to less-stable economies, but it also encourages peace in the world. Tourism has been credited in part for the opening of China's borders, liberalizing Eastern Europe and reducing international conflicts. When people from different countries have the opportunity to meet each other and exchange ideas, inhibitions and misconceptions are destroyed.

Ecotravel operators and nonprofit organizations offer programs that enable people of different cultures to interact with people who share common interests. Such interests range from a mutual concern for the environment to a shared interest in a certain sport or activity. Groups of travelers from several different countries can combine their efforts through programs, research projects or rallies for environmental action and world peace.

Several people-to-people programs are arranged by religious organizations, educational institutions and environmental groups. The range of these programs is expansive and includes many options such as student exchanges, hosting a foreign traveler in your home, trekking through small villages or staying in international dormitories or hostels. Often it is possible to arrange such projects individually. *The Green Travel Sourcebook* (John Wiley & Sons, New York) offers a selection of trips and contacts for other options.

ORGANIZATIONS PROMOTING ECOTRAVEL

The Ecotourism Society is a nonprofit organization dedicated to conserving natural environments and sustaining the well-being of local people through responsible travel. Founded in 1990 to serve tourism and conservation professionals, the society also provides discounts to members on a host of ecotourism publications and resources. It also publishes a quarterly newsletter and other information on visitor management, impacts, literature, gear and contacts for ecosystems worldwide. Along with the World Wildlife Fund, the Nature Conservancy, the Sierra

: *To produce one pat of butter, how much water is required?*

ECOTRAVEL

Club and the National Audubon Society, the society is working to develop "green evaluations" of tour operators and lodges, planned for publication in 1994. See "Monitoring Ecotourism," page 306. Contact the Ecotourism Society, 801 Devon Place, Alexandria, VA 22314, (703) 549-8979.

The Center for Responsible Tourism was formed in 1984 to involve North Americans in ethical issues about travel in the third world. It advocates travel that benefits both host and guest and harms neither. A nonprofit organization, the Center for Responsible Tourism produces a quarterly newsletter called "Responsible Traveling" for its contributors. For information, send a stamped, self-addressed envelope to the Center for Responsible Tourism, P.O. Box 827, San Anselmo, CA 94979, (510) 843-5506.

The Adventure Travel Society (ATS) is the only trade association in the world that addresses the issues related to nature-based tourism and adventure travel. ATS organizes the World Congress on Adventure Travel and Ecotourism. For more information, contact the society at 6551 S. Revere Parkway, Ste. 160, Englewood, CO 80111, (303) 649-9016.

Cultural Survival defends the rights of indigenous people and ethnic minorities on five continents. The 20-year-old organization works with native people around the world to help them adapt new technologies to their traditional systems of production. Cultural Survival also publishes an informative quarterly magazine. Contact Cultural Survival, Inc., 215 First St., Cambridge, MA 02142, (617) 621-3818. Three back issues of the quarterly focus on tourism and its impacts. "The Tourist Trap: Who's Getting Caught," and parts one and two of "Tourism and Development: Breaking Out of the Tourist Trap" are available from the above address for $5 each, prepaid.

Lonely Planet guides are dependably clear-eyed and fiesty, and pay plenty of attention to environmental, cultural and social issues in the numerous countries they cover. Contact Lonely Planet Publications, Embarcadero W., 155 Filbert St., Ste. 251, Oakland, CA 94607.

Pelican Publishing Company, Gretna, LA, $12.95.

This guide details for travelers all of the major attractions on the islands, as well as some less-visited sites of interest.

Sacred Sites: A Guidebook to Sacred Centers and Mysterious Places in the United States, edited by Frank Joseph, Llewellyn Publications, St. Paul, MN, $16.95.

Providing physical descriptions and locations of 27 sacred sites across the country, this travel guide is sure to provide for a unique ecoventure.

Resources

The Buzzworm Magazine Guide to Ecotravel, by the editors of Buzzworm Magazine, Buzzworm Books, Boulder, CO, $11.95.

The editors of *Buzzworm* have chosen 100 exciting ecotravel trips from all over the world. Each trip is described in detail, with a map of the area and information on the trip outfitter.

The Green Travel Sourcebook, by Daniel Grotta and Sally Wiener Grotta, John Wiley & Sons, New York, $14.95.

Subtitled "A Guide for the Physically Adventurous, the Intellectually Curious, or the Socially Aware," this book lists travel outfitters, organizations that sponsor "green" trips and much more.

A: *100 gallons*
Source: 50 Simple Things You Can Do to Save the Earth, *Earth Works Press*

EARTH JOURNAL

Directory to EcoTravel Outfitters

*P*ick your trip, and pack your bags! Here are dozens of ecotravel outfitters ready to take you on the eco-adventure of your dreams. Whether you enjoy hiking, kayaking, scuba diving or birdwatching, there's an ecotravel trip for you.

1. **ABOVE THE CLOUDS TREKKING**
   P.O. Box 398
   Worcester, MA 01602-0398
   (508) 799-4499; (800) 233-4499
   Types of Adventure: Culturally oriented trekking and walking trips. **Destinations:** Nepal, Bhutan, Pakistan, India, Norway, Austria, France, Great Britain, Ireland, Poland, Madagascar, Kenya, Zimbabwe, Botswana, Argentina, Costa Rica. **Special Emphasis:** Walking trips through remote mountainous areas; coming into contact with local people who are unaccustomed to outsiders; culturally sensitive. **Average Cost:** $2,000/person for land package for a 3-week trip.

2. **ACTION WHITEWATER ADVENTURES**
   P.O. Box 1634
   Provo, UT 84603
   (800) 453-1482
   Types of Adventure: Whitewater rafting trips. **Destinations:** Main Salmon River (Idaho), Middle Fork of the Salmon (Idaho), Colorado River/Grand Canyon, American River (California). **Special Emphasis:** Quality rafting vacations. **Average Cost:** $900/person/week, $180/person/2 days.

3. **ADVENTURE CENTER**
   1311 63rd St., Ste. 200
   Emeryville, CA 94608
   (510) 654-1879
   Types of Adventure: Wildlife safaris, treks, overland expeditions, cycling, train journeys, cultural trips, camel trips. **Destinations:** Worldwide—over 100 countries, including Africa, Asia, Latin America, South Pacific, Europe, Middle East. **Special Emphasis:** Cultural interaction with local peoples, flexible itineraries, international groups, low environmental impact. **Average Cost:** $500-$7,000 depending on itinerary length. Trips from 1 week to 6 months.

4. **AFRICATOURS**
   210 Post St., #911
   San Francisco, CA 94108
   (415) 391-5788
   Types of Adventure: Educational, photo safaris, canoeing, rafting, hiking, mountain climbing, wildlife observations. **Destinations:** Botswana, Kenya, South Africa, Malawi, Tanzania, Zimbabwe, Zambia. **Special Emphasis:** Wildlife observation. **Average Cost:** $2,500.

5. **ALASKA DISCOVERY, INC.**
   234 Gold St.
   Juneau, AK 99801
   (907) 586-1911; Fax (907) 586-2332
   Types of Adventure: Sea kayaking, canoeing and rafting. **Destinations:** Glacier Bay, Admiralty Island, southeast Alaska, Brooks Range in the Arctic National Wildlife Refuge, Tatshenshini/Alsek rivers. **Special Emphasis:** Low-impact, comfortable camping. **Average Cost:** $1,100 for 5-day tour, $1,350 for 7-day tour, $2,200 for 10-day Arctic trip, $1,700 for 10-day Tatshenshini raft trip.

6. **ALASKA WILDLAND ADVENTURES**
   P.O. Box 389
   Girdwood, AK 99587
   (907) 783-2928; (800) 334-8730
   Types of Adventure: Alaska natural history safaris, senior safaris, wilderness expeditions. **Destinations:** Denali National Park, Kenai Fjords National Park, Kenai National Wildlife Refuge, Chugach National Forest, seacoast glaciers, Inside Passage. **Special Emphasis:** Natural history tours that take travelers beyond the conventional bus tours and cruises. Trips are informative and experience-oriented with group size of 18. **Average Cost:** $1,950 for 7 days, $2,395 for 10 days, $2,795 for 12 days.

7. **ALL ADVENTURE TRAVEL, INC.**
   P.O. Box 4307
   Boulder, CO 80306
   (303) 499-1981
   Types of Adventure: Biking, hiking, nature safaris, expedition cruises. **Destinations:** US, Canada, Europe, South America, Asia, Africa. **Special Emphasis:** Soft adventure, cultural interaction. **Average Cost:** $150-$250/day.

8. **AMAZON TOURS & CRUISES**
   8700 W. Flagler, Ste. 190

 Which US state has the most national parks?

Miami, FL 33174
(305) 227-2266; (800) 423-2791
Types of Adventure: Amazon river trips—from expeditions to air-conditioned ships in the upper and mid-Amazon areas. **Destinations:** Iquitos, Peru; Leticia, Colombia; Tabatinga and Manaus, Brazil. **Special Emphasis:** FIT and small groups. **Average Cost:** $70/day.

9. **AMAZONIA EXPEDITIONS**
   1824 NW 102nd Way
   Gainesville, FL 32606
   (904) 332-4051

Types of Adventure: Expeditions in Amazon rainforest. **Destinations:** Amazon rainforest. **Special Emphasis:** Custom itineraries for personal interests and needs. **Average Cost:** $1,375/person for 2 weeks.

10. **AMERICAN WILDERNESS EXPERIENCE, INC. (AWE!)**
    P.O. Box 1486
    Boulder, CO 80306
    (303) 444-2622; (800) 444-0099

Types of Adventure: Horse packing, llama trekking, whitewater rafting, backpacking, canoeing, sailing, mountain biking, sea kayaking, natural history, trekking, snorkeling and diving. **Destinations:** Rocky Mountain West, desert Southwest, Alaska, Hawaii, Canada, Mexico, New England, Florida, Virgin Islands, Australia, New Zealand, Peru and Belize. **Special Emphasis:** Small group, ecologically sensitive backcountry travel. **Average Cost:** $74-$175/person for 6 days, domestic; $100-$175/person for 12 days, international.

11. **ANCIENT FOREST ADVENTURES**
    Central Oregon Environmental Center
    16 NW Kansas Ave.
    Bend, OR 97701
    (503) 385-8633; (800) 551-1043

Types of Adventure: Interpretive trips into the ancient forests of Oregon by foot, snowshoe and cross-country skiing. **Destinations:** Ancient or old growth forests in Oregon; from the coast to the dry east side of the state. **Special Emphasis:** Ecological management. Discovering the magic and the mystery of the ancient forests through Native American lore, individual quiet time and night walks. **Average Cost:** $525/week for resort-based trips, $350/week for camping.

12. **APPALACHIAN MOUNTAIN CLUB**
    P.O. Box 298-ET
    Gorham, NH 03581
    (603) 466-2721

Types of Adventure: Guided and unguided hikes, outdoor skills workshops, nature and wildlife exploration, environmental education programs for all ages. **Destinations:** White Mountain National Forest (New Hampshire), Berkshires (Massachusetts), Catskills (New York). **Special Emphasis:** Promoting the protection, enjoyment and wise use of our natural resources. **Average Cost:** $200.

13. **ASIAN PACIFIC ADVENTURES**
    826 S. Sierra Bonita Ave.
    Los Angeles, CA 90036
    (213) 935-3156; (800) 825-1680

Types of Adventure: Bicycling, in-depth cultural and environmental trips, hiking, jungle trekking, mountain biking, overland expeditions, photography, rafting, safaris, wildlife. **Destinations:** Bali, Borneo, China, Himalayas, India, Indonesia, Japan, Malaysia, Nepal, Pakistan, Southeast Asia, Thailand, Tibet, Vietnam, Laos and Cambodia. **Special Emphasis:** Environmental and cultural adventures to spectacular, remote areas in small groups that allow in-depth experiences with minimal impact on the environment and indigenous cultures. **Average Cost:** $150/person/day.

14. **BAIKAL REFLECTIONS, INC.**
    13 Ridge Rd.
    Fairfax, CA 94930
    (415) 455-0155; (800) 927-2797

Types of Adventure: Ocean kayaking, backpacking, cycling, wildlife, sailing. **Destinations:** Lake Baikal and other parts of the Commonwealth of Independent States (former Soviet Union). **Special Emphasis:** Educational, ethics-based, environment-centered, indigenous cultures. **Average Cost:** $1,700 land cost.

15. **BAJA EXPEDITIONS**
    2625 Garnet Ave.
    San Diego, CA 92109
    (619) 581-3311; (800) 843-6967
    Fax (619) 581-6542

Types of Adventure: Whale watching, sea kayaking, Sea of Cortez cruises, mountain biking, wilderness sailing, scuba diving. **Destinations:** Baja, Mexico and Costa Rica. **Special Emphasis:** Natural history and adventure travel. **Average Cost:** $150/day (includes air).

16. **BICYCLE AFRICA**
    4887-B Columbia Dr. S.
    Seattle, WA 98108
    (206) 628-9314

Types of Adventure: Culturally sensitive, environmentally friendly people-to-people bicycle tours. Easy to moderate difficulty, with extra side trips for the active. **Destinations:** Tunisia, Zimbabwe, Botswana, Kenya, Benin, Togo, Ghana,

Alaska, with 23
Source: The Daily Camera, *Boulder, Colorado*

Mali, Senegal, Gambia. **Special Emphasis:** Small group bicycle tours to untouristed areas promoting cultural and environmental understanding while benefiting local economy. **Average Cost:** $990-$1,290 plus airfare, for 2 weeks.

17. **BICYCLE ROMANTIC SCOTLAND, PETER COSTELLO, LTD.**
P.O. Box 23490
Baltimore, MD 21203
(410) 783-1229; Fax (410) 539-3250
Types of Adventure: Van-supported, 6-day bicycle tours. Hiking and trekking. **Destinations:** Scotland—Sir Walter Scott's fictionalized historic Scottish "Borders." **Special Emphasis:** Challenging, varied trips, passionate countryside, glorious castles, stately houses, charming market towns, quiet rivers, uncluttered byways. **Average Cost:** $1,050.

18. **BILL DVORAK KAYAK & RAFTING EXPEDITIONS, INC.**
17921-Z, US Hwy. 285
Nathrop, CO 81236
(719) 539-6851; (800) 824-3795
Types of Adventure: Whitewater rafting, kayaking, canoeing, custom fishing, horseback riding, mountain biking, instructional seminars, classical music journeys, international expeditions. **Destinations:** Colorado, Utah, Wyoming, Arizona, New Mexico, Texas, Mexico, Hawaii, New Zealand, Australia. **Special Emphasis:** Dvorak Expeditions fosters a participatory spirit in whitewater expeditions as well as supporting river conservation and a socially responsible, low-impact style of travel. **Average Cost:** $32-$81 for half-day to full-day trips, $190-$1,929 for 2-day to 12-day trips.

19. **BIOLOGICAL JOURNEYS**
1696 Ocean Dr.
McKinleyville, CA 95521
(707) 839-0178; Fax (707) 839-4656
Types of Adventure: Small group, educational adventure travel. **Destinations:** Alaska, Baja California, Australia, the Amazon and the Galápagos. **Special Emphasis:** Endangered wildlife, threatened habitats and conservation activism. **Average Cost:** $1,295-$5,300.

20. **BOLDER ADVENTURES, INC.**
P.O. Box 1279
Boulder, CO 80306
(303) 443-6789
Types of Adventure: Natural history and in-depth cultural programs, both small group and independent travel arrangements. **Destinations:** Southeast Asia, including Thailand, Indonesia, Laos, Cambodia, Vietnam, Malaysia, Burma.

Special Emphasis: Activity-oriented programs that get you close to the people and natural beauty of Southeast Asia. **Average Cost:** $1,500-$3,000.

21. **BORTON OVERSEAS**
5516 Lyndale Ave. S.
Minneapolis, MN 55419
(612) 824-4415; (800) 843-0602
Types of Adventure: African safaris, Kilimanjaro climb, glacier and mountain hiking, rafting, whale watching, skiing, birdwatching, dog and reindeer sledding. **Destinations:** Africa, Scandinavia, Lapland, Greenland, Spitsbergen. **Special Emphasis:** Environmentally and culturally sensitive trips. Individual and small groups. **Average Cost:** $400-$4,000.

22. **CANADIAN RIVER EXPEDITIONS**
3524 W. 16th Ave.
Vancouver, BC, V6R 3C1, Canada
(604) 738-4449
Types of Adventure: Wilderness expeditions to remote regions in western Canada and Alaska. Natural history exploration of rivers and coastlines. **Destinations:** Tatshenshini and Alsek rivers, British Columbia and Alaska. Firth River, Chilko River, Chilcotin River and Fraser River. **Special Emphasis:** Wilderness and natural history. Small groups with expert naturalists. Each trip focuses on a different landscape and ecosystem. **Average Cost:** $1,200 for 12 days, fully guided, all-inclusive.

23. **CANYONLANDS FIELD INSTITUTE**
P.O. Box 68
Moab, UT 84532
(801) 259-7750
Types of Adventure: Ecology studies collected on various wilderness expeditions. **Destinations:** Utah, Colorado, Arizona, New Mexico. **Special Emphasis:** Monitoring the flora and fauna of the western US. **Average Cost:** $40-$750.

24. **CASCADE RAFT COMPANY**
P.O. Box 6
Garden Valley, ID 83622
(800) 292-7238
Types of Adventure: River trips and river treks. **Destinations:** Payette River and Main Salmon River (Idaho), Nepal, Costa Rica. **Special Emphasis:** Environmentally sensitive river trips and treks. **Average Cost:** $80-$120/day.

25. **CLEARWATER CANOE OUTFITTERS & LODGE**
355 Gunflint Trail
Grand Marais, MN 55604

: *For how long does most nuclear waste generated in nuclear power plants remain toxic?*

(218) 388-2254; (800) 527-0554
Types of Adventure: Wilderness canoe/camping trips or historic lodge/B&B, secluded cabins. Hiking, mountain biking, guided kayak tours on Lake Superior. **Destinations:** Minnesota's Boundary Waters Canoe Area, Canada's Quetico Park in Ontario, Isle Royale on Lake Superior. **Special Emphasis:** Quality equipment and Kevlar canoes, vegetarian menu options, personalized routing and unique sailboat camping package. **Average Cost:** $40-$115/day.

26. **CO-OP AMERICA TRAVEL-LINKS**
    14 Arrow St.
    Cambridge, MA 02138
    (617) 628-2667

Types of Adventure: Personally designed tours for individuals and small groups. **Destinations:** Caribbean, Central America, Africa, China, Hawaii. **Special Emphasis:** How to be a socially responsible traveler. **Average Cost:** $1,200.

27. **COSTA RICA EXPERTS, INC.**
    3540 NW 13th St.
    Gainesville, FL 32609
    (904) 377-7111; (800) 858-0999

Types of Adventure: Natural history adventure, birding, botany. **Destinations:** Costa Rica. **Special Emphasis:** Rainforests, birds, botany. **Average Cost:** $406-$1,875.

28. **COUNTRY WALKERS**
    P.O. Box 180
    Waterbury, VT 05676
    (802) 244-1387

Types of Adventure: Walking vacations. **Destinations:** US, New Zealand, Ireland. **Special Emphasis:** Environmental preservation and natural history tours led by local specialists. **Average Cost:** $299-$2,499.

29. **EARTHWATCH**
    680 Mt. Auburn St., Box 403
    Watertown, MA 02272
    (617) 926-8200; Fax (617) 926-8532

Types of Adventure: Scientific expeditions. **Destinations:** Worldwide. **Special Emphasis:** Environmental education, monitoring global change, conservation of endangered habitats and species, cultural exploration, fostering world health and international cooperation. **Average Cost:** $1,400.

30. **ECOSUMMER EXPEDITIONS**
    1516 Duranleau St.
    Vancouver, BC, V6H 3S4, Canada
    (604) 669-7741

Types of Adventure: Sea kayaking, trekking, rafting, canoeing, photo/nature expeditions. **Destinations:** Baja, Belize, Bahamas, British Columbia Pacific coast, Yukon, High Arctic, Costa Rica. **Special Emphasis:** Keeping the "adventure" in wilderness travel with a high level of cultural and historical interpretation. **Average Cost:** $2,000 for 10 days.

31. **ECOTOURS OF HAWAII**
    P.O. Box 2193
    Kamuela, HI 96743
    (808) 885-7759

Types of Adventure: Hiking, bicycling, kayaking and sailing. **Destinations:** Hawaiian islands. **Special Emphasis:** Low-impact adventures with security, education and appreciation of natural beauty. **Average Cost:** $45-$150/day.

32. **F & H TRAVEL CONSULTING**
    2441 Janin Way
    Solvang, CA 93463
    (805) 688-2441

Types of Adventure: Rainforest lodge trips, jeep safaris, birdwatching, fishing, trans-Pantanal tour. **Destinations:** Amazon and Pantanal, Brazil. **Special Emphasis:** Ecological tours of Brazil. **Average Cost:** $4,000.

33. **FANTASY ADVENTURES OF EARTH, INC.**
    P.O. Box 368
    Lincolndale, NY 10540
    (914) 248-5107

Types of Adventure: Wilderness adventures, including horseback riding, mountain biking, high mountain treks, whitewater challenges, sailing and scuba diving. **Destinations:** US, Canada, Peru, New Zealand, Belize, Alaska and Hawaii. **Special Emphasis:** Personal challenges, custom trips, team building, nature photography, family and single trips. **Average Cost:** $75-$125/day, wilderness trips; $90-$175/day, dude ranch program.

34. **FISHING AND FLYING**
    P.O. Box 2349
    Cordova, AK 99574
    (907) 424-3324

Types of Adventure: Hiking, fishing, rafting, mountaineering, sightseeing. **Destinations:** Alaska's Gulf Coast, Wrangell-St. Elias National Park, Glacier Bay National Park, Tongass and Chugach national forests. **Special Emphasis:** Wildlife viewing, natural wonders, off-the-beaten-track. **Average Cost:** Varies.

35. **FOUR CORNERS SCHOOL OF OUTDOOR EDUCATION**
    East Route
    Monticello, UT 84535

A: *Up to 220,000 years*
Source: Greenpeace

(801) 587-2156
Types of Adventure: Rafting, backpacking, cross-country skiing, hiking. **Destinations:** Southwestern US—Utah, Arizona, Colorado and New Mexico. **Special Emphasis:** Natural and human history of the Colorado Plateau, wilderness preservation. **Average Cost:** $500-$800.

36. **FRIENDS OF THE EARTH ECOTOURS**
    4-8-15 Naka Meguro
    Meguro-ku, Tokyo 153, Japan
    (81-3) 3760-3644; Fax (81-3) 3760-6959

Types of Adventure: Birding and wildlife tours led by professional naturalists, with a special emphasis upon deeping the participant's understanding and appreciation of the ecosystem visited. **Destinations:** Japan, Hawaii, Siberia, Vietnam, Antarctica. **Special Emphasis:** Friends of the Earth is a nonprofit environmental organization. All profits from the tours go to conservation of the area visited and for the preservation of Japan's forests and wetlands. Small groups, personal interaction with international conservationists, trips strive to "tread as lightly" as possible. **Average Cost:** $800-$10,000.

37. **GALAPAGOS NETWORK**
    7200 Corporate Center Dr., Ste. 404
    Miami, FL 33126-7972
    (305) 592-2294; (800) 633-7972

Types of Adventure: Yacht cruises in the Galápagos Islands aboard a 20-passenger, first-class yacht, including crew and naturalist guide. **Destinations:** Ecuador and the Galápagos Islands. **Special Emphasis:** Wildlife and natural history tours, birdwatching, yacht charters (motor and sail), environmental education. **Average Cost:** $600/3 nights, $800/4 nights, $1,400/7 nights.

38. **GEO EXPEDITIONS**
    P.O. Box 3656
    Sonora, CA 95370
    (800) 351-5041

Types of Adventure: Tented safaris, walking safaris in Africa, yacht cruises in the Galápagos Islands. **Destinations:** Kenya, Tanzania, Rwanda, Zaire, Botswana, Madagascar, Seychelles Islands, Galápagos Islands. **Special Emphasis:** Natural history tours to off-the-beaten-path areas, geared to small groups and led by qualified naturalist guides. **Average Cost:** $3,400-$7,500.

39. **GLACIER BAY SEA KAYAKING**
    Box 26
    Gustavus, AK 99826
    (907) 697-2257

Types of Adventure: Sea kayaking, camping. **Destinations:** Alaska—Glacier Bay National Park. **Special Emphasis:** Sea kayak rentals and transportation to remote areas of Glacier Bay. **Average Cost:** $500/3 days, 2 people.

40. **GREAT EXPEDITIONS INC.**
    5915 West Blvd.
    Vancouver, BC, V6M 3X1, Canada
    (604) 263-1505; (604) 263-1476

Types of Adventure: Natural history and soft adventure tours. **Destinations:** Zimbabwe, Botswana, Galápagos, Costa Rica, Indonesia, Thailand, Nepal, British Columbia, Yukon, Alaska. **Special Emphasis:** Natural history of the various areas being visited, the flora and fauna and how they are being affected by exploitation of the natural resources. **Average Cost:** $1,000-$6,500.

41. **GREAT PLAINS WILDLIFE INSTITUTE**
    P.O. Box 7580
    Jackson, WY 83001
    (307) 733-2623

Types of Adventure: North American wildlife safaris with groups of 6 people. **Destinations:** Northwestern Wyoming (the Yellowstone ecosystem). **Special Emphasis:** Wildlife and scenery. **Average Cost:** $1,635/person.

42. **HIGH DESERT ADVENTURES**
    757 E. South Temple, #201
    Salt Lake City, UT 84102
    (801) 355-5444; (800) 345-RAFT

Types of Adventure: Rafting, mountain biking, hiking, combination trips with two or more of these activities. **Destinations:** Grand Canyon, Canyonlands, Dinosaur National Park, southwestern Colorado, River of No Return Wilderness in Idaho. **Special Emphasis:** Environmentally sensitive and personally challenging adventures in Western wilderness areas. **Average Cost:** $230-$1,818.

43. **HIMALAYA TREKKING & WILDERNESS EXPEDITIONS**
    1900 Eighth St.
    Berkeley, CA 94710
    (510) 540-8040; (800) 777-TREK

Types of Adventure: Trekking, river rafting, river cruises in Siberia, safaris, cultural touring. **Destinations:** India, Nepal, Tibet, Pakistan, China (Silk Road), Bhutan, Sikkim, Siberia, Ladakh. **Special Emphasis:** Low-impact, small group sizes (6 to 8), ecological awareness information, sensitivity to conditions along route. **Average Cost:** $2,000 land cost.

44. **HIMALAYAN JOURNEYS**
    757 E. South Temple, #201
    Salt Lake City, UT 84102

Q: *On the University of Puget Sound campus, what percentage of students said they would participate if a recycling program were started?*

ECOTRAVEL

(800) 345-7238
Types of Adventure: Trekking and overland journeys in the Himalayan region. **Destinations:** Nepal, Ladakh, Tibet, Bhutan and Sikkim. **Special Emphasis:** Trekking and overland journeys in the Himalayas, with insight into the mountains' natural environment and the human culture of the region. **Average Cost:** $1,850.

45. **HIMALAYAN TRAVEL, INC.**
112 Prospect St.
Stamford, CT 06901
(800) 225-2380
Types of Adventure: Trekking, hiking, wildlife safaris, jungle expeditions, natural history cruises. **Destinations:** Russia, Peru, Galápagos, Nepal, India, Tibet, Bhutan, Pakistan, Thailand, Kenya, Tanzania, Rwanda, Egypt, Morocco, Israel, Turkey, Switzerland, France, Spain, Greece, Britain, Eastern Europe. **Special Emphasis:** Trekking in Nepal, camping safaris in East Africa, hiking in Europe. **Average Cost:** $500/week plus airfare.

46. **INNERASIA EXPEDITIONS**
2627 Lombard St.
San Francisco, CA 94123
(415) 922-0448; (800) 777-8183
Types of Adventure: Travel for environmental nonprofit organizations (e.g., Nature Conservancy, Audubon), commercial treks and overland tours. **Destinations:** Asia, Europe, South America and the Pacific. **Special Emphasis:** Educational travel. **Average Cost:** $2,400.

47. **INTERNATIONAL EXPEDITIONS, INC.**
One Environs Park
Helena, AL 35080
(205) 428-1700; (800) 633-4734
Types of Adventure: Natural history travel to 30 destinations worldwide, rainforest workshops in Costa Rica and the Amazon. **Destinations:** Amazon, Belize, Costa Rica, Galápagos, Panama, Venezuela, Hawaii, Alaska, Kenya, Tanzania, India, Malaysia, Australia, China. **Special Emphasis:** Rainforests. **Average Cost:** $2,500.

48. **INTERNATIONAL OCEANOGRAPHIC FOUNDATION**
4600 Rickenbacker Causeway
Miami, FL 33149
(305) 361-4697
Types of Adventure: Whale watching, natural history, sailing, scuba diving, cultural studies. **Destinations:** Baja California, Alaska, Arctic regions, Amazon, Costa Rica, South Africa, Florida coast, Mississippi islands, Egypt. **Special Emphasis:** Oceanography, cultures of coastal cities, marine mammals, reef studies. **Average Cost:** $2,300-$5,000.

49. **ISLAND PACKERS**
1867 Spinnaker Dr.
Ventura Harbor, CA 93001
(805) 642-1393
Types of Adventure: Recreation and research-oriented travel. **Destinations:** Channel Islands, California. **Special Emphasis:** Preservation of the islands, flora, fauna and marine life. **Average Cost:** $37 for a one-day trip.

50. **JAPAN & ORIENT TOURS, INC./J&O HOLIDAYS**
3131 Camino del Rio N., Ste. 1080
San Diego, CA 92108
(619) 282-3131; (800) 877-8777
Types of Adventure: Trekking (Nepal, Burma, Thailand, New Zealand), adventure ecotravel (Papua New Guinea), whitewater rafting (Nepal), diving (South Pacific and Southeast Asia), bicycling (Nepal). **Destinations:** Orient and South Pacific. **Special Emphasis:** Custom-designed independent adventure travel with special emphasis on Nepal and Papua New Guinea. **Average Cost:** $1,500-$5,000/person.

51. **JOURNEY TO THE EAST, INC.**
P.O. Box 1334
Flushing, NY 11352-1334
(718) 358-4034; (800) 366-4034
Types of Adventure: Hiking, bicycling and bus trips that emphasize contact with the indigenous cultures of the various ethnic groups in China, Tibet and Mongolia. **Destinations:** China, Tibet, Mongolia, Nepal and Asia. **Special Emphasis:** Opportunity to meet indigenous people and participate in their festivals. Remote locations, small groups, photography, scenery and affordable prices. **Average Cost:** Varies.

52. **JOURNEYS**
4011 Jackson Rd.
Ann Arbor, MI 48103
(800) 255-8735
Types of Adventure: Trekking, hiking, wildlife safaris, family trips, worldwide nature and culture explorations. **Destinations:** Nepal, Ladakh, Sikkim, Papua New Guinea, Australia, Indonesia, Japan, Thailand, Bhutan, Kenya, Tanzania, Botswana, Namibia, Zimbabwe, Madagascar, Hawaii, Norway, Latin America. **Special Emphasis:** Small groups, expert local guides, cross-cultural emphasis, first-class service, remote destinations. **Average Cost:** $80-$150/day.

 99 percent
Source: The Student Environmental Action Guide, *EarthWorks Press*

53. **LITTLE ST. SIMONS ISLAND**
P.O. Box 1078
St. Simons, GA 31522
(912) 638-7472
Types of Adventure: Country inn on private barrier island wilderness preserve. **Destinations:** Little St. Simons Island. **Special Emphasis:** Interpretive natural history programs, birding, barrier island ecology, canoeing, fishing, horseback riding. **Average Cost:** $175/person, all-inclusive.

54. **MAINE SPORT OUTFITTERS**
P.O. Box 956
Rockport, ME 04856
(207) 236-8797; (800) 722-0826
Types of Adventure: Sea kayaking, wilderness canoe trips, outdoor school, fishing trips, island ecology, seascape photography, youth day camps. **Destinations:** All of Maine's coastal waters, islands and inland waterways. **Special Emphasis:** Sea kayaking instruction and tours, island camping and exploring, natural history topics. **Average Cost:** $75 for day trips, $100/day for extended trips, all-inclusive.

55. **MARINE ADVENTURE SAILING TOURS**
945 Fritz Cove Rd.
Juneau, AK 99801
(907) 789-0919
Types of Adventure: Custom-designed trips for groups up to 6 in Glacier Bay and trips from southeast Alaska to Puget Sound. **Destinations:** Southeast Alaska, British Columbia, Puget Sound. **Special Emphasis:** Glacier Bay, West Chichagof Wilderness Area, Tracy Arm Wilderness Area. Family trips. **Average Cost:** $125-$500/day.

56. **MOUNTAIN TRAIL HORSE CENTER INC.**
RD 2, Box 53
Wellsboro, PA 16901
(717) 376-5561
Types of Adventure: Overnight horseback camping trips in Pennsylvania wilderness, ride and cross-country skiing combination trips, ride and raft combinations. **Destinations:** Pennsylvania Grand Canyon country and state forest in north-central Pennsylvania. **Special Emphasis:** Adventure trips with personalized service. Minimum-impact camping, wildlife sightings and information, fall foliage, mountain laurel, rides to country inn. **Average Cost:** $60-$125/day.

57. **MOUNTAIN TRAVEL/SOBEK, THE ADVENTURE COMPANY**
6420 Fairmount Ave.
El Cerrito, CA 94530-3606
(510) 527-8100; (800) 227-2384
Types of Adventure: Environmental-education trips exploring the hottest environmental issues in remote areas of the world. Tours operated with noted environmental organizations, including Natural Resources Defense Council, Rainforest Action Network and National Wildlife Federation. **Destinations:** Worldwide, including the Amazon rainforest, Lake Baikal (former USSR), Alaska, Africa, Poland, Hawaii, Antarctic, Indonesia, Galápagos Islands. **Special Emphasis:** Interaction with indigenous cultures such as Xiavane tribe in Brazil, working with indigenous people to develop ecotourism programs in their countries. **Average Cost:** $2,000-$3,500.

58. **NANTAHALA OUTDOOR CENTER**
41 Hwy. 19 W.
Bryson City, NC 28713
(704) 488-2175, Fax (704) 488-2498
Types of Adventure: Whitewater kayaking, canoeing and rafting trips. **Destinations:** US, Central and South America, Nepal and Malaysia. **Special Emphasis:** Preservation of natural river habitats. **Average Cost:** $1,500-$2,000

58. **NICHOLS EXPEDITIONS**
497 N. Main
Moab, UT 84532
(801) 259-7882; (800) 635-1792
Types of Adventure: Mountain biking, trekking, sea kayaking, backpacking, rafting. **Destinations:** Canyonlands National Park, Grand Canyon, Salmon River and hot springs (Idaho), Alaska, Baja, Thailand. **Special Emphasis:** Low-impact camping and wilderness skills, environmental ecology and specific, sport-related skills taught. **Average Cost:** $50-$100/day.

60. **OSPREY EXPEDITIONS**
P.O. Box 209
Denali National Park, AK 99755
(907) 683-2734
Types of Adventure: Deluxe wilderness raft expeditions, 2-12-day trips in Alaska. **Destinations:** Copper River, Chitina River, Talkeetna River, Yanert Fork, South Fork Kuskoywim, Tazlina River, Nenana River. **Special Emphasis:** Wilderness, wildlife, minimal-impact camping, small group, personal attention, first-class service. **Average Cost:** $175/person/day, all-inclusive.

61. **OUTER EDGE EXPEDITIONS**
45500 Pontiac Trail, Ste. B
Wallod Lake, MI 48390
(313) 624-5140; (800) 322-5235.
Types of Adventure: One- to ten-person adventure expeditions with close wildlife

: *For every glass bottle that is recycled, how much energy is saved?*

encounters in an ecologically sound fashion. Kayaking among whales, swimming with pink dolphins, traveling in a dugout or on a camel, trekking with goats, mushing a dogsled. **Destinations:** Amazon jungle, Australia, British Columbia, Canadian Rockies, Great Barrier Reef, Machu Picchu, Peru, Vancouver Island. **Special Emphasis:** Very small groups, unusual and remote wilderness locations, traveling lightly with minimal environmental impact, natural history education. **Average Cost:** $90-$220/day.

62. **OVERSEAS ADVENTURE TRAVEL**
    349 Broadway
    Cambridge, MA 02139
    (617) 876-0533; (800) 221-0814

Types of Adventure: Trekking, photo safaris, family adventure, summit climbing. **Destinations:** Tanzania, Costa Rica, Galápagos, Venezuela, Botswana, Morocco, Amazon, Egypt, Chile, Bolivia, South Africa, Zambia. **Special Emphasis:** Natural history tours, minimal-impact trips. **Average Cost:** $1,200-$3,200 for a 2-week trip.

63. **RARA AVIS S.A.-RAINFOREST LODGE AND RESERVE**
    P.O. Box 8105-1000
    San José, Costa Rica
    (506) 53-0844

Types of Adventure: Guided nature walks in a private rainforest. **Destinations:** Costa Rica. **Special Emphasis:** Rainforest ecology, research and preservation issues are discussed by naturalist guides. **Average Cost:** Varies.

64. **REI ADVENTURES**
    P.O. Box 1938
    Sumner, WA 98390
    (206) 891-2631; (800) 622-2236

Types of Adventure: Bicycling, kayaking, climbing, trekking and walking. **Destinations:** Asia, US, New Zealand, Australia, South America, Central America, Europe, Commonwealth of Independent States (former Soviet Union). **Special Emphasis:** Active adventure travel in small groups with hands-on experience. **Average Cost:** $1,000-$1,500.

65. **ROCKY MOUNTAIN RIVER TOURS**
    P.O. Box 2552-BW
    Boise, ID 83701
    (208) 345-2400

Types of Adventure: Wilderness paddle rafting and kayaking. **Destinations:** Idaho's Middle Fork Salmon. **Special Emphasis:** No motorized boats, no crowds, small groups, all equipment provided. **Average Cost:** $745 for 4 days, $1,145 for 6 days.

66. **SAFARI CONSULTANTS, LTD.**
    4N211 Locust Ave.
    West Chicago, IL 60185
    (708) 293-9288; (800) 762-4027

Types of Adventure: Hard and soft adventure travel throughout Africa. **Destinations:** Botswana, Namibia, Zimbabwe, Zambia, South Africa, Kenya, Tanzania. **Special Emphasis:** Firsthand exposure to learning about the effects of tourism, game management and wildlife preservation under the direction of government and public institutions. **Average Cost:** $2,200 for a 13-day safari.

67. **SAFARICENTRE**
    Box 309
    Manhattan Beach, CA 90266
    (310) 546-4411

Types of Adventure: Overland safaris, photo safaris, trekking, camping, river expeditions. **Destinations:** Africa, Asia, Latin America, Oceania, North America. **Special Emphasis:** Nature and adventure safaris worldwide. **Average Cost:** $75-$200/day.

68. **SAILING YACHT "STRANGER"**
    8145 Oak Park Rd.
    Orlando, FL 32819
    (407) 578-8792

Types of Adventure: Sailing, snorkeling, coral reef ecology. **Destinations:** US and Virgin Islands. **Special Emphasis:** Education, participation, relaxation, discussion of life's big questions. **Average Cost:** $80-$100/person/day, including food and drink.

69. **SILVER CLOUD EXPEDITIONS**
    P.O. Box 1006-B
    Salmon, ID 83467
    (208) 756-6215

Types of Adventure: Wilderness whitewater trips, wilderness steelhead fishing trips. **Destinations:** River of No Return Wilderness trips, Salmon River. **Special Emphasis:** Client participation, natural history, evening campfire programs, wilderness ethics, low-impact camping. **Average Cost:** $600-$700 for summer trips, $900-$1,000 for steelhead fishing trips.

70. **SLICKROCK ADVENTURES**
    P.O. Box 1400
    Moab, UT 84532
    (801) 259-6996

Types of Adventure: Belize sea kayaking, Mexico river trips. **Destinations:** Belize barrier reef, Jatate and Usumacinta rivers in Lacandon

A: *Enough to light a 100-watt bulb for four hours*
  Source: The Denver Post 1991 Colorado Recycling Guide

rainforest, Mexico. **Special Emphasis:** Wilderness setting, local cultures, unique geography, tropical rainforest. **Average Cost:** $120/day.

71. **SOUTHWIND ADVENTURES, INC.**
P.O. Box 621057
Littleton, CO 80162-1057
(303) 972-0701
Types of Adventure: Minimum-impact trekking, camping, rafting, climbing, jungle expeditions, cultural exchanges, mountain biking, bird and wildlife viewing. **Destinations:** South America, Venezuela, Ecuador, Peru, Bolivia, Argentina, Chile. **Special Emphasis:** Experiencing and protecting nature and native cultures. **Average Cost:** $1,200-$1,700 for 13 to 17 days, varies according to destination and length of trip.

72. **SPECIAL EXPEDITIONS, INC.**
720 Fifth Ave.
New York, NY 10019
(212) 765-7740; (800) 762-0003
Types of Adventure: Adventure voyages aboard comfortable expedition ships accommodating less than 80 passengers, plus tented safaris to Africa and other land expeditions to far-flung destinations worldwide. Expeditions are environmentally responsible, educating travelers about the places visited with an expert staff of naturalists, historians, geologists and archaeologists. **Destinations:** Alaska, Baja California, Columbia and Snake rivers, US Southwest, Caribbean, Belize, Costa Rica, Amazon, Orinoco River, Galápagos, British Isles, Arctic Norway, Western Europe, Baltics, Russia, Egypt, Morocco, East Africa, Australia, Papua New Guinea. **Special Emphasis:** Focus is on nature's wonders, wildlife and diverse cultures. **Average Cost:** $275-$550/day.

73. **TAMU SAFARIS**
P.O. Box 247
West Chesterfield, NH 03466
(802) 257-2607; (800) 766-9199
Types of Adventure: Customized natural history and cultural tours to off-the-beaten-track areas where comfortable camping and wilderness lodges are combined with exciting cross-cultural activities. **Destinations:** Botswana, Kenya, Zimbabwe, Tanzania, Madagascar, Seychelles Islands. **Special Emphasis:** Ecologically and socially responsible travel for individuals and groups, including wildlife and birdwatching tours, relaxed safari camps, wilderness lodges, canoe trips, photography, cross-cultural explorations, family trips, personalized service, reasonable costs. **Average Cost:** Varies.

74. **TAYLOR-CASSLING, LTD., TRAVEL PROFESSIONALS**
4880 River Bend Rd.
Boulder, CO 80301
(303) 442-8585
Types of Adventure: Trekking, photo safaris, rafting, bicycling, climbing. **Destinations:** Worldwide. **Special Emphasis:** Complete planning, inclusive travel, minimum impact. **Average Cost:** $1,250/person, inclusive.

75. **UNIVERSITY RESEARCH EXPEDITIONS PROGRAM**
University of California
Berkeley, CA 94720
(510) 642-6586
Types of Adventure: Joining scientists in researching crucial problems that threaten life on Earth. **Destinations:** Worldwide. **Special Emphasis:** Demystifying the scientific method and involving as many people as possible in the rewards of discovery. **Average Cost:** $1,400.

76. **VICTOR EMANUEL NATURE TOURS, INC.**
P.O. Box 33008
Austin, TX 78764
(512) 328-5221; (800) 328-VENT
Types of Adventure: Birding tours, natural history tours, photo tours. **Destinations:** Worldwide. **Special Emphasis:** Birds, nature. **Average Cost:** $150/day.

77. **VOYAGERS INTERNATIONAL**
P.O. Box 915
Ithaca, NY 14851
(607) 257-3091; (800) 633-0299
Types of Adventure: Natural history, nature photography, birding tours led by outstanding naturalists and professional photographers. **Destinations:** Worldwide, Africa, Galápagos, Costa Rica, New Zealand, Australia, Ireland, Hawaii, Nepal, Antarctica, Alaska, Indonesia, India, Belize. **Special Emphasis:** Outdoor photography, birding, natural history, small groups, low-impact trips. **Average Cost:** $2,000-$5,000.

78. **WHITE MAGIC UNLIMITED**
P.O. Box 5506
Mill Valley, CA 94942
(415) 381-8889; (800) 869-9874
Types of Adventure: Rafting, trekking, diving, jungle safaris, mountain biking, whale watching, cultural tours, natural history, archaeology. **Destinations:** Alaska, Grand Canyon, Middle and Main Salmon rivers, Mexico, Guatemala, Costa Rica, Chile, Belize, Nepal, India, Thailand, Turkey, Albania, US Southwest, California. **Special**

Q: *Which country is home to at least 150,000 living species found nowhere else in the world?*

ECOTRAVEL

Emphasis: Actualizing dreams on Earth. **Average Cost:** $500-$2,500.

79. **WILDERNESS ALASKA**
    P.O. Box 113063
    Anchorage, AK 99511
    (907) 345-3567
    Types of Adventure: Scheduled and custom backpacking, rafting, base camp trips. **Destinations:** Brooks Range, Alaska, Arctic National Wildlife Refuge. **Special Emphasis:** Small groups in remote wilderness unfolding the unique natural history of the Brooks Range. **Average Cost:** $1,800.

80. **WILDERNESS SOUTHEAST**
    711 Sandtown Rd.
    Savannah, GA 31410
    (912) 897-5108
    Types of Adventure: Wilderness camping, canoeing/kayaking, hiking, backpacking, snorkeling, natural history education. **Destinations:** Okefenokee Swamp, Everglades, Great Smoky Mountains, Cumberland Island, Florida springs, British Virgin Islands, Bahamas, Belize, Costa Rica, Amazon Basin and the Pantanal. **Special Emphasis:** Natural history, ecology. **Average Cost:** $70-$100/day.

81. **WILDERNESS TRAVEL**
    801 Allston Way
    Berkeley, CA 94710
    (510) 548-0420; (800) 368-2794
    Types of Adventure: Adventure travel, natural history, wildlife and cultural expeditions around the world, 100 different itineraries, over 300 departures, hiking, trekking, adventure cruises, jungle and safari trips. **Destinations:** South America, Galápagos, Asia, Pacific and Pacific Rim, Africa, Europe. **Special Emphasis:** Small groups with expert leaders, unrivaled customer service, innovative itineraries. **Average Cost:** $2,000/person/land cost.

82. **WILDLAND ADVENTURES**
    3516 NE 155th St.
    Seattle, WA 98155
    (206) 365-0686; (800) 345-4453
    Fax (206) 363-6615
    Types of Adventure: Wildlife safaris, trekking, jungle expeditions, hiking, conservation, worldwide nature and culture explorations to Earth's wild places and exotic cultures since 1978. Small groups, uncommon itineraries, personal cross-cultural interaction, pre-trip planning assistance and personal attention. **Destinations:** Andes, Amazon, Himalayas, Africa, Costa Rica, Belize, Mexico, Alaska, Turkey.

Special Emphasis: Trips support conservation and community development projects through our nonprofit, affiliated conservation organization, The Earth Preservation Fund. **Average Cost:** $1,500/land cost.

83. **WOODSWOMEN**
    25 W. Diamond Lake Rd.
    Minneapolis, MN 55419
    (612) 822-3809; (800) 279-0555
    Types of Adventure: Trekking Nepal with native women sherpanis, exploring Costa Rica with local naturalist, cruising the Galápagos Islands. **Destinations:** Nepal, Africa, Ecuador, Switzerland, Mexico, Costa Rica, Ireland, US, New Zealand. **Special Emphasis:** Women's adventure travel for women of all ages, skill levels and backgrounds. **Average Cost:** $765-$5,695.

84. **ZEGRAHM EXPEDITIONS**
    1414 Dexter Ave. N., Ste. 327
    Seattle, WA 98109
    (206) 285-4000
    Types of Adventure: Cultural expeditions, diving, expedition cruises, natural history expeditions, photography expeditions. **Destinations:** Antarctica, Arctic, Asmat, Galápagos, Amazon, Cambodia, Vietnam, Laos, Botswana. **Special Emphasis:** Small groups with expert leadership. **Average Cost:** $2,000-$8,000.

A: *Madagascar*
Source: Time magazine

VIII
ECOCONNECTIONS

"Women in thin raincoats, drinking tea from paper cups" is how Joan Didion once described feminist activists. "People with their sleeves rolled up, enjoying the sun, wind and rain and the beauty of the Earth" might be a good description of environmental activists. We've listed dozens of volunteer projects that need your hard work and hard thinking—plus environmental education opportunities, national and international awards, and an exciting list of "green" groups to join.

How to Write Congress

"Ask and ye shall receive" is a phrase to remember when you're dealing with elected representatives. If you don't *ask* for specific legislative action, you have almost no chance of receiving it. If you *do* ask—and lots of others ask, too—you have an excellent chance of getting a bill passed or derailed. Here are some tips for writing your elected representatives.

1. Refer to the bill by name if you are writing about specific legislation. Let your representatives know *exactly* how you feel about the bill, and how his or her vote will affect *your* future votes.

2. When you're writing about a specific bill, include local news, personal experiences or other information that supports your stand. If you are writing to a senator, he or she may be unaware of events that have happened in your part of the state. Include local news clips.

3. Be sure to include your own name and complete address, so the elected official can reply to your letter. Feel free to ask questions—and expect a response.

4. Address your letter properly. The form for members of the House is: The Honorable (full name of representative), House of Representatives, Washington, DC 20515. For members of

Volunteer!

Would you like to watch whales and catalog their behavior . . . clear brush and build trails . . . dive in and study sea turtles or dolphins? Volunteers are needed everywhere, every day. Mother Earth needs you!

WRITE THE RIVER

The MERRIMACK RIVER WATERSHED COUNCIL, a citizens' conservation organization dedicated to the restoration and protection of the Merrimack River and its watershed in New Hampshire and Massachusetts, needs volunteers to assist the Manager of Development. Duties include fundraising, writing grants and proposals, research and networking with corporations and small businesses. Volunteers will become familiar with development activities required to fund water quality studies, water monitoring projects, trails and purchase of scenic lands. Contact Caron Soond, MERRIMACK RIVER WATERSHED COUNCIL, 694 Main St., West Newbury, MA 01985, (508) 363-5777.

SKI GREEN

The GREEN MOUNTAIN NATIONAL FOREST seeks volunteers to work as nordic ski trail patrollers, backcountry rangers, photographers and construction workers. Volunteers are also needed for positions in wildlife and fisheries management, archaeology and other fields. Many jobs include a stipend. Although extremely limited, housing is available. Contact Roxanne Ramah, Volunteer Coordinator, GREEN MOUNTAIN NATIONAL FOREST, P.O. Box 519, Rutland, VT 05702-0519, (802) 773-0300.

WILDLIFE PROJECTS

The 51 affiliated offices of the NATIONAL WILDLIFE FEDERATION seek volunteers to work on various environmental projects conducted by each group. Projects may include lobbying, coordinating events, maintaining correspondence, working on youth camps and grassroots mobilization. Contact in writing, Affiliate and Regional Programs, NATIONAL WILDLIFE FEDERATION, 1400 16th St. NW, Washington, DC 20036.

: *Which US city has undertaken a project designed to convert six acres of landfill into grass meadows and young woodlands?*

ECOCONNECTIONS

UNDERWATER DOCUMENTS
CEDAM INTERNATIONAL needs volunteers to document marine species, map underwater topography, participate in cleanup dives and take underwater photographs. Land projects are also offered. Members can participate for share-of-cost donation. Contact CEDAM INTERNATIONAL, 1 Fox Rd., Croton, NY 10520, (914) 271-5265.

WOLF NATURALISTS
The WILD CANID SURVIVAL AND RESEARCH CENTER AND WOLF SANCTUARY offers volunteer positions ranging from fundraisers to naturalists. The center also offers an Adopt-a-Wolf program. Contact WILD CANID SURVIVAL AND RESEARCH CENTER, P.O. Box 760, Eureka, MO 63025, (314) 938-6490.

BIG BIRDS
TEXAS PARKS AND WILDLIFE DEPARTMENT is looking for volunteers who are interested in leading birding tours or providing other interpretive services to state park visitors. Expertise in birding, wildflowers, geology and history are especially requested. Indicate park preference on a letter of interest and send it to Kevin Good, TEXAS PARKS AND WILDLIFE DEPARTMENT, 4200 Smith School Rd., Austin, TX 78744.

PARK MANAGEMENT
ROCK CREEK PARK, in the city of Washington, DC, seeks volunteers to assist with historical and natural interpretation, resource management and office work. The nearly 1,800-acre park has 28 miles of trails and numerous other recreational activities. Benefits include housing and training. Must be 18 years old and have a current driver's license. Own transportation helpful. Contact Volunteer Coordinator Joe Burns, ROCK CREEK PARK, 5000 Glover Rd. NW, Washington, DC 20015, (202) 426-6832.

ANIMAL TALK
The New York Zoological Society seeks volunteers to provide wildlife education at the CENTRAL PARK ZOO. Volunteers will teach zoo visitors about animals by delivering mini-talks at selected exhibits and talking to visitors about wildlife and the environment in the Wildlife Conservation Center and at Education

the Senate, the form is: The Honorable (full name of senator), United States Senate, Washington, DC 20510. To write to the president or vice president, the address is: The White House, Washington, DC 20500. For state officials, you can write to: (full name of state official), State Capitol, (city, state and zip code of your capital city).

5. A new book, *Dear Mr. President*, by Marc Davenport (Carol Publishing, New York, 1991), gives you 100 already-written letters on topics ranging from deforestation to global warming to solar power, and more. If you're pressed for time, you can tear out or photocopy these letters, sign them with your name, and mail them to your representatives. Ideally, you can use these letters as a starting point, to give you ideas and facts for the letters *you* write. *Dear Mr. President* includes a list of "green" groups that can give you more information on environmental topics, as well as a complete index of legislative addresses.

6. Remember: A democratic nation is governed "of the people, by the people, for the people." You don't have to be an expert on a topic to have a valuable opinion on that topic. Maybe you have no idea what energy bills are pending, but you feel strongly that nuclear power plants are dangerous and that wind or solar power is safer; write a letter saying exactly that. Maybe you

A: *Hackensack, New Jersey*
Source: Garbage magazine

have no idea what kind of automobile bills are pending, but you strongly support higher miles-per-gallon standards, plus tax rebates for people who buy electric cars. Write that!

7. Put your money where your mouth is. Instead of just pointing out that "nuclear power creates dangerous waste," ask that your tax dollars be spent on solar power research. Instead of complaining that there is not enough wilderness, ask that your tax money be spent to purchase and support wilderness areas. What you are willing to pay for is what you will get.

Eco-Awards

Environmental awards are a win-win situation. Even if your group or effort doesn't win an award, everyone on the planet wins when people work to protect the Earth. Here are some recent winners, plus information on how to contact administrators for each award:

Agricultural Conservation Awards
American Farmland Trust, 1920 N St. NW, Ste. 400, Washington, DC 20036, (202) 659-5170.

Established in 1984 by the American Farmland Trust to honor individuals and organizations making outstanding contributions to farmland preservation, the 1992 awards were given to groups in ten different categories.

à la Cart. Volunteers are also needed to help with a variety of special zoo events. Weekday volunteers work one day per week; weekend volunteers work one day every other week. Volunteers must be 18 or older and will receive their training in a special course taught by Central Park Zoo Staff. Contact April Rivkin, CENTRAL PARK ZOO, 830 Fifth Ave., New York, NY 10021, (212) 439-6538.

FRESH AS A MARSH

JOHN HEINZ NATIONAL WILDLIFE REFUGE at Tinicum cares for the largest remaining freshwater tidal marsh in Pennsylvania, which is home to a variety of plants and animals typical of a wetland environment. Nature guides are wanted to lead walks on weekends, scout educational groups for teachers and complete various other duties. Some knowledge of plant or animal identification and natural history is needed. Contact Jackie Burns, JOHN HEINZ NATIONAL WILDLIFE REFUGE, Ste. 104, Scott Plaza 2, Philadelphia, PA 19113, (215) 521-0662.

TURTLE GROUNDS

SEA TURTLE RESTORATION PROJECT needs help to support the first wildlife refuge proposed for an endangered species. The establishment of the Archie Carr Refuge would protect some of Florida's east coast beaches from development and create a sanctuary for the world's second-largest loggerhead sea turtle population and for 40 percent of all green turtles that nest in Florida. Contact SEA TURTLE RESTORATION PROJECT, Earth Island Institute, 300 Broadway, Ste. 28, San Francisco, CA 94133, (415) 788-3666.

WORLDWIDE VOLUNTEERS

FOUNDATION FOR FIELD RESEARCH offers volunteer research positions worldwide in many fields. Areas of study include marine biology and mammalogy, archaeology, botany and more. Volunteer positions are available in Grenada, Mexico, Africa, US and elsewhere. Fee required. Contact FOUNDATION FOR FIELD RESEARCH, P.O. Box 2010, Alpine, CA 91903, (619) 445-9264.

GULLY-BUSTERS

CLEVELAND NATIONAL FOREST invites volunteers to conduct nature walks, staff the visitor center, help

Q: *How much toxic waste is produced by the US military every minute?*

ECOCONNECTIONS

build a new hiking trail and serve as mounted or foot wilderness rangers. Participants are also needed for special projects such as tree pruning, gully-busting, wildlife projects and fence building. Contact Laura Lambert, CLEVELAND NATIONAL FOREST, 3348 Alpine Blvd., Alpine, CA 91901, (619) 445-6235.

HAWK MIGRATION

HAWKWATCH INTERNATIONAL seeks field participants to assist with migration counts, capture and banding programs, and nest surveys of hawks and other birds of prey in several Western states. Office workers are also needed. Experienced applicants preferred. Volunteers are expected to participate for four to 15 weeks. Send letter of application, resume and three references to Chuck Goodmacher, HAWKWATCH INTERNATIONAL, P.O. Box 35706, Albuquerque, NM 87176-5393, (505) 255-7522.

COLORADO CONSERVATION

The COLORADO ENVIRONMENTAL COALITION invites volunteers to work on a variety of issues, including public lands, rivers, water quality, timber and environmental health. The coalition has been working to stop large dam construction on various rivers throughout Colorado and is working to control the impact of oil and gas drilling. The coalition's environmental health committee currently needs volunteers to assist with an ongoing conservation project. Contact Dorothy Cohen, COLORADO ENVIRONMENTAL COALITION, 777 Grant St., Ste. 606, Denver, CO 80203, (303) 837-8701.

UTAH AND YOU

UTAH STATE PARKS AND RECREATION offers statewide volunteer opportunities. Volunteers may choose a park to work in. Activities may include interpretation, guiding visitors, research, maintenance and more. Contact Mary L. Tullius, Division of Parks and Recreation, 1636 W. North Temple, Salt Lake City, UT 84116, (801) 538-7220.

EAGLE WATCH

Volunteer opportunities are available at the ROCKY MOUNTAIN ARSENAL. Volunteers will help with a variety of activities, including guiding tours, planning events and coordinating programs at the Visitor

Among this year's winners were a large Virginia corporation that restored the land and wildlife habitat on a historic 1,800-acre farm near Jamestown, and two citizen action groups working in a California wine region to save threatened farmlands.

America's Corporate Conscience Awards

Council on Economic Priorities, 30 Irving Place, New York, NY 10003, (212) 420-1133.

Church & Dwight, US West, General Mills, Prudential Insurance, Supermarkets General, Donnelly Corp., Tom's of Maine, Lotus and Conservatree won the 1992 awards sponsored by the Council on Economic Priorities. These companies were given awards based on categories ranging from environmental stewardship to charitable contributions and community outreach.

Arizona Heirloom Fruit & Nut Regis-TREE

Native Seeds Search, 2509 N. Campbell, #325, Tucson, AZ 85719, (602) 327-9123.

The Regis-TREE program, initiated in 1991, promotes the conservation of Arizona's useful perennial plants by honoring people who have planted or cared for plants that enrich both our diets and our lives. In 1992, 11 awards were given to a variety of individual and group efforts to preserve plants such as wild chiles and apricots that can faithfully bear fruit in harsh conditions at 6,000 feet.

A: *More than one ton*
Source: The Threat at Home, *Beacon Press*

Automotive Solutions Competition
Council for Solid Waste Solutions, 1275 K St. NW, Ste. 400, Washington, DC 20005, (202) 371-5319.

The Council for Solid Waste Solutions, an environmental task force of the plastics industry, and the Michigan Materials and Processing Institute, an association of automobile manufacturers, presented the Automotive Solutions Competition prize to a team of three undergraduate industrial engineering students at Iowa State University. The three students—Jim Fraise, Heather Foley and Greg Kent—designed a facility that would produce two marketable primary materials from recovered automobile bumpers.

Best Bets Awards
The Center for Policy Alternatives, 1875 Connecticut Ave. NW, Ste. 710, Washington, DC 20009, (202) 387-6030.

The Center for Policy Alternatives presented its fifth annual Best Bets awards to ten state actions, including those of two governors. Governor Bill Clinton (Dem.) of Arkansas was honored for legislation to recover the profit from environmental violations. Governor Michael Castle (Rep.) of Delaware received his award for promoting "green" activity by business.

"Best Recycling Story" Award
The American Paper Institute, 485 Madison

Center and the Eagle Watch. Contact the US Fish and Wildlife Service, ROCKY MOUNTAIN ARSENAL, Bldg. 613, Commerce City, CO 80022-2180, (303) 289-0232.

NATIONAL OPPORTUNITIES
STUDENT CONSERVATION ASSOCIATION organizes groups of volunteers to work in cooperation with public management agencies and in other programs. Participants have the opportunity to do a large variety of tasks, such as interpretation and visitor services, resource management, recreation planning, forestry, range and wildlife management, trail construction and maintenance, backcountry and river patrols and research in archaeology, geology and hydrology. Projects are available around the country. Contact STUDENT CONSERVATION ASSOCIATION, P.O. Box 550, Charlestown, NH 03603, (603) 543-1700.

GONE FISHING
TROUT UNLIMITED needs volunteers to help local chapters with Embrace-A-Stream projects. Volunteers may assist in a variety of duties, including monitoring cold-water habitats to document damage, restoring deteriorated lakes, improving salmonid habitats and educating the public through workshops and seminars. Contact TROUT UNLIMITED, 800 Follin Lane SE, Ste. 250, Vienna, VA 22180-4959, (703) 281-1100.

SHORES OF CONSERVATION
The LAKE MICHIGAN FEDERATION seeks volunteers for the Shorekeepers program. Volunteers will monitor stretches of Lake Michigan shoreline for pollution, lake level changes, erosion, wildlife and recreation usage. Contact LAKE MICHIGAN FEDERATION, 59 E. Van Buren St., Ste. 2215, Chicago, IL 60605, (312) 939-0838.

PRESERVE A PARK
The AMERICAN HIKING SOCIETY needs volunteers to help preserve America's parks and forests. Volunteers spend 10 days on a variety of projects ranging from trail maintenance in New Mexico to bridge building in Wyoming. Most of the work sites are in remote and primitive areas. Volunteers must be at least 16 years old, experienced hikers and physically fit. A $40 registration fee is required. For free information about the program, send a stamped,

 What percentage of all polyester carpet made and sold in the US today contains recycled plastic?

ECOCONNECTIONS

self-addressed envelope to AMERICAN HIKING SOCIETY, Volunteer Vacations, P.O. Box 86, Dept. AHS/VV, North Scituate, MA 02060.

ANIMAL RIGHTS

PROGRESSIVE ANIMAL WELFARE SOCIETY (PAWS) seeks volunteers to help in a variety of areas, including animal care, animal rights, wildlife rehabilitation, education programs, building maintenance and general office assistance. Orientation meetings for new volunteers are held regularly. Contact PROGRESSIVE ANIMAL WELFARE SOCIETY/PAWS, P.O. Box 1037, Lynnwood, WA 98046, (206) 743-3845.

SEA CREATURES

CENTER FOR MARINE CONSERVATION offers volunteer programs in marine habitat protection, sea turtle, marine mammal and seabird conservation, marine debris and entanglement, fisheries management and conservation. Volunteers are also needed to help with mailings and office work. Applicants should send a resume and a two- to five-page writing sample to Volunteer Coordinator, CENTER FOR MARINE CONSERVATION, 1725 DeSales St. NW, Ste. 500, Washington, DC 20036, (202) 429-5609.

BUTTERFLIES IN THE STOMACH

Volunteers are needed to participate in the monarch monitoring project of the CALIFORNIA MONARCH STUDIES program. Field surveyors record roosting habitats and tag butterflies at monarch overwintering sites along the California coast. Site location and training provided. Contact Walt Sakai, CALIFORNIA MONARCH STUDIES, Santa Monica College, 1900 Pico Blvd., Santa Monica, CA 90405-1628, (213) 450-5150, ext. 9713.

LOONY INFO

LONG POINT BIRD OBSERVATORY seeks volunteers to survey lakes in the summer and record information about the breeding of loons. Volunteer surveyors watch for pairs of loons and nesting behavior, observe the appearance of newly hatched chicks and record how many chicks have survived the summer. Volunteers will receive a survey kit with instructions and a report form to be returned at the end of the season. Contact Canadian Lakes Loon Survey, LONG

Ave., New York, NY 10022-5896, (212) 371-2200.

This contest, created by the American Paper Institute to recognize the achievements of dedicated paper recyclers and the importance of paper recycling, offers prizes of $500 nationally and $250 regionally. This year's top prize went to a Michigan mother who mounted a recycling crusade in schools around her state after suffering serious burns in a household trash fire.

Chevron Conservation Awards
Chevron Corp., P.O. Box 7753, San Francisco, CA 94120, (415) 894-2457.

This program, established in 1954, recognizes outstanding individuals and organizations that serve as environmental role models. This year, 25 individuals were honored in three categories—Professional, Citizen Volunteer and Nonprofit Organization/Public Agency.

City and State Environmental Achievement Award
City and State Magazine, 740 N. Rush St., Chicago, IL 60611-2590, (312) 649-5347.

This environmental achievement award, given by *City and State* magazine, is presented at the annual meeting of the National League of Cities. The 1992 winner was the City of Los Angeles, for its outstanding curbside recycling, waste reduction, source reduction and public education programs.

A: *35 percent*
Source: The Denver Post *1991 Recycling Guide*

Corporate Wildlife Stewardship Award
Department of the Interior, 1849 C St. NW, Washington, DC 20240. Contact Bob Walker, (202) 208-3171.

The award, established in 1990 to recognize corporations making outstanding contributions to conservation and management of the environment (including wildlife), was given to the Kansas Power and Light Company in 1992.

Direct Marketing Association Robert Rodale Environmental Award
Direct Marketing Association, 11 W. 42nd St., New York, NY 10036-8096, (212) 768-7277.

Established in 1991 to recognize environmental achievement in the direct marketing field, this award is named in honor of the late Robert Rodale, founder of Rodale Press. In 1992, the $10,000 awards went to the Fingerhut Company (for non-environmentally related companies) and to Seventh Generation (for environmentally related companies).

Earl A. Chiles Award
High Desert Museum, 59800 South Hwy 87, Bend, OR 97702 (503) 382-4754.

The Chiles Award was established by the High Desert Museum in 1983 in honor of Oregon philanthropist Earl A. Chiles. It recognizes individuals whose accomplishments help preserve high deserts. Leontine Nappe of Reno, Nevada, won for building a

POINT BIRD OBSERVATORY, P.O. Box 160, Port Rowan, Ontario N0E 1M0, Canada, (519) 586-3531.

CRY WOLF

The MEXICAN WOLF COALITION needs volunteers to help with an activist network to promote wolf recovery in Arizona, New Mexico and Texas. Participants work as part of a grassroots educational program to inform the public of the conditions facing the Mexican wolf. Volunteers write letters to agencies and Congress, participate in fundraisers and provide educational talks to schools and community groups. Contact Bobbie Holaday, Preserve Arizona's Wolves, 1413 E. Dobbins Rd., Phoenix, AZ 85040, (602) 268-1089, or Susan Larson, MEXICAN WOLF COALITION, 207 San Pedro NE, Albuquerque, NM 87108, (505) 265-5506.

AMERICAN BEAUTY

KEEP AMERICA BEAUTIFUL offers various volunteer opportunities, including educating the public through school and community presentations, and planning and directing solid waste management, recycling and composting programs. Other responsibilities include working with public land stewardship programs like Adopt-A-Beach, park maintenance, trail signing and maintenance, and sensitive land protection. Keep America Beautiful also offers unlimited opportunities in communications, public relations and special projects, depending on the needs of the local office and affiliate concerns. Contact Volunteer Coordinator, KEEP AMERICA BEAUTIFUL, 9 W. Broad St., Stamford, CT 06902, (203) 323-8987.

ELK HABITS

ROCKY MOUNTAIN ELK FOUNDATION needs volunteers to assist in evaluating conservation project ideas, promoting elk habitat preservation and enhancement, and organizing fundraising banquets to benefit the elk and other wildlife. Contact ROCKY MOUNTAIN ELK FOUNDATION, Attn. Field Operations, P.O. Box 8249, Missoula, MT 59807, (800) CALL-ELK.

STREAM KEEPERS

The ADOPT-A-STREAM FOUNDATION seeks volunteers to join its network of Stream Keepers, individuals

Q: *What percentage of the world's grain is fed to livestock?*

involved in monitoring, enhancing and watchdogging the streams of Washington state. Volunteers are also needed to manage the Stream Keeper database and assist with fundraising activities. Send a self-addressed, stamped envelope to Workshop Info, ADOPT-A-STREAM FOUNDATION, P.O. Box 5558, Everett, WA 98206, (206) 388-3487.

CHARTERS FOR DEER

Volunteer with one of WHITETAILS UNLIMITED's already-existing local chapters or start your own. Duties will include fundraising, membership building, and supporting and promoting the program of the national organization. Whitetails Unlimited addresses issues such as loss of habitat, poaching or illegal kills, disease and parasites, overpopulation and deer/car accidents. Contact WHITETAILS UNLIMITED, Dave Hawkey, Chapter Coordinator, P.O. Box 422, Sturgeon Bay, WI 54235, (414) 743-6777.

WHALE'S TALES

The AMERICAN CETACEAN SOCIETY sponsors a gray whale census. Census volunteers are trained to observe the behavior and migratory patterns of the Pacific gray whale. Positions are available from December through May. Volunteers will also record the number of whales that pass by the Palos Verdes Peninsula. Contact Alisa Schulman, AMERICAN CETACEAN SOCIETY, P.O. Box 2698, San Pedro, CA 90731, (310) 519-8963.

WORKING FOR THE EARTH FIRST

EARTH FIRST! seeks volunteers for its national office and its local groups. Volunteer opportunities in newsletter writing, education, research, demonstrations, tabling and computers are available. Contact EARTH FIRST!, Mike Roselle, P.O. Box 5176, Missoula, MT 59806, (406) 728-8114.

GATEKEEPING

HAWK MOUNTAIN SANCTUARY ASSOCIATION keeps volunteers occupied throughout the year carrying out duties such as collecting admissions, working in the bookstore, gatekeeping and general maintenance, including trail work and intermittent special projects. During the winter months, volunteers may work on a special project or a direct mail campaign

consensus between conservation and sport groups to secure water rights for wildlife in the Lahontan Valley of western Nevada.

"Earth Teacher" Award
Amway Corporation, 7575 E. Fulton St., Ada, MI 49355, (616) 676-5178; Time Magazine, Rockefeller Center, Time Life Bldg., New York, NY 10020, (212) 522-5196.

Amway Corp. and *Time* magazine presented their annual "Earth Teacher" Award to nine teachers in 1992. Each teacher won $10,000 in educational materials for his or her school, in recognition of the teacher's environmental activities. The nine winners (selected by the Alliance for Environmental Education) also received a $500 gift and have been invited to attend the United Nations Environment Programme's Global Youth Forum in June 1993 in New York City, as guests of Amway Corp. and *Time*.

Edward Abbey Ecofiction Award
Buzzworm Magazine, 2305 Canyon, Ste. 206, Boulder, CO 80302.

Established in 1990 by *Buzzworm: The Environmental Journal* and Patagonia, Inc., to recognize outstanding novel-length fiction in the tradition of Edward Abbey. Recently published and about-to-be-published novels are eligible for consideration. The 1992 winner was Kent Nelson's novel *Language in the Blood* (Gibbs-Smith, Layton,

A: *Nearly 40 percent*
Source: Worldwatch *magazine*

Utah, 1992). The winner receives a $2,000 cash prize and a gold-plated monkey wrench, plus publication of an excerpt from the winning novel in the July/August issue of *Buzzworm*.

Endowed Chair at Dartmouth
Dartmouth College, 38 N. Main St., Hanover, NH 03755-3623, (603) 646-3661.
  Professor James F. Hornig has been named Dartmouth Professor of Chemistry and of Environmental Studies, an endowed chair made possible by an anonymous $1.5 million endowment to Dartmouth. Hornig has played a key role in making environmental studies a permanent program at Dartmouth.

Environmental Women of Action
Women's Sports Foundation, 342 Madison Ave., New York, NY 10173, (800) 227-3988.
  Co-sponsored by the Women's Sports Foundation and Tampax (Tambrands), the Environmental Women of Action awards honor 50 women—one from each state—who work to save the environment. The 50 winners and descriptions of their work are published in a paperback book each year; 1 million copies of the book are given away with boxes of Tampax. The winners are honored at an awards dinner each year.

Frederick Law Olmsted Medal
American Society of

and help with routine office tasks such as typing and filing. Contact Sue Wolfe, HAWK MOUNTAIN SANCTUARY ASSOCIATION, Rte. 2, Box 191, Kempton, PA 19529, (215) 756-6961.

THE PLANET IS FULL OF OPPORTUNITY
PLANET DRUM FOUNDATION'S Green City Pioneer Volunteer Referral Network seeks volunteers in the Bay Area. This comprehensive volunteer network provides information on hands-on environmental volunteer opportunities. Activities include recycling, habitat restoration, tree planting and wild animal care. Planet Drum's Referral Network provides a unique service through which volunteers can find work and organizations can list opportunities. Contact PLANET DRUM FOUNDATION, P.O. Box 31251, San Francisco, CA 94131, (415) 285-6556.

TALKING ANIMALS
LINCOLN PARK ZOO seeks volunteers to lead tours, give talks, conduct animal observations and perform a variety of duties. Limited handling of small animals is involved. Zoo volunteers are required to make a one-year commitment of four to six hours per week. Training is provided. Contact THE LINCOLN PARK ZOOLOGICAL SOCIETY, 2200 N. Cannon Dr., Chicago, IL 60614, (312) 294-4676.

WILDERNESS ASSISTANCE
SOUTHERN UTAH WILDERNESS ALLIANCE seeks volunteer assistance from lawyers, geologists, botanists, surveyors, photographers and artists. Professional and technical skills are required, along with a three-month commitment. Contact Susan Tixier, SOUTHERN UTAH WILDERNESS ALLIANCE, 1471 S. 1100 E., Salt Lake City, UT 84105-2423, (801) 486-3161.

CONSERVING CAREERS
The CONSERVATION CAREER DEVELOPMENT PROGRAM offers year-round volunteer activities in the fields of wildlife and resource management, forestry, planning, park and recreation administration, history and archaeology. Small stipend provided to participants for summer fieldwork. Contact CONSERVATION CAREER DEVELOPMENT PROGRAM, 1800 N. Kent St., Ste. 1260, Arlington, VA 22209, (703) 524-2441.

 Where is the site of the worst nuclear accident in US history?

WILD STUDIES

WILDLANDS STUDIES offers a worldwide field studies program to help protect wildlife and preserve wilderness environments. Activities may include a firsthand search for endangered timber wolves in the Northern Cascades, on-site investigation of whale behavior in Canada, examinations of threatened Rocky Mountain wilderness land in Montana, Wyoming and Colorado, and studies of Hawaii's environments and cultures. Opportunities for field studies in Nepal, New Zealand, Thailand and Alaska are offered as well. Participants can earn three to 14 university credits. Contact Crandall Bay, Director, WILDLANDS STUDIES, 3 Mosswood Cir., Box B, Cazadero, CA 95421, (707) 632-5665.

THE MARINE MOVEMENT

THE COASTAL SOCIETY offers a variety of volunteer positions around the US. Positions include work in marine policy, marine law, coastal management, living resource management, special area management, marine research and literature research. For information and placement contact Tom Bigford, THE COASTAL SOCIETY, P.O. Box 2081, Gloucester, MA 01930-2081, (508) 281-9209.

THIRD WORLD EXCHANGE

The ALTERNATIVE DEVELOPMENT EXCHANGE, a project of the environmental organization Earth Island Institute, seeks skilled volunteers for third world sustainable development projects such as reforestation, alternative energy and potable water systems. The organization responds to requests from nongovernment organizations and community project leaders for urgent short-term technical advice, facilitating consultation with experts, searching technical literature and networking with other projects that have faced similar problems. The Alternative Development Exchange also collects and distributes tools, instruments, reference and education materials, seeds and other materials requested by overseas projects. Send resume highlighting language and technical skills and third world experience to ALTERNATIVE DEVELOPMENT EXCHANGE, Earth Island Institute, 300 Broadway, Ste. 28, San Francisco, CA 94133, (415) 788-3666.

Landscape Architects, 4401 Connecticut Ave. NW, Washington, DC 20008, (202) 686-ASLA.

The American Society of Landscape Architects has given the Frederick Law Olmsted Medal, awarded for dedication to environmental stewardship, to the Nature Conservancy. The American Society of Landscape Architects was founded in 1899.

Goldman Environmental Prize
The Lippin Group, 6100 Wilshire Blvd., Ste. 400, Los Angeles, CA 90048-5111, (213) 965-1990.

The Goldman Environmental Prizes are given annually to six of the world's leading environmentalists. The 1992 winners were: Colleen McCrory of Canada, who has led campaigns to protect Canada's temperate rainforests; Jeton Anjain of the Marshall Islands, who organized the evacuation of Rongelop Atoll, an area contaminated by US nuclear bomb tests; Medha Patkar of India, a leader in the protests against a series of dams that would displace 100,000 villagers; Christine Jean of France, who has led opposition to dams in the Loire valley; Carlos Alberto Ricardo of Brazil, who works for indigenous people's rights in Amazonia; and Wadja Mathieu Egnankou of Ivory Coast, a scientist struggling to protect West Africa's coastal forests.

Good Earthkeeping Seal Award
Environmental Institute,

 Three Mile Island, Pennsylvania
Source: Union of Concerned Scientists

3520 NW 13th St., Gainesville, FL 32609, (904) 375-2221.

The Environmental Institute, a national non-profit organization dedicated to helping businesses improve their environmental responsibility, awarded Conservatree Paper Company its 1992 Good Earthkeeping Seal Award. Conservatree is the nation's largest wholesaler specializing exclusively in environmentally sound paper products.

Government Wildlife Protection Award

US Fish and Wildlife Service, Dept. of the Interior, 1849 C St. NW, Washington, DC 20240, (202) 208-5634.

The Corporate Wildlife Stewardship Award is the highest honor the US Fish and Wildlife Service can bestow. Nippon Telegraph and Telephone Corporation won the award for its innovative miniature satellite transmitter that allows biologists to track the migration of Wrangel Island geese.

Grassroots Peace Award

Peace Development Fund, 44 N. Prospect St., P.O. Box 270, Amherst, MA 01004, (413) 256-8306.

This award, a unique national citation honoring groups that think globally and act locally, is given by the Amherst-based Peace Development Fund and carries a $10,000 stipend. This year the award goes to an upstate New York

GO GREYHOUND

USA DEFENDERS OF GREYHOUNDS, previously known as Indiana Retired Greyhounds As Pets, seeks volunteers to provide a permanent home for greyhounds no longer needed, wanted or valued by the greyhound racing industry. Volunteers are also needed for public awareness and grassroots education activities. Contact Volunteer Coordinator, USA DEFENDERS OF GREYHOUNDS, P.O. Box 111, Camby, IN 46113, (317) 244-0113.

FORMING A HABITAT

The GAIA INSTITUTE of the Cathedral of St. John the Divine seeks volunteers to assist in its efforts to establish the Microbial Habitats Resource Center. The institute is currently gathering and categorizing information concerning the microbial populations within specific habitats. Volunteers must have access to scientific journals and libraries and will be selected to represent specific geographical locations throughout the world. The information gathered will be used to create a specialized database. Contact GAIA INSTITUTE, Microbial Habitats Resource Center, c/o I.D.E.A., Inc., 22 Forest St., Providence, RI 02906, (401) 273-7093.

PLANTING TREES

California's TREEPEOPLE welcomes volunteers of all ages to help plant trees throughout Los Angeles. The group also offers citizen forester training to support those who wish to lead planting projects in their communities. Contact TREEPEOPLE, 12601 Mulholland Dr., Beverly Hills, CA 90210, (818) 753-4600.

WATER WORKS

FRIENDS OF THE RIVER seeks volunteers to help research water policy, plan river trips and help with general office work. Contact FRIENDS OF THE RIVER, Ft. Mason Ctr., Bldg. C, San Francisco, CA 94123, (415) 771-0400.

ENVIRONMENTAL PEACE

The PEACE CORPS needs environmental and forestry volunteers, among others. Volunteers are needed for assignments in university forestry education, parks and wildlife management, environmental education, community forestry extension and related fields in

: *In an effort to slow global warming, which country gives its sheep anti-flatulence pills to reduce methane gas emissions?*

ECOCONNECTIONS

Poland, Nepal, Chile, Tanzania, Papua New Guinea and other worldwide locations. Volunteers may work at senior levels or in villages, depending on their skills. Expenses are paid. Returned volunteers receive a stipend and preferential treatment for US government jobs. Special scholarships and fellowships are available at over 50 colleges and universities for former volunteers. Each applicant must be a US citizen and have a bachelor's degree in an environmental subject. Contact PEACE CORPS, 1990 K St. NW, Washington, DC 20526, (800) 424-8580, ext. 2293.

TURTLE TALES
CARETTA RESEARCH PROJECT is looking for volunteers to help with a sea turtle study. Volunteers will tag nesting sea turtles on a Georgia barrier island and protect the turtle nests from predators during one-week shifts. Room, board and transportation costs are covered. Contact Robert Moulis, CARETTA RESEARCH PROJECT, 4405 Paulsen St., Savannah, GA 31405, (912) 355-6705.

ENVIRONMENTAL LITERACY
RENE DUBOS CENTER FOR HUMAN ENVIRONMENTS needs a volunteer program assistant to organize resources for network centers and clearinghouses and to enlist support for a program on environmental literacy. Contact Sara Peracca, RENE DUBOS CENTER FOR HUMAN ENVIRONMENTS, 100 E. 85th St., New York, NY 10028, (212) 249-7745.

PET RESEARCH
Volunteer opportunities are available with PEOPLE FOR THE ETHICAL TREATMENT OF ANIMALS, including media relations, grassroots assistance, work with research projects and aiding in investigation and program policy issues. Contact PEOPLE FOR THE ETHICAL TREATMENT OF ANIMALS, P.O. Box 42516, Washington, DC 20015, (301) 770-PETA.

TEACH THE WORLD
WORLD TEACH seeks volunteers to teach environmental education and English in Costa Rica, Namibia, Thailand, South Africa, Poland, Kenya, Ecuador or China. Volunteers of all ages will teach for a year, term or summer. Most programs provide housing and a small salary. Student loans can be deferred while in the

multiracial youth movement called Young & Teen Peacemakers. The group publishes the country's only youth-written magazine on peacemaking, which is read by 20,000 children in the US and 15 other countries.

Green Stars
The Environmental Action Coalition, 625 Broadway, New York, NY 10012, (212) 677-1601.
The Environmental Action Coalition honored Bill Howell, Tessa Huxley, Thomas Jorling and George Whalen for their recognizable environmental procedures ranging from implementing the use of recycled plastics for traffic cones, to advocating widespread use of water-conserving plumbing fixtures, to leadership in landmark environmental legislation and lifelong environmental activism.

Heinz National Recycling Awards
The United Conference of Mayors, 1220 I St. NW, Washington, DC 20006, (202) 293-7330.
H.J. Heinz Company, in cooperation with the United States Conference of Mayors, presented its second annual National Recycling Award to two cities: Los Angeles (population 3.5 million) and Newton, Massachusetts (population 83,236). Each citiy received a $10,000 check to be spent on citizen education for recycling.

A: *New Zealand*
Source: The Colorado Daily, Boulder, Colorado

International Corporate Environmental Achievement

The World Environment Center, 419 Park Ave. S., Ste 1800, New York, NY 10016.

The 1992 World Environment Center Gold Medal for International Corporate Environmental Achievement was presented to Procter & Gamble Company for its dedication to "developing and implementing innovative approaches to meeting environmental needs around the world."

John M. Collier Forest History Journalism Award

Forest History Society, 701 Vickers Ave., Durham, NC 27701, (919) 682-9319.

Sponsored by the Forest History Society, the 1992 John M. Collier Journalism Award was given to T. H. Watkins for his article in *American Heritage* magazine (June 1991). The story traced the early forestry activities of conservation leader Gifford Pinchot.

John Muir Award

The Sierra Club, 730 Polk St., San Francisco, CA 94109, (415) 776-2211.

The Sierra Club presents an annual John Muir Award (named after the 19th-century environmentalist) to a person the club deems the outstanding environmentalist of the year. The 1992 winner was James C. Catlin, for his work in protecting and promoting Utah wilderness areas.

program. Volunteers must have a bachelor's degree. A fee to cover the cost of travel, health insurance, placement, orientation, field support and program administration is required. Contact WORLD TEACH, Harvard Institute for International Development, 1 Eliot St., Cambridge, MA 02138-5705, (617) 495-5527.

CONCERNED WRITERS

CONCERN, INC., seeks volunteers and interns to assist with the research and writing of CONCERN'S Community Action Series, an overview of environmental issues. Knowledge of environmental issues, strong research and writing skills, and good interpersonal communication required. Word processing and database skills are desirable. Contact CONCERN, INC., 1794 Columbia Rd. NW, Washington, DC 20009, (202) 328-8160.

AUTHOR, AUTHOR

FOOD AND WATER INC. seeks volunteer research assistants to help compile a consumers' guide to safe food. Volunteers must have good writing and communications skills. Contact Lori Williams, FOOD AND WATER INC., RR I, Box 30, Old School House Common, Marshfield, VT 05658, (802) 426-3700.

WILD CANADA

WILDERNESS COMMITTEE needs volunteers to fill positions in areas including computer programming, slide-show coordinating, public education and research assistance. Volunteers are also needed to assist in the construction of trails and boardwalks at Boise Creek. Contact Kerry Dawson, WILDERNESS COMMITTEE, 20 Water St., Vancouver, British Columbia V6B 1A4, Canada, (604) 683-8220.

WATER RESEARCH

CLEAN WATER ACTION, a national citizens' organization working for clean and safe water, needs volunteers to assist with computer data entry and donor research. Contact Barbara Schecter, Director of Development, CLEAN WATER ACTION, 1320 18th St. NW, Ste. 300, Washington, DC 20036, (202) 457-1286.

HOME BUILDING

HABITAT FOR HUMANITY has a wide range of volunteer positions available. Opportunities include

 How many dolphins are legally allowed to be killed by the US tuna fleet each year?

project directors, habitat construction supervisors and workers, clerical and secretarial positions, fundraising and development positions, volunteer coordinators, international media relations positions, writers and business managers. A three-month commitment is requested. Room and board are provided for habitat construction workers. Contact Hilary Cook, HABITAT FOR HUMANITY, 121 Habitat St., Americus, GA 31709, (912) 924-6295.

ON GOLDEN PONDS

Volunteer writers are needed to submit their work to the NATIONAL POND SOCIETY. The society is looking for stories about pond keeping, childhood pond essays and stories about the wildlife surrounding ponds. Contact Karla Sperling, NATIONAL POND SOCIETY, P.O. Box 449, Acworth, GA 30101, (404) 975-0277.

ANTI-POLLUTION

SEACOAST ANTI-POLLUTION LEAGUE has a variety of volunteer positions available. Duties include educational projects in elementary and high schools, publicity work compiling press releases on the costs of nuclear power, membership development, fundraising, grant writing and general office help. Contact June Daigneault, SEACOAST ANTI-POLLUTION LEAGUE, 5 Market St., Portsmouth, NH 03801, (603) 431-5089.

EARTH SOUNDS

EARTH ON THE AIR radio, a grassroots organization, seeks volunteer submissions of interviews, music and sound collages that deal with environmental and social justice issues. Membership is acquired through work or pay. Prior radio experience is not required. Contact Susan Gleason, EARTH ON THE AIR, P.O. Box 45883, Seattle, WA 98145-0883, (206) 526-0551.

SAY NO TO PESTICIDES

NATIONAL COALITION AGAINST THE MISUSE OF PESTICIDES seeks volunteers to fill a variety of positions. Volunteers are needed to research pesticide uses on fields and in homes. Other available opportunities include fundraising and public relations work, scientist assistantships and newsletter writing. Contact Sarah Sullivan, NATIONAL COALITION

Julie and Spencer Penrose Award
El Pomar Foundation, 10 Lake Circle, Colorado Springs, CO 80906, (719) 577-5709.
The Denver Museum of Natural History won the Julie and Spencer Penrose Award for outstanding nonprofit organization in Colorado. Founded in 1900, the museum is the fifth-largest natural history museum in the US, with approximately 1.35 million visitors each year.

Marjory Stoneman Douglas Award
Faultless Starch/Bon Ami Company, 1025 W. Eighth St., Kansas City, MO 64104, (816) 842-1230.
This award, named in honor of Marjory Stoneman Douglas, who devoted many years to preserving the fragile ecosystem of the Florida Everglades, recognizes outstanding efforts resulting in protection of a unit or a proposed unit of the National Park System. This year's recipient was Isaac C. Eastvold, founder and president of Friends of the Albuquerque Petroglyphs, a group dedicated to preventing the destruction of ancient rock art on a 17-mile-long escarpment near Albuquerque, New Mexico.

National Conservation Achievement Award
National Wildlife Federation, 1400 16th St. NW, Washington, DC 20036-2266, (202) 797-6850.
Established in 1965, the National Wildlife Federation's National

A: *20,500*
Source: Greenpeace *magazine*

Conservation Achievement Awards have been presented to individuals and organizations providing leadership in spreading the conservation message and protecting natural resources. This year the award was given to physicist Amory Lovins for his many accomplishments in the field of energy policy, both domestically and internationally.

National Environmental Awards Council
National Environmental Awards Council, P.O. Box 362, East Haddam, CT 06423-0362, (203) 434-8666.

The National Association for Humane and Environmental Education's Adopt-A-Teacher program has been awarded a Certificate of Environmental Achievement for its success in advancing the cause of environmental protection, while serving as a model that can be replicated around the country.

National Park Foundation
National Park Foundation, 1101 17th St. NW, Ste., 1102, Washington, DC 20036, (202) 785-4500.

The National Park Foundation sponsored over $140,000 in grants to benefit priority projects in the parks. The 11 Winter 1992 awards range from a challenge grant to Saguaro National Monument, Arizona, for a wheelchair-accessible nature trail, to an innovative environmental education program for

AGAINST THE MISUSE OF PESTICIDES, 701 E St. SE, Ste. 200, Washington, DC 20003, (202) 543-5450.

WETLANDS WORK
THE SAVE THE DUNES COUNCIL needs volunteers to assist in influencing federal legislation that will protect vital dunes, wetlands and prairies and add over 1,000 acres to the Indiana Dunes National Lakeshore. Contact Tom Anderson, SAVE THE DUNES COUNCIL, 444 Barker Rd., Michigan City, IN 46360, (219) 879-3937.

TRAILING ALONG
The APPALACHIAN TRAIL CONFERENCE is seeking volunteer trail crews. Crews operate in most trail states from Maine to Georgia, building and maintaining the Appalachian Trail in some of the most scenic locations in the East. A one- to two-week commitment is required. For more information contact Crews, ATC, BZ-92B, P.O. Box 10, Newport, VA 24128, (703) 544-7388.

FOREST FRIENDS
SAVE AMERICA'S FORESTS seeks volunteers for all types of activism, including political, press and creative fundraising. Volunteers are also needed to work from their homes or travel across the country to organize congressional districts and build the coalition throughout the US. Contact SAVE AMERICA'S FORESTS, 4 Library Ct. SE, Washington, DC 20003, (202) 544-9219.

AFRICAN GREEN
Through planting and maintaining trees, TREES FOR AFRICA's mission is to conserve southern Africa's environment. The group encourages and supports individual planting projects. Under its advisement, corporations and industries have responded by sponsoring planting in disadvantaged communities for recreational developments. TREES FOR AFRICA also sponsors agroforestry and woodlot projects in rural areas. Environmental education centers have been established to provide ongoing training, field experience and logistic support for greening projects. TREES FOR AFRICA needs international volunteers to assist with these projects. Contact TREES FOR AFRICA, P.O. Box 18, Johannesburg 2000, South Africa, (27 11) 802 4867.

Q: *According to EPA estimates, over the next 50 years, how many Americans are expected to contract skin cancer as a result of ultraviolet exposure?*

ECOCONNECTIONS

EARTH EFFORT
The EARTH ISLAND INSTITUTE seeks volunteers in its San Francisco office for administrative duties and office support (reception and mailings). People are also needed to staff events in the San Francisco Bay area. Contact the Volunteer Coordinator, EARTH ISLAND INSTITUTE, 300 Broadway, Ste. 28, San Francisco, CA 94133, (415) 788-3666.

CARING FOR COLORADO
VOLUNTEERS FOR OUTDOOR COLORADO organizes citizens to work on public land improvement projects. All volunteers are led by trained crew leaders, so no experience is necessary. For more information or to sign up for projects, call VOLUNTEERS FOR OUTDOOR COLORADO, 1410 Grant St., B105, Denver, CO 80203, (303) 830-7792.

FRIENDS OF THE FOREST
Volunteers are sought by FRIENDSHIP PROJECTS, INC., to help with research, administrative activities, fundraising and on-site professional assistance for the preparation of a reforestation project in Madagascar. Contact FRIENDSHIP PROJECTS, INC., P.O. Box 1003, Dutch Harbor, AK 99692.

DOLPHIN VENTURE
FLORIDA MARINE CONSERVATION CORPS offers research volunteer positions with the Caribbean Dolphin Research Project team. Volunteers participate in at-sea studies of wild dolphin biology, sociology and physiology in Florida and the Caribbean Islands. Fee required. Space limited. Contact FLORIDA MARINE CONSERVATION CORPS, 160 Elaine Rd., West Palm Beach, FL 33413, (407) 683-9647.

DEDICATED TO DUNES?
GREAT SAND DUNES NATIONAL MONUMENT needs volunteers for visitor center operations and information services, a computer operator, librarian, photographer, interpreter and maintenance crew and curatorial workers. Benefits include training, supervision, housing and college credit (if arranged). Contact Libbie Landreth, GREAT SAND DUNES NATIONAL MONUMENT, 11500 Hwy. 150, Mosca, CO 81146, (719) 378-2312.

elementary schoolchildren at Apostle Islands National Lakeshore, Wisconsin.

NAPCOR PET Plastic Recycling Awards
NAPCOR, 4828 Parkway Plaza Blvd., Ste 260, Charlotte, NC 28217, (704) 357-3250.

The National Association for Plastic Container Recovery (NAPCOR) presented its annual awards recognizing corporate and community organizations for the advancement of the recycling of soft drink bottles and other containers made of polyethylene terephthalate (PET) plastic. The 1992 winners were the national solid waste company, Browning-Ferris Industries, Inc., and the community recycling program in Louisville, Kentucky.

Oliver Award
National Groundwater Association, 6375 Riverside Dr., Dublin, OH 43017, (614) 761-1711.

The National Groundwater Association honored John McMahan with the 1992 Oliver Award, recognizing outstanding contributions to the groundwater industry. The Association's Science Award went to Dr. John Bredehoeft of the US Geological Survey. Jerry Bronicel and Robert Stringer shared the Groundwater Advocate Award.

Polartec Performance Challenge Awards
Polartec, 450 Seventh Ave., New York, NY 10123, (212) 563-0404.

A:
12 million
Source: Greenpeace

Polartec's Performance Challenge project, launched in August 1991 in support of international expeditions and individual pursuits, has awarded $25,000 in prize money to eight individuals who best exemplify original approaches to adventure, outdoor sports, environmental issues and vanishing cultures.

Rainforest Champion Award
Rainforest Alliance, 270 Lafayette St., Ste. 512, New York, NY 10012, (212) 941-1900.
  The Rainforest Alliance established the Rainforest Champion Award this year to honor individuals who have made a significant contribution to the limitation of tropical deforestation. The first award was given to Dr. Thomas E. Lovejoy, who has initiated several important projects to save rainforest, including creating the Minimum Critical Size of Ecosystems Project near Manaus, Brazil.

Rene Dubos Environmental Awards
Rene Dubos Center for Human Environments, Inc., 100 E. 85th St., New York, NY 10028, (212) 249-7745.
  The Rene Dubos Environmental Awards, named for the renowned scientist, were established in 1984 by the Rene Dubos Center, a nonprofit education and research organization. The awards are presented each May at a dinner, and proceeds

PROTECT THE COLLECTION

The JOHN MUIR NATIONAL HISTORIC SITE needs volunteers to lead guided walks of the Muir home and to garden, catalog, research and do nontechnical maintenance on the collection. Two-month commitment required. Funding may be arranged on a person-by-person basis. Contact Thaddeus Shay, NATIONAL PARK SERVICE, 4202 Alhambra Ave., Martinez, CA 94553, (510) 228-8860.

PARK PROJECTS

REDWOOD NATIONAL PARK seeks volunteers to assist with information and interpretation, trail work, library and curatorial work, photography, clerical and resource management. A two-month commitment is required. Benefits include dorm housing, training and experience. Contact Robin Galea, REDWOOD NATIONAL PARK, 1111 Second St., Crescent City, CA 95531, (707) 464-6101.

HANDLING THE PANHANDLE

IDAHO PANHANDLE NATIONAL FOREST is seeking volunteers for various positions. Some of these include trail crew workers, timber crew members, district volunteer coordinators, maintenance workers/caretakers, survey crew members, inventory crew members, campground and picnic ground hosts, and office workers. Other positions also available. Contact Debra Gallegos, IDAHO PANHANDLE NATIONAL FOREST, 1201 Ironwood Dr., Coeur d'Alene, ID 83814, (208) 765-7265.

OPEN SPACE

MIDPENINSULA REGIONAL OPEN SPACE volunteer program offers a variety of restoration, construction, resource enhancement and trail maintenance projects. Crew leadership training courses are offered, as well as group service projects, on 28 separate space preserves. Other opportunities include a district docent program including 60 hours of training, and individual fieldwork. The open space region encompasses close to 35,000 acres. Contact Volunteer Coordinator, MIDPENINSULA REGIONAL OPEN SPACE PROGRAM, 330 Distel Cir., Los Altos, CA 94022, (415) 691-1200.

 Q: *Of Czechoslovakian countryside wells, what percentage are not contaminated?*

SEEKING REFUGE

ARANSAS NATIONAL WILDLIFE REFUGE seeks volunteers for winter and year-round positions. Winter volunteers will guide bird tours and trail hikes and help operate the visitor center. Year-round opportunities include computer work, administrative assistance, and trail and general maintenance. Contact Smokey Cranfill, ARANSAS NATIONAL WILDLIFE REFUGE, P.O. Box 100, Austwell, TX 77950, (512) 286-3559.

ECO-RETAIL

RECREATIONAL EQUIPMENT, INC., offers volunteer opportunities through its annual service projects in communities served by its 35 retail stores. The company's service projects focus on enhancing outdoor recreation opportunities through trail building and maintenance, campground rehabilitation and enhancement, shoreline cleanups and other community projects. For information regarding service projects in your area, contact RECREATIONAL EQUIPMENT, INC., Public Affairs, Service Projects, P.O. Box 1938, Sumner, WA 98390-0800, (206) 395-5957 or (800) 999-4734.

CREW COORDINATORS

The PACIFIC CREST TRAIL CONFERENCE needs up to 100 volunteer crew leaders. Training in trail construction, maintenance and other skills provided. Each crew leader coordinates the efforts of several volunteers. Contact Louise B. Marshall, PACIFIC CREST TRAIL CONFERENCE, P.O. Box 2040, Lynnwood, WA 98036, (206) 771-7208.

RECREATION RESEARCH

OUTDOOR RECREATION AND WILDERNESS RESEARCH is seeking enthusiastic individuals who can interview visitors at national forests. The information gathered is used to plan for future recreation uses on public lands. Fieldwork involves traveling to interview sites (use your own vehicle), camping (use your own tent or trailer), interviewing visitors and reporting progress weekly. A daily per diem is provided. Travel and lodging will be reimbursed. All ages welcome. Contact Jerry Coker, SE Forest Experiment Station, OUTDOOR RECREATION AND

from the dinner ceremony support the development of educational resources for environmental literacy and education. The 1992 award winners were Will D. Carpenter, W. Parker Mauldin, David Maybury-Lewis, Pia Maybury-Lewis and Chauncey Starr.

Renew America Environmental Achievement Award
Renew America, 1400 16th St. NW, Ste. 710, Washington, DC 20036, (202) 232-2252.

These awards, established to recognize and promote successful projects around the country to serve as environmental models, were given this year to 28 of the country's leading environmental organizations.

River Conservationist of the Year
Perception, Inc., 1110 Powdersville Rd., Easley, SC 29640, (803) 859-7518.

Scootch Pankonin of Washington, DC, has been named the 11th annual recipient of the River Conservationist of the Year Award, presented by Perception, Inc. Pankonin has been working with several organizations to assess the Glen Canyon Dam and Grand Canyon water situation.

Robert C. Barnard Environmental Science and Engineering Scholarship
American Association for the Advancement of Science, 1333 H St.

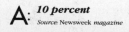
A: *10 percent*
Source: Newsweek *magazine*

NW, Washington, DC 20005, (202) 326-6431.

Dr. Eric Andrew Pani, a geoscientist at Northeast Louisiana University, was recently awarded the Robert C. Barnard Environmental Science and Engineering Scholarship for his work on pollution venting from the mixed layer into the free troposphere. The $3,000 research grant is given annually to an outstanding participant in the American Association for the Advancement of Science.

Samuel C. Johnson Environmental Stewardship Award
S.C. Johnson & Son, Inc., 1525 Howe St., Racine, WI 53403-5011, (414) 631-2000.

This first annual award was founded to recognize employees, retirees or employee spouses connected to S.C. Johnson & Son, Inc., who have demonstrated their personal commitment to a better world through voluntary efforts to improve, preserve and sustain the environment for future generations. This year's award was given to several individuals worldwide who have made contributions to improving the environments where they live.

Steel Can Recycling Institute Scholarship Program
Steel Can Recycling Institute, Foster Plaza X, 680 Andersen Dr., Pittsburgh, PA 15220, (800) 876-SCRI.

The competition is designed to help promote

WILDERNESS RESEARCH, Carlton/Green Sts., Athens, GA 30601, (404) 546-2451.

WORK FOR WALNUT

WALNUT CANYON NATIONAL MONUMENT is looking to fill volunteer positions in park operations, resource management and park maintenance. Varied responsibilities include interpretation, tours, inventory, trail work, cleaning, painting and carpentry. Benefits include free housing (limited). Bicycle commuting to Flagstaff is possible. Contact Kim Watson, Wupatki/Sunset Crater Volcano, Walnut Canyon National Monument, 2717 N. Steves Blvd., #3, Flagstaff, AZ 86004, (602) 556-7134.

T.R.A.I.L. BOSS

BOY SCOUTS OF AMERICA is seeking volunteers for its T.R.A.I.L. Boss program. The goal is to teach volunteer leaders specialized skills for training and leading volunteer crews involved in conservation projects. Benefits include environmental education and greater stewardship of cultural and natural resources. Contact Robert J. Pruden, Director of Boy Scout Camping and Conservation, BOY SCOUTS OF AMERICA, 1325 Walnut Hill Lane, P.O. Box 152079, Irving, TX 75015.

ILLINOIS NATURE

ILLINOIS NATURE PRESERVE COMMISSION needs interns to assist natural areas preservation specialists with researching, identifying and securing protection status for Illinois natural lands. Undergraduate training in natural or biological sciences, or experience in real estate or law, is necessary. Benefits include training, supervision and college credit if pre-arranged. Contact Dr. Brian D. Anderson, ILLINOIS NATURE PRESERVE COMMISSION, 524 S. Second St., Springfield, IL 62701-1787, (217) 785-8686.

ENVIRONMENTAL WRITERS

The ENVIRONMENTAL AND ENERGY STUDY INSTITUTE is looking for interns with demonstrated research and writing skills as well as good verbal skills. Specifically, programs that interns will work on include marketing of the weekly publication, water quality, water efficiency, international environmental

: *How much energy does it take to supply pork to US consumers for consumption?*

issues, global climate change, energy efficiency and development. Generally, interns serve for approximately three or four months, but student schedules can be accommodated. Contact Karen Park, Intern Coordinator, ENVIRONMENTAL AND ENERGY STUDY INSTITUTE, 122 C St. NW, Ste. 700, Washington, DC 20001, (202) 628-1400.

GLACIAL GUIDANCE

THE GLACIER INSTITUTE seeks student interns for a season to help run an outdoor education program. The institute offers outdoor classes for all ages and interests, including one- to five-day residential environmental programs for first- through ninth-grade students. Contact THE GLACIER INSTITUTE, P.O. Box 1457, Kalispell, MT 59903, (406) 756-3911.

PUBLIC PROJECTS

CENTER FOR SCIENCE IN THE PUBLIC INTEREST offers internships to a small number of qualified students in undergraduate, graduate, law and medical schools each summer and during the school year. Some of the areas involving interns are the consumer health/nutrition project, nutrition action healthletter, legal affairs office, alcohol and public policy, Americans for safe food and management/administration. Internships usually last ten weeks. Interns may qualify for a stipend if the budget permits. Application materials should include a cover letter, a resume, a writing sample, two letters of recommendation and an official transcript of courses and grades. Contact Intern Coordinator, Center for Science, 1875 Connecticut Ave. NW, Ste. 300, Washington, DC 20009, (202) 332-9110.

SEA TURTLE SAVERS

The EARTH ISLAND INSTITUTE has intern opportunities for their Sea Turtle Restoration project. Responsibilities could include assisting with special research projects, administrative work, database entry, outreach coordination, fundraising, writing and editing, and correspondence with members and international environmental organizations. A commitment of at least eight hours per week is required. Contact Kathy Neilson, 300 Broadway, Ste. 28, San Francisco, CA 94133, (415) 788-3666.

solid waste management education in schools throughout the country, and to heighten public awareness of the importance of recycling and solid waste management. This year, five students were selected from each of the seven states involved in the initial year of the competition, and each will receive a $1,000 college scholarship to use at the school of their choice.

Stephen Tyng Mather Award
Faultless Starch/Bon Ami Company, 1025 W. Eighth St., Kansas City, MO 64104, (816) 842-1230.

This award, named for the first director of the National Park Service, recognizes a Park Service employee who has risked his or her job or career for the principles and practices of good stewardship. This year's recipient is Christian L. Shaver, chief of the Policy, Planning, and Permit Review Branch of the National Park Service's Air Quality Division.

United Nations Environmental Award
United Nations Environment Programme, 1889 F St. NW, Washington, DC 20006, (202) 289-8456.

Northern Telecom Ltd. won the 1992 North American Environmental Leadership Award for outstanding environmental achievement. In 1991, Northern Telecom became the world's first large electronics company to fulfill its

A: *15 million calories of energy for every pound*
Source: Worldwatch *magazine*

commitment to eliminate ozone-depleting CFC-113 solvents from its manufacturing and research operations.

Vision for America Award
Keep America Beautiful, Inc., 9 W. Broad St., Stamford, CT 06902, (203) 323-8987.

Keep America Beautiful, Inc., presented the 1992 Vision for America Award to Kmart Corporation CEO Joseph E. Antonini. Keep America Beautiful said Kmart has "dedicated itself to providing a healthier environment for America" with programs ranging from battery and tire recycling to elementary school education outreach.

We Care for America Grants
S.C. Johnson & Son, Inc., 1525 Howe St., Racine, WI 53403-5011, (414) 631-2000.

S.C. Johnson & Son, Inc., awards approximately $500,000 in "We Care for America" grant funds each year. The awards are administered independently by World Wildlife Fund. Grants have gone to such diverse groups as Ocean Arks International, Colorado State University, Pima Trails Association and the North Carolina Natural Heritage Foundation.

Windstar Foundation Award
Windstar Foundation, 2317 Snowmass Creek Rd., Snowmass, CO 81654, (303) 927-4777.

The Windstar Award

WILDLIFE REHABILITATION

VOLUNTEER FOR WILDLIFE, INC., needs people to work at its Wildlife Rehabilitation and Education Center on Long Island. Volunteers will assist in the rescue, medical care, rehabilitation and release of injured or displaced animals. Other opportunities include monitoring the wildlife advisory telephone, assisting education programs and contributing office skills. This is a local program with no housing. Contact Stephanie Brunetta, VOLUNTEER FOR WILDLIFE, INC., 27 Lloyd Harbor Rd., Huntington, NY 11743, (516) 423-0982.

FOSTER THE FUND

Internships are available with the Brower Fund, an EARTH ISLAND INSTITUTE project that provides short-term support and guidance for innovative efforts to conserve, preserve and restore the environment. Two to three afternoons for at least three months are required. Duties include office support and administration, research, writing, public outreach, event organizing and fundraising. Skills needed include Mac literacy and previous research or fundraising experience. Possible stipend. Send resume and cover letter to Jerry Langman, EARTH ISLAND INSTITUTE, 300 Broadway, Ste. 28, San Francisco, CA 94133, (415) 788-3666.

GLOBAL EFFICIENCY

INTERNATIONAL ENVIRONMENT AND DEVELOPMENT SERVICE seeks volunteers who are experts from industry and academia to meet the increased demand for its pro bono services in Asia, Eastern and Central Europe and the near East. Some of the issues confronted are industrial health, safety and environmental management, waste minimization, pollution prevention, emergency preparedness and energy and resource efficiency. Contact World Environment Center, 419 Park Ave. S., Ste. 1800, New York, NY 10016, (212) 683-4700.

PHILIPPINE ECOLOGY

The LOWER IFUGAO ENVIRONMENTAL AWARENESS PROGRAM is an environmental education program recently launched in Ifugao Province of the Philippines and run by the indigenous tribal group Ifugao. A large component of the program is

 How much energy does it take to supply rice, potatoes, fruit or vegetables to US consumers for consumption?

environmental education in four rural high schools. The Ifugao program seeks educational materials emphasizing forest protection and sustainable agriculture, and international advice and ideas for the training of community developers working on the program. Contact John Klock, Botany Dept., Ohio University, 218 R.T.E.C. Bldg., Athens, OH 45701, (614) 593-1126.

KID POWER

KIDS FOR A CLEAN ENVIRONMENT needs child volunteers to help with general office duties such as typing, filing and newsletter work. The group's newsletter tells kids about groups and organizations that invite child volunteers to help with activities. Contact Trish Poe, KIDS FOR A CLEAN ENVIRONMENT, P.O. Box 158254, Nashville, TN 37215, (615) 331-7381.

GARDENS AND GROUNDS

The AMERICAN HORTICULTURAL SOCIETY needs volunteers to assist with horticultural and non-horticultural activities. Volunteers will help maintain and enhance the gardens and grounds, staff the Gardeners' Information Service and help with the National Backyard Compost Demonstration Park. Participants also assist with public relations and a variety of other activities. Contact Maureen Heffernan, AMERICAN HORTICULTURAL SOCIETY, 7931 E. Boulevard Dr., Alexandria, VA 22308 (703) 768-5700.

BIG CITY TURTLES

The NEW YORK TURTLE AND TORTOISE SOCIETY needs volunteers to help with turtle care and rehabilitation. Volunteers will be trained in turtle handling and will work to find temporary homes for the turtles until they are released back into the wild. Participants will also act as watchdogs to report any violations of legislation that protects turtles and tortoises. Contact Lori Cramer, NEW YORK TURTLE AND TORTOISE SOCIETY, 163 Amsterdam Ave., Ste. 365, New York, NY 10023, (212) 459-4803.

is presented annually to a global citizen whose personal and professional life exemplifies Windstar's commitment to the creation of a sustainable, peaceful future. Professor Phil Lane, Jr., of Yankton Sioux and Chickasaw heritage, received this year's $10,000 award. Professor Lane is recognized for his leadership in human and community relations and for his commitment to the Four Worlds Development Project, which seeks to eliminate alcohol and drug abuse in North American Native communities by the year 2000.

Windstar Youth Award
Windstar Foundation, 2317 Snowmass Creek Rd., Snowmass, CO 81654, (303) 927-4777.
This $2,500 scholarship, sponsored by Half Price Books of Dallas, is awarded each year to a young person whose actions contribute to an environmentally sustainable and peaceful future. This year's recipient, Erika Peña of Laredo, Texas, received her award after founding the Kids for the Earth Recycling Club. The club members organized the first curbside recycling program in Laredo, participated in Earth Day events and organized an annual reuseable toy drive at Christmas to benefit the children in their sister city, Nuevo Laredo, Mexico.

A: *1 million calories of energy for every pound*
Source: Worldwatch *magazine*

EARTH JOURNAL

Directory to Environmental Groups

As David Maybury-Lewis points out in his book, Millennium, humans have lived in tribes for thousands of years. Do you have a tribe? No? Here are some you can join. We asked each group to send us information on their purpose, current emphasis, membership, funding and expenditures. (The letters "dnd" next to a heading mean the organization "did not disclose" that information.)

THE ACID RAIN FOUNDATION, INC.
1410 Varsity Dr.
Raleigh, NC 27606
(919) 828-9443
Fax: (919) 515 3593
Purpose: To foster greater understanding of global atmospheric issues by raising the level of public awareness, supplying educational resources and supporting research. **Current Emphasis:** Acid rain, global atmosphere, recycling and forest ecosystems. **Members:** dnd. **Fees:** $35. **Funding:** Membership, 2.5%; Programs, 44%; Direct public support, 53.5%. **Annual Revenue:** $100,000. **Usage:** Administration, 24%; Fundraising, 9%; Programs, 67%. **Volunteer Programs:** Education, development, library and marketing.

AFRICAN WILDLIFE FOUNDATION
1717 Massachusetts Ave. NW
Washington, DC 20036
(202) 265-8394
Fax: (202) 265-2361
Purpose: Working directly with Africans at all government and private levels in over 25 countries since 1961, AWF's staff promotes, establishes and supports grassroots and institutional programs in conservation education, wildlife management training and management of threatened conservation areas. **Current Emphasis:** Public awareness campaign to encourage ivory boycotts and support of Mountain Gorilla Project in Rwanda. **Members:** 100,000. **Fees:** No minimum for membership; $15 to receive newsletter. **Funding:** Membership, 60%; Corporate, 15%; Foundation/Donor, 25%. **Annual Revenue:** $4,676,000. **Usage:** Administration, 7%; Fundraising, 10%; Programs, 83%. **Volunteer Programs:** "How You Can Help" letter lists ways in which interested people can help curtail the demand for ivory.

ALLIANCE FOR ENVIRONMENTAL EDUCATION
P.O. Box 368
The Plains, VA 22171
(703) 253-5812
Fax: (703) 253-5811
Purpose: To serve as an advocate for a quality environment through education and advanced communication, cooperation and exchange among organizations. **Current Emphasis:** The alliance is establishing a network of interactive environmental education centers based in colleges, universities and institutions across America. **Members:** 300 organizations, millions of individuals. **Fees:** $125-$2,500. **Funding:** Membership, 10%; Corporate, 20%; Other, 70%. **Annual Revenue:** $550,000. **Usage:** Administration, 20%; Fundraising, 10%; Programs, 70%. **Volunteer Programs:** Board membership, task forces, special committees, regional advisory councils and internships.

AMERICAN ASSOCIATION OF ZOOLOGICAL PARKS AND AQUARIUMS
7970-D Old Georgetown Rd.
Bethesda, MD 20814
(301) 907-7777
Fax: (301) 907-2980
Purpose: A professional organization representing 160 accredited zoos and aquariums in North America. The primary goal is to further wildlife conservation and education and to enforce a code of ethics for all individual members and zoological institutions. **Current Emphasis:** Wildlife conservation through captive propagation. **Members:** 6,000. **Fees:** $35 for associates. **Funding:** Membership, 65%; Corporate, 2%. **Annual Revenue:** dnd. **Usage:** Administration, 7%; Membership, 31%; Conservation, 30%. **Volunteer Programs:** None.

Q: *What percentage of daily garbage can be attributed to packaging?*

ECOCONNECTIONS

AMERICAN FORESTS
P.O. Box 2000
Washington, DC 20013
(202) 667-3300
Fax: (202) 667-7751

Purpose: To protect forests and community trees and restore damaged ecosystems through action, education, research and policy advocacy. **Current Emphasis:** 1) To further the understanding and management of national forests as ecosystems and restore forest health, especially in the West. 2) Global ReLeaf grassroots campaign mobilizes people, businesses and governments to plant and care for trees—saving energy, slowing global warming and protecting wildlife. **Members:** 108,000. **Fees:** Subscribing memberships from $24; Contributing memberships from $15. **Funding:** Membership, 32%; Grants, 49%; Other, 19%. **Annual Revenue:** $3,977,761. **Usage:** Fundraising, 13%; Programs, 51%; Other, 36%. **Volunteer Programs:** Call for more information on Global ReLeaf Projects in your area.

AMERICAN GEOGRAPHICAL SOCIETY
156 Fifth Ave., Ste. 600
New York, NY 10010-7002
(212) 242-0214
Fax: (212) 989-1583

Purpose: To expand and disseminate geographical knowledge through publications, awards, travel programs, lectures and consulting, with a strong emphasis on ecology and environmental issues abroad and in the US. **Current Emphasis:** Publication of *The Geographical Review* and *Focus* magazine, and the AGS Newsletter, provision of educational travel program, lecturers to educational and business audiences, and award program to encourage research. **Members:** 1,800. **Fees:** $25. **Funding:** Membership, 8%; Corporate, 15%; Foundation/Donor, 7%; Other, 70%. **Annual Revenue:** $565,000. **Usage:** Administration, 28%; Programs, 72%. **Volunteer Programs:** Volunteer positions available.

AMERICAN HIKING SOCIETY
P.O. Box 20160
Washington, DC 20041-2160
(703) 385-3252
Fax: (703) 754-9008

Purpose: To protect and promote hiking trails in America. Over 100 club affiliates provide for information exchange within the trails community. **Current Emphasis:** Development of the first transcontinental hiking trail, coordinating first National Trail Day (set for June 1993), effectively educating members of US Congress regarding value of trails, maintaining a public information service to trail users and managers regarding facilities, organizations and best use of trails to protect the environment. **Members:** 5,000. **Fees:** $25. **Funding:** Membership, 38%; Corporate, 5.5%; Other, 56.5%. **Annual Revenue:** $250,000. **Usage:** Administration, 22%; Fundraising, 3.5%; Other, 74.5%. **Volunteer Programs:** Encourages volunteers in trail building and maintenance through work trips called "Volunteer Vacations" and by publishing a directory of volunteer opportunities on public lands.

AMERICAN HORSE PROTECTION ASSOCIATION, INC.
1000 29th St. NW, Ste. T-100
Washington, DC 20007
(202) 965-0500
Fax: (202) 965-9621

Purpose: Dedicated entirely to the welfare of equines, wild and domestic, by fighting for the humane treatment of horses through litigation, investigation and public awareness of proper and humane horse care. **Current Emphasis:** Preserving and protecting horses and burros; preventing abuse of horses in competition; solving problems of neglect and mistreatment of horses; promoting safe and humane equine transportation. **Members:** 8,000. **Fees:** $20. **Funding:** dnd. **Annual Revenue:** $260,000. **Usage:** dnd. **Volunteer Programs:** None.

AMERICAN HUMANE ASSOCIATION
63 Inverness Dr. E.
Englewood, CO 80112
(303) 792-9900
Fax: (303) 792-5333

Purpose: To prevent the neglect, abuse, cruelty and exploitation of children and animals and to assure that their interests and well-being are fully, effectively and humanely guaranteed by an aware and caring society. **Current Emphasis:** dnd. **Members:** 30,000. **Fees:** $15-$50. **Funding:** dnd. **Annual Revenue:** $4,900,000. **Usage:** Fundraising, 21%; Programs, 79%. **Volunteer Programs:** None.

AMERICAN RIVERS
801 Pennsylvania Ave. SE, Ste. 400
Washington, DC 20003
(202) 547-6900
Fax: (202) 543-6142

Purpose: Nonprofit conservation organization leading the effort to protect and restore the nation's river systems. The organization has effectively preserved over 10,000 river miles for

A: *65 percent*
Source: Garbage magazine

clean water, threatened fish and wildlife, recreation and scenic beauty. Concerns include dams, diversions, channelizations and adverse development. **Current Emphasis:** Wild and Scenic river system; hydropower relicensing; endangered aquatic species protection (especially salmon); instream flow issues. **Members:** 18,000. **Fees:** Begin at $20. **Funding:** Membership, 45%; Corporate, 12%; Foundation/Donor, 38%; Other, 5%. **Annual Revenue:** $1,500,000. **Usage:** Administration, 8%; Fundraising, 14%; Programs, 78%. **Volunteer Programs:** River activist network; internships.

ANIMAL PROTECTION INSTITUTE OF AMERICA
2831 Fruitridge Rd.
Sacramento, CA 95822
(916) 731-5521
Fax: (916) 731-4467

Purpose: To eliminate fear, pain and suffering inflicted on animals and to preserve threatened species. **Current Emphasis:** Publications, animal welfare issues, education and legislative issues. **Members:** 120,000+. **Fees:** dnd. **Funding:** dnd. **Annual Revenue:** dnd. **Usage:** dnd. **Volunteer Programs:** Teachers are encouraged to participate in educating students using API "Know a Teacher" literature.

THE ASSOCIATION OF FOAM PACKAGING RECYCLERS
1025 Connecticut Ave. NW, Ste. 515
Washington, DC 20036
(800) 944-8448
Fax: (202) 331-0538

Purpose: With more than 45 plant locations nationwide to serve as central collection points, AFPR brings the recycling of foam packaging to every major metropolitan area. This establishes the basis to produce protective foam packaging made with recycled content. **Current Emphasis:** Encouraging the reuse of loose-fill foam packaging and the recycling and reprocessing of molded foam packaging. **Members:** dnd. **Fees:** dnd. **Funding:** Corporate, 100%. **Annual Revenue:** dnd. **Usage:** dnd. **Volunteer Programs:** None.

BEYOND BEEF
1130 17th St. NW, Ste. 300
Washington, DC 20036
(202) 775-1132
Fax: (202) 775-0074

Purpose: International public interest coalition working to educate people about the harm caused by beef consumption. The main goal is to encourage people to reduce their individual beef consumption. **Current Emphasis:** Nationwide public speaking tour promoting the campaign. **Members:** dnd. **Fees:** dnd. **Funding:** dnd. **Annual Revenue:** dnd. **Usage:** dnd. **Volunteer Programs:** Yes.

BIG BLUE FOUNDATION, INC.
8446 Melrose Place
Los Angeles, CA 90069
(213) 852-1414
Fax: (213) 658-5853

Purpose: Harnessing media power to publicize and positively affect environmental issues is the mission of the Big Blue Foundation, a nonprofit, environmental media group dedicated to the protection of our planet. **Current Emphasis:** Producing a national multimedia campaign to promote the permanent protection of America's national forests. **Members:** dnd. **Fees:** dnd. **Funding:** Corporate, 10%; Foundation/Donor, 90%. **Annual Revenue:** dnd. **Usage:** dnd. **Volunteer Programs:** Film production internships.

CENTER FOR ENVIRONMENTAL INFORMATION
46 Prince St.
Rochester, NY 14607
(716) 271-3550
Fax: (716) 271-0606

Purpose: Established to provide timely, accurate and comprehensive information on environmental issues. CEI has developed a multifaceted program of publications, education programs and information services. **Current Emphasis:** Environmental education, ethics, laws, communication and global environmental change. **Members:** 700. **Fees:** $25. **Funding:** Membership, 50%; Corporate, 50%. **Annual Revenue:** $600,000. **Usage:** Administration, 5%; Fundraising, 10%; Programs, 85%. **Volunteer Programs:** Library services, conferences and program coordination, and publications.

CENTER FOR HOLISTIC RESOURCE MANAGEMENT
5820 Fourth St. NW
Albuquerque, NM 87107
(505) 344-3445
Fax: (505) 344-9079

Purpose: Community development based on a proven process of goal setting and decision making that helps communities restore their well-being and the natural resources on which they depend. **Current Emphasis:** Expanding the number of individuals capable of offering training in holistic resource management.

 How many disposable cups does the average American college student throw out each year?

ECOCONNECTIONS

Members: 1,500. **Fees:** $35. **Funding:** Membership, 4%; Corporate, 11%; Foundation/Donor, 22%; Programs, 63%. **Annual Revenue:** $900,000. **Usage:** Administration, 16%; Fundraising, 6%; Programs, 78%. **Volunteer Programs:** Currently looking for volunteers interested in becoming trainers in holistic resource management.

CENTER FOR MARINE CONSERVATION
1725 DeSales St. NW, Ste. 500
Washington, DC 20036
(202) 429-5609
Fax: (202) 872-0619

Purpose: Center for Marine Conservation is the leading nonprofit organization dedicated solely to the conservation of marine wildlife and habitats. Focusing on five major goals: conserving marine habitats; preventing marine pollution; fisheries conservation; protecting endangered species; and promoting and educating about marine biodiversity. **Current Emphasis:** All of the above. **Members:** 110,000+. **Fees:** $20. **Funding:** Membership, 50%; Corporate, 7%; Foundation/ Donor, 34%; Government, 9%. **Annual Revenue:** $3,600,000. **Usage:** Administration, 10%; Fundraising, 13.5%; Programs, 54%; Other, 22.5%. **Volunteer Programs:** National Beach Cleanup in September. Some internships at national and regional offices.

CENTER FOR PLANT CONSERVATION
Missouri Botanical Garden
P.O. Box 299
St. Louis, MO 63166-0299
(314) 577-9450
Fax: (314) 664-0465

Purpose: To conserve rare and endangered native plants through research, cultivation and education at botanical gardens and arboreta in the US. Through 25 affiliated gardens and arboreta, the center establishes off-site germ plasm collections in the National Collection of Endangered Plants. **Current Emphasis:** Five priority regions: Hawaii, Florida, California, Texas and Puerto Rico. **Members:** dnd. **Fees:** dnd. **Funding:** Donations, 100%. **Annual Revenue:** $865,000. **Usage:** Administration, 17%; Fundraising, 13%; Programs, 70%. **Volunteer Programs:** None.

CENTER FOR SCIENCE INFORMATION
4252 20th St.
San Francisco, CA 94103
(415) 553-8178
Fax: (415) 861-4908

Purpose: To educate decision-makers and journalists about the environmental applications of biotechnology. **Current Emphasis:** All of the above. **Members:** dnd. **Fees:** dnd. **Funding:** Corporate, 5%; Foundation/Donor, 95%. **Annual Revenue:** $150,000. **Usage:** Administration, 10%; Fundraising, 5%; Programs, 85%. **Volunteer Programs:** None.

CITIZEN'S CLEARINGHOUSE FOR HAZARDOUS WASTE (CENTER FOR ENVIRONMENTAL JUSTICE)
P.O Box 6806
Falls Church, VA 22040
(703) 237-2249

Purpose: To assist communities to fight environmental threats through grassroots efforts. **Current Emphasis:** Contaminated sites campaign. Convicting the EPA for child abuse for not cleaning up these sites. **Members:** 22,000. **Fees:** $25. **Funding:** Membership, 31%; Grants, 10%; Foundation/Donor, 51%; Other, 8%. **Annual Revenue:** $850,000. **Usage:** Administration, 12%; Programs, 88%. **Volunteer Programs:** Volunteer positions and paid field internships are available.

CLEAN WATER ACTION
1320 18th St. NW, Ste. 300
Washington, DC 20036
(202) 457-1286
Fax: (202) 457-0287

Purpose: National citizens' organization working for clean and safe water at an affordable cost; control of toxic chemicals; protection and conservation of wetlands, groundwater and coastal waters; safe solid waste management; public health; and environmental safety of all citizens. **Current Emphasis:** Citizen organizing and education to effect environmental change and safety. **Members:** 600,000. **Fees:** $24. **Funding:** Membership, 98%; Foundation/Donor, 2%. **Annual Revenue:** $9,000,000. **Usage:** Administration, 7%; Fundraising, 27%; Programs, 66%. **Volunteer Programs:** Consumer education programs, community organizing.

COLORADO RIVER STUDIES OFFICE
P.O. Box 11568
Salt Lake City, UT 84147
(801) 524-5491
Fax: (801) 524-5499

Purpose: Preparation of environmental impact statement on Glen Canyon Dam and office overseeing related environmental studies. **Current Emphasis:** Glen Canyon Dam environmental impact statement and related scientific

A: *About 500*
Source: The Student Environmental Action Guide, *EarthWorks Press*

studies. **Members:** Not a membership organization. **Fees:** N/A. **Funding:** dnd. **Annual Revenue:** dnd. **Usage:** dnd. **Volunteer Programs:** None.

CONSERVATION INTERNATIONAL
1015 18th St. NW, Ste. 1000
Washington, DC 20036
(202) 429-5660
Fax: (202) 887-5188

Purpose: To conserve ecosystems and biological diversity and the ecological processes that support life on Earth. **Current Emphasis:** Works with partner organizations and local people in tropical and temperate countries, particularly the "megadiversity" countries containing over half of all species, to develop and implement ecosystem conservation projects. **Members:** 50,000. **Fees:** $25. **Funding:** Membership, 7%; Corporate, 4%; Foundation/Donor, 83%; Other, 6%. **Annual Revenue:** $9,821,536. **Usage:** Administration, 10%; Fundraising, 6%; Programs, 84%. **Volunteer Programs:** None.

THE COUSTEAU SOCIETY, INC.
870 Greenbrier Circle, Ste. 402
Chesapeake, VA 23320
(804) 523-9335
Fax: (804) 523-2747

Purpose: Dedicated to the protection and improvement of the quality of life. Founded in 1973 by Captain Jacques Cousteau and Jean-Michel Cousteau in the belief that an informed and alerted public can best make the choices to insure a healthy and productive world. The society produces television films, books, membership publications and articles, and offers lectures and a summer field study program. **Current Emphasis:** Worldwide petition campaign for the United Nations to adopt a Bill of Rights for Future Generations, acknowledging the right to an uncontaminated and undamaged Earth; rediscovery of the world filming and research expeditions. **Members:** Over 300,000 worldwide. **Fees:** Individual, $20; Family, $28. **Funding:** Nonprofit, membership-supported. **Annual Revenue:** $17,986,508. **Usage:** Administration, 8%; Fundraising, 12%; Programs, 73%; Other, 7%. **Volunteer Programs:** Chesapeake headquarters only.

DEFENDERS OF WILDLIFE
1244 19th St. NW
Washington, DC 20036
(202) 659-9510
Fax: (202) 833-3349

Purpose: A national nonprofit organization, Defenders utilizes public education, litigation and advocacy of progressive public policies aimed at protecting the diversity of wildlife and preserving the habitat critical to its survival. **Current Emphasis:** Specific projects include: restoring the gray wolf to its former range in Yellowstone National Park; preventing entanglement of marine mammals in plastic debris and discarded fishnets; working with congress to develop a bill to strengthen our National Wildlife Refuge system; and combating the trade of wild-caught birds. **Members:** 82,000. **Fees:** Individual, $20. **Funding:** dnd. **Annual Revenue:** $4,345,902. **Usage:** Administration, 13%; Programs, 64%; Membership, 10%; Other, 13%. **Volunteer Programs:** Defenders has an activist network consisting of more than 6,000 individuals. Volunteers are also welcome to assist staff at national headquarters or at any one of the four regional offices.

DUCKS UNLIMITED, INC.
One Waterfowl Way
Memphis, TN 38120
(901) 754-4666
Fax: (901) 753-2613

Purpose: To conserve and enhance wetland ecosystems throughout North America. **Current Emphasis:** dnd. **Members:** 510,000. **Fees:** $20. **Funding:** dnd. **Annual Revenue:** $63,000,000. **Usage:** Administration, 3.8%; Fundraising, 17.9%; Programs, 76.3%; Other, 2%. **Volunteer Programs:** Wetland habitat conservation and enhancement projects.

EARTH ISLAND INSTITUTE
300 Broadway, Ste. 28
San Francisco, CA 94133
(415) 788-3666
Fax: (415) 788-7324

Purpose: To develop innovative projects for the conservation, preservation and restoration of the global environment. **Current Emphasis:** *Earth Island Journal,* the International Marine Mammal Project, the Sea Turtle Restoration Project, Baikal Watch, Urban Habitat Program and International Green Circle, among others. **Members:** 35,000. **Fees:** Individual, $25; student, $15. **Funding:** Membership, 55%; Foundation/Donor, 35%; Other, 10%. **Annual Revenue:** $1,300,000. **Usage:** dnd. **Volunteer Programs:** Volunteer programs and internships available in most projects.

 What source of air pollution is responsible for one-third of all air pollutants, as well as being the fastest growing source of air pollution in the world?

ECOCONNECTIONS

EARTHWATCH
P.O. Box 403N
680 Mt. Auburn St.
Watertown, MA 02172
(800) 776-0188, (617) 926-8200
Fax: (617) 926-8532

Purpose: Sending volunteers to work with scientists around the world who are working to save rainforests and endangered species, preserve archaeological finds and study pollution effects. **Current Emphasis:** Working to research and create management plans to help alleviate crucial environmental problems. **Members:** 70,000. **Fees:** $25/year for Earthwatch membership. Includes six issues of *Earthwatch* magazine. **Funding:** Membership, 80%; Corporate, 10%; Foundation/Donor, 10%. **Annual Revenue:** dnd. **Usage:** Administration, 16%; Fundraising, 4%; Programs, 80%. **Volunteer Programs:** Work in 46 countries with scientists from around the world on projects ranging from two to three weeks. Volunteers contribute a share of the cost to the project, ranging from $800 to $2,200.

ENVIRONMENTAL ACTION
6930 Carroll Ave., Ste. 600
Takoma Park, MD 20912
(301) 891-1100
Fax: (301) 891-2218

Purpose: To protect our resources for present and future generations by encouraging pollution prevention and conservation of natural resources. Environmental Action believes an unpolluted environment is a fundamental human right. **Current Emphasis:** Includes solid waste disposal, toxic waste disposal, community right to know, lobbying and energy issues. **Members:** 20,000. **Fees:** $25/year. **Funding:** Membership, 40%; Foundation/Donor, 60%. **Annual Revenue:** $1,300,000. **Usage:** Administration, 3%; Fundraising, 17%; Programs, 80%. **Volunteer Programs:** In some areas. Internships are available.

ENVIRONMENTAL DATA RESEARCH INSTITUTE, INC.
797 Elmwood Ave.
Rochester, NY 14620
(716) 473-3090, (800) 724-1857
Fax: (716) 473-0968

Purpose: Established in 1989 to provide the environmental community with information on funding. EDRI maintains a large database on environmental grants and has recently published the first edition of *Environmental Grantmaking Foundations 1992.* This 490-page directory is an in-depth guide to 250 private and community foundations that give grants for environmental programs. In addition to contact information, profiles include history and philosophy, officers and directors, financial data, environmental program details, sample grants, application procedures, emphases and limitations. Indexes document foundation location, recipient location, activity location. A 59-page supplemental index includes an alphabetical index of 472 environmental topics (such as global warming and endangered species) and activities (such as education and advocacy); and an index to officers, trustees, directors and contacts. **Current Emphasis:** Analysis of funding by topic (biodiversity, energy, toxics, etc.), activity (research, advocacy, etc.), geographic region and scope. **Members:** dnd. **Fees:** dnd. **Funding:** Foundation/Donor, 100%. **Annual Revenue:** dnd. **Usage:** dnd. **Volunteer Programs:** Some internships are available.

ENVIRONMENTAL DEFENSE FUND
257 Park Ave. S.
New York, NY 10010
(212) 505-2100
Fax: (212) 505-2375

Purpose: To link science, economics and law to create innovative, economically viable solutions to today's environmental problems. **Current Emphasis:** Solid waste management, global climate change, tropical rainforest deforestation and toxin control. **Members:** 200,000. **Fees:** $20. **Funding:** Membership, 53%; Corporate, 1%; Other, 46%. **Annual Revenue:** $18,500,000. **Usage:** Administration, 3%; Fundraising, 13%; Programs, 82%; Other, 2%. **Volunteer Programs:** Summer internships are available in various departments in all six offices.

THE ENVIRONMENTAL EXCHANGE
1930 18th St. NW, Ste. 24
Washington, DC 20009
(202) 387-2182
Fax: (202) 588-9422

Purpose: To facilitate grassroots environmental projects through the exchange of information from organizations to the public about working environmental projects. **Current Emphasis:** Communicating effective initiatives in air pollution reduction, transportation alternatives, and toxic waste reduction. **Members:** dnd. **Fees:** dnd. **Funding:** dnd. **Annual Revenue:** $150,000. **Usage:** dnd. **Volunteer Programs:** Volunteer internships.

A: *The automobile*
Source: Greenpeace

ENVIRONMENTAL SUPPORT CENTER, INC.
1875 Connecticut Ave. NW, Ste. 340
Washington, DC 20009
(202) 328-7813
Fax: (202) 265-9419
Purpose: The center operates programs to strengthen regional, state, local and grassroots organizations working on environmental issues. ESC pays most of the cost of contracting with professionals to provide training and technical assistance to those groups in fundraising, organizational development and strategic planning. ESC obtains equipment and services for groups. **Current Emphasis:** In addition to the programs listed above, the center also helps groups make use of workplace solicitation as a fundraising tool. **Members:** dnd. **Fees:** dnd. **Funding:** Foundation/Donor, 100%. **Annual Revenue:** $650,000. **Usage:** dnd. **Volunteer Programs:** None.

FISH AND WILDLIFE REFERENCE SERVICE
5430 Grosvenor Ln., Ste. 110
Bethesda, MD 20814
(301) 492-6403
Fax: (301) 564-4059
Purpose: A computerized information retrieval system and clearinghouse providing fish and wildlife management research reports. **Current Emphasis:** Fish and wildlife management and protection of endangered species. **Members:** 9,000. **Fees:** Some user fees. **Funding:** dnd. **Annual Revenue:** $10,000. **Usage:** dnd. **Volunteer Programs:** Up to five nonpaid interns.

FRIENDS OF THE EARTH
218 D St. SE
Washington, DC 20003
(202) 544-2600
Fax: (202) 543-4710
Purpose: Global environmental advocates dedicated to the conservation, protection and rational use of the Earth. Engaged in lobbying in Washington, DC, and various state capitals and in disseminating public information on a wide variety of environmental issues. FOE publishes an award-winning newsletter, *Friends of the Earth*, and is affiliated with 47 other Friends of the Earth groups around the world. **Current Emphasis:** Ozone depletion, agricultural biotechnology, toxic chemical safety, groundwater protection, nuclear weapons production wastes, tropical deforestation and various international projects. **Members:** 50,000. **Fees:** Individuals, $25; Student/low-income/senior, $15. **Funding:** dnd. **Annual Revenue:** $2,800,000. **Usage:** dnd. **Volunteer Programs:** Internships available. Volunteer work also available.

THE FUND FOR ANIMALS
200 W. 57th St.
New York, NY 10019
(212) 246-2096
Fax: (212) 246-2633
Purpose: To oppose cruelty to animals—whether wild or domestic—wherever and whenever it occurs and to preserve biodiversity. **Current Emphasis:** To oppose all sport hunting, to limit the breeding of domestic animals and to preserve rare species and ecosystems. **Members:** 200,000. **Fees:** $20. **Funding:** Membership, 98%; Foundation/Donor, 2%. **Annual Revenue:** $2,000,000. **Usage:** Administration, 20%; Fundraising, 10%; Programs, 70%. **Volunteer Programs:** Washington, DC, office internships.

GRAND CANYON TRUST
1400 16th St. NW, #300
Washington, DC 20036
(202) 797-5429
Fax: (202) 797-5411
Purpose: To advocate the preservation and wise management of the natural resources of the Colorado Plateau and the Grand Canyon. **Current Emphasis:** Glen Canyon Dam operations, air pollution in the canyon and Utah wilderness designation. **Members:** 5,000 **Fees:** $25. **Funding:** Membership, 10%; Foundation/Donor, 48%; Other, 42%. **Annual Revenue:** $780,000. **Usage:** Administration, 6%; Fundraising, 9%; Programs, 85%; Membership, 10%. **Volunteer Programs:** Washington, DC, and Flagstaff, AZ, offices use volunteers.

GREAT OLD BROADS FOR WILDERNESS
P.O. Box 520307
Salt Lake City, UT 84152-0307
(801) 539-8208
Purpose: A nationwide group of women aged 45 and over who enjoy the wilderness and want to take an active part in preserving and protecting it; dedicated specifically to the growth and protection of the National Wilderness Preservation System. **Current Emphasis:** Supporting wilderness bills currently before congress and on the way to congress. **Members:** 800. **Fees:** $15. **Funding:** Membership, 100%. **Annual Revenue:** dnd. **Usage:** Programs, 100%. **Volunteer Programs:** None.

 In an independent study by Consumers Union, what percentage of US swordfish samples had mercury levels exceeding safety guidelines set by the FDA?

GREATER YELLOWSTONE COALITION
P.O. Box 1874
Bozeman, MT 59771
(406) 586-1593
Fax: (406) 586-0851

Purpose: To ensure the preservation and protection of the Greater Yellowstone Ecosystem and the quality of life it sustains. The Greater Yellowstone Ecosystem is one of the largest essentially intact ecosystems in the temperate zones of the Earth. **Current Emphasis:** Concerned with inappropriate oil and gas development on national forest lands, logging, mining, grazing and excess development. **Members:** 4,500+ individuals, 90 member organizations. **Fees:** Basic, $25; Patron, $500. **Funding:** Membership, 45%; Corporate, 5%; Foundation/Donor, 50%. **Annual Revenue:** $725,000. **Usage:** Administration, 14%; Fundraising, 15%; Programs, 71%. **Volunteer Programs:** Internships available.

GREENPEACE USA
1436 U St. NW
Washington, DC 20009
(202) 462-1177
Fax: (202) 462-4507

Purpose: Greenpeace is an international environmental organization dedicated to protecting the environment and all the life it supports. **Current Emphasis:** Greenpeace campaigns to free the Earth of nuclear and toxic pollution, protect marine ecology and end atmospheric destruction. The *Greenpeace* newsletter is published on a quarterly basis. **Members:** Supporters: 1.8 million in the US, over 4 million worldwide. **Fees:** A $20 donation is requested to receive the newsletter. **Funding:** Corporate, 1%; Foundation/Donor, 1%; Supporters, 98%. **Annual Revenue:** $50,000,000. **Usage:** Administration, 3%; Fundraising, 19%; Programs, 78%. **Volunteer Programs:** Volunteers and interns accepted in all regional offices.

HAWKWATCH INTERNATIONAL, INC.
P.O. Box 35706
Albuquerque, NM 87176-5706
(505) 255-7622
Fax: (505) 255-7832

Purpose: The conservation of birds of prey and their habitats in the western US through research and public education. Supports seven field projects to monitor trends and migration patterns of migratory raptors in the Rocky Mountain West. **Current Emphasis:** Standardized counts of migrating raptors at strategic observation points and large-scale capture and banding program. **Members:** 2,800. **Fees:** Individual, $20; Family, $30. **Funding:** Membership, 35%; Grants, 30%; Programs, 20%; Other, 15%. **Annual Revenue:** $243,000. **Usage:** Administration, 10%; Fundraising, 10%; Programs, 80%. **Volunteer Programs:** Spring and fall research and education internships. Volunteer banders, writers and outreach volunteers also needed.

HEARTWOOD
P.O. Box 402
Paoli, IN 47454
(812) 723-2430
Fax: (812) 723-2430

Purpose: Heartwood is an association of grassroots groups, individuals and businesses dedicated to the health and well-being of the native forests of the Central Hardwood region, and its plant and animal inhabitants, including the humans. The Central Hardwood region extends from the Appalachian Mountains to the Ozarks, and from the Great Lakes to the Tall Grass Prairie. We believe that the principal role of public lands is the conservation of functioning ecological systems and the wealth of genetic information they contain. **Current Emphasis:** Challenging public forest destruction through direct action and the appeals process; political organizing; and research and education. **Members:** Member groups, businesses and individuals in 19 states. **Fees:** dnd. **Funding:** dnd. **Annual Revenue:** Under $10,000. **Usage:** Programs, 100%. **Volunteer Programs:** A wide variety of volunteer opportunities available.

THE HUMANE SOCIETY OF THE UNITED STATES
2100 L St. NW
Washington, DC 20037
(202) 452-1100
Fax: (202) 778-6132

Purpose: To promote the humane treatment of animals and foster respect, understanding and compassion for all creatures. **Current Emphasis:** Reducing the overbreeding of dogs and cats and promoting responsible pet ownership; addressing critical environmental issues in terms of their impact on animals and humans; protecting endangered wildlife and marine mammals and their habitat; halting the cruelty of the international trade in wildlife, especially exotic birds and elephant ivory; promoting the use of non-animal alternatives for research, testing and experimentation; campaigning for or against federal, state and local local legislation that affects animal protection and monitoring its enforcement; working with

A: *40 percent*
Source: Time *magazine*

animal-control agencies and local humane societies to establish effective and humane programs; conducting workshops, symposia and seminars to train professionals and others in animal-related work. **Members:** 1,600,000. **Fees:** dnd. **Funding:** dnd. **Annual Revenue:** $15,142,844. **Usage:** dnd. **Volunteer Programs:** Contact your local organization for area activities.

INFORM
381 Park Ave. S.
New York, NY 10016
(212) 689-4040
Fax: (212) 447-0689

Purpose: Environmental research and education organization that identifies and reports on practical actions for the preservation and conservation of natural resources and public health. Current research focuses on such critical environmental issues as hazardous waste reduction, garbage management, urban air quality and land and water conservation. Approximately six reports published per year. **Current Emphasis:** Toxics in everyday products; waste reduction planning for municipalities; reducing and recycling business waste; and garbage incineration. **Members:** 1,000. **Fees:** From $25. **Funding:** Membership, 22%; Corporate, 18%; Other, 60%. **Annual Revenue:** $1,500,000. **Usage:** Research and education, 100%. **Volunteer Programs:** Occasional availability in clerical and communications.

INSTITUTE FOR CONSERVATION LEADERSHIP
2000 P St. NW, Ste. 413
Washington, DC 20036
(202) 466-3330
Fax: (202) 659-3897

Purpose: To serve the entire conservation/environmental community with leadership training and organizational development programs. To help build volunteer involvement, increase organizational leadership, help establish state networks and improve individual leadership skills and abilities. Our goal is to increase the number and effectiveness of volunteer organizations and leaders in the entire community. **Current Emphasis:** Week-long, individual training sessions, state networking conferences, board-of-directors training and long-range planning facilitation. **Members:** None. **Fees:** Vary. **Funding:** Foundation/Donor, 70%; Other, 30%. **Annual Revenue:** $270,000. **Usage:** Administration, 30%; Fundraising, 10%; Programs, 60%. **Volunteer Programs:** None.

INSTITUTE FOR EARTH EDUCATION
Cedar Cove
Greenville, WV 24945
(304) 832-6404

Purpose: IEE develops and disseminates educational programs that help people build an understanding of, appreciation for and harmony with the Earth and its life. Through its worldwide network, the institute conducts workshops, provides a seasonal journal, hosts international and regional conferences, supports local groups, distributes an annual catalog and publishes books and program materials. **Current Emphasis:** Earth Education program development and support for teachers and leaders. **Members:** 2,000+. **Fees:** Personal, $25; Professional, $35; Affiliate, $50; Sponsor, $100. **Funding:** dnd. **Annual Revenue:** $300,000+. **Usage:** dnd. **Volunteer Programs:** Available through international sharing centers.

INTERNATIONAL ALLIANCE FOR SUSTAINABLE AGRICULTURE
1701 University Ave. SE
Minneapolis, MN 55414
(612) 331-1099
Fax: (612) 379-1527

Purpose: IASA works for the worldwide realization of sustainable agriculture—food systems that are ecologically sound, economically viable, socially just and humane. The alliance focuses on three goals: 1) building a strong sustainable agriculture industry and movement; 2) widespread understanding of and participation in sustainable agriculture; 3) universal adoption of policies that implement sustainable agriculture. **Current Emphasis:** 1993 International Sustainable Agriculture Conference to be held at the University of Minnesota; publication of a national directory on humane sustainable agriculture; and updating *Planting a Future: A Resource Guide to Sustainable Agriculture in the Third World.* **Members:** 800. **Fees:** $10-$1,000. **Funding:** Membership, 6%; Corporate, 23%; Foundation/Donor, 71%. **Annual Revenue:** $187,000. **Usage:** Administration, 13%; Fundraising, 2%; Programs, 85%. **Volunteer Programs:** Opportunities are available in every area of current emphasis described above and through formal committee and volunteer structure. Call for more information.

 How much animal waste do US livestock industry cattle produce each year?

INTERNATIONAL FUND FOR ANIMAL WELFARE
411 Main St.
Yarmouth Port, MA 02675
(508) 362-4944
Fax: (508) 362-5841
Purpose: An international animal welfare organization dedicated to protecting wild and domestic animals from cruelty. **Current Emphasis:** Preservation of harp and hood seals in Canada; dog and cat abuse in the Philippines and South Korea; elephants in Africa; the use of animals in laboratory testing for the cosmetics industry; and whales and other marine mammals around the world. **Members:** 1,000,000 worldwide. **Fees:** dnd. **Funding:** Donors. **Annual Revenue:** $4,916,491. **Usage:** Expenses equal 108% of donations: Administration, 18%; Fundraising, 17%; Programs, 73%. **Volunteer Programs:** Pilot whale stranding network on Cape Cod for residents of Massachusetts.

INTERNATIONAL PRIMATE PROTECTION LEAGUE
P.O. Box 766
Summerville, SC 29484
(803) 871-2280
Fax: (803) 871-7988
Purpose: Dedicated to the conservation and protection of apes, monkeys and prosimians, maintenance of a gibbon sanctuary and support of overseas projects. Includes a quarterly newsletter, *The IPPL News*. **Current Emphasis:** Uncovering illegal trafficking in primates, and support of primate sanctuaries overseas. **Members:** 12,000. **Fees:** $20. **Funding:** Membership, 75%; Corporate, 1%; Other, 24%. **Annual Revenue:** $250,000. **Usage:** Administration, 20%; Fundraising, 10%; Programs, 70%. **Volunteer Programs:** None.

IZAAK WALTON LEAGUE OF AMERICA
1401 Wilson Blvd., Level B
Arlington, VA 22209
(703) 528-1818
Fax: (703) 528-1836
Purpose: Established in 1922, the Izaak Walton League of America is a national conservation organization whose 56,000 members work to protect and enjoy America's soil, air, woods, waters and wildlife. **Current Emphasis:** Clean water, energy efficiency, clean air, wildlife habitat protection, improved public lands management and outdoor ethics. **Members:** 56,000. **Fees:** $20 individual. **Funding:** Membership, 36%; Corporate, 22%; Foundation/Donor, 36%; Other, 6%. **Annual Revenue:** $2,000,000. **Usage:** Administration, 11%; Fundraising, 13%; Programs, 76%. **Volunteer Programs:** Editorial internships, Save Our Streams internship.

THE JANE GOODALL INSTITUTE
P.O. Box 41720
Tucson, AZ 85717
(602) 325-1211, (800) 999-CHIMP
Fax: (602) 325-0020
Purpose: Ongoing support and expansion of field research on wild chimpanzees and studies of chimpanzees in captive environments. The institute is dedicated to publicizing the unique status and needs of chimpanzees to ensure their preservation in the wild and their physical and psychological well-being in captivity. **Current Emphasis:** Field research activities at Gombe Stream Research Centre in Tanzania; the ChimpanZoo study of captive chimpanzees in zoos or other captive colonies in the US, and conservation activities targeting wild and captive chimpanzees, including those in biomedical research laboratories. **Members:** 4,000. **Fees:** From $30. **Funding:** Membership, 20%; Corporate, 2%; Lecture tour, 39%; Other, 39%. **Annual Revenue:** $599,800. **Usage:** Administration, 15%; Fundraising, 15%; Programs, 70%. **Volunteer Programs:** University of Southern California Goodall Fellowship.

THE LAND AND WATER FUND OF THE ROCKIES
2260 Baseline, Ste. 200
Boulder, CO 80302
(303) 444-1188
Fax: (303) 786-8054
Purpose: "Legal Aid for the Environment," to provide free legal aid to grassroots environmental groups in Arizona, Colorado, Idaho, Montana, New Mexico, Utah and Wyoming. LAW Fund staff attorneys and a regional network of local volunteer attorneys supply advice and counsel, and will litigate for client groups. Founded in 1990. **Current Emphasis:** Public lands, water and toxics, energy efficiency. **Members:** 250. **Fees:** Regular, $25; Student/senior/limited-income, $15; Organization, $100; Special, $50-$1,000. **Funding:** Membership, 3%; Foundation/Donor, 96%; Other, 1%. **Annual Revenue:** $450,000. **Usage:** Administration, 25%; Fundraising, 1%; Programs, 74%. **Volunteer Programs:** Pro Bono Attorney Program, "Adopt-A-Forest," opportunities for technical experts.

A: *158 million tons*
Source: Worldwatch *magazine*

LAND TRUST ALLIANCE
900 17th St. NW, Ste. 410
Washington, DC 20006-2596
(202) 785-1410
Fax: (202) 785-1408
Purpose: A national organization of local and regional land conservation groups that provides programs and services to help land trusts reach their full potential, fosters public policies supportive of land conservation, and builds public awareness of land trusts and their goals. **Current Emphasis:** Providing educational materials and technical assistance for land trusts and other land conservation professionals. **Members:** 925. **Fees:** From $30. **Funding:** dnd. **Annual Revenue:** $865,000. **Usage:** Administration, 23%; Fundraising, 3%; Programs, 74%. **Volunteer Programs:** LTA can put individuals in touch with land trusts across the country that seek volunteer assistance.

LEAGUE OF CONSERVATION VOTERS
1707 L St., NW, Ste. 550
Washington, DC 20036
(202) 785-8683
Fax: (202) 835-0491
Purpose: The League of Conservation Voters is the 21-year-old national, nonpartisan, political arm of the environmental community. The league's goal is to change the balance of power in the US Congress to reflect the pro-environmental concerns of the American public. **Current Emphasis:** Endorsing and supporting candidates for election to the US House and Senate. **Members:** 50,000. **Fees:** $25. **Funding:** Membership, 35%; Foundation/Donor, 65%. **Annual Revenue:** $2,000,000. **Usage:** Administration, 4%; Fundraising, 21%; Programs, 75%. **Volunteer Programs:** Six-month paid internship focusing on political/candidate research.

LEAGUE TO SAVE LAKE TAHOE
989 Tahoe Keys Blvd., Ste. 6
S. Lake Tahoe, CA 96150
(916) 541-5388
Fax: (916) 541-5454
Purpose: Dedicated to preserving the environmental balance, scenic beauty and recreational opportunities of the Lake Tahoe Basin. Subsidiary organizations include League to Save Lake Tahoe Charitable Trust. **Current Emphasis:** Reversing the water- and air-quality decline at Lake Tahoe. **Members:** 4,000. **Fees:** $35. **Funding:** dnd. **Annual Revenue:** $250,000. **Usage:** Administration, 40%; Programs, 60%. **Volunteer Programs:** None.

LIGHTHAWK
P.O. Box 8163
Santa Fe, NM 87504
(505) 982-9656
Fax: (505) 984-8381
Purpose: To use and encourage the advantages of flight to shed light on and correct environmental mismanagement and empower others to do the same. The goal is to greatly enhance humankind's capacity to sustain biological diversity, intact ecosystems and ecological processes that support life on Earth. **Current Emphasis:** Working to protect America's national forest system, particularly the last vestiges of our once-vast Pacific Northwest rainforests. **Members:** 4,000. **Fees:** $35, $100 and up. **Funding:** dnd. **Annual Revenue:** $1,000,000. **Usage:** Administration, 10%; Fundraising, 2%; Programs, 88%. **Volunteer Programs:** Volunteer aircraft owner/pilots with at least 1,000 hours flight time in their own aircraft needed.

MANOMET BIRD OBSERVATORY
P.O. Box 1770
Manomet, MA 02345
(508) 224-6521
Fax: (508) 224-9220
Purpose: Nonprofit, membership-supported environmental research and education institute. MBO is dedicated to promoting informed conservation policy through long-term research on natural systems throughout the Americas. MBO also conducts a college-accredited Field Biology Training Program for hands-on experience in field research; undergraduate and beginning graduate students can apply. The education program develops elementary and secondary school environmental curricula. **Current Emphasis:** Fisheries conservation and management issues in the northwest Atlantic; tropical and deciduous forest ecology and management; international shorebird migration and identification of critical wetlands habitats; estuarine biomonitoring in New York and Massachusetts with management implications; long-term monitoring, ecology and conservation of migrant songbird populations; endangered avian species biology and conservation. **Members:** 2,000. **Fees:** Regular, $25; Student and Seniors, $15. **Funding:** dnd. **Annual Revenue:** $1,500,000. **Usage:** Research/education programs, 100%. **Volunteer Programs:** Opportunities available in all programs.

 : *How many people are born in the US each hour?*

ECOCONNECTIONS

NATIONAL ARBOR DAY FOUNDATION
211 N. 12th St., Ste. 501
Lincoln, NE 68508
(402) 474-5655
Fax: (402) 474-0820
Purpose: An education organization dedicated to tree planting and environmental stewardship. **Current Emphasis:** Programs such as Trees for America, Tree City USA, Conservation Trees, Celebrate Arbor Day and the National Arbor Day Complex. **Members:** 1,000,000. **Fees:** $10. **Funding:** Membership, 57%; Other, 43%. **Annual Revenue:** $16,110,483. **Usage:** Administration, 1%; Fundraising, 12%; Programs, 87%. **Volunteer Programs:** None.

NATIONAL ASSOCIATION OF BIOLOGY TEACHERS
11250 Roger Bacon Dr., #19
Reston, VA 22090
(703) 471-1134
Fax: (703) 435-5582
Purpose: Dedicated exclusively to the concerns of biology teachers. Publishes *The American Biology Teacher,* a nationally recognized journal that highlights research findings, innovative teaching strategies, laboratory exercises and reviews of publications, computer programs and videos. **Current Emphasis:** Projects under way include middle school teacher training, biotechnology labs and equipment loan programs, alternative use of animals in the classroom and elementary education environmental curriculum. **Members:** 7,000. **Fees:** $38. **Funding:** Membership, 40%; Other, 60%. **Annual Revenue:** $700,000. **Usage:** Administration, 33%; Fundraising, 1%; Programs, 66%. **Volunteer Programs:** None.

NATIONAL ASSOCIATION OF CONSERVATION DISTRICTS
P.O. Box 855
League City, TX 77574-0855
(800) 825-5547
Fax: (718) 332-5259
Purpose: The mission of NACD is to advance the interests of the nation's conservation districts and provide needed services to further the conservation, management and orderly development of natural resources. **Current Emphasis:** Conservation, management and orderly development of America's natural resources. **Members:** 4,000. **Fees:** $10 to $100. **Funding:** dnd. **Annual Revenue:** dnd. **Usage:** dnd. **Volunteer Programs:** None.

NATIONAL ASSOCIATION OF INTERPRETATION
P.O. Box 1892
Fort Collins, CO 80522
(303) 491-6434
Fax: (303) 491-2255
Purpose: A professional organization serving the needs and interests of interpreters employed by agencies and organizations concerned with natural and cultural resources, conservation and management. **Current Emphasis:** Representing all those whose job it is to convey the meanings and relationships between people and their natural, cultural and recreational world. **Members:** 2,800. **Fees:** Student, $25, individual, $40. **Funding:** Membership, 90%; Corporate, 10%. **Annual Revenue:** $132,000. **Usage:** Administration, 38%; Programs, 62%. **Volunteer Programs:** dnd.

NATIONAL AUDUBON SOCIETY
700 Broadway
New York, NY 10003
(212) 979-3000
Purpose: National Audubon Society is a grassroots environmental organization dedicated to protecting wildlife and its habitats. Audubon's 600,000 members and staff of scientists, lobbyists, lawyers, policy analysts and educators work through state and regional offices and participate in policy research, lobbying, litigation and citizen action to protect and restore various habitats throughout the Americas. **Current Emphasis:** Ancient Forests, Wetlands, Endangered Species, Arctic National Wildlife Refuge, Platte River, Everglades and the Adirondack Park. **Members:** 600,000 adult and family chapter members. 550,000 students enrolled in the Audubon Adventures program in 17,500 classrooms around the country. **Fees:** $20. **Funding:** Membership, 35%; Bequests/Contributions, 37%; Other, 28%. **Annual Revenue:** $37,000,000. **Usage:** Administration, 8%; Fundraising, 9%; Programs, 72%; Membership, 11%. **Volunteer Programs:** Annual Christmas Bird Count and Breeding Bird Census; Audubon Activist Network and involvement through chapters, state and regional offices.

NATIONAL AUDUBON SOCIETY EXPEDITION INSTITUTE
P.O. Box 365
Belfast, ME 04915
(207) 338-5859
Fax: (207) 338-1037

A: *480*
Source: Children's Television Workshop

Purpose: Graduate, undergraduate and high school education program that offers year-long and semester expeditions providing an alternative to traditional education and emphasizing environmental education. Exciting list of courses offered. **Current Emphasis:** Varies. **Members:** 80-100 per year. **Fees:** Semester, $7,038; year, $12,000. **Funding:** Membership, 90%; Foundation/Donor, 10%. **Annual Revenue:** dnd. **Usage:** dnd. **Volunteer Programs:** None.

NATIONAL COALITION AGAINST THE MISUSE OF PESTICIDES
701 E St. SE, Ste. 200
Washington, DC 20003
(202) 543-5450

Purpose: To serve as a national network committed to pesticide safety and the adoption of alternative pest management strategies that reduce or eliminate dependency on toxic chemicals. **Current Emphasis:** To effect change through local action, assisting individuals and community-based organizations in this endeavor. **Members:** dnd. **Fees:** $25. **Funding:** Membership, 33%; Grants, 67%. **Annual Revenue:** dnd. **Usage:** dnd. **Volunteer Programs:** Internships available.

NATIONAL TOXICS CAMPAIGN
1168 Commonwealth Ave.
Boston, MA 02134
(617) 232-0327
Fax: (617) 232-3945

Purpose: Preventing pollution, protecting public health, guidance for communities, laboratory testing for air and water pollution, pressuring corporations to clean up and manufacture safer products. **Current Emphasis:** dnd. **Members:** 75,000. **Fees:** $25-$36. **Funding:** dnd. **Annual Revenue:** dnd. **Usage:** dnd. **Volunteer Programs:** Yes.

NATIONAL WILDFLOWER RESEARCH CENTER
2600 FM 973 N.
Austin, TX 78725-4201
(512) 929-3600
Fax: (512) 929-0513

Purpose: To promote the preservation and reestablishment of native wildflowers, grasses, shrubs and trees in North America. **Current Emphasis:** To encourage ecological stability through public education on the use of native plants in landscaping, and incorporating native flora in the repair of the environment. **Members:** 16,500. **Fees:** From $25. **Funding:** dnd. **Annual Revenue:** dnd. **Usage:** dnd. **Volunteer Programs:** Active volunteer program as well as public relations and marketing internships through the University of Texas.

NATIONAL WILDLIFE FEDERATION
1400 16th St. NW
Washington, DC 20036
(202) 797-6800
Fax: (202) 797-6646

Purpose: To be an effective conservation education organization promoting the responsible use of natural resources and protection of the global environment. The federation distributes periodicals and education materials, sponsors outdoor nature programs, lobbies congress and litigates environmental disputes in an effort to conserve fisheries, wildlife and natural resources. **Current Emphasis:** Endangered species and forest protection, wetlands and water resource conservation, grazing and mining reform, biotechnology and toxic pollution sunsetting. **Members:** 5,300,000. **Fees:** $15-$20. **Funding:** Membership, 47%; Foundation/Donor, 14%; Educational materials, 39%. **Annual Revenue:** $92,000,000. **Usage:** Programs, 68%; Membership, 21%; Administration/Fundraising, 11%. **Volunteer Programs:** The National Wildlife Federation has affiliate organizations in 51 states and the US Virgin Islands.

NATIVE FOREST COUNCIL
P.O. Box 2171
Eugene, OR 97402
(503) 688-2600
Fax: (503) 461-2156

Purpose: To protect, preserve and restore America's forests. **Current Emphasis:** National education, citizen action and accountability and media. **Members:** 5,000. **Fees:** $25 US, $50 non-US. **Funding:** Membership, 35%; Corporate, 5%; Foundation/Donor, 60%. **Annual Revenue:** $300,000. **Usage:** Administration, 9%; Fundraising, 7%; Programs, 84%. **Volunteer Programs:** Journalism, advertising, business, public relations, environment and computer science.

NATURAL RESOURCES DEFENSE COUNCIL
40 W. 20th St.
New York, NY 10011
(212) 727-2700
Fax: (212) 727-1773

Purpose: To protect America's natural resources and improve the quality of the human environment. NRDC combines legal action, scientific research and citizen education

---

 Which country has only 10 percent of its natural coral reefs remaining, due to extensive dynamiting, polluting and collecting?

in an environmental protection program. **Current Emphasis:** Major accomplishments have been in the area of energy policy and nuclear safety, air and water pollution, urban transportation issues, pesticides and toxic substances, forest protection, global warming and the international environment. **Members:** 170,000. **Fees:** $10. **Funding:** Membership, 37%; Foundation/Donor, 31%; Major gifts, 22%; Other, 10%. **Annual Revenue:** $16,000,000. **Usage:** Administration, 9%; Fundraising, 13%; Programs, 78%. **Volunteer Programs:** Legal internships as well as internships for graduate and undergraduate college students in several offices.

THE NATURE CONSERVANCY
1815 N. Lynn St.
Arlington, VA 22209
(703) 841-5300

Purpose: International organization whose mission is to preserve plants, animals and natural communities that represent the diversity of life on Earth by protecting the lands and waters they need to survive. Manages a system of more than 1,300 nature sanctuaries in all 50 states. Assists in-country nongovernmental organizations to do the same throughout Latin America and the Pacific. **Current Emphasis:** "Last Great Places" initiative is establishing ecosystem conservation models to demonstrate that large-scale biodiversity protection can also accommodate human economic and cultural needs. **Members:** 680,000. **Fees:** $25. **Funding:** Membership, 71.8%; Corporate, 12%; Foundation/Donor, 16.2%. **Annual Revenue:** $172,990,000. **Usage:** Administration, 6.6%; Fundraising, 3.9%; Programs, 85%; Membership, 4.5%. **Volunteer Programs:** Positions available through state chapters.

NORTH AMERICAN NATIVE FISHES ASSOCIATION
123 W. Mount Airy Ave.
Philadelphia, PA 19119
(215) 247-0384

Purpose: To bring together people interested in fishes native to this continent for scientific purposes or aquarium study; to encourage increased scientific appreciation and conservation of native fishes through observation, study and research; to assemble and distribute information about native fishes. **Current Emphasis:** dnd. **Members:** 400. **Fees:** $11-$15. **Funding:** dnd. **Annual Revenue:** dnd. **Usage:** dnd. **Volunteer Programs:** None.

NORTHERN ALASKA ENVIRONMENTAL CENTER
218 Driveway
Fairbanks, AK 99701
(907) 452-5021
Fax: (907) 452-3100

Purpose: Dedicated to the protection of the quality of the Alaskan environment through action and education. NAEC covers areas north of the Alaska Range and works closely with government agencies on land-use issues such as the Arctic National Wildlife Refuge, implementation of the Alaska Lands Act, oil and gas leasing and placer gold-mining. **Current Emphasis:** Wilderness designation for the coastal plain of Arctic National Wildlife Refuge and arctic development issues. **Members:** 750. **Fees:** Individual, $25; Family, $35. **Funding:** Membership, 55%; Grants, 30%; Events, 15%. **Annual Revenue:** $90,000. **Usage:** Administration, 17%; Fundraising, 11%; Programs, 72%. **Volunteer Programs:** Opportunities for research internships and a wide variety of volunteer programs.

NUCLEAR INFORMATION AND RESOURCE SERVICE
1424 16th St. NW, Ste. 601
Washington, DC 20036
(202) 328-0002
Fax: (202) 462-2183

Purpose: To serve as a networking and information clearinghouse for environmental activists concerned with nuclear power and waste issues; to provide citizens with the information and tools necessary to challenge nuclear facilities and policies; to work for increased energy efficiency and toward a sustainable, renewable energy future. **Current Emphasis:** Challenging "low-level" radioactive waste policy; publication of energy audit manual for towns and universities; working to prevent a new generation of nuclear reactors. **Members:** 1,200. **Fees:** Any-size contribution; $35/year for biweekly newsletter. **Funding:** Membership, 20%; Foundation/Donor, 80%. **Annual Revenue:** $300,000. **Usage:** Administration, 12%; Fundraising, 5%; Programs, 83%. **Volunteer Programs:** Intern applications accepted year-round and $100/week stipend offered.

POPULATION CRISIS COMMITTEE
1120 19th St. NW, #550
Washington, DC 20036
(202) 659-1833
Fax: (202) 293-1795

A: *The Philippines*
Source: Newsweek magazine

Purpose: To stimulate public awareness, understanding and action toward reducing population growth rates. Advocates universal, voluntary access to family planning services to achieve world population stabilization. **Current Emphasis:** Relationship between population growth and environmental degradation, contraceptive availability. **Members:** dnd. **Fees:** dnd. **Funding:** dnd. **Annual Revenue:** $4,000,000. **Usage:** Administration, 9%; Fundraising, 5%; Programs, 86%. **Volunteer Programs:** None.

POPULATION-ENVIRONMENT BALANCE
1325 G St. NW, Ste. 1003
Washington, DC 20005-3104
(202) 879-3000
Fax: (202) 879-3019

Purpose: A national, nonprofit membership organization dedicated to education and advocacy of measures that would encourage population stabilization in the US, in order to safeguard our environment. **Current Emphasis:** Environmental protection, birth control availability/research, local growth control and immigration policy. **Members:** 5,000. **Fees:** $25. **Funding:** Membership, 35%; Foundation/Donor, 65%. **Annual Revenue:** $650,000. **Usage:** Administration, 25%; Fundraising, 10%; Programs, 65%. **Volunteer Programs:** Volunteer positions regularly available.

PROTECT OUR WOODS
P.O. Box 352
Paoli, IN 47454
(812) 678-4303

Purpose: Protect Our Woods comprises groups of landowners and concerned individuals working toward preserving and protecting the forests of Indiana and the wild and rural areas of the state. **Current Emphasis:** Improving forest management on both public and private lands, protecting rivers and resisting development of rural Indiana. **Members:** 800. **Fees:** Individual, $15; Family/Household, $25; Woodland Owner, $35. **Funding:** Membership, 100%. **Annual Revenue:** $12,000. **Usage:** Administration, 5-10%; Programs, 90-95%. **Volunteer Programs:** Volunteer positions are available.

RAILS-TO-TRAILS CONSERVANCY
1400 16th St. NW, Ste. 300
Washington, DC 20036
(202) 797-5400
Fax: (202) 797-5411

Purpose: Converting thousands of miles of abandoned railroad corridors to public trails for walking, bicycling, horseback riding, cross-country skiing, wildlife habitat and nature appreciation. **Current Emphasis:** Linking major metropolitan areas via rail-trails and established greenways. **Members:** 50,000. **Fees:** $18. **Funding:** Membership, 42.4%; Grants, 14%; Contributions, 30.5%; Other, 13.1%. **Annual Revenue:** $2,023,949. **Usage:** Administration, 7.7%; Fundraising, 11.1%; Programs, 40.4%; Membership, 27.5%; Public information, 13.3%. **Volunteer Programs:** Six-month paid internships are available at the national office in Washington, DC, and volunteers can serve at chapter offices and on specific projects.

RAINFOREST ACTION NETWORK
450 Sansome St., Ste. 700
San Francisco, CA 94111
(415) 398-4404
Fax: (415) 398-2732

Purpose: Nonprofit activist organization working nationally and internationally to save the world's rainforests. Works internationally in cooperation with other environmental and human rights organizations on major campaigns to protect rainforests. RAN works with over 150 rainforest action groups nationwide. **Current Emphasis:** Protecting rainforests in Hawaii, Amazonia and Southeast Asia. **Members:** 31,000. **Fees:** $15 minimum. **Funding:** Membership, 70%; Foundation/Donor, 10%; Special projects, 20%. **Annual Revenue:** dnd. **Usage:** Administration, 15%; Fundraising, 10%; Programs, 75%. **Volunteer Programs:** Internship program is available and volunteers are needed in all locations.

RAINFOREST ALLIANCE
270 Lafayette St., Ste. 512
New York, NY 10012
(212) 941-1900
Fax: (212) 941-4986

Purpose: Dedicated to the conservation of the world's tropical forests, the Rainforest Alliance aims primarily to develop and promote sound alternatives to the activities that cause tropical deforestation—opportunities for people to utilize tropical forests without destroying them. Also involved in public education and building new constituencies for conservation. **Current Emphasis:** Timber project (including "Smart Wood" certification); Periwinkle Project (medicinal plant information); Edelstein Fellowship for Medicinal Plant Research in Brazil; Committee for Conservation and Higher Education; Tropical Conservation Media Center; Kleinhans Fellowship for Non-Timber Forest Products,

 Nuclear power provides what percentage of US energy?

Banana Project, Amazon Rivers Conservation Project. **Members:** 15,000. **Fees:** $25. **Funding:** Membership, 20%; Corporate, 15%; Foundation/Donor, 65%. **Annual Revenue:** $1,300,000. **Usage:** Administration, 9%; Fundraising, 11%; Programs, 80%. **Volunteer Programs:** Variety of programs.

RARE CENTER FOR TROPICAL BIRD CONSERVATION
1529 Walnut St., Third Fl.
Philadelphia, PA 19102
(215) 568-0420
Fax: (215) 568-0516

Purpose: RARE Center is a small organization doing innovative work to preserve threatened habitats and ecosystems in Latin America and the Caribbean. RARE focuses on endangered birds because of their value as environmental indicators and rallying points for conservation initiatives. **Current Emphasis:** Conservation education and applied research. **Members:** 1,000. **Fees:** $30. **Funding:** Membership, 40%; Foundation/Donor, 60%. **Annual Revenue:** $300,000. **Usage:** Administration, 20%; Fundraising, 15%; Programs, 65%. **Volunteer Programs:** Occasionally.

REEF RELIEF
P.O. Box 430
Key West, FL 33041
(305) 294-3100
Fax: (305) 293-9515

Purpose: To preserve and protect the living coral reef of the Florida Keys. **Current Emphasis:** Regional public education and outreach program on why and how to protect the coral reef, maintenence of over 100 reef mooring buoys at seven Key West-area reefs. **Members:** 1,800. **Fees:** $20/year. **Funding:** Membership, 29%; Foundation/Donor, 9%; Fundraising, 45%; Other, 17%. **Annual Revenue:** $150,000. **Usage:** Fundraising, 9%; Programs, 60%; Membership, 16%; Rent and utilities, 15%. **Volunteer Programs:** Daily, weekly, monthly, summer and semester.

THE RENE DUBOS CENTER FOR HUMAN ENVIRONMENTS, INC.
100 E. 85th St.
New York, NY 10028
(212) 249-7745

Purpose: An independent education and research organization founded by the eminent scientist/humanist Rene Dubos to focus on the humanistic and social aspects of environmental problems. The center's mission is to develop creative policies for the resolution of environmental conflicts and to help decision-makers and the general public formulate new environmental values. **Current Emphasis:** Forums, publications, multimedia computer applications and other related activities to increase environmental literacy in schools, the work place and the community. **Members:** dnd. **Fees:** dnd. **Funding:** dnd. **Annual Revenue:** dnd. **Usage:** dnd. **Volunteer Programs:** dnd.

RENEW AMERICA
1400 16th St. NW, Ste. 710
Washington, DC 20036
(202) 232-2252

Purpose: Committed to renewing America's community spirit through environmental success. **Current Emphasis:** Identification, verification, and promotion of successful environmental programs in 20 categories ranging from air pollution reduction to wildlife conservation. These programs (more than 1,600 in 1992) are published in the annual *Environmental Success Index,* distributed nationwide to policymakers, business leaders, industry, government and media. **Members:** 6,000. **Fees:** $25. **Funding:** Membership, 22%; Foundation/Donor, 73%; Publication, 5%. **Annual Revenue:** $600,000. **Usage:** Administration, 23%; Programs, 44%; Membership, 22%; Reports, 11%. **Volunteer Programs:** Internship positions available.

RHINO RESCUE USA, INC.
1150 17th St. NW, Ste. 400
Washington, DC 20036
(202) 293-5305
Fax: (202) 223-0346

Purpose: To save the rhinoceros from extinction by funding rhino sanctuaries and research into rhino conservation and working to end the illegal trade of rhino horn. **Current Emphasis:** Funding rhinoceros sanctuaries and the research needed to effectively manage and increase remaining rhino populations. **Members:** dnd. **Fees:** dnd. **Funding:** Individual donors and foundations, 100%. **Annual Revenue:** $22,000. **Usage:** Administration, 15%; Programs, 85%. **Volunteer Programs:** None.

ROCKY MOUNTAIN RECYCLERS ASSOCIATION
P.O. Box 224
Denver, CO 80214-1896
(303) 441-9445

Purpose: To improve economic conditions for the Rocky Mountain recycling industry; to organize this industry to achieve economic

A: *About 6 percent*
Source: Nuclear Information and Resource Service

policy reform; and to assist groups and individuals with recycling concerns. **Current Emphasis:** All of the above. **Members:** 10. **Fees:** $18-$500. **Funding:** Membership, 25%; Corporate, 25%; Foundation/Donor, 50%. **Annual Revenue:** $70,000. **Usage:** dnd. **Volunteer Programs:** Volunteer positions are available.

THE RUFFED GROUSE SOCIETY
451 McCormick Rd.
Coraopolis, PA 15108
(412) 262-4044
Fax: (412) 262-9207

Purpose: Dedicated to improving the environment for ruffed grouse, American woodcock and other forest wildlife. **Current Emphasis:** Direct assistance in cooperation with land managers in creating and improving young-forest habitat on public lands. **Members:** 25,000. **Fees:** $20. **Funding:** Membership, 55%; Banquet, 32%; Other, 13%. **Annual Revenue:** $2,108,000. **Usage:** Administration, 33%; Programs, 67%. **Volunteer Programs:** Assisting with local banquets and chapters.

THE SACRED EARTH NETWORK
426 Sixth Ave.
Brooklyn, NY 11215
(718) 768-8569
Fax: (718) 768-2858

Purpose: The Sacred Earth Network seeks to reacquaint people with their fundamental connection with the natural world through any nonviolent means necessary. To enhance ties with former Soviet environmentalists through the establishment of an electronic mail network, translating "cutting edge" solutions to environmental problems from English to Russian and Russian to English. **Current Emphasis:** To continue more translations, workshops and seminars in deep ecology. Protecting and preserving the forests of the Siberian wilderness. **Members:** 1,100. **Fees:** $25. **Funding:** Membership, 50%; Foundation/Donor, 50%. **Annual Revenue:** $71,000. **Usage:** Administration, 15%; Programs, 85%. **Volunteer Programs:** No volunteer programs have been established yet.

SAVE THE DUNES COUNCIL, INC.
444 Barker Rd.
Michigan City, IN 46360
(219) 879-3937

Purpose: Preservation and protection of the Indiana Dunes for public use and enjoyment by working for the control of air, water and waste pollution affecting the National Lakeshore and northwest Indiana area. Involved in shoreline erosion and shoreline policy issues affecting the Indiana Lake Michigan shoreline, wetlands preservation and groundwater protection. **Current Emphasis:** Water and air pollution, wetlands, shoreline management, parkland purchase and planning and development issues affecting the Indiana Dunes National Lakeshore. **Members:** 1,800. **Fees:** Senior and student, $15; Individual, $20; Couple, $35; Life, $500. **Funding:** Membership, 80%; Events, 20%. **Annual Revenue:** $25,000. **Usage:** Fundraising, 1%; Programs, 99%. **Volunteer Programs:** Volunteers operate the Dunes Shop, whose revenue supports the council.

SAVE-THE-REDWOODS LEAGUE
114 Sansome St., Room 605
San Francisco, CA 94104
(415) 362-2352
Fax: (415) 362-7017

Purpose: To purchase redwood groves and watershed lands for protection in public parks; to support reforestation, research and educational programs. **Current Emphasis:** All of the above. **Members:** 50,000. **Fees:** $10. **Funding:** Membership, 100%. **Annual Revenue:** $2,000,000. **Usage:** Programs, 100%. **Volunteer Programs:** None.

SAVE THE WHALES, INC.
P.O. Box 2397
Venice, CA 90291
(310) 392-6226
Fax: (310) 392-8968

Purpose: To educate children and adults about marine mammals, their environment and their preservation. Save the Whales is beginning educational programs via a mobile unit (Whales on Wheels) that will bring lectures and hands-on materials to schoolchildren. **Current Emphasis:** Education through lectures, newsletters (four times a year) to world-wide membership, letter-writing campaigns, media appearances and support of marine mammal research in the wild. **Members:** 1,000. **Fees:** Adults, $20; Children, $10; Classroom, $30. **Funding:** dnd. **Annual Revenue:** dnd. **Usage:** dnd. **Volunteer Programs:** None.

SCENIC AMERICA
21 Dupont Cir. NW
Washington, DC 20036
(202) 833-4300
Fax: (202) 833-4304

Purpose: To preserve and enhance the scenic quality of America's communities and countryside. Provides information and technical

 How many glass bottles and jars do Americans throw away every two weeks?

assistance on scenic byways, tree preservation, economics of aesthetic regulation, billboard and sign control, scenic areas preservation, growth management and all forms of aesthetic regulation. **Current Emphasis:** Scenic highways, billboard control, tree preservation, economics of aesthetic regulation. **Members:** 2,000. **Fees:** dnd. **Funding:** Membership, 20%; Foundation/Donor, 65%; Other, 15%. **Annual Revenue:** $450,000. **Usage:** Administration, 9%; Fundraising, 10%; Programs, 81%. **Volunteer Programs:** Internships in environmental journalism, conservation advocacy, tree preservation, aesthetic regulation and land-use regulation.

SEA SHEPHERD CONSERVATION SOCIETY
1314 Second St.
Santa Monica, CA 90301
(310) 394-3198
Fax: (310) 390-0360

Purpose: To protect marine animals and marine habitats. **Current Emphasis:** Prevention of killing of dolphins by the tuna industry in the tropical Pacific; protecting pilot whales in the Faeroe Islands; enforcing a moratorium on whaling; rescue of whales and marine mammals in distress and of dolphins in international waters. **Members:** 20,000. **Fees:** Donations. **Funding:** Membership, 100%. **Annual Revenue:** $500,000. **Usage:** Programs, 100%. **Volunteer Programs:** Sea Shepherd is entirely a volunteer organization.

SIERRA CLUB
730 Polk St.
San Francisco, CA 94109
(415) 776-2211
Fax: (415) 776-0350

Purpose: To explore, enjoy and protect the wild places of the Earth; to practice and promote the responsible use of the Earth's ecosystems and resources; to educate and enlist humanity to protect and restore the quality of the natural and human environment; and to use all lawful means to carry out these objectives. **Current Emphasis:** Old growth forest protection; global warming/auto fuel efficiency/energy policy; Arctic National Wildlife Refuge protection; Bureau of Land Management wilderness/desert/national parks protection; toxic waste regulations; international development lending; tropical forest preservation. **Members:** 648,000. **Fees:** Individual, $35; Student/Senior, $15. **Funding:** Membership, 38%; Other, 62%. **Annual Revenue:** $38,000,000. **Usage:** Administration, 14%; Fundraising, 11%; Programs, 30%; Membership, 16%; Influencing public policy, 29%. **Volunteer Programs:** Extensive opportunities available throughout the country.

SIERRA CLUB LEGAL DEFENSE FUND
180 Montgomery St., Ste. 1400
San Francisco, CA 94104
(415) 627-6700
Fax: (415) 627-6740

Purpose: To provide legal representation to environmental organizations on matters involving land use, public lands, pollution, endangered species and wildlife habitats. **Current Emphasis:** Currently involved in many endangered species cases that involve the use of public lands. New offices have opened in Louisiana and Florida and are currently involved in pollution liability and groundwater rights cases respectively. Also attempting to protect native Ecuadorians and rainforest lands from oil exploration. **Members:** 150,000. **Fees:** $10. **Funding:** Membership, 75%; Foundation/Donor, 20%; Other, 5%. **Annual Revenue:** $9,000,000. **Usage:** Administration, 25%; Programs, 75%. **Volunteer Programs:** Law internships available in most offices with a small stipend provided.

SINAPU
P.O. Box 3243
Boulder, CO
(303) 494-3710

Purpose: Educating about and advocating for the reintroduction and protection of wolves in Colorado. Sinapu means "wolves" in the Ute language. **Current Emphasis:** Showing a slide show promoting wolf reintroduction and pressing federal agencies to take steps toward reintroduction. **Members:** 50. **Fees:** $15-$35. **Funding:** Membership, 100%. **Annual Revenue:** $3,000. **Usage:** Programs, 100%; Education and advocacy, 100%. **Volunteer Programs:** Sinapu functions totally on volunteer labor.

SOIL AND WATER CONSERVATION SOCIETY
7515 NE Ankeny Rd.
Ankeny, IA 50021-9764
(515) 289-2331
Fax: (515) 289-1227

Purpose: The mission of the Soil and Water Conservation Society is to advocate the conservation of soil, water and related natural resources. **Current Emphasis:** dnd. **Members:** 12,000. **Fees:** First-time, $30; Regular, $44. **Funding:** Membership, 32%; Corporate,

A: *Enough to fill both of New York's 1,350-foot World Trade Center's twin towers*
Source: Environmental Defense Fund

1%; Foundation/Donor, 20%. **Annual Revenue:** dnd. **Usage:** dnd. **Volunteer Programs:** Journalism intern, public affairs specialist intern, Soil Conservation Service—Earth Team.

SOUTHWEST NETWORK FOR ENVIRONMENTAL AND ECONOMIC JUSTICE
P.O. Box 7399
Albuquerque, NM 87194
(505) 242-0416
Fax: (505) 242-5609
Purpose: Recognizing the direct link between economic and environmental issues, the Southwest Network for Environmental and Economic Justice was formed to bring together activists, grassroots organizations of people of color, and native nations in the Southwest to broaden regional strategies and perspectives on environmental degradation and other social, racial and economic injustices. **Current Emphasis:** Campaigns include EPA: accountability campaign for environmental justice; High Tech Industry: workplace hazards; Free Trade: *maquiladoras,* US-Mexico border. **Members:** Over 70 organizations. **Fees:** Range from $10 to $500. **Funding:** dnd. **Annual Revenue:** dnd. **Usage:** dnd. **Volunteer Programs:** None.

SOUTHWEST ORGANIZING PROJECT
211 10th St. SW
Albuquerque, NM 87102
(505) 247-8832
Fax: (505) 247-9972
Purpose: Southwest Organizing Project's mission is to empower the disenfranchised in the Southwest to realize racial and gender equality and social and economic justice. **Current Emphasis:** Environmental justice in communities composed of people of color. **Members:** 50. **Fees:** $2/month. **Funding:** Membership, 10%; Foundation/Donor, 90%. **Annual Revenue:** $150,000. **Usage:** Administration, 15%; Fundraising, 15%; Programs, 70%. **Volunteer Programs:** None.

SOUTHWEST UTAH WILDERNESS ALLIANCE
1471 South 1100 E.
Salt Lake City, UT 84105
(801) 486-3181
Fax: (801) 486-4233
Purpose: To obtain wilderness protection and proper management for federal public lands in southern Utah and for related water, wildlife, archaeological resources, etc. **Current Emphasis:** dnd. **Members:** 10,000. **Fees:** $25 year. **Funding:** Membership, 60%; Other, 40%. **Annual Revenue:** $600,000. **Usage:** Administration, 10%; Fundraising, 10%; Programs, 80%. **Volunteer Programs:** Occasionally.

STUDENT CONSERVATION ASSOCIATION INC.
P.O. Box 550
Charlestown, NH 03603
(603) 543-1700
Fax: (603) 543-1828
Purpose: Since 1957, SCA has provided educational opportunities for student and adult volunteers to assist with the stewardship of our public lands and natural resources while gaining experience that enhances career directions or personal goals. Volunteers serve in national parks, national forests, wildlife refuges and other public or private conservation areas nationwide. **Current Emphasis:** Through its Conservation Career Development Program, encouraging youth (particularly people of color and women) to pursue careers in conservation/natural resource management. Volunteer assistance for ecological restoration; publishing *Earth Work,* a monthly magazine for current and future conservation professionals. **Members:** 12,000. **Fees:** $15 and up. **Funding:** Membership, 20%; Corporate, 4%; Foundation/Donor, 12%; Other, 64%. **Annual Revenue:** $3,800,000. **Usage:** Administration, 19%; Fundraising, 14%; Programs, 65%; Other, 2%. **Volunteer Programs:** The heart of SCA's activity is two programs: Resource Assistant Program (year-round) for college students/other adults, and the High School Program (summer) for students 16-18. No fee; most expenses paid; 1,500 openings per year. Listings and applications available.

THORNE ECOLOGICAL INSTITUTE
5398 Manhattan Cir.
Boulder, CO 80303
(303) 499-3647
Fax: (303) 499-8340
Purpose: TEI is committed to educating individuals about the conservation of the environment. **Current Emphasis:** Educational programs for adults, children and community. **Members:** 400. **Fees:** $25-$1,000. **Funding:** Membership, 10%; Corporate, 70%; Foundation/Donor, 20%. **Annual Revenue:** $225,000. **Usage:** dnd. **Volunteer Programs:** Both internships and volunteer opportunities are available.

 According to the American Lung Association, vehicle pollution causes how much in health care costs for Americans each year?

ECOCONNECTIONS

20/20 VISION
30 Cottage St.
Amherst, MA 01002
(413) 549-4555
Fax: (413) 549-0544
Purpose: 20/20 Vision seeks to revitalize democracy by creating persistent, strategic citizen action to persuade decision-makers to protect the Earth by reducing militarism and preserving the environment. For $20 a year, 20/20 Vision subscribers receive brightly colored, monthly postcards describing the most important 20-minute action they can take. **Current Emphasis:** To stop the military's assault on the environment, protect the ozone layer, preserve ecosystems, stabilize climate and increase funding for the environment. **Members:** 11,000. **Fees:** $20/year. **Funding:** Membership, 35%; Foundation/Donor, 65%. **Annual Revenue:** $420,000. **Usage:** Administration, 20%; Fundraising, 15%; Programs, 65%. **Volunteer Programs:** Interns are needed in both Amherst and Washington, DC, offices.

UNION OF CONCERNED SCIENTISTS
26 Church St.
Cambridge, MA 02238
(617) 547-5552
Fax: (617) 864-9405
Purpose: The Union of Concerned Scientists is an independent nonprofit organization of scientists and other citizens concerned about the impact of advanced technology on society. UCS focuses on energy policy, global environmental problems and arms control. **Current Emphasis:** Global warming, national energy policy, renewable energy, transportation, nuclear power safety and the relationship between population growth and environmental problems. **Members:** 100,000. **Fees:** dnd. **Funding:** Membership, 70%; Foundation/Donor, 30%. **Annual Revenue:** $4,000,000. **Usage:** Administration, 3%; Fundraising, 21%; Programs, 76%. **Volunteer Programs:** Three- to five-month paid internships are available in both the Cambridge and Washington, DC, offices. Many volunteer opportunities are available as part of the Scientists Action Network.

UNITED NATIONS ENVIRONMENT PROGRAMME
2 United Nations Plaza, Room DC2-303
New York, NY 10017
(212) 963-8093
Fax: (212) 963-7341
Purpose: Created in 1972, the United Nations Environment Programme, with its headquarters in Nairobi, Kenya, has a professional staff of approximately 200 worldwide, serving as the environmental agency of the United Nations system. UNEP helps formulate and coordinate environmental policies at the municipal, national, regional and international levels by working with scientific agencies, the private and public sectors, nongovernmental organizations legal institutions and others. **Current Emphasis:** Current emphasis includes the Global Environmental Facility (in cooperation with the World Bank and UNDP); building environmental accounting; acting as the secretariat to the ozone layer, hazardous wastes and biological diversity conventions; and building national environmental management capacities. **Members:** dnd. **Fees:** dnd. **Funding:** Grants, 100%. **Annual Revenue:** dnd. **Usage:** dnd. **Volunteer Programs:** A limited number of internship programs are available.

UNIVERSITY RESEARCH EXPEDITIONS PROGRAM
University of California
Berkeley, CA 94720
(510) 642-6586
Fax: (510) 642-6791
Purpose: To promote public involvement in ongoing, worldwide, scientific research and educational activities in the environmental, natural and social sciences. **Current Emphasis:** Projects in cooperation with scientists from developing nations, focused on preserving the Earth's resources and improving people's lives. **Members:** 400 participants per year. **Fees:** Participants make a tax-deductible donation to the university to support the research and cover their expenses. **Funding:** dnd. **Annual Revenue:** dnd. **Usage:** dnd. **Volunteer Programs:** Volunteer participants needed from all walks of life to serve two to three weeks as field assistants on research teams.

WILD CANID SURVIVAL AND RESEARCH CENTER/WOLF SANCTUARY
P.O. Box 760
Eureka, MO 63025
(314) 938-5900
Purpose: Environmental education. Captive breeding of the endangered red and Mexican wolves for the purpose of reintroduction into the wild. **Current Emphasis:** Providing as much breeding space as possible for the rare Mexican gray wolf. **Members:** 2,800. **Fees:** From $25 up. **Funding:** Membership, 80%; Corporate, 10%; Foundation/Donor, 10%.

A: *Approximately $93 billion*
Source: Campaign for New Transportation Priorities

Annual Revenue: $115,000. Usage: dnd. Volunteer Programs: Volunteers assist with fundraising, educational programs and administration.

THE WILDERNESS SOCIETY
900 17th St. NW
Washington, DC 20006
(202) 833-2300
Fax: (202) 429-3959

Purpose: Protecting wildlands and wildlife; safeguarding the integrity of our federal public lands, national forests, wildlife refuges, national seashores, recreation areas and public domain lands. **Current Emphasis:** Arctic wildlife refuge, national forest policy, national parks, endangered species protection and the economics of public land use. **Members:** 300,000. **Fees:** New, $15; Renewal, $30. **Funding:** Membership, 50%; Grants, 16%; Foundation/Donor, 16%; Other, 18%. **Annual Revenue:** $10,932,448. **Usage:** Administration, 10%; Fundraising, 6%; Programs, 72%; Recruiting, 12%. **Volunteer Programs:** None.

WILDLIFE CONSERVATION INTERNATIONAL
c/o New York Zoological Society
Bronx, NY 10460
(718) 220-6891
Fax: (718) 364-7963

Purpose: To help preserve the Earth's biological diversity and valuable ecosystems. With 140 projects in 45 countries, WCI addresses conflicts between humans and wildlife and explores locally sustainable solutions. **Current Emphasis:** Tropical rainforests; African elephant and rhino; Tibetan plateau and conservation training in developing countries. **Members:** 60,000. **Fees:** $25. **Funding:** Membership, 60%; Corporate, 10%; Foundation/Donor, 30%. **Annual Revenue:** $7,000,000. **Usage:** Administration, 5%; Programs, 95%. **Volunteer Programs:** Grants for graduate students and professionals in wildlife sciences upon proposal submission.

WILDLIFE DAMAGE REVIEW
P.O. Box 2541
Tucson, AZ 85702-2541
(602) 882-4218

Purpose: Founded to bring public scrutiny to the government agency Animal Damage Control (ADC), that controls predators, often by lethal means, that inhibit the livestock industry. Wildlife Damage Review wants to enlighten the public about ADC's goals and methods with the hope of eventually eliminating the agency. **Current Emphasis:** Encouraging public action against ADC. **Members:** dnd. **Fees:** dnd. **Funding:** Grants, 100%. **Annual Revenue:** $18,000. **Usage:** Programs, 70%; Membership, 30%. **Volunteer Programs:** Volunteer network is expanding. Demands individual initiative for regional work with support from Wildlife Damage Review.

THE WILDLIFE SOCIETY
5410 Grosvenor Ln.
Bethesda, MD 20814-2197
(301) 897-9770
Fax: (301) 530-2471

Purpose: The Wildlife Society is a nonprofit scientific and educational organization dedicated to conserving and sustaining wildlife productivity and diversity through resource management and to enhancing the scientific and technical capability and performance of wildlife professionals. **Current Emphasis:** Providing current scientific and management information on wildlife resources and enhancing the professionalism of wildlife managers. **Members:** 8,900. **Fees:** From $33; Students from $17. **Funding:** Membership, 37%; Publications and contributions, 46%; Other, 17%. **Annual Revenue:** $900,000. **Usage:** Administration, 8%; Fundraising, 1%; Programs, 67%; Membership, 24%. **Volunteer Programs:** A three- to six-month wildlife policy internship program with stipend is available.

WOMEN'S ENVIRONMENT AND DEVELOPMENT ORGANIZATION
845 Third Ave., 15th Fl.
New York, NY 10022
(212) 759-7982
Fax: (212) 759-8647

Purpose: Women's Environment and Development Organization's goal is to organize women's visibility as participants, policymakers and leaders in fate-of-the-Earth decisions. As half the world's population, women must have an equal say in post-Earth Summit and governmental policymaking. **Current Emphasis:** Organizing regional and community "Women for a Healthy Planet" groups; Community Health Report Card projects. **Members:** dnd. **Fees:** dnd. **Funding:** Foundation/Donor, 50%; Other, 50%. **Annual Revenue:** dnd. **Usage:** dnd. **Volunteer Programs:** Yes.

WORLD RESOURCES INSTITUTE
1709 New York Ave. NW, Ste. 700
Washington, DC 20006
(202) 638-6300
Fax: (202) 638-0036

Q: *How much of an explosive arsenal do the combined atomic weapons of the world's nuclear countries collectively possess?*

Purpose: A research and policy institute helping governments, the private sector, environmental and development organizations, and others address a fundamental question: How can societies meet human needs and nurture economic growth while preserving the natural resources and environmental integrity on which life and economic vitality ultimately depend? **Current Emphasis:** Forests, biodiversity, economics, technology, institutions, climate, energy, pollution, resource and environmental information, governance. **Members:** Not a membership organization. **Fees:** N/A. **Funding:** Private foundations, governmental and intergovernmental institutions, 100%. **Annual Revenue:** dnd. **Usage:** dnd. **Volunteer Programs:** None.

WORLD SOCIETY FOR THE PROTECTION OF ANIMALS
P.O. Box 190
Boston, MA 02130
(617) 522-7000
Fax: (617) 522-7077

Purpose: International animal protection/wildlife conservation organization frequently working in less-developed countries where animal protection societies are nonexistent or need support to prevent animal suffering. World Society for the Protection of Animals is one of the few international animal protection organizations with consultative status at the UN. **Current Emphasis:** To enact animal protection legislation; "Libearty" campaign to protect the world's bears; disaster relief program to aid animal victims of natural and human-caused disasters anywhere in the world; humane transport and slaughter of livestock; animal spectacles; humane education. **Members:** 11,000/US. **Fees:** $20. **Funding:** Corporate, 6%; Individual funding and bequests, 78%; Other, 16%. **Annual Revenue:** $747,000. **Usage:** Administration, 29.5%; Fundraising, 3.7%; Programs, 65.9%; Other, 0.9%. **Volunteer Programs:** Occasionally.

WORLD WILDLIFE FUND
1250 24th St. NW, Ste. 400
Washington, DC 20037
(202) 293-4800
Fax: (202) 293-9211

Purpose: Working worldwide to preserve endangered wildlife and wildlands by encouraging sustainable development and the preservation of biodiversity, particularly in the tropical forests of Latin America, Africa and Asia. **Current Emphasis:** Conservation of tropical rainforests, preserving biological diversity. **Members:** 1,000,000. **Fees:** $15. **Funding:** Membership, 62%; Corporate, 2%; Other, 36%. **Annual Revenue:** dnd. **Usage:** Administration, 15%; Programs, 85%. **Volunteer Programs:** Not available.

XERCES SOCIETY
10 SW Ash St.
Portland, OR 97204
(503) 222-2788
Fax: (503) 222-2763

Purpose: An international, nonprofit organization dedicated to invertebrates and the preservation of critical biosystems worldwide. The society is committed to protecting invertebrates as the major component of biological diversity. Because invertebrates sustain biological systems, invertebrate conservation means preserving ecosystems and ecological functions as well as individual species. **Current Emphasis:** Conservation science for reserve design in Madagascar and Jamaica; in-country conservation training and public education; public policy initiatives protecting invertebrates; protecting Pacific Northwest old growth forests. **Members:** 3,700. **Fees:** $15 to $1,000. **Funding:** Membership, 34%; Corporate, 7%; Foundation/Donor, 59%. **Annual Revenue:** $275,000. **Usage:** dnd. **Volunteer Programs:** None.

ZERO POPULATION GROWTH
1400 16th St. NW, Ste. 320
Washington, DC 20036
(202) 332-2200
Fax: (202) 332-2302

Purpose: To achieve a sustainable balance between the Earth's population, its environment and its resources. Primary activities include publishing newsletters and research reports, developing in-school population education programs and coordinating local and national citizen action efforts. **Current Emphasis:** Urbanization and local growth issues, global warming, sustainability, transportation, family planning and other key population concerns. **Members:** 45,000. **Fees:** Student/senior, $10; Individual, $20. **Funding:** Membership, 80%; Grants, 20%. **Annual Revenue:** $1,300,000. **Usage:** Administration, 11%; Fundraising, 9%; Programs, 80%. **Volunteer Programs:** Action Alert Network, Roving Reporter, Growthbusters.

A: *Enough to equal four tons of TNT for every man, woman and child alive today*
Source: Greenpeace

Directory to Environmental Education Programs

This directory features environmental education for every age—kids' camps, high school programs, colleges and graduate schools, adult education programs, and lots of exciting, hands-on learning experiences.

Education Programs

ACID RAIN FOUNDATION, INC.
1410 Varsity Dr.
Raleigh, NC 27606
(919) 828-9443
Emphasis: Acid rain, air quality, global change, air pollutants and forests. **Programs:** Information, education services, curriculum. **Type:** Outdoor education, natural resources. **Cost:** Varies. **Age Emphasis:** All ages. **Application Deadline:** None.

ADIRONDACK OUTDOOR EDUCATION CENTER
Camp Chingachgook
Pilot Knob Rd., HC01-Box 35
Kattskill Bay, NY 12844
(518) 656-9462
Emphasis: Outdoor recreation on Lake George. **Programs:** High Ropes, Team Building, Lake & Wetland Exploration. **Type:** Outdoor education, expedition/experiential, natural science/history, residential. **Cost:** $20 per person per day, 1-5 days. **Age Emphasis:** All ages. **Application Deadline:** March 15.

APPALACHIAN MOUNTAIN CLUB
Pinkham Notch Camp
P.O. Box 298
Gorham, NH 03581
(603) 466-2721
Emphasis: Wide variety of environmental/outdoor and conservation education programs. A new residential school program promotes environmental awareness and land stewardship. **Programs:** Guided hikes, field seminars (college credit available), workshops, residential school programs, mountain leadership school, program for inner-city youth. **Type:** Outdoor education, natural science/history, residential. **Cost:** Varies, from free to $350/week. **Age Emphasis:** All ages. **Application Deadline:** None.

AUDUBON NATURALIST SOCIETY OF THE CENTRAL ATLANTIC STATES
8940 Jones Mill Rd.
Chevy Chase, MD 20815
(301) 652-9188
Emphasis: Environmental education classes and programs, family activities, field trips. **Programs:** Natural History Field Studies adult education program, children's programs, adult programs. **Type:** Expedition/experiential, natural science/history. **Cost:** Varies. **Age Emphasis:** All ages. **Application Deadline:** None.

BEAR MOUNTAIN OUTDOOR SCHOOL, INC.
Rte. 250
Hightown, VA 24444
(703) 468-2700
Emphasis: Environmental education, ecology awareness workshops, eco-construction workshops, organic gardening workshops. **Programs:** Rural Living Skills, Homebuilding Skills, Natural History/Science, Mountain Ecology Camps for Youth. **Type:** "Learning Vacations" (April-October)—Week-long and weekend residential at 4,200-foot altitude lodge in the Potomac Highlands. **Cost:** Varies from $175-$600. Family retreats and ecology camps (call for rates). **Age Emphasis:** All ages. **Application Deadline:** Call or write Bear Mountain for brochure and program schedule.

B.O.C.E.S./OUTDOOR ENVIRONMENTAL EDUCATION
P.O. Box 604
Smithtown, NY 11787
(516) 360-3652
Emphasis: In-service teacher training for integrating environmental education into traditional curricula; residential service for teachers and their classes; in-depth specialized topics and publications. Focus on hands-on, experiential learning. **Programs:** Day use, residential and special services. **Type:** Teacher training programs, school nature tours. **Cost:** Varies. **Age Emphasis:** All ages. **Application Deadline:** Service contracts are due by April 15.

Q: *What percentage of US energy consumption is from either petroleum or natural gas?*

BOK TOWER GARDENS
P.O. Box 3810
Lake Wales, FL 33859-3810
(813) 676-1408
Emphasis: Horticulture, Florida history, natural history, art and/or music. **Programs:** Adult Education Program; adult day and overnight trips to unique places; lectures; workshops; "A Garden Classroom" for fourth-graders. **Type:** School tours, member and non-member nature history field courses. **Cost:** 4th-graders, free; adult costs vary. **Age Emphasis:** All ages. **Application Deadline:** None.

CAL-WOOD ENVIRONMENTAL EDUCATION RESOURCE CENTER
P.O. Box 347
Jamestown, CO 80455
Emphasis: Conservation education, focus on natural resource management; international leadership development (Mexico, Brazil, Nigeria). **Programs:** Tailored to meet needs of individual groups. Center contracts with schools, organizations, etc. (no openings for individuals). **Type:** Outdoor education, expedition/experiential, natural science/history, natural resources, residential. **Cost:** Negotiable by contract. **Age Emphasis:** All ages. **Application Deadline:** 3 to 6 months in advance; one year in advance for large programs.

CAPE OUTDOOR DISCOVERY
P.O. Box 1565
Buzzards Bay, MA 02532
(508) 224-3040
Emphasis: To increase environmental awareness by providing students with positive experiences in a natural environment. Emphasis on marine environments and ecological lifestyles. **Programs:** Group Challenges, whalewatching, local history, residential programs, day programs, *Cape Cod Environmental Data* (publication). **Type:** Outdoor education, natural science/history, natural resources, residential. **Cost:** Varies, from $65-$135 per student, depending on the length of stay (school group rates). **Age Emphasis:** Youth. **Application Deadline:** None.

CARRIE MURRAY OUTDOOR EDUCATION CAMPUS
1901 Ridgetop Rd.
Baltimore, MD 21207
(410) 396-0808
Emphasis: Outdoor education. **Programs:** Year-round outdoor education program for children and adults; raptor rehabilitation and education program; adventure education trips and workshops; summer nature camp. **Type:** Outdoor education, expedition/experiential, natural science/history. **Cost:** $1-$5 for classes and workshops. **Age Emphasis:** All ages. **Application Deadline:** None.

CITY OF AURORA, UTILITIES DEPT.
1470 S. Havana, Ste. 400
Aurora, CO 80012
(303) 695-7381
Emphasis: Water conservation via xeriscape, or indoors with retrofit or renovation; water awareness video for grades K-12; co-published "Landscaping for Water Conservation: Xeriscape!" **Programs:** Audiovisual resources. **Type:** School education, residential. **Cost:** Varies. **Age Emphasis:** All ages. **Application Deadline:** None.

CITY OF BOULDER ADVENTURE PROGRAM
P.O. Box 791
Boulder, CO 80306
(303) 441-4401
Emphasis: Instruction and trips in rock climbing, kayaking, mountain biking, canoeing, skiing and general mountaineering with a focus on minimum-impact camping and experiential education. **Programs:** Rock climbing, canoeing, mountain biking and kayaking trips in Colorado and Canyonlands, the youth/teen program and adult program. **Type:** Outdoor education, expedition/experiential, natural science/history. **Cost:** Varies, $15-$400. **Age Emphasis:** All ages. **Application Deadline:** Courses run year-round.

COALITION FOR EDUCATION IN THE OUTDOORS
Box 2000, SUNY College at Cortland
Cortland, NY 13045
(607) 753-4971
Emphasis: The coalition is a network of affiliated businesses, professional organizations, institutions, centers and groups in support of outdoor/environmental education. **Programs:** Conferences, symposia, quarterly newsletter. **Type:** Outdoor education, expedition/experiential. **Cost:** Graduated scale based on membership category. **Age Emphasis:** Adults. **Application Deadline:** None.

A: *63 percent*
Source: Union of Concerned Scientists

COLORADO MOUNTAIN COLLEGE
P.O. Box 10001 BW
Glenwood Springs, CO 81602
(303) 945-8691
Emphasis: Outdoor education and environmental courses. Academic classroom work combined with work, skills and adventure/expeditions. **Programs:** Environmental technology, outdoor semester in the Rockies, wilderness studies. **Type:** Certificate, degree, residential campuses in Glenwood Springs, Leadville and Steamboat Springs. **Cost:** $9,000/year. for out-of-state student comprehensive cost. **Age Emphasis:** College student. **Application Deadline:** Rolling admissions for most programs, limited enrollment for outdoor semester each term.

COLORADO OUTWARD BOUND SCHOOL
945 Pennsylvania
Denver, CO 80203
(303) 837-0880
Emphasis: Year-round wilderness expeditions since 1961. **Programs:** Mountaineering, whitewater rafting and kayaking, desert and canyon exploration. Special health services and professional development programs. **Type:** Outdoor education, expedition/experiential, natural science/history. **Cost:** $50/credit hour in-state; $150/credit hour out of state. **Age Emphasis:** Adults. **Application Deadline:** August 31.

NATURE QUEST
1400 16th St. NW
Washington, DC 20036-2266
(800) 245-5484
Emphasis: NatureQuest is the National Wildlife Federation's certified training program for camp directors, nature and science counselors, naturalists and outdoor educators. Three-day workshops expose participants to the "Quest" model, then participants design sites and needs. **Programs:** NatureQuest. **Type:** Outdoor education training. **Cost:** $1,980 program fee. **Age Emphasis:** Adult. **Application Deadline:** One month prior to session; sessions are in March and April.

CREATIVE MEDIA/ENVIRONMENTAL EDUCATION/NORTHWEST TRAVEL & STUDY
9730 Manitou Place
Bainbridge Island, WA 98110
(206) 456-1854
Emphasis: Environmental education program planning; workshops for educators. Consultation. **Programs:** Environmental education, wetlands, recycling, water quality education. **Type:** Outdoor education, natural science/history, natural resources. **Cost:** Free. **Age Emphasis:** All ages. **Application Deadline:** None.

DAHLEM ENVIRONMENTAL EDUCATION CENTER
7117 S. Jackson Rd.
Jackson, MI 49201
(517) 782-3453
Emphasis: Comprehensive field study curriculum for P-6th graders; special events, workshops, weekend programs for teachers, adults and families; summer ecology day camp; county-wide "Bring Back the Bluebirds Project;" hosts Bluebird Festival and Wildlife Art Show each March. **Programs:** Internships for college students and recent graduates (interpretation, wildlife biology, research). **Type:** Natural science/history, natural resources. **Cost:** Varies. **Age Emphasis:** All ages. **Application Deadline:** Intern applications accepted seasonally.

ECOLOGIC
P.O. Box 1514
Antigonish, Nova Scotia, B2G 2L8, Canada
(902) 863-5984
Emphasis: Designs and facilitates environmental education programs for teachers, industry, community groups, government. Specializes in participatory workshops with emphasis on global issues. **Programs:** Diverse and varied; designed specifically for individual clients. **Type:** Expedition/experiential, natural science/history. **Cost:** Varies. **Age Emphasis:** All ages. **Application Deadline:** None.

EDUDEX ASSOCIATES, KETTLE MORAINE DIVISION
604 Second Ave.
West Bend, WI 53095
(414) 334-4978
Emphasis: Natural and human resources organization designs, delivers and evaluates outdoor, conservation and environmental experiential programs on an international level. Assists individuals, groups, classes, etc. **Programs:** Programs prepared by foresters, soil scientists, wildlife biologists, education staff. **Type:** Outdoor education, expedition/experiential, natural science/history, natural resources. **Cost:** Based on ability and need. **Age Emphasis:** All ages. **Application Deadline:** None.

 What percentage of American students believe that they and their classmates do not know enough about environmental problems and solutions?

ECOCONNECTIONS

EDWARDS CAMP AND CONFERENCE CENTER
P.O. Box 16
East Troy, WI 53120
(414) 642-7466
Emphasis: Environmental education in YMCA resident camp setting. Stresses diversity of ecosystems, lesson plans and curriculum, recreation opportunities. **Programs:** Wildlife, aquatic, forest/prairie, environmental awareness, weather, astronomy, recreation and sports, low rope course/group initiatives. **Type:** Outdoor education. **Cost:** Two nights, seven meals: $52.45, cabins; $59.40, lodges. **Age Emphasis:** Youth. **Application Deadline:** Arrange with camp prior to stay.

FOUR CORNERS SCHOOL OF OUTDOOR EDUCATION
East Rte.
Monticello, UT 84535
(801) 587-2859
Emphasis: Educational field programs in the Southwest. Focus on archaeology, geology, cultural studies, as well as photography and writing workshops, backpacking, river rafting and base camp experiences on the Colorado Plateau. **Programs:** Rock Art, Dig the Past, Geography of the San Juan River, Kaiparowits Plateau Survey, etc. **Type:** Degree program, outdoor education, expedition/experiential, natural science/history, natural resources, residential. **Cost:** Most programs, $375-$875. **Age Emphasis:** All ages. **Application Deadline:** Varies.

GLEN HELEN OUTDOOR EDUCATION CENTER
1075 Rte. 343
Yellow Springs, OH 45387
(513) 767-7648
Emphasis: Residential programs for local schools; nature interpretation focusing on non-competitive games. **Programs:** Two simultaneous programs, a residential environmental education experience for (mostly) 6th graders, and a training program for naturalists/interns. Interns get stipend, room, board and graduate or undergraduate credits. **Type:** Outdoor education, expedition/experiential, raptor rehabilitation, natural science/history, natural resources, residential. **Cost:** None. **Age Emphasis:** All ages. **Application Deadline:** None.

GREAT SMOKY MOUNTAINS INSTITUTE AT TREMONT
Rte. 1, Box 700
Townsend, TN 37882
(615) 448-6709
Emphasis: Wide variety of programs emphasizing cultural and natural resources of Great Smoky Mountains National Park. **Programs:** School programs, adult workshops, teacher training, camps, elder hostel. **Type:** Outdoor education, natural science/history, natural resources, residential. **Cost:** Varies, $20-$30/day. **Age Emphasis:** All ages. **Application Deadline:** None.

GUIDED DISCOVERIES, INC.
P.O. Box 1360
Claremont, CA 91711
(714) 625-6194
Emphasis: Marine biology, island ecology, physical science, astronomy. **Programs:** Catalina Island Marine Institute at Toyon Bay, Cherry Cove and Astrocamp. **Type:** Outdoor education. **Cost:** Toyon: 3 days/$125, 5 days/$225, weekends/$110. Cherry Cove: 3 days/$125, 5 days/$225. Astro: 3 days/$100, 5 days/$195, weekends/$82. **Age Emphasis:** Youth. **Application Deadline:** None.

HOUSEHOLD HAZARDOUS WASTE PROJECT
1031 E. Battlefield, Ste. 214
Springfield, MO 65807
Emphasis: Award-winning community education program to identify and minimize household hazardous products. Emphasis on safety, storage disposal, promotion of safer alternatives. **Programs:** Training workshops, learning materials, audiovisual programs, *Guide to Hazardous Products Around the Home* (personal action manual). Telephone information request line. **Type:** N/A. **Cost:** Training and course materials are $60 out-of-state, $50 in-state. **Age Emphasis:** All ages. **Application Deadline:** Training course offered in fall and spring.

HOUSTON INDEPENDENT SCHOOL DISTRICT'S OUTDOOR EDUCATION CENTER
Cullen OEC
Rte. 3, Box 135D1
Trinity, TX 75862
(409) 594-2274
Emphasis: Designed to serve the children of Houston as an extension of the classroom, to

A: *90 percent*
Source: The Student Environmental Action Guide, *EarthWorks Press*

promote concern for the environment and to provide interaction among children of different economic, racial and cultural backgrounds. **Programs:** 4-day programs, September-May. **Type:** Outdoor education, residential. **Cost:** No cost to students. **Age Emphasis:** Youth. **Application Deadline:** None.

HUNTSMAN MARINE SCIENCE CENTRE
Brandy Cove Rd.
St. Andrews, New Brunswick E0G 2X0, Canada
Emphasis: Marine science education for students, teachers and the public. Experience-based programs that access the Bay of Fundy ecosystem (one of the most productive on the planet) and facilities of the Science Centre. **Programs:** Intro to Marine Biology and Intro to Marine Vertebrates (high school), Sea Trek (adults). **Type:** Outdoor education, expedition/experiential. **Cost:** Varies, $450-$550 Canadian. **Age Emphasis:** Youth and adult. **Application Deadline:** May 31.

ISAAC W. BERNHEIM FOUNDATION
Bernheim Arboretum and Research Forest
Hwy 245
Clermont, KY 40110
(502) 543-2451
Emphasis: Environmental education, arboretum, art classes, research. **Programs:** Research Institute for Education and the Environment. **Type:** Outdoor education. **Cost:** Weekdays, free. Weekends and holidays, $5 per vehicle. **Age Emphasis:** All ages. **Application Deadline:** None.

JOY OUTDOOR EDUCATION CENTER
Box 157
Clarkeville, OH 45113
Emphasis: Outdoor, environmental education for elementary and high schools, summer camp for disadvantaged children, corporate team-building program. **Programs:** Camp Joy; Venture Out! **Type:** Outdoor education, residential. **Cost:** $90-$200 for adults. **Age Emphasis:** All ages. **Application Deadline:** None.

KEEWADEN ENVIRONMENTAL EDUCATION CENTER
RR 1, Box 88
Salisbury, VT 05769
(802) 352-4187
Emphasis: Natural communities and human impact and responsibility. Local history, including Native Americans, early settlers, early industries. **Programs:** Spring and fall at Lake Dunmore; winter at Seyon Ranch in Groton, VT. **Type:** Outdoor education, natural science/history, natural resources, residential. **Cost:** $30-$90, depends on length of stay. **Age Emphasis:** Youth. **Application Deadline:** None.

KORTRIGHT CENTER
5 Shoreham Dr.
Downsview, Ontario M3N 1S4, Canada
(416) 832-2289
Emphasis: Outdoor education center. **Programs:** Field study programs on the environment and natural history. Topics include water, wildlife, energy conservation and forestry. **Type:** Experiential, natural science/history, natural resources. **Cost:** $2-$45. **Age Emphasis:** All ages. **Application Deadline:** None.

LEGACY INTERNATIONAL
Rte. 4, Box 265
Bedford, VA 24523
(703) 297-5982
Emphasis: Youth leadership training. Youth for Environment and Service (YES) provides youths with an opportunity to build bridges of understanding between people of diverse cultures, races and backgrounds. **Programs:** Youth Environmental Services, Dialogue, Global Issues and Summer Youth Leadership Training, English as a Foreign Language, Performing Arts for Social Change. **Type:** Experiential, natural resources, residential. **Cost:** $1,480-$3,500. Length: 6 weeks. **Age Emphasis:** All ages. **Application Deadline:** Ongoing, apply early.

MAINE CONSERVATION SCHOOL
P.O. Box 188
Bryant Pond, ME 04219
(207) 665-2068
Emphasis: Environmental education through nature studies and outdoor learning activities with emphasis on management issues. **Programs:** School programs, teacher workshop, adult programs, community events and outreach programs/workshops. **Type:** Outdoor education, natural science/history, natural resources, residential. **Cost:** $25 per day. **Age Emphasis:** All ages. **Application Deadline:** None.

 What percentage of American students said they would take environmental action if they had more information on what to do?

MINNESOTA VALLEY NATIONAL WILDLIFE REFUGE
3815 E. 80th St.
Bloomington, MN 55425-1600
(612) 854-5200
Emphasis: Environmental education, interpretive programming, training for teachers, activities and events to create environmentally literate citizens. **Programs:** Volunteer Wildlife Interpreters course, teacher workshops, environmental education for school groups, seasonal interpretive programs and events. **Type:** Natural-resource-based, environmental education, recreation. **Cost:** No fee. **Age Emphasis:** All ages. **Application Deadline:** None.

MOHICAN SCHOOL IN THE OUT-OF-DOORS, INC.
21882 Shadley Valley Rd.
Danville, OH 43014
(614) 599-9753
Emphasis: Curriculum enrichment emphasizes resident outdoor environmental education. **Programs:** Resident programs and field trips for grades 4-7. Adult workshops. **Type:** Outdoor education, natural science/history, natural resources, residential. **Cost:** $63.50/half week, $118/full week. **Age Emphasis:** Middle school, grades 4-7. **Application Deadline:** None.

NATIONAL AUDUBON EXPEDITION INSTITUTE
P.O. Box 365
Belfast, ME 04915
(207) 338-5859
Programs: A nonprofit traveling education program that offers year-long and semester expeditions providing students an alternative to traditional education and an understanding of nature's spiritual dimension. Requires the desire to be challenged academically and an open attitude toward global environmental issues. **Type:** High school, B.S. and M.S. in environmental education. **Cost:** $12,000/year, $7,038/semester. **Age Emphasis:** Youth, college student. **Application Deadline:** None.

NEWFOUND HARBOR MARINE INSTITUTE AT SEACAMP
Rte. 3, Box 170
Big Pine Key, FL 33043
(305) 872-2331
Emphasis: Combines direct, hands-on experience with conceptual learning. Offers marine environmental education courses to increase understanding of the ocean and natural ecosystems in a diverse tropical community. **Programs:** Coral reef ecology, shallow bay ecology, mangrove ecology, coastal ecology, ichthyology, tropical island botany, ornithology. **Type:** Outdoor education, expedition/experiential, residential, marine science. **Cost:** Average cost, $90/day inclusive. **Age Emphasis:** 10 and up. **Application Deadline:** None.

THE NIZHONI SCHOOL FOR GLOBAL CONSCIOUSNESS
Rte. 14, Box 203
Santa Fe, NM 87505
(505) 473-4848
Emphasis: Spiritual education for those who seek to lead multidimensional lives, discover themselves and their unique gifts through a path of global consciousness. **Programs:** Academy for Ecology, Energy and Global Business, School for Language and Communication and the Divinity School. **Type:** International boarding/day school founded by Chris Griscom. **Cost:** $16,200/boarding school, $10,200/day school. **Age Emphasis:** All ages. **Application Deadline:** None.

NORTH CASCADES INSTITUTE
2105 Hwy 20
Sedro Woolley, WA 98284
(206) 856-5700
Emphasis: Natural and cultural history of the Pacific Northwest. Most seminars cover natural history, but art, photography and writing workshops are also offered. Seminars are experiential and field-based. **Programs:** Varied; everything from wildflowers and mushrooms to North Coast Native American carving. **Type:** Outdoor education, natural science/history **Cost:** Varies, $35-$300. **Age Emphasis:** All ages. **Application Deadline:** First come, first served.

PARADIGM VENTURES
29 Azul Loop
Santa Fe, NM 87505
(505) 988-1813
Emphasis: Personal and professional development. Customized programs utilizing Southwest wilderness areas. Ropes course programs. **Programs:** Customized to client needs and abilities. **Type:** Expedition/experiential. **Cost:** Varies. **Age Emphasis:** All ages. **Application Deadline:** None.

A: *84 percent*
Source: The Student Environmental Action Guide, *EarthWorks Press*

ROCKY MOUNTAIN SEMINARS
Rocky Mountain National Park
Estes Park, CO 80517
(303) 586-2371, ext. 294
Emphasis: Uses natural resources of Rocky Mountain National Park as an education tool. **Programs:** Wide variety of topics, from photography to geology and ecology of the park. **Type:** Outdoor education, expedition/experiential, natural science/history. **Cost:** $150/week-long seminars, $80/weekend, $40/one day. **Age Emphasis:** All ages. **Application Deadline:** None; classes are open until they fill.

ROGER TORY PETERSON INSTITUTE
110 Marvin Pkwy.
Jamestown, NY 14701
(716) 665-2473
Emphasis: Nature-in-education programs for teachers and nature education professionals. **Programs:** Nature Journals (workshop for middle school teachers to facilitate integration of nature and whole language curriculum); conference for teachers and nature educators; Nature Educator Awards (annual $1,000 awards to outstanding nature educators); "Birds, Bats, and Butterflies" (a quarterly nature study leaflet for parents and teachers). **Type:** Natural science/history. **Cost:** Varies. **Age Emphasis:** Adults who work with youth. **Application Deadline:** None.

SAFARI CLUB INTERNATIONAL
4800 W. Gates Pass Rd.
Tucson, AZ 85745
(602) 620-1220
Emphasis: Each 10-day workshop provides challenging experiences in wildlife ecology, management and conservation, together with instruction in firearm safety, fly tying, wilderness survival archery, outdoor interpretive techniques, Project WILD and outdoor ethics. **Programs:** American Wilderness Leadership School. **Type:** Degree program, outdoor education, natural science/history, natural resources, residential. **Cost:** $600. **Age Emphasis:** Adult. **Application Deadline:** April 30.

SCHOOL FOR INTERNATIONAL TRAINING/WORLD LEARNING, INC.
P.O. Box 676
Brattleboro, VT 05302-0676
(802) 257-7751; (800) 451-4465
Emphasis: Programs give students a firsthand experience of the environment in different parts of the world and point out how it is being jeopardized. **Programs:** Madagascar/Brazil, Amazon Studies and Ecology; Australia, Natural and Human Environment; Dominican Republic, Caribbean Area Studies/Development; Kenya, Coastal Studies; Tanzania, Wildlife Ecology and Conservationism; Ecuador, Comparative Ecology. 40 programs total. **Type:** Expedition/experiential, natural resources, residential. **Cost:** $8,300-$10,300. **Length:** 15 weeks. **Age Emphasis:** College student. Summer programs for high school students. **Application Deadline:** October 15, May 15.

SOLO
RFD 1, Box 163
Conway, NH 03818
(603) 447-6711
Emphasis: Wilderness and emergency medical programs. **Programs:** Backcountry Medicine, Wilderness First Responder, Wilderness EMT. **Type:** Outdoor education, expedition/experiential, residential. **Cost:** Varies, $75-$1,325. **Age Emphasis:** Over 18 years old. **Application Deadline:** None.

THAMES SCIENCE CENTER
Gallows Lane
New London, CT 06320
(203) 442-0391
Emphasis: Environmental science curriculum focuses on local watersheds. Teachers and students conduct lab and field experiments. **Programs:** Watershed Worlds, Global Views, Regional Perspectives. **Type:** Outdoor education, expedition/experiential, natural science/history, natural resources, residential. **Cost:** Open to classroom teachers, stipend provided. **Age Emphasis:** Middle school—adults. **Application Deadline:** None.

THORNE ECOLOGICAL INSTITUTE
5398 Manhattan Circle
Boulder, CO 80303
(303) 499-3647
Emphasis: TEI is committed to educating individuals about the environment. **Programs:** Boulder and Denver Natural Science Schools, Annual Symposium: Issues and Technology in the Management of Impacted Wildlife, Viewpoints, Environmental Stewardship 2000.

 The US livestock industry occupies how much of the country's land for feed crops, pasture and range?

ECOCONNECTIONS

Type: Natural science/history, natural resources. **Cost:** Free-$120. **Length:** 1 day to 2 weeks. **Age Emphasis:** All ages. **Application Deadline:** None.

TREES FOR TOMORROW
P.O. Box 609
Eagle River, WI 54521
(715) 479-6456
Emphasis: Stresses people management and preservation of natural resources. **Programs:** Natural resources education. *Northbound*, a quarterly publication. School subscription—$3/year or with membership. **Type:** Natural resources. **Cost:** $78.50/3-day program (school rate). **Age Emphasis:** All ages. **Application Deadline:** None.

UNITED STATES BOTANIC GARDEN
245 First St. SW
Washington, DC 20024
(202) 226-4082
Emphasis: Displays of plants from around the world, including many rare and endangered plants. Epiphytic House; collection of orchids and begonias; new dinosaur garden; cactus house; four flower shows a year, each with a different theme. **Programs:** Horticultural classes. **Type:** Natural science. **Cost:** Free. **Age Emphasis:** All ages. **Application Deadline:** None.

URBAN OPTIONS
135 Linden St.
East Lansing, MI 48823
(517) 337-0422
Emphasis: Nonprofit community information resource for greater Lansing area. **Programs:** Demonstration house, environmental information service, weatherization programs, tool lending, library, newsletter, classes. **Type:** Educational programs and services on energy and the environment. **Cost:** Varies from free-$200. **Age Emphasis:** All ages. **Application Deadline:** None.

WESLEY WOODS UNITED METHODIST OUTDOOR MINISTRIES, INC.
329 Wesley Woods Rd.
Townsend, TN 37882
(615) 448-2246
Emphasis: Outdoor experiential and environmental education. **Programs:** Outdoor/Environmental Education, Residential Summer Camp, Youth Trips & Adventure Programs, Elder Hostel. **Type:** Outdoor education, expedition, experiential, natural science, history, natural resources, residential. **Cost:** $51/half week, $89/week. **Age Emphasis:** All ages. **Application Deadline:** None.

WILD BASIN WILDERNESS PRESERVE
805 N. Capitol of Texas Hwy.
Austin, TX 78746
(512) 327-7622
Emphasis: Preservation of 227-acre nature preserve located near urban area; education in all aspects of the environment for adults and children. **Programs:** Wild Basin Environmental Education (for schoolchildren); Moonlighting, stargazing, etc. (for adults). **Type:** Outdoor education, natural science/history, natural resources. **Cost:** Most programs at a minimal cost. **Age Emphasis:** All ages. **Application Deadline:** None.

WILDERNESS SOUTHEAST
711 Sandtown Rd.
Savannah, GA 31410
(912) 897-5108
Emphasis: Nonprofit school of the outdoors. Small groups camp in diverse wilderness areas for a close-up look at remarkable ecosystems. Leisurely pace and flexible itineraries. Relax and unwind, stretch both mind and muscles. **Programs:** Natural history. **Type:** Outdoor education, expedition/experiential, natural science/history. **Cost:** $70-$125/day. **Age Emphasis:** All ages. **Application Deadline:** None.

WILDERNESS TRANSITIONS, INC.
70 Rodeo Ave.
Sausalito, CA 94965
(415) 331-5380; (415) 332-9558
Emphasis: Wilderness vision quest trips. Small groups share two days in base camp, then individuals live alone for three days to rediscover one's self and one's connectedness to the Earth and all living beings. **Programs:** Vision Quest (includes instruction in nature lore and wilderness living). **Type:** Outdoor education, expedition/experiential. **Cost:** $460/week. **Age Emphasis:** All ages over 15. **Application Deadline:** Varies.

WOODSWOMEN, INC.
25 W. Diamond Lake Rd.
Minneapolis, MN 55419
(612) 822-3809
Emphasis: Supportive and challenging learning opportunities in the context of safe,

A: *An area of land equivalent to about half of the continental US*
Source: Worldwatch *magazine*

enjoyable outdoor and wilderness travel experiences and leadership development. Courses provide healthy living, options, skills, development, new perspectives on natural history. **Programs:** Adventure travel for women. **Type:** Outdoor education, expedition/experiential, natural science/history. **Cost:** $18-$100/day. **Age Emphasis:** Adult women/college women. **Application Deadline:** None.

THE YELLOWSTONE INSTITUTE
Box 117
Yellowstone National Park, WY 82190
(307) 344-2294

Emphasis: Enhancement of people's appreciation of the park through the presentation of various natural and human history topics. **Programs:** College credit/summer, fall, and winter field study classes. **Type:** Backcountry expedition/experiential, natural science/history, natural resources, residential. **Cost:** $35-$100/day. **Length:** 2-6 days. **Age Emphasis:** All ages. **Application Deadline:** None.

YMCA OF GREATER NEW YORK, GREENKILL OUTDOOR ENVIRONMENTAL EDUCATION CENTER
YMCA Greenkill
P.O. Box B
Huguenot, NY 12746
(914) 856-4382

Emphasis: Environmental education resident experience for elementary through high school students, focusing on environmental science, outdoor skills and history. Stresses interrelationship and cooperation. **Programs:** Environmental Science, Outdoor Skills, History, Challenge Education. **Type:** Outdoor education, natural science/history, residential. **Cost:** $85-$130 per person. **Age Emphasis:** Youth. **Application Deadline:** None.

YMCA OF THE ROCKIES—ESTES PARK CENTER
2515 Tunnel Rd.
Estes Park, CO 80511
(303) 586-3341, ext. 1106

Emphasis: Curriculum instruction, hands-on experiences and exploration of the natural world provided through discovery and linked to interdisciplinary approach. **Programs:** Beaver Walk, Kinship, Star Challenger, Nocturnal Enchantment, Adventurer (ropes), Orienteering, Survival, Earthwalker, Nature Graphics, Writers' Camp. **Type:** Outdoor education. **Cost:** $18/lodging and food. **Age Emphasis:** All ages. **Application Deadline:** None.

YOSEMITE ASSOCIATION
P.O. Box 230
El Portal, CA 95318
(209) 379-2646

Emphasis: Outdoor natural history seminars. College credit, family trips, backpacking trips for beginners through advanced. **Programs:** History, botany, geology, Native American studies, art, photography. Over 75 classes. **Type:** Outdoor education. **Cost:** $45-$350 (some include meals and lodging). **Age Emphasis:** All ages, plus families. **Application Deadline:** Write for free brochure.

YOSEMITE NATIONAL INSTITUTE
Golden Gate National Recreation Area,
Bldg. 1033
Sausalito, CA 94965
(415) 332-5771

Emphasis: Environmental learning experiences for teachers, students and the general public. Three campuses: Yosemite National Park, Olympic National Park and Golden Gate National Park. **Programs:** Residential environmental education. **Type:** Outdoor education, experiential, natural science/history, natural resources, residential. **Cost:** Varies, $100-$200 per week, including all meals, lodging, instruction. **Age Emphasis:** All ages. **Application Deadline:** None.

YUKON CONSERVATION SOCIETY
P.O. Box 4163
Whitehorse, Yukon YIA 3T3, Canada
(403) 668-5678

Emphasis: Advocacy, education and research on conservation issues in the Yukon Territory. **Programs:** Natural appreciation, travel guides, green consumer guide. **Type:** Natural science/history, natural resources. **Cost:** Varies. **Age Emphasis:** All ages. **Application Deadline:** None.

Nature Centers

ANDORRA NATURAL AREA
Old Northwestern Ave.
Philadelphia, PA 19118
(215) 685-9285

Emphasis: Interpretive programs centering on natural and woodland stream areas of Fairmount Park, one of the world's largest urban parks. **Programs:** Bird walks, campfires, interpretive hikes, maple sugaring program. **Type:** Outdoor education. **Cost:** Most programs are free. **Age Emphasis:** All ages. **Application Deadline:** None.

Q: *Which world event was responsible for mobilizing international cooperation in the future phasing out of production of CFCs?*

ECOCONNECTIONS

ANITA PURVES NATURE CENTER
1505 N. Broadway
Urbana, IL 61801
(217) 384-4062
Emphasis: Nature Center and adjacent Busey Woods provide a natural resource for individuals, families and school groups. Center exhibits highlight animals, birds and natural phenomena. **Programs:** Vary seasonally. **Type:** Expedition/experiential, natural science/history, natural resources. **Cost:** Free or varied fees; most under $20. **Age Emphasis:** All ages. **Application Deadline:** Generally open, but depends on program.

BANDELIER NATIONAL MONUMENT
HCR 1, Box 1, Ste. 15
Los Alamos, NM 87544-9701
(505) 672-3861
Emphasis: Bandelier National Monument was established to preserve Native American ruins in a beautiful, wild setting. **Programs:** Native American culture, past and present; plants; wildlife; geology; wilderness; etc. **Type:** Natural science/history, natural resources. **Cost:** $5 per carload, good for 7 days. **Age Emphasis:** All ages. **Application Deadline:** Groups call ahead.

CABLE NATURAL HISTORY MUSEUM
P.O. Box 416
Cable, WI 54821
(715) 798-3890
Emphasis: Natural history of the northern Great Lakes region. **Programs:** Workshops, lectures, field trips, etc. **Type:** Outdoor education, expedition/experiential, natural science/history, natural resources. **Cost:** Varies, from $1 to $600. **Age Emphasis:** All ages. **Application Deadline:** None.

CALLAWAY GARDENS
P.O. Box 2000
Pine Mountain, GA 31822-2000
(706) 663-5140
Emphasis: Wide variety of natural history and gardening programs. **Programs:** Summer Recreation Program, Summer Internships. **Type:** Outdoor education, natural science/history. **Cost:** Varies. **Age Emphasis:** All ages; internships available for college students. **Application Deadline:** February 1, spring. March 1, summer.

CAMP ALLEN
Rte. 1, Box 426
Navasota, TX 77868
(409) 825-7175
Emphasis: To provide educational experiences that encourage students to actively participate in solving and preventing the problems facing their generation and future operations. **Programs:** The Discovery Program. **Type:** Outdoor education. **Cost:** Varies, $25-$40/day. **Age Emphasis:** All ages. **Application Deadline:** None.

CAPE COD MUSEUM OF NATURAL HISTORY
P.O. Box 1710, Rte. 6A
Brewster, MA 02631
(508) 896-3867
Emphasis: Nonprofit education and research science center focused on the natural environment of Cape Cod. To promote better understanding and appreciation for the environment and the means to sustain it. **Programs:** Environmental education. **Type:** Marine labs, outdoor education, teacher education, kids' ecology club, natural science/history, natural resources, nature camp. **Cost:** From $5 per year to $66 per class. **Age Emphasis:** All ages. **Application Deadline:** None.

CENTER FOR ALASKAN COASTAL STUDIES
P.O. Box 2225
Homer, AK 99603
(907) 235-6667
Emphasis: Intertidal studies and forest ecology, northwest coast rainforest. **Programs:** Natural history day tours. **Type:** Outdoor education, expedition/experiential, natural science/history, natural resources, residential. **Cost:** $49/adult, $39/seniors, $29/children. **Age Emphasis:** All ages. **Application Deadline:** None.

CINCINNATI ZOO AND BOTANICAL GARDEN
3400 Vine St.
Cincinnati, OH 45220
(513) 281-4700
Emphasis: Extensive captive breeding and research, emphasis on local solutions to global environmental problems. One of the largest captive collections of endangered species. **Programs:** College internships, college courses, high school vocational program,

A: *The Montreal Treaty of 1987*
Source: Newsweek magazine

accelerated nature study for gifted students, members' programs, multimedia environmental materials. **Type:** Outdoor education, expedition/experiential, natural science/history, natural resources. **Cost:** Varies. **Age Emphasis:** All ages. **Application Deadline:** None.

CLAY PIT PONDS STATE PARK PRESERVE
83 Nielsen Ave.
Staten Island, NY 10309
(718) 967-1976
Emphasis: 260-acre natural area on Staten Island. Once the site of a clay mining operation, the park contains a mixture of unique habitats such as wetlands, sandy barrens, fields, etc. Managed to retain unique ecology and to provide educational programs. **Programs:** Year-round nature programs. **Type:** Outdoor education, natural science/history, natural resources. **Cost:** Free. **Age Emphasis:** All ages. **Application Deadline:** None.

COLORADO OUTDOOR EDUCATION CENTER, SANBORN WESTERN CAMPS
2000 Old Stage Rd.
Florissant, CO 80816
(719) 748-3341; (719) 748-3475
Emphasis: The Nature Place (team building for business and educational groups). **Programs:** High Trails, Outdoor Education, The Nature Place, Sanborn Western Camps. **Type:** Outdoor education, expedition/experiential, natural science/history, residential. **Cost:** Depends on session. **Age Emphasis:** All ages. **Application Deadline:** Depends on program.

DALLAS MUSEUM OF NATURAL HISTORY
P.O. Box 150439
Dallas, TX 75315
(214) 670-8458
Emphasis: Regional museum focusing on the natural history of Texas. Collects, preserves and interprets the record of the natural world, of humans and their environment. Also, research and teaching. **Programs:** Weekend discovery centers and Saturday science courses. Also, summer classes and week-long day camps, outreach programs, teacher courses (accredited). **Type:** Natural science/history, natural resources. **Cost:** Some programs are free; others range from $4-$30. **Age Emphasis:** All ages. **Application Deadline:** Varies, but registration is required for all courses.

EAGLE'S NEST CAMP
633 Summit St.
Winston Salem, NC 27101
(919) 761-1040 (winter)
(704) 877-4349 (summer)
Emphasis: Wilderness experience, natural history, art, music, community life, cultural exchange. **Programs:** Summer camp, Hante mountaineering, whitewater, Paleo man experiment, Longhouse journey. **Type:** Outdoor education, expedition/experiential, natural science/history, natural resources, residential. **Cost:** $390/week. **Age Emphasis:** Youth. **Application Deadline:** Spring.

ELGIN PUBLIC MUSEUM
225 Grand Blvd.
Elgin, IL 60120
(708) 741-6655
Emphasis: Museum of natural history, including anthropology, botany, geology, paleontology, zoology. Emphasis on North America, specifically the Midwest. Discovery Room. **Programs:** Preschool through high school environmental education and natural history. Adult lectures. **Type:** Outdoor education, natural science/history. **Cost:** Admission fee, $1/adults, $.25/children and seniors. Group fee charged for group programs. **Age Emphasis:** Youth. **Application Deadline:** None.

GREENBURGH NATURE CENTER
Dromore Rd. off Central Ave.
Scarsdale, NY 10583
(914) 723-3470
Emphasis: Natural history exhibits, live animal museum, programs for all ages, college and high school internships (residential available), special events, environmental education, cultural programs. **Programs:** Plant and Animal Adaptation, To Build or Not to Build, etc. **Type:** Outdoor education, natural science/history, natural resources. **Cost:** Varies. **Age Emphasis:** All ages. **Application Deadline:** None.

THE HAWAII NATURE CENTER
2131 Makiki Heights Dr.
Honolulu, HI 96822
(808) 955-0100
Emphasis: Discovery of nature through one-of-a-kind, hands-on experiences for children. Outdoor field trips for K-5 grades. Family nature adventures. Guided interpretative hikes.

Q: *How many people will die in 1992 from cancers resulting from radiation exposure from nuclear testing?*

Field sites: Oahu and Maui. **Programs:** Numerous school and community programs. **Type:** Outdoor education, expedition/experiential, natural science/history. **Cost:** Varies. **Age Emphasis:** All ages. **Application Deadline:** None.

THE HIGH DESERT MUSEUM
59800 S. Hwy. 97
Bend, OR 97702
(503) 382-4754
Emphasis: Natural and cultural history of the Intermountain West. **Programs:** Field excursions; fall, spring, summer classes; speaker series. **Type:** Outdoor education, expedition/experiential, natural science/history, natural resources, wildlife and landscape photography. **Cost:** $5/Seniors, $5.50/Adults, $2.75/children (5-12); free/members and under 5. **Age Emphasis:** All ages. **Application Deadline:** None.

HOUSTON MUSEUM OF NATURAL SCIENCE
1 Hermann Circle Dr.
Houston, TX 77030
(713) 639-4629
Emphasis: Resource center for the public and teachers. Ecological principles behind environmental issues. **Programs:** Teacher workshops (state-mandated environmental education training site), lectures, in-service programs, children's classes. **Type:** Natural science/history, natural resources. **Cost:** Minimal. **Age Emphasis:** Adults and teachers. **Application Deadline:** None.

HULBERT OUTDOOR CENTER
RR #1, Box 91A
Fairlee, VT 05045
(802) 333-9840
Emphasis: Year-round residential and wilderness trip programs, team-building, natural history, photography, outdoor skill development. **Programs:** Youth groups, adult groups, schools, corporations, elder hostels. **Type:** Outdoor education, expedition/experiential, natural science/history, residential. **Cost:** Varies from $30-$250/person/day. **Age Emphasis:** All ages. **Application Deadline:** None.

ISLAND INSTITUTE
4004 58th Pl. SW
Seattle, WA 98116
(206) 938-0345
Emphasis: Marine science immersion experiences; sea kayaking, snorkel/SCUBA, whale-watching, island explorations in the San Juan Islands, boat charters. **Programs:** Advanced marine mammal ecology course, whale camp, boat camp. **Type:** Degree program, expedition/experiential, natural science/history, natural resources, residential. **Cost:** $695 per week. **Age Emphasis:** All ages. **Application Deadline:** Register by May 15.

KEYSTONE SCIENCE SCHOOL
P.O. Box 8606
Keystone, CO 80435
(303) 468-5824
Emphasis: Field science focuses on ecosystems of the Central Rockies; montane; subalpine and alpine ecology; forest ecology; aquatic ecology; winter ecology; snow physics; geology; mining; cultural history; wildlife biology. **Programs:** School groups, teacher workshops, elder hostels, summer discovery camp. **Type:** Outdoor education, natural science/history, residential. **Cost:** $46 per day. **Age Emphasis:** All ages. **Application Deadline:** Usually filled 9-12 months in advance.

KOKEE NATURAL HISTORY MUSEUM
P.O. Box 100
Kekaha, HI 96752
(808) 335-9975
Emphasis: Natural history of Kauai and Hawaiian archipelago. Museum is open 365 days per year/free admission. **Programs:** Forest-Wise Earth Education Camp, Earth-Wise Symposium and Festival. **Type:** Outdoor education, expedition/experiential, natural science/history, residential. **Cost:** Varies, from free to $250/week. **Age Emphasis:** All ages. **Application Deadline:** None.

LAC LAWRANN CONSERVANCY
1115 S. Main St.
West Bend, WI 53095
(414) 335-5080.
Emphasis: Natural sciences, environmental education programs for the general public and on an appointment basis, regularly scheduled youth, adult and school programs during the fall and spring. **Programs:** Project Wild, Living Lightly in the City, Nature Scope, Dream Chasers or other national programs. **Type:** Outdoor education, expedition/experiential. **Cost:** Free on site, at cost off site. **Age Emphasis:** All ages. **Application Deadline:** None.

A: *430,000 worldwide*
Source: Greenpeace

LAKE ERIE NATURE AND SCIENCE CENTER
28728 Wolf Rd.
Bay Village, OH 44140
(216) 871-2900
Emphasis: Provides nature experiences and stimulates environmental action through education. **Programs:** 25 physical and natural science classes (K-8) plus various play-school and adult programs. **Type:** Expedition/experiential, natural science/history. **Cost:** $3/child. **Age Emphasis:** Youth. **Application Deadline:** None.

MARINE MAMMAL STRANDING CENTER
P.O. Box 773
Brigantine, NJ 08203
(609) 266-0538
Emphasis: Responsible for rescuing, rehabilitating and releasing marine mammals and turtles in New Jersey. The center has worked with over 850 animals since 1978, and these animals are the basis for the educational program. **Programs:** Ocean's Barometer, Songs, Stories and Birds. **Type:** Natural science/history, natural resources. **Cost:** $1/person on site; fees vary by mileage away from site, from one hour to full day. **Age Emphasis:** All ages. **Application Deadline:** None.

NORTH PARK VILLAGE NATURE CENTER
5801 N. Pulaski
Chicago, IL 60646
Emphasis: Nature Discovery school field trip program; maple syrup education program and festival; Arbor Day activities; public programs for natural history education, bird walks, wildflower identification, etc.; teacher workshops. **Programs:** Natural history field trips and public programming. **Type:** Outdoor education, expedition/experiential, natural science/history, natural resources. **Cost:** Most programs are free. **Age Emphasis:** All ages. **Application Deadline:** Reservations are often neccessary.

OGLEBAY INSTITUTE NATURE AND ENVIRONMENTAL EDUCATION DEPARTMENT
Oglebay Park; A.B. Brooks Nature Center
Wheeling, WV 26003
(304) 242-6855
Emphasis: Outdoor environmental education. **Programs:** None. **Type:** Junior Nature Camp—one- or two-week resident camp sessions to study and enjoy natural history, out door recreation, and canoe tripping. **Cost:** $275/both weeks, $140/one week. **Age Emphasis:** 11-15, students. **Application Deadline:** August.

OGLEBAY INSTITUTE NATURE AND ENVIRONMENTAL EDUCATION DEPARTMENT
Oglebay Park; A.B. Brooks Nature Center
Wheeling, WV 26003
(304) 242-6855
Emphasis: Family camping. **Programs:** Varies according to season. **Type:** Terra Alta Mountain Nature Camp—one- or two-week resident camp sessions conducted in the Appalachians of West Virginia from a rustic base camp. Geological formations and flora and fauna unique to a boreal forest will be explored. **Cost:** $300/two weeks, $155/one week. **Age Emphasis:** 16 and above. **Application Deadline:** June.

POINT BONITA YMCA OUTDOOR EDUCATION CONFERENCE CENTER
Bldg. 981
Fort Barry, Sausalito, CA 94965
(415) 331-9622
Emphasis: Hands-on learning experiences about coastal ecology, environments and habitats. Lighthouse and military history resources nearby. **Programs:** N/A. **Type:** Outdoor education, expedition/experiential, natural science/history, natural resources, residential. **Cost:** $34 lodging, 3 meals, day and evening program. 1-5 days. **Age Emphasis:** Youth. **Application Deadline:** None.

POTOMAC OVERLOOK REGIONAL PARK & NATURE CENTER
2845 Marcey Rd.
Arlington, VA 22207
(703) 528-5406
Emphasis: Interrelationships between people and the rest of nature. **Programs:** Examples: The Human and Natural Heritage of Potomac Overlook Regional Park; Canoe the Marsh. **Type:** Outdoor education, expedition/experiential, natural science/history, natural resources. **Cost:** Most are free; $5/person charge for group canoe trips. **Age Emphasis:** All ages. **Application Deadline:** Usually 2 weeks prior to program.

ROCKY MOUNTAIN NATURE ASSOCIATION
Rocky Mountain National Park
Estes Park, CO 80517
(303) 586-3565, ext. 258

 How many plastic bottles do Americans use every hour?

Emphasis: Field seminars on the geology, ecology, and natural history of Rocky Mountain National Park. **Programs:** Hiking, lectures, group discussions, bird watching, photography and plant identification. **Type:** Experiential, natural science/history, natural resources, residential. **Cost:** $40-$160; 1-7 days. **Age Emphasis:** All ages. **Application Deadline:** None.

SAN FRANCISCO BAY BIRD OBSERVATORY
P.O. Box 247
Alviso, CA 95002
(408) 946-6548

Emphasis: Research and education focuses on marshes, sloughs and salt ponds of the south San Francisco Bay. Volunteers run research and education programs. **Programs:** Varied. **Type:** Outdoor education, expedition/experiential, natural science/history. **Cost:** Varies, $30-$100. Tours (camping) range from $35-$500. **Age Emphasis:** All ages. **Application Deadline:** None.

SEACAMP
Rte. 3, Box 170
Big Pine Key, FL 33043
(305) 872-2331

Emphasis: The marine science program is the heart of Seacamp. **Programs:** Activities include scuba, windsurfing, sailing, arts and crafts, snorkeling, fishing and photography. **Type:** Outdoor education, expedition/experiential, natural science/history, natural resources, residential, marine science. **Cost:** $1,900. **Age Emphasis:** 12-17. **Application Deadline:** First come, first served; usually booked by March.

SLIDE RANCH
2025 Shoreline Hwy.
Muir Beach, CA 94965
(415) 381-6155

Emphasis: Programs focus on human dependence upon the Earth's resources to provide food, clothing, shelter and energy. The wildlands, ocean shore and farmstead provide an outdoor classroom for exploring life, water and energy cycles. **Programs:** School and family programs, intern program (interns live and work at Slide Ranch gaining practical experience leading programs for children, seniors, disabled, family groups). **Type:** Outdoor education, expedition/experiential, natural science/history, natural resources, residential, farm. **Cost:** Volunteers must commit for 6 months; room and board provided. School and Family sessions, $10/person/day. **Age Emphasis:** All ages. **Application Deadline:** End of October/(spring), end of April/(summer); send for application.

SOMERSET COUNTY PARK COMMISSION'S ENVIRONMENTAL EDUCATION CENTER
190 Lord Stirling Rd.
Basking Ridge, NJ 07920
(908) 766-2489

Emphasis: Conservation and preservation of our natural resourses. **Programs:** Environmental education programs for schools and the general public. **Type:** Outdoor education, expedition/experiential, natural science/history, natural resources. **Cost:** Varies. **Age Emphasis:** All ages. **Application Deadline:** None.

SONORAN ARTHROPOD STUDIES, INC.
P.O. Box 5624
Tucson, AZ 85703
(602) 883-3945

Emphasis: Environmental and science education focusing on insects and other arthropods with an emphasis on interrelationships and biodiversity. Conservation is high priority, especially reduction of pesticide use through increased knowledge. **Programs:** Large variety, from lectures to expeditions. **Type:** Outdoor education, expedition/experiential, natural science/history. **Cost:** Varies, $10-$75 for workshops and day trips, $5-$1,500 for expeditions. **Age Emphasis:** All ages. **Application Deadline:** Continuous through year. Members notified through publications, public through media.

TREE HILL, JACKSONVILLE NATURE CENTER
7152 Lone Star Rd.
Jacksonville, FL 32211
(904) 724-4646

Emphasis: Tree Hill is a nature preserve featuring two nature trails, a garden parcourse, two science laboratories and a natural history museum. Field experience in a "nature classroom" focuses on forest ecology. **Programs:** Environmental education for elementary grades in Duval County; forest ecology for Scouts; senior citizen programs; Science Day Camp. **Cost:** Contracts with agencies; some programs fee-based. **Age Emphasis:** All ages. **Application Deadline:** None.

A: *2.5 million*
Source: 50 Simple Things You Can Do to Save the Earth, *EarthWorks Press*

WYOMING GAME AND FISH DEPARTMENT
5400 Bishop Blvd.
Cheyenne, WY 82006
(307) 777-4543
Emphasis: Wildlife conservation; education/interpretive education; wildlife and wildland viewing experiences in Wyoming under the "Wyoming Wildlife—Worth the Watching" program. **Programs:** Visitor centers, viewing areas, nature areas tour guide. **Type:** Outdoor education. **Cost:** No charge at visitor centers and other facilities. **Age Emphasis:** All ages. **Application Deadline:** None.

University Programs

ANTIOCH NEW ENGLAND GRADUATE SCHOOL
103 Roxbury St.
Keene, NH 03431
(603) 357-3122
Emphasis: Graduate professional training programs for the environmental field. Includes resource management, environmental administration, environmental communications, environmental education and teacher certification in biology, general science and elementary science. **Programs:** Master of Science in Environmental Studies, Master of Science in Resource Management and Administration. **Type:** Degree program, outdoor education, expedition/experiential, natural science/history, natural resources. **Cost:** 50-credit, Resource Management M.S.: $14,700. 40-credit, Environmental Studies M.S.: $11,750. **Age Emphasis:** Adults. **Application Deadline:** Entry points are September, January and June.

BARD COLLEGE—GRADUATE SCHOOL OF ENVIRONMENTAL STUDIES
Annandale Rd.
Annandale, NY 12504
(914) 758-7483
Emphasis: Provides an interdisciplinary education that leads to an effective understanding of current environmental problems and the ability to put that understanding to use. **Programs:** Master of Science in Environmental Studies. **Type:** Three summer sessions degree program; source and fieldwork; natural science/social science mix. **Cost:** Tuition $4,950; some financial aid available. **Age Emphasis:** Adults. **Application Deadline:** April 1.

BASTYR COLLEGE
144 NE 54th
Seattle, WA 98105
(206) 523-9585
Emphasis: Extensive clinical training of practitioners of natural therapeutics. **Programs:** Naturopathic medicine, acupuncture, nutrition. **Type:** Degree program, residential. **Cost:** 4-year, $10,000/year; 3-year, $7,000/year; 2-year, $5,000/year. **Age Emphasis:** Adults. **Application Deadline:** February 1, April 1.

BAYLOR UNIVERSITY—INSTITUTE OF ENVIRONMENTAL STUDIES
P.O. Box 97266
Waco, TX 76798-7266
(817) 755-3405
Emphasis: Renewable energies; hazardous waste management; air and water pollution; sustainable agriculture; permitting and regulations; environmental ethics. **Programs:** Bachelor of Science; Bachelor of Arts; Master of Science; Master of Environmental Studies. **Type:** Degree program, natural sciences/history, natural resources. **Cost:** $12,000/year for B.A./B.S. program; $7,250/year for M.S. (negotiated with director for full support). **Application Deadline:** 90 days before semester begins.

BELOIT COLLEGE, DEPT. OF BIOLOGY
700 College Ave.
Beloit, WI 53511
(608) 363-2287
Emphasis: An ecologic approach to concepts and issues, rooted in a liberal arts tradition with strong emphasis on practical experience, individualized programs, management and research. **Programs:** Environmental biology. **Type:** Private residential degree program. **Cost:** $14,000/year; financial aid available. Length of program: 4 years. **Age Emphasis:** Adult, college student. **Application Deadline:** None.

BRADFORD WOODS
5040 State Rd. 67N
Martinsville, IN 46151
(317) 342-2915
Emphasis: Outdoor recreation. **Programs:** Outdoor education, challenge education, summer camp, leadership development, internships. **Type:** Degree program, outdoor education, expedition/experiential, residential. **Cost:** Varies. **Age Emphasis:** All ages. **Application Deadline:** None.

: *In which country do more than 40 percent of the people bicycle to work?*

ECOCONNECTIONS

THE CITY COLLEGE OF NEW YORK C.U.N.Y., SCHOOL OF EDUCATION
Mr. Joseph Fonseca
138 St. and Convent Ave., R6204
New York, NY 10031
(212) 650-6236
Emphasis: Science and social studies. **Programs:** An interdisciplinary graduate program leading to an M.A. degree and New York state teacher certification. **Type:** Degree program. **Cost:** $82/credit; 30-credit program. **Age Emphasis:** Adults. **Application Deadline:** For fall matriculation, March 1; for spring, October 1.

CORNELL LABORATORY OF ORNITHOLOGY
159 Sapsucker Woods Rd.
Ithaca, NY 14850
(607) 254-BIRD
Emphasis: International center for study, appreciation and conservation of birds. Develops, applies and shares tools for understanding birds and protecting bird populations. **Programs:** Home study course in bird biology, field courses in bird study and bird song recording, educational birding tours. **Type:** Expedition/experiential, natural science/history. **Cost:** Home study course, $135; field courses, $100 and up; tours, $1,500 and up. **Age Emphasis:** All ages. **Application Deadline:** Write for information.

CUEST (CENTER FOR UNDERSTANDING ENVIRONMENTS, SCIENCE AND TECHNOLOGY)
Box 770
Northern State University
Aberdeen, SD 57401
(605) 622-2627
Emphasis: CUEST is an environmental education resource center for teachers that offers environmental education workshops, preservice and in-service teacher training, audiovisual and printed matter on environmental issues, and activities and speakers for community outreach. **Programs:** Environmental education, undergraduate level classes and a collection of environmental education resourse materials. **Type:** Degree program (undergraduate and continuing education), natural resources. **Cost:** $76.45 (on campus), $65.00 (off campus)/undergraduate credit hour, $96.45 (on campus), $85.00 (off campus)/graduate credit hour. **Age Emphasis:** All ages. **Application Deadline:** None.

DE ANZA COLLEGE—ENVIRONMENTAL STUDIES/BIOLOGY
21250 Stevens Creek Blvd.
Cupertino, CA 95014
(408) 864-8525
Emphasis: Promotes environmental education, environmental biology careers and environmental awareness. **Programs:** A.A. degree in environmental studies, tours of environmental studies area of the De Anza College. **Type:** Outdoor education, expedition/experiential, natural science/history, natural resources, residential. **Cost:** Varies. **Age Emphasis:** All ages. **Application Deadline:** None.

EASTERN KENTUCKY UNIVERSITY—DEPT. OF ENVIRONMENTAL HEALTH SCIENCE
219 Dizney Bldg.
Richmond, KY 40475-3135
(606) 622-1939
Emphasis: Environmental health. **Programs:** B.S. in Environmental Health Science. **Type:** Degree program. **Cost:** Tuition plus fees. **Age Emphasis:** Adults. **Application Deadline:** None.

GARRETT COMMUNITY COLLEGE
P.O. Box 151
McHenry, MD 21541
(800) 695-4221; (301) 387-6666
Emphasis: Prepares students for careers in wildlife management, soil and water conservation, fisheries management, water quality monitoring and other natural resources areas. **Programs:** Natural Resources and Wildlife Technology, A.A. Degree (Degree in Associate Arts). **Type:** Degree program, outdoor education, expedition/experiential, natural science/history, natural resources. **Cost:** $44/credit (MD and WV residents), $111/credit (out of state). **Age Emphasis:** Adult. **Application Deadline:** Must register before classes begin in September.

ILLINOIS STATE UNIVERSITY—DEPT. OF HEALTH SCIENCES
5220 Health Sciences
Normal, IL 61761-6901
(309) 438-8329
Emphasis: Majors available in Environmental Health or Health Education. Combinations of the two degrees are available as a major and minor. **Programs:** Bachelor of Science in Health

A: *Denmark*
Source: Protect Our Planet Calendar, Running Press Book Publisher

Science. **Type:** Degree program. **Cost:** Tuition, fees and books are approximately $1,500 per semester. **Age Emphasis:** Adults. **Application Deadline:** Submit as early as possible.

IOWA STATE UNIVERSITY—ENVIRONMENTAL STUDIES PROGRAM
201 Bessey Hall
Ames, IA 50011
(515) 294-4787; (515) 294-4911
Emphasis: Comprehensive program of departmental and multi-disciplinary environmental courses in the natural and social sciences and humanities. **Programs:** Environmental studies major and minor. **Type:** Degree program. **Cost:** Tuition: $1,044/semester in-state, $3,428 semester qout-of-state. **Age Emphasis:** Adults. **Application Deadline:** None.

JORDAN COLLEGE ENERGY INSTITUTE
155 Seven Mile Rd. NW
Comstock Park, MI 49321
(616) 784-7595
Emphasis: Renewable energy technology. **Programs:** Degree programs in renewable energies, applied environmental technology, energy management, energy-efficient construction and design, solar retrofit technology and business courses available. **Type:** Degree program (B.S.). **Cost:** $120/credit hour maximum, financial aid available. Length of program: 1-year certificates, 2-year associate degrees, 4-year bachelor's. **Age Emphasis:** Adult college student. **Application Deadline:** None.

MCNEESE STATE UNIVERSITY—DEPT. OF BIOLOGICAL AND ENVIRONMENTAL SCIENCES
P.O. Box 92000-0655
Lake Charles, LA 70609-2000
(318) 475-5674
Emphasis: Studies in biology, microbiology, chemistry and mathematics to aid students in finding creative solutions to the complex problems of our environment. Graduate students specialize in air and water quality studies. **Programs:** B.S. and M.S. in Environmental Science and Biological Sciences. **Type:** Degree program. **Cost:** Tuition varies. **Age Emphasis:** Adults. **Application Deadline:** Fall registration, August 15; spring registration, December 15.

MURRAY STATE UNIVERSITY—DEPT. OF ELEMENTARY AND SECONDARY EDUCATION
Wells Hall
Murray, KY 42071
(502) 762-2747
Emphasis: Teacher training, pre-service, in-service, community and public school outreach, curriculum development and dissemination, research, resource room. **Programs:** Environmental Outreach, Center for Environmental Education Resource Room, Western Kentucky Environmental Consortium. **Cost:** None. **Age Emphasis:** All ages. **Application Deadline:** None.

NEW YORK UNIVERSITY
Professor Millard Clements
Department of Teaching and Learning
239 Greene St., Rm. 635
Washington Square
New York, NY 10003
(212) 998-5495
Emphasis: Non-science interdisciplinary M.A. program based on core courses in social, philosophical and organizational aspects of the environment, with half the credits in electives. Internship required. **Programs:** M.A. in Environmental Conservation Education. **Type:** Degree program, interdisciplinary. **Cost:** $484/credit; 37 credits. **Age Emphasis:** Adults. **Application Deadline:** Rolling admissions.

NEW YORK UNIVERSITY—MANAGEMENTS INSTITUTE
44 W. Fourth St.
New York, NY 10012
(212) 998-0200
Emphasis: Business and the environment. **Programs:** Managing for the Future, Environmental Economics; Recycling: How to Make it Work; Environmental Marketing; Environmental Law for the Layperson; Environmentally Sound Product Development, etc. **Type:** Continuing Education. **Cost:** $250-$300. Length: 6-12 weeks. **Age Emphasis:** Adult. **Application Deadline:** None.

NORTHLAND COLLEGE
1411 Ellis Ave.
Ashland, WI 54806
(715) 682-1699
Emphasis: Environmental education, field experience, internships. **Programs:** N/A.

: *Of the 34 most widely used lawn pesticides, how many has the EPA fully assessed for health effects?*

Type: Degree program, outdoor education, expedition/experiential, natural science/history, natural resources, residential. **Cost:** $12,810 per academic year, financial aid available. **Age Emphasis:** College student. **Application Deadline:** None.

OBERLIN COLLEGE—ENVIRONMENTAL STUDIES PROGRAM
Rice Hall-31
Oberlin, OH 44074
(216) 775-8409
Emphasis: Interdisciplinary program of study (natural science, social science, humanities) leading to a B.A. degree. **Programs:** Environmental Studies Major. **Type:** Degree program. **Cost:** Tuition: $17,600/year. **Age Emphasis:** Adult. **Application Deadline:** None.

OHIO SEA GRANT EDUCATION PROGRAM
Ohio State University
059 Ramseyer Hall
29 W. Woodruff Ave.
Columbus, OH 43210-1085
(614) 292-1078
Emphasis: Curriculum development and teacher education in Great Lakes education. **Programs:** Great Lakes Education Workshop. **Type:** Natural science/history, natural resources. **Cost:** $30 materials fee, plus university tuition. **Age Emphasis:** Teachers. **Application Deadline:** None.

OPPORTUNITIES IN SCIENCE, INC.
P.O. Box 1176
Bemidji, MN 56601
(218) 751-1110
Emphasis: Self-contained course in ecology for adults and secondary students focuses on Environmental Education. **Programs:** WILDWAYS: Understanding Wildlife Conservation. **Type:** Degree program. **Cost:** $450 for text and computer disks. **Age Emphasis:** Adult. **Application Deadline:** None.

PRESCOTT COLLEGE
220 Grove Ave.
Prescott, AZ 86301
(602) 778-2090
Emphasis: A 4-year liberal arts college with an environmental mission. Entire curriculum has a strong environmental emphasis. Curriculum is interdisciplinary, emphasizing fieldwork throughout the Southwest. **Programs:** Environmental Studies, Human Development, Humanities, Outdoor Leadership and Cultural and Regional Studies. **Type:** Degree program, outdoor education, expedition/experiential, natural science/history, natural resources. **Cost:** $8,600/year. **Age Emphasis:** Adult. **Application Deadline:** Fall, May 15; spring, November 15.

PURDUE UNIVERSITY—ENVIRONMENTAL ENGINEERING DEPT.
1284 Civil Engineering
W. Lafayette, IN 47907-1284
(317) 494-2194
Emphasis: Industrial Wastewater Treatment, Drinking Water, Air Toxics/Noise, Municipal/Hazardous Waste, Physical/Chemical Treatment, Environmental Chemistry, Groundwater Remediation, Water Quality Management. **Programs:** Master of Science, MSCE, Ph.D. **Type:** Degree programs. **Cost:** $90/credit in-state, $269/credit out-of-state. **Age Emphasis:** Adults. **Application Deadline:** None.

RAMAPO COLLEGE OF NEW JERSEY
505 Ramapo Valley Rd.
Mahwah, NJ 07430
(201) 529-7500
Emphasis: To train individuals to address environmental problems by balancing scientific knowledge and activity to communicate across social, political and economic boundaries. **Programs:** Degree programs in environmental science and environmental studies offered. **Type:** Natural history, undergraduate degree (B.S.). **Cost:** $900 per semester, $3,300 total. Length of program: 4 years. **Age Emphasis:** College student. **Application Deadline:** None.

SHAVER'S CREEK ENVIRONMENTAL CENTER—PENN STATE UNIVERSITY
School of HR and RM
203 S. Henderson Bldg.
University Park, PA 16802
(814) 863-2000
Emphasis: On- and off-site natural history programs for schools, scouts, general public, raptor rehabilitation, teacher workshops, cooperation course, high ropes source, extensive intern program and Penn State student training courses. Residential Outdoor School for 5th and 6th graders. **Programs:** Season Discovery Walks, Traveling Road Shows, Group

A: *Two*
Source: The Student Environmental Action Guide, *EarthWorks Press*

Initiatives, Outdoor School, Maple Sugaring, Summer Day Camps. **Type:** Outdoor education, natural science/history, residential. **Cost:** Varies. **Age Emphasis:** All ages. **Application Deadline:** None.

SONOMA STATE UNIVERSITY—DEPT. OF ENVIRONMENTAL STUDIES AND PLANNING
1801 E. Cotati Ave.
Rohnert Park, CA 94928
(707) 664-2306
Emphasis: Bachelor's degree programs in environmental education, environmental conservation and restoration, environmental education, environmental technology and planning. **Programs:** Bachelor's degree in Environmental Studies and Planning. **Type:** Degree program, natural science/history, natural resources, residential. **Cost:** $551/semester. **Age Emphasis:** Adults. **Application Deadline:** Varies according to time of desired enrollment.

SOUTHERN VERMONT COLLEGE
Monument Ave.
Bennington, VT 05201
(802) 442-5427
Emphasis: Environmental studies. **Programs:** 2-year associate and 4-year bachelor programs in environmental studies. **Type:** N/A. **Cost:** Tuition, room and board: $10,600/year. **Application Deadline:** None.

STATE UNIVERSITY OF NEW YORK AT CORTLAND
P.O. Box 2000, Park Ctr.
Cortland, NY 13045
(607) 753-4941
Emphasis: The College at Cortland offers undergraduate and graduate courses and degree options in outdoor education, environmental education and interpretation, and outdoor pursuits. The college maintains three field campuses, including historic "great camp" in the Adirondacks. Cortland is also headquarters for CEO, the Coalition for Education in the Outdoors. **Type:** Degree program, outdoor education, expedition/experiential, natural science/history, natural resources, residential. **Cost:** $45/hour in-state undergraduates, $90 graduate students. **Age Emphasis:** Adults. **Application Deadline:** None.

TETON SCIENCE SCHOOL
Box 68
Kelly, WY 83011
(307) 733-4765
Emphasis: Experiential natural science education programs in the greater Yellowstone ecosystem for people of all ages. Natural science education using research as a tool. School group programs, adult seminars, 4 accredited college courses, high school field ecology. **Programs:** Ecology, Field Ornithology, Environmental Ethics, Winter Ecology. **Type:** Degree program, experiential, natural history/science, natural resources, residential. **Cost:** $40-$50/day. **Length:** 1 day-6 weeks. **Age Emphasis:** All ages. **Application Deadline:** None.

THIEL COLLEGE
75 College Ave.
Greenville, PA 16125
(412) 589-2000
Emphasis: B.A. in environmental science with emphasis in land analysis, includes introduction to GIS and land-use planning. **Programs:** B.A. **Type:** Degree program. **Cost:** $14,458/year, including room and board (tuition: $9,700). **Age Emphasis:** Adults. **Application Deadline:** The spring prior to fall semester.

UNIVERSITY OF ARIZONA—NATURAL RESOURCES CONSERVATION WORKSHOPS
RNR/BSE 325
Tucson, AZ 85721
(602) 621-7269
Emphasis: For educators, camp directors, interpretive naturalists and school teachers (K-12). Basic natural resource conservation. **Programs:** Natural resources conservation workshop for educators. **Type:** Degree program, outdoor education, natural science/history, natural resources. **Cost:** $200 plus tuition for university credit. **Age Emphasis:** Adults. **Application Deadline:** May 15, 1993.

UNIVERSITY OF TAMPA—DIVISION OF SCIENCE AND MATH
401 W. Kennedy Blvd.
Tampa, FL 33606
(813) 253-3333
Emphasis: Bachelor's degree in biology with a specialized environmental science track.

 Of the 32 most common lawn pesticides lacking full EPA health effects assessment, how many are feared to cause birth defects, gene mutation or cancer?

ECOCONNECTIONS

Programs: Degree program includes several ecologically oriented biology and chemistry courses, and recommends work in scientific writing, economics, statistics, political science and interdisciplinary studies. **Type:** Degree program. **Cost:** University tuition. **Age Emphasis:** Adults. **Application Deadline:** None.

UNIVERSITY OF UTAH—RED BUTTE GARDENS AND ARBORETUM
390 Wakara Way
Salt Lake City, UT 84108
(801) 581-8936
Emphasis: Environmental education (K-12), general horticulture and gardening techniques, native plants and ecosystems, nature programs. **Programs:** Examples—Walks in the Wasatch, plants for residential landscapes, garden lecture series, paper making, etc. **Type:** Outdoor education, expedition/experiential, natural science/history. **Cost:** Varies, from free workshops to $90/quarter university credit. **Age Emphasis:** All ages. **Application Deadline:** Ask for quarterly program announcement.

UNIVERSITY OF VERMONT—DEPT. OF BOTANY
Marsh Life Science Bldg.
Burlington, VT 05405-0086
(802) 656-2930
Emphasis: Integrative field science, environmental problem solving, strong communication skills. **Programs:** Field Naturalist Program. **Type:** Degree program. **Cost:** Fellowship. Length: 2 years. **Age Emphasis:** College student. **Application deadline:** March 1.

WILLIAMS COLLEGE—CENTER FOR ENVIRONMENTAL STUDIES
Kellogg House, P.O. Box 632,
Williamstown, MA 01267
(413) 597-2346
Emphasis: Environmental policy, planning, ethics, science. **Programs:** Undergraduate degree with concentration in Environment Studies. **Type:** Degree program. **Cost:** $23,095/year for tuition, room and board at residential liberal arts college. **Age Emphasis:** Adults. **Application Deadline:** None.

YUKON COLLEGE
Box 2799
Whitehorse, Yukon Y1A 5K4, Canada
(403) 668-8778
Emphasis: Explore the Yukon's magnificent heritage while reflecting upon the environment, society, role of education. **Programs:** Two-week program in July—Environmental Studies and Education. **Type:** Expedition/experiential, philosophical, natural science/history. **Cost:** Call for current price. **Age Emphasis:** Adults. **Application Deadline:** June.

Youth

BETSY-JEFF PENN 4-H EDUCATIONAL CENTER
804 Cedar Lane
Reidsville, NC 27320
(919) 349-9445
Emphasis: Interdisciplinary approach to exploring the environment. Environmental education program for grades 2-8; global awareness seminars (high school); teaching in a living classroom (for teachers). **Programs:** Pond ecology, forest communities, wildlife adaptations. **Type:** Outdoor education. **Cost:** Varies, $65/3 days, $120/5 days. **Age Emphasis:** Youth. **Application Deadline:** None.

CHEWONKI FOUNDATION, INC.
RR 2, Box 1200
Wiscasset, ME 04578
(207) 882-7323
Emphasis: Residential and day environmental education programs for school groups; summer camp; wilderness expeditions for youth, families and adults. **Programs:** Summer camp for boys, ages 8-14; coed wilderness experience in inland Maine, coastal Maine and Quebec, ages 14-18; The Maine Coast Semester (personal and academic challenge for 11th graders interested in environmental issues and natural science); family and adult wilderness expeditions. **Type:** Outdoor education, expedition/experiential, natural science/history, residential/day, academic semester. **Cost:** Varies, from $12/day to $8,250/semester. **Age Emphasis:** All ages. **Application Deadline:** Depends on program.

CLEARING MAGAZINE—ENVIRONMENTAL EDUCATION PROJECT
19600 S. Molalla Ave.
Oregon City, OR 97045
(503) 656-0155

A: *More than 20 percent*
Source: The Student Environmental Action Guide, *Earth Works Press*

Emphasis: Provides high-quality resource materials, teaching ideas, and information for parents and teachers (K-12) interested in providing environmental education. **Programs:** *Clearing Magazine: Environmental Education in the Pacific Northwest.* **Type:** Magazine. **Cost:** $15/year (5 issues). **Age Emphasis:** All ages. **Application Deadline:** None.

CORNELL WASTE MANAGEMENT INSTITUTE
468 Hollister Hall
Ithaca, NY 14853
(607) 255-1187
Emphasis: Solid waste, recycling and composting education for K-12. **Programs:** Audiovisual resources, games, workbooks, posters, videos, computer disks. **Type:** Natural resources. **Cost:** Varies. **Age Emphasis:** Youth. **Application Deadline:** None.

EBERSOLE ENVIRONMENTAL EDUCATION AND CONFERENCE CENTER
3400 Second St.
Wayland, MI 49348
(616) 792-6294
Emphasis: Environmental education programs for school-age children. Site includes fen, lake, climax hardwood forest, prairie. **Programs:** Varied; discovery-oriented curriculum; off-site wilderness canoe trip; 2-week summer camp; 10 days in Canadian wilderness. **Type:** Outdoor education, natural science/history, natural resources, residential. **Age Emphasis:** Youth. **Cost:** Varies, $50-$120. **Application Deadline:** None.

ECHO HILL OUTDOOR SCHOOL
13655 Bloomingneck Rd.
Worton, MD 21678
(301) 348-5880
Emphasis: Helping children learn about the environment and themselves in residential programs designed to heighten awareness and appreciation of the natural world. **Programs:** Outdoor School, Explore, Quicksilver Project. **Type:** Outdoor education, experiential, natural science/history, natural resources, residential. **Cost:** Free; 3-5 days. **Age Emphasis:** Youth. **Application Deadline:** None.

EDUCATIONAL DEVELOPMENT SPECIALISTS
5505 E. Carson St., Ste. 250
Lakewood, CA, 90713
(310) 420-6814

Emphasis: Natural resource and energy source education programs for grades K-12. **Programs:** "Think Earth" for grades K-6; "Energy Source" for K-12. Program is private; local government or business sponsors exist in some areas. **Type:** Natural resources. **Cost:** Varies. **Age Emphasis:** Youth. **Application Deadline:** None.

EXPEDITION YELLOWSTONE!
National Park Service
P.O. Box 168
Yellowstone National Park, WY 82190
(307) 344-7381
Emphasis: Curriculum and story-book about Yellowstone Park, targeted to grades 4-6. **Programs:** Expeditions to the park in spring and fall. Learning activities cover geology, history, wildlife; suitable for classroom or outdoors. **Type:** Outdoor education. **Cost:** Write for information. **Age Emphasis:** Youth. **Application Deadline:** Write for information.

GEOTHERMAL EDUCATION OFFICE
664 Hilary Dr.
Tiburon, CA 94920
(800) 866-4GEO; (415) 435-4544
Emphasis: K-12 and adult education about geothermal energy and renewables, with environmental orientation. **Programs:** Materials only; no programs. User-friendly lay information. **Type:** Natural resources. **Cost:** Single sets of information, $3. Minimal cost for classroom sets. **Age Emphasis:** All ages. **Application Deadline:** None.

THE GREEN SCENE
University of Arizona School for Renewable Natural Resources
BioScience E., Rm. 325
Tucson, AZ 85721
Emphasis: Joint effort of the Wilderness Society, University of Arizona and USDA Forest Service. **Programs:** The Green Scene. **Type:** Curriculum for middle schools (grades 4-8) on forest ecology and wilderness. **Cost:** $5. **Age Emphasis:** Youth. **Application Deadline:** None.

HIDDEN VILLA ENVIRONMENTAL EDUCATION PROGRAM
26870 Moody Rd.
Los Altos Hills, CA 94022
(415) 948-4690

Q: *How many disposable diapers are thrown away by Americans every year?*

Emphasis: Providing children with firsthand experiences with nature, emphasizing the interrelatedness of all living things. **Programs:** Farm Tours, Farm and Wilderness Exploration. **Type:** Experiential, natural science/history. **Cost:** $4-$12/person. **Age Emphasis:** Preschool to grade 6. **Application Deadline:** Reservations September-May.

HIDDEN VILLA ENVIRONMENTAL EDUCATION INTERNSHIP PROGRAM
26870 Moody Rd.
Los Altos Hills, CA 94022
(415) 948-4690

Emphasis: Providing young adults with skills in teaching environmental education and organic farming. **Programs:** 9-month internship (sometimes available for one semester). **Type:** Experiential, natural science/history, organic farming. **Cost:** No cost. Pays $100/month. **Age Emphasis:** College age or beyond. **Application Deadline:** None.

HIDDEN VILLA ENVIRONMENTAL EDUCATION SUMMER CAMP PROGRAM
26870 Moody Rd.
Los Altos Hills, CA 94022
(415) 948-4690

Emphasis: Building relationships among young people of diverse backgrounds and between young people and our natural world. **Programs:** One- or two-week sessions: Day Camp, Resident Camp, Farm and Wilderness Camp, Bay to Sea Backpack Camp, Community Leadership Training Camp. **Type:** Day and residential, outdoor, organic farming, experiential, camp. **Cost:** $160-$700. Scholarships available. **Age Emphasis:** 7-18. **Application Deadline:** Spring.

K.E.E.P.—KERN ENVIRONMENTAL EDUCATION PROGRAM
Star Rte. 1, Box 311
Posey, CA 93260
(805) 536-8403

Emphasis: Provides environmental education to Kern County grade 6 students. Emphasis on wildlife, plants, water and human impacts on the environment. **Programs:** N/A. **Type:** Experiential, natural science/history, natural resources, residential. **Cost:** $141/child; 5 days. **Age Emphasis:** 11-13. **Application Deadline:** None.

KEYSTONE SCIENCE SCHOOL
P.O. Box 8606
Keystone, CO 80435-7998
(303) 468-5824

Emphasis: Promotes a scientific understanding of nature and our relationship to the natural environment. **Programs:** Elder Hostel, Discovery Camp, Project Wild 2. **Type:** Outdoor education, experiential, natural science/history, natural resources, residential. **Cost:** $46/day; 2-14 days. **Age Emphasis:** K-12. **Application Deadline:** None.

KIWANIS CAMP WYMAN
600 Kiwanis Dr.
Eureka, MO 63025
(314) 938-5245

Emphasis: Environmental education, adventure education, outdoor education. Summer camp, day camp, retreat facilities, outreach basis. **Programs:** Sunship Earth, customized programs. **Type:** Outdoor education, expedition/experiential, natural science/history, natural resources, residential. **Cost:** $6-$60/day. **Age Emphasis:** All ages. **Application Deadline:** None.

LEARNING FORUM
1725 S. Hill St.
Oceanside, CA 92054
(619) 722-0072

Emphasis: Accelerated learning programs for self-confidence, personal growth, self-esteem and motivation. Learn how-to-learn skills. **Programs:** Super Camp and College Forum. **Type:** Residential. **Cost:** $1,495. **Age Emphasis:** 12-24. **Application Deadline:** None.

THE NATIONAL ASSOCIATION FOR HUMANE AND ENVIRONMENTAL EDUCATION; YOUTH EDUCATION DIVISION OF THE HUMANE SOCIETY OF THE US
P.O. Box 362
East Haddam, CT 06423-1736
(203) 434-8666

Emphasis: Humane and environmental education, focus on teaching tools, reading, science and health, includes science lessons and writing assignments. **Programs:** NAHEE'S Adopt-A-Teacher, *Kind* (Kids in Nature's Defense) *News*. **Type:** Elementary education, residential. **Cost:** $18. Length: 9-month school year. **Age Emphasis:** Youth. **Application Deadline:** None.

---

 18 billion, enough to stretch to the moon and back seven times
Source: 50 Simple Things You Can Do to Save the Earth, *EarthWorks* Press

NATIONAL WILDLIFE FEDERATION—LEADERSHIP TRAINING
1400 16th St. NW
Washington, DC 20036-2266
(800) 245-5484
Emphasis: Leadership Training Program teaches leadership skills to teens ages 14-17 who aspire to lead and teach younger children in an outdoor setting. Participants assist counselors at the National Wildlife Federation's Wildlife Camp. **Programs:** Trainees assist with crafts, recreation activities, evening programs, daily camp activities. **Type:** Outdoor environmental education, residential camp. **Cost:** $550. **Age Emphasis:** Youth, ages 14-17. **Application Deadline:** Sessions usually fill by March or April.

NATIONAL WILDLIFE FEDERATION—TEEN ADVENTURE
1400 16th St. NW
Washington, DC 20036-2266
(800) 245-5484
Emphasis: Teen Adventure is for ages 14-17. Teens discover nature by being active members of the natural world in wilderness areas seldom disturbed by human activity. While hiking and backpacking, participants navigate trails with maps and compasses, set up overnight campsites using minimum impact techniques and study wildlife biology, ecosystems, land management, geology and Native American culture. **Programs:** Eastern Teen Adventure, Black Mountain, NC; Western Teen Adventure, Estes Park, CO. **Type:** Outdoor environmental education, residential camp. **Cost:** $600. **Age Emphasis:** Youth, ages 14-17. **Application Deadline:** Sessions usually filled by March or April.

NATIONAL WILDLIFE FEDERATION—WILDLIFE CAMP
1400 16th St. NW
Washington, DC 20036-2266
(800) 245-5484
Emphasis: Wildlife Camp is for children ages 9-13 to develop an understanding of the natural world and foster an attitude of environmental citizenship. Campers participate in Quests and MiniQuests that cover areas such as plant ecology, lakes and streams, birds, wilderness survival, Earth Savers and outdoor challenges. **Programs:** Eastern Wildlife Camp in Asheville, NC; Western Wildlife Camp in Boulder, CO. **Type:** Outdoor environmental education, residential camp. **Cost:** $500-$600. **Age Emphasis:** Youth, ages 9-13. **Application Deadline:** Sessions usually filled by March or April.

THE SCIENCE CENTER AND ENVIRONMENT PARK OF FORSYTH COUNTY
400 Hanes Mill Rd.
Winston Salem, NC 27105
(919) 767-6730
Emphasis: Programs for school groups and adults; addresses issues of environmental awareness and protection. **Programs:** Living Links, Endangered Species, Recycling, Rainforest. **Type:** Outdoor education, natural science/history. **Cost:** Varies, call for info. **Age Emphasis:** All ages. **Application Deadline:** None.

ORANGE COUNTY OUTDOOR SCIENCE SCHOOL
1833 Mentone Blvd.
Mentone, CA 92359
(714) 794-1988
Emphasis: Natural science program for grades 5-6. **Programs:** 5-day residential program. **Type:** Natural science. **Cost:** $175/week/student. **Age Emphasis:** Grades 5-6. **Application Deadline:** None.

RIVERBEND ENVIRONMENTAL EDUCATION CENTER
P.O. Box 2
Gladwyne, PA 19035
(215) 527-5234.
Emphasis: Hands-on environmental education for children of all ages, school teachers, group leaders. **Programs:** 14 programs focusing on aspects of the environment. **Type:** Outdoor education, natural science/history, natural resources. **Cost:** $3/per child. **Age Emphasis:** All ages. **Application Deadline:** None.

RIVERBEND ENVIRONMENTAL EDUCATION CENTER—CAMP GREEN HERON
P.O. Box 2
Gladwyne, PA 19035
(215) 527-5234
Emphasis: Summer day camp focusing on environmental themes. **Programs:** Games, songs, crafts, explorations. **Type:** Outdoor education, natural science/history. **Cost:** $45-$90/week. **Age Emphasis:** Ages 2-11. **Application Deadline:** None.

Q: *Approximately how many American cities and towns already have curbside recycling programs?*

SOUTH SLOUGH NATIONAL ESTUARINE RESERVE
P.O. Box 5417
Charleston, OR 97420
(503) 888-5558

Emphasis: Education about estuaries, ecologically oriented program. **Programs:** Treasures of the South Slough, Secret of the Medallion, Estuary: An Ecosystem and a Resource, Lore of the South Slough. Also, custom activities and interpretive programs. Groups must schedule ahead. **Type:** Outdoor education, experiential, natural science/history, natural resources. **Cost:** Free. **Age Emphasis:** Youth. **Application Deadline:** None.

TRAILSIDE DISCOVERY PROGRAMS
519 W. Eighth, Ste. 201
Anchorage, AK 99501
(907) 561-5437

Emphasis: Nature camps and outdoor education programs for young people ages 4-18. Located in the amazing wild lands of south-central Alaska. **Programs:** Marine science, ornithology, alpine ecology, spirit keepers, Alaskan Quest trips. **Type:** Outdoor education, expedition/experiential, natural science/history, natural resources, residential. **Cost:** $75-$400, depending on program. **Age Emphasis:** Youth. **Application Deadline:** None.

UNIVERSITY OF GEORGIA ENVIRONMENTAL EDUCATION PROGRAM
Rock Eagle 4-H Center
350 Rock Eagle Rd. NW
Eatonton, GA 31024-9599
(404) 485-2831

Emphasis: Residential environmental education for grades 3-8. **Programs:** Outdoor education at Jekyll Island, Tybee Island, Rock Eagle and Wahsega 4-H centers. **Type:** Outdoor education, natural science/history, expedition/experiential, natural resources, residential. **Cost:** $50/per person. **Age Emphasis:** Youth. College students can apply for internships. **Application Deadline:** None.

WATER POLLUTION CONTROL FEDERATION
601 Wythe St.
Alexandria, VA 22314-1994
(800) 666-0206

Emphasis: Protects water resources and maintains water quality through Water Environment Curriculum Series. **Programs:** Surface Water Video, Waste Water Treatment, The Groundwater Video Adventure, Saving Water/The Conservation Unit. **Type:** Natural resources. **Cost:** $49 per program includes teacher guide and 20 student guides. **Age Emphasis:** Youth. **Application Deadline:** None. Send SASE for information and brochure.

Y.O. ADVENTURE CAMP, INC.
HC 01, P.O. Box 555
Mountain Home, TX 78058-9705
(512) 640-3220

Emphasis: Outdoor skills, environmental education activities, challenge/adventure activities. **Programs:** Outdoor Awareness Program. **Type:** Outdoor education. **Cost:** $55/day. **Age Emphasis:** Youth. **Application Deadline:** Apply 6 months in advance; for summer programs, 2-3 months in advance.

ZERO POPULATION GROWTH, INC.
1400 16th St. NW, Ste. 320
Washington, DC 20036
(202) 332-2200.

Emphasis: Teacher training workshops for hands-on activities in population education for K-12; development of teaching materials for population education. **Programs:** ZPG teacher training workshops. **Type:** Hands-on workshop. **Cost:** Varies. **Age Emphasis:** Teachers. **Application Deadline:** None.

A: *More than 2,500*
Source: Environmental Defense Fund

PART 4
ECOVOICE

EARTH DIGEST IX

Reading essays is like composting: You mix together bits of tomato and orange peel and tea leaves, and you end up with rich fertilizer in which flowers can blossom. Or—in your mind's "compost heap"—you mix together Freeman Dyson's essay on Gaia, and Vaclav Havel's wise words on the environment, and the other essays we've included, and you end up with a rich base in which new ideas can grow.

Rio and the New Millennium

- By Vaclav Havel -

Prague—The United Nations Conference on Environment and Development is taking place two years after the collapse of the Communist totalitarian system. That system, one of the most monstrous in history, destroyed not only people and their souls but nature as well.

I live in a country that suffers from serious environmental problems and is one of the greatest polluters in Europe. A large part of our forests is dying, one would shrink from dipping a finger in rivers, and there are areas where most people almost cannot breathe; in those areas, people die younger than elsewhere and children are born ill. Some parts of my country have turned into something like a lunar landscape.

I can give a number of reasons why Czechoslovakia finds itself in this condition. Our economy, subordinated for decades to the strategic and military interests of the former Soviet Union, was a one-way street: It was directed toward production growth, regardless of the quality and marketability of products, energy consumption or effects on the environment.

The then-ruling regime took the per capita output of cement and steel as evidence of its own indispensability, as a symptom of prosperity and social development. That system, based on ruthless exploitation of the past and the future at the expense of the present, ingeniously took advantage of the fact that environmental consciousness was nonexistent or suppressed and consigned to the periphery of public concern.

The main thing was to give the people decent wages and enough to eat in order to keep them from rebelling. Natural resources were squandered; investments in efficient, modern technology were lacking and free discussion on the consequences of such conduct was not allowed. *"Après nous le déluge"* was the underlying principle.

But that is still not the main problem. These are but the consequences of something that goes deeper than that—man's attitude toward the world, toward nature, toward other humans, toward being itself.

These are the consequences of Marxist ideology—the consequences of the arrogance of modern man, who believes he understands everything and knows everything, who names himself master of nature and the world (who is the only one who understands them)—for whose sake this planet is in existence. Such was the thinking of man who refused to recognize anything above him, anything higher than himself.

Unwittingly, even the term "environment" may be a product of this anthropocentrism. It implies that whatever is not human is just something that envelops man—surroundings that are inferior to him and that he should tend and develop in his own image. Nothing but the arrogance of an alleged master of the world and superior proprietor of reason could have produced the erroneous concept that life, the economy—the whole world—can be managed from one single center by one single planner.

The Rio de Janeiro conference is taking place at an unusual time.

 Q: *How many plastic bottles do Americans use every hour?*

Communism has fallen and a bipolar division of the world into a West and an East has ceased to exist. It is being said more and more often that a new polarization may be developing between the rich countries in the North and the poor ones in the South.

This dichotomy reflects the theme of the Rio conference—environment and development—in a rather unfortunate way. The real theme of the conference is neither the environment alone nor development in itself: It is the combination of the two, with emphasis placed on the word "and."

Yet many view the two things separately, as if the states of the North cared first and foremost about environment, while the states of the South sought primarily development. In the midst of the painful quest for a new world order, striking a balance between these points of view appears to be a nearly superhuman task.

The states of the South find it difficult to overcome their mistrust of the North. They believe that the northern countries should finally understand that today's patterns of production and consumption, besides not being sustainable, are the principal cause of the threat facing the global ecosystem, and that the northern states therefore have to accept substantial blame for environmental degradation in the poorer countries.

They have a right to expect that the northern countries will change their profligate way of life and will help find a way to sustainable development for other countries as well. If they do not find the understanding they expect, they will feel frustrated.

The northern states, in turn, point out that they are already giving the South considerable financial assistance, and that this assistance is not used efficiently or distributed fairly—and sometimes is even lost in the safes of those who do not need it at all and who actually impede changes that would provide for the growth needed by the countries concerned. Moreover, the northern countries are confronted with a worldwide recession and do not show much willingness to mobilize new or additional financial resources.

Many post-Communist countries, Czechoslovakia in particular, find themselves in a special position on the North-South issue. No more than 50 years ago, my country was one of the world's most advanced states. We had modern industry, a well-educated and skilled population and a model welfare policy. Forty years of Communist rule was enough to bring Czechoslovakia economically down to the level of certain southern countries. We have had a bitter experience of dramatic decline for which we hardly find a parallel in modern history.

We have many problems in common with the so-called third world and are turning to the most advanced states for assistance. Meanwhile, our way of thinking is still that of the advanced industrial North, and we have a fairly good chance of making up for the loss soon. Maybe that is why we are able to see things from both sides; maybe that is why we know that an isolated course of action suggested by one side or the other would not be the solution.

The only solution is indicated by the word "and." It lies in the combination of economic growth and respect for the environment. This has been the essence of the concept of sustainable development that was put forward five years ago by the Brundtland Commission (named for Gro Harlem Brundtland, the Norwegian prime minister) and that is on the agenda of

A: **2.5 million**
Source: 50 Simple Things You Can Do to Save the Earth, EarthWorks Press

the Rio conference. In other words, the key is to maintain economic development, yet to do it in a way that would be in keeping with the needs of both man and nature.

This is not just a technological, economic or ecological task. This tremendous challenge has a moral and spiritual dimension. The past era has taught us, survivors of the totalitarian regime, one very good lesson—man cannot command wind and rain, as a propaganda song once promised in my country.

Man is not an omnipotent master of the universe, allowed to do with impunity whatever he thinks, or whatever suits him at the moment. The world we live in is made of an immensely complex and mysterious tissue about which we know very little and which we must treat with utmost humility.

Vaclav Havel is the former president of Czechoslovakia.

Reprinted by permission from the New York Times, "Rio and the New Millenium," June 3, 1992.

Moving from a Material World

- By Susan Okie -

Kisumu, Kenya—My 8-year-old son, Peter, was playing soccer on the patio with our housekeeper's 9-year-old son, Alfred. The "ball" they were kicking back and forth was not really a ball but a wad of newspaper taped together into a rough sphere. Peter's toys had not yet arrived at our new home in Kenya, so Alfred taught him a time-honored African technique for making a plaything out of readily available materials. Peter quickly discovered that if he collected sticks, string and feathers, the security guard sent by the American embassy to protect our house would help him make a set of bows and arrows.

For me, newly arrived in Africa with my family, the first and most overwhelming lesson of my new life has been that my ordinary, middle-class American lifestyle is embarrassingly excessive. Things I have long taken for granted as necessities—running water, reliable electric power, shoes, a separate bedroom for my children, affordable schools, health care when I need it—are beyond the reach of millions of Africans. Even a job that pays a subsistence wage is impossible for many Kenyans to find in a country with declining economic growth rates and a labor force that is growing at 4 percent a year. An American family like ours, settling in for a two- or three-year stay, is like a new business opening in a small town: a precious source of jobs. People come up to us on the streets of our new neighborhood to ask for employment.

The most jarring reminder of our new status as rich *wazungu* (Swahili for "whites") is the panoply of security measures urged upon us by the embassy. In Nairobi, both expatriates and well-off Kenyans live like caged birds, with high walls, guards, and gratings over every window. We were issued a shortwave radio so that we could call the Marines at the embassy to send reinforcements in case of a break-in. At a security briefing during our first week in

Q: *If the average commuter passenger load in the US were increased by just one person per day, how much gasoline would the US save daily?*

town, we heard harrowing stories about bands of machete-wielding burglars besieging houses at three in the morning, carrying off stereos and computers while the terrified tenants hid in upstairs bedrooms. The crimes were rarely personal or violent, we were assured, and people seldom got hurt. But with an annual inflation rate of 25 percent, crime is on the rise because too many Kenyans cannot afford to feed their families.

We confront poverty more frequently and at closer range than in the US and must accustom ourselves to this. In Washington, I had reached an uneasy equilibrium with the homeless men and women whom I passed on the walk from subway to office. I tried to meet their gaze, returned their greetings, gave a little money if I thought their pleas were sincere. But how was I to respond to the ragged boys who pressed their faces to my car window on Nairobi streets, demanding a shilling every time I stopped at a red light? They looked about Peter's age, and they were there all day, every day. Were they homeless, orphaned by AIDS or other diseases? Or were they kept out of school by parents who sent them onto the streets to beg, as a taxi driver assured me? Waiting on a traffic island in downtown Nairobi one day in a crowd of well-dressed Kenyans, I looked down and saw a child who looked about 5 years old sleeping all alone on the pavement by my feet, a tin can for coins by his side. When the light changed, everyone stepped around him and crossed the street.

My most intimate glimpses of everyday African life come from my housekeeper, a 43-year-old woman who cooks and cleans for us in return for a room, partial board and a monthly salary of about $70—well above the average per capita income of about $360 a year. Determined to keep her four children in school, she approached us at the beginning of the academic year to ask us to lend her the fees. Schools, she said, got more expensive every year and soon only rich people would be able to afford them.

When we asked why her husband—a cook in another household—was not contributing, she explained that he had another wife who did not work and he must use his earnings to support that wife's children. The school fees for the year totaled more than our housekeeper's annual salary. We decided to figure out how much she could afford to pay back, lent her that amount and made her a gift of the rest.

That assuaged my conscience for a while, but I stopped feeling virtuous when our household goods arrived from the US in nine enormous wooden crates carried by three trucks. In Maryland, it had not been a bigger-than-average amount for a family of four, and my husband had persuaded me that we would adjust to life in Africa faster if we had familiar things around us. In Kenya, it seemed obscene. My heart sank as I unpacked boxes of books, cases of Cheerios and peanut butter, cartons of movies on videotape. The carpenter and electrician working in our house stopped to watch Peter's ecstatic reunion with his Nintendo set. My husband had been partially right—it was comforting to have such things. Yet we had lived without them for four months, as if we had given up our worldly goods. Now we could no longer hide the fact that we were rich wazungu.

A: *33 million gallons*
Source: Campaign for New Transportation Priorities

My housekeeper's reaction was another lesson on the African approach to life. "Nine crates is a lot!" she exclaimed. "You are going to have a lot of wood. Maybe I can have some of it to make furniture."

Susan Okie is a science writer for the Washington Post *who is on leave in Africa.*

Reprinted by permission from the Washington Post, "Moving From a Material World," February 2, 1992.

Green Guilt and Ecological Overload

- By Theodore Roszak -

I am listening to a lecture by Helen Caldicott, the environmental activist. Dr. Caldicott is in top form, holding forth with her usual bracing mixture of caustic wit and prophetical urgency. All around me, an audience of the faithful is responding with camp-meeting fervor, cheering her on as she itemizes a familiar checklist of impending calamities: acid rain, global warming, endangered species.

She has even come up with a fresh wrinkle on one of the standard environmental horrors, nuclear energy. Did we know, she asks, that nuclear energy is producing scores of anencephalic births in the industrial shantytowns along the Mexican border? "Every time you turn on an electric light," she admonishes us, "you are making another brainless baby."

Dr. Caldicott's presentation is meant to instill unease. In my case, she is succeeding, though not in the way she intends. She is making me worry, as so many of my fellow environmentalists have begun to make me worry— not simply for the fate of the Earth, but for the fate of this movement on which so much depends. As much as I want to endorse what I hear, Dr. Caldicott's effort to shock and shame just isn't taking. I am as sympathetic a listener as she can expect to find, yet rather than collapsing into self-castigation, as I once might have, I find myself going numb.

Is it possible that green guilt, the mainstay of the movement, has lost its ethical sting?

Despite my reservations, I do my best to go along with what Dr. Caldicott has to say—even though I suspect (as I think most of her audience does) that there is no connection between light bulbs and brainless babies. The increase in anencephalic births around the US-dominated *maquiladoras* on the Mexican border probably has more to do with the dumping of toxic wastes. And isn't that bad enough?

Still, I remind myself that the important thing is to spread the alarm. And Dr. Caldicott is an inspired alarmist. I try to bear with her habit of moral hyperbole because I know what she is up against—especially in the US. She is struggling to move a mountain of official complacency.

Judging by the leisurely pace at which the world's political and corporate leaders seem prepared to "phase in" such reforms as emissions controls and "phase out" such dangers as ozone-depleting chemicals, it is clear

Q: *How much will the clean-up of leaky storage facilities for nuclear debris cost US taxpayers?*

that they do not share my sense of urgency—even when it comes to such biospheric imperatives as the greenhouse effect.

But if politicians and corporate leaders have been remiss in registering the emergency, the public at large may not share their insouciance. Over the past two decades, environmentalists have done a good job of scaring and shaming people. They have been so effective that the movement may be in danger of crippling the public's capacity to take action.

If we were to compile all the warnings of all the ecology groups, there would be little that we in the industrial world could do that would not be lethal, wicked or both. From the dioxin-laced coffee filters we use in the morning to the electric blankets we cover ourselves with at night, we are besieged by deadly hazards.

Worse still, many of those hazards make us unwitting accessories to crimes against the biosphere. My eyeglasses, for example. How could I have guessed that the frames are made from an endangered species, the hawksbill turtle?

There is no question in my mind that these problems are as serious as environmentalists contend. It is simply that there are so many of them and each comes at us crying, "Me first! Me first!" In part, the ecological overload arises from the haphazard way in which the movement operates. The pattern too much resembles those "disease of the month" telethons that leave us wondering if there is anybody still alive out there.

Only a few groups, like the Worldwatch Institute, deal with the planetary habitat as a whole, seeking to assign the issues some priority. Otherwise, the biosphere has been Balkanized into a landscape of disaster areas. Scores of groups compete for public attention and funds, each fixed on a single horror. Hunger, pollution, the ozone layer, the topsoil, the rainforests, the whales, the wolves, the spotted owls. In politics, a thousand people each demanding that we do a different right thing may add up to one big bad thing: public rejection.

That result is all the more likely when environmentalists take to scolding their audience. As Jeremy Burgess, a science writer and supporter of environmentalism, asks: "Is it just me, or does everyone feel guilty for being alive too? . . . Eventually, and probably soon, we shall all be reduced to creeping about in disgrace, nervous of our simplest pleasures."

Environmentalists are not unaware of the problem. One group has tried to make light of the matter, calling itself the Voluntary Human Extinction Movement. Under the slogan "The Answer to All Our Problems," its founder observes that "the extinction of *Homo sapiens* would mean survival for millions, if not billions, of other Earth-dwelling species."

The humorous touch is welcome, but it may come too late. As we enter the 1990s, environmentalists are beginning to see who is reaping the political benefit of the guilty frustration they have so diligently disseminated. A fanatical anti-environmental backlash now under way is stripping away the ecologists' most important asset: their claim to public virtue.

Until now, the business community has been forced to handle the movement with care. Rather than confront it directly, they resorted to "greenwashing," trying to take the side of the angels. They redefined their products

A: *$200 billion, more than twice the price of an Apollo moon landing*
Source: Greenpeace

in eco-friendly terms, ran advertisements featuring frisky animals and Edenic landscapes and claimed, usually deceptively, to be defending the biosphere at every turn.

But now a new tactic has emerged: environmental hardball. Corporations are sponsoring citizens' groups that purport to speak for hikers, hunters, anglers and dirt-bike riders who merely want to enjoy the simple, God-given pleasures of nature. The Big Three automakers, for example, have created the Coalition for Vehicle Choice. Through the Alliance for America, lumber and mining corporations portray ecologists as bullying spoilsports and hasten to champion the little guys, helpless victims of the elitist environmental organizations.

For conservatives, a green scare is replacing the red menace. The Competitive Enterprise Institute, which promotes "free market ecomanagement," announces, "There is an intellectual war taking place between pro-market and anti-market forces to which business contributing a vigorous defense of its social role." The institute claims that "anti-human" ecologists believe that "every consumer product and every consumer action is inherently anti-environmental."

At its extreme, this rhetoric can be venomous. Ron Arnold, a lobbyist for the Alliance for America, describes the environmental movement as "the perfect boyman" and admits that his goal is to "destroy" it. Another critic, George Reisman, a professor of economics at Pepperdine University, condemns environmentalism as every bit as menacing to capitalism as Bolshevism or Nazism. The movement's contention that non-human nature possesses intrinsic value is a thin cover for its true goal, Reisman has written, which is "nothing less than *the undoing of the Industrial Revolution*, and the return to the poverty, filth and misery of earlier centuries."

It is clearly time for the environmental movement to draw up a psychological impact statement. Have we pushed scare tactics and guilt trips as far as they can take us? At the least, the problem is one of effective public relations. Shame has always been one of the worst and most unpredictable motivators in politics; it too easily laps over into resentment. Call people's entire way of life into question and what you are apt to get is defensive rigidity.

Jan Beyea of the National Audubon Society wisely cautions: "Environmentalists need to be very careful to watch their own psychological state. Many of my friends . . . get such a psychological reward from being in the battle, the good guys against the bad guys, that they lose sight of what they are trying to do." Beyea wants to replace the "politics of blame" with what he calls the "politics of vision," by which he means "showing people practical ways that they can do better."

The popularity of John Javna's *50 Simple Things You Can Do to Save the Earth* stems from the fact that the book provides its readers with some small chance to act—though hardly enough to satisfy more radical environmentalists.

The response of *Earth Island Journal* was to catalog 50 difficult things. The list begins:
1. Dismantle your car.
2. Become a total vegetarian.
3. Grow your own vegetables.
4. Have your power lines disconnected.
5. Don't have children.

The intention is not entirely humorous.

But there is a philosophical issue

According to a recent study, what percentage of North America's fish are rare or imperiled due to habitat destruction, pollution and overfishing?

here that goes deeper than public relations. Every political movement is grounded in a vision of human nature. What do people need, what do they fear, what do they love? The question of motivation determines everything that follows in a political program. Start from the assumption that people are greedy brutes, and the tone of all you say will be one of contempt.

Issues like these become all the more pertinent for a movement that requires such sweeping change. Environmentalism may not require the total "undoing of the Industrial Revolution," but it does involve inventing new concepts of wealth and well-being that challenge many of the values on which Western politics has been based for two centuries.

The agenda for change on this scale will be the work of generations. But here and now something basic has to be decided. In its task of saving life on Earth, does this movement believe it has anything more to draw on than the ethical resolution of a small group of overworked, increasingly vituperative activists who feel they may have to be entrusted with more and more domineering control over the conduct of daily life? Or is there an ecological dimension to the human personality that is both "natural" and universal?

I believe there is a sense of connectedness with nature as rooted in the psyche as Freud once believed the libido to be. When we experience this shared identity person to person, we call it love. More coolly and distantly felt between the human and not-human, it is called compassion. In either case, the result is spontaneous loyalty.

Those of us who presume to act as the planet's guardians must decide if we believe such a bond exists between ourselves and the planet that gives us life. There have been few movements as internally diverse and contentious as environmentalism.

Meanwhile, on the world scene, as the Earth Summit in Rio de Janeiro demonstrated, environmentalism is fast becoming the catchall for every form of third world discontent. The movement runs the risk of disintegrating into an angry chaos of conflicting agendas. In the nasty give-and-take of daily politics, it is sometimes difficult to realize that the essential motivation of most environmentalists is a spontaneous love of the natural beauties.

On the far side of the Earth Summit, we might do well to begin asking what environmental politics connects with in people that is generous, joyous, freely given and noble.

Theodore Roszak, author of The Voice of the Earth, *is professor of history at California State University, Hayward.*

Reprinted by permission from the New York Times, "Green Guilt and Ecological Overload," June 9, 1992.

A: *33 percent*
Source: National Wildlife *magazine*

Snapshots of a World Coming Apart

- By Eduardo Galeano -

"*We can be like them,*" proclaims the giant neon sign on the road to development. The third world will become the first world. It will be rich and happy, as long as it behaves itself and does what it's told.

But "what cannot be, cannot be, and besides, is impossible," as the bullfighter Pedro el Gallo said so well. If the third world produced and squandered as much as the rich countries, our planet would perish.

Already acid rain kills our forests and lakes. Toxic waste poisons our rivers and seas. In the South, agroindustry rips both trees and humans from their roots. With delirious enthusiasm, humankind is sawing the branch on which it is seated.

The average American consumes as much as 50 Haitians. Of course, this statistic does not represent the likes of Baby Doc Duvalier or the average resident of Harlem, but we must ask ourselves anyway: What would happen if the 50 Haitians consumed as many cars, as many televisions, as many refrigerators or as many luxury goods as the one American?

Nothing. Nothing would ever happen again. We would have to change planets. Ours, which is already close to catastrophe, couldn't take it.

The precarious equilibrium of the world depends on the perpetuation of injustice. So that some can consume more, people must continue to consume less. To keep people in their place, the system produced armaments. Incapable of fighting poverty, the system fights the poor.

"Life is something that happens while you're doing something else," John Lennon used to say. In our era, we no longer work to live. We live to work. Some work ever harder in order to satisfy their basic necessities. Others work ever harder in order to squander.

In Latin America, the eight-hour workday pertains to the realm of abstract art. Moonlighting, rarely reflected in statistics, is a way of life for people who have no other way to escape hunger. But in the places where development is at its apex, should humans work like ants?

To be is to have, says the system. But in the end, things are the masters of people. The automobile, for example, not only takes up space but also time. Much of the workday pays for the commute to the workplace. Cars, portable telephones, televisions, VCRs and personal computers, all conceived to save or to pass time, actually appropriate time.

Over the last 20 years, the workday has grown longer in the US. The number of Americans suffering from stress has doubled. According to the World Health Organization, the US consumes almost half of all tranquilizers sold on the planet.

In Latin America, there are two things that not even the richest can buy: clean air and silence. In Brazil, Volkswagen and Ford produce cars with emission controls for export to the US and Europe. For Brazil, they also produce cars—without controls.

 Q: *What portion of newspapers are never recycled?*

Likewise, Argentina refines lead-free gasoline for export and poison for domestic use. Cars are free to cough up gobs of lead. From the car's point of view, lead increases octane and profits. From the human's point of view, lead damages the brain and the nervous system. Cars, masters of the cities, don't listen to the importunate.

In June of 1989, Santiago, Chile, rivaled São Paulo, Brazil, for the world title of biggest polluter. Chile's newly elected government imposed a few light measures on the 800 tons of noxious gas that becomes part of the city's air each day. Motorists and businesses strongly objected. The right to pollute is a fundamental attraction for foreign investment, almost as important as the right to pay miserable salaries.

Walking in a large Latin American city is a high-risk activity. Staying at home is, too. The city as prison: Those who are not prisoners of necessity are prisoners of fear. Those who have something live in fear of the next holdup. Those who have a lot live holed up in fortresses.

According to the *New York Times*, the police have killed more than 40 children on the streets of Guatemala City. The bodies of child beggars, thieves and rubbish diggers were found without tongues, without ears, thrown out like trash. Public opinion creators apologize every day for the crime. Last year in Buenos Aires, an engineer shot two young thieves who had stolen the cassette player from his car. Bernardo Neustadt, an influential Argentine journalist, told a local television station: "I would have done the same thing." In a savage capitalist society, the right to own is more important than the right to live.

Since Christopher Columbus, Latin America has lived the development of foreign capitalism as its own tragedy. Now we're starting over. The tragedy repeats itself as a farce. A dwarf plays the role of a child. It is a caricature of development. The "Bolivian Miracle," for example, occurred thanks to drug profits. The tin rush is over, and, along with tin, the mines and the most militant Bolivian unions will fall. The village of Llallagua doesn't have water but does have a parabolic television antenna on top of Mt. Calvario.

The "Chilean Miracle," product of General Pinochet's magic wand, is now sold to former Eastern-bloc countries like snake oil. In Chile, the food supply has increased. So has starvation. Did failure go to Pinochet's head? In 1970, 25 percent of Chileans were poor; today it's 45 percent.

Numbers attest but do not repent. Human dignity is a cost-benefit calculation. The sacrifice of the poor is the "social cost" of progress.

The West is euphoric in triumph. The collapse of Communism gives the West the perfect alibi: In the East it was worse. Were the two systems any different? The West sacrifices justice in the name of liberty on the altar of the goddess productivity. The East used to sacrifice freedom in the name of justice on the altar of the goddess productivity.

In the South, let's ask ourselves if this goddess deserves our lives.

Uruguay's Eduardo Galeano is a writer who has long chronicled the tensions between the third world and first world. This essay was translated by Kevin O'Donnell. Galeano is author of The Book of Embraces, *W.W. Norton, New York, 1991.*

Reprinted by permission from In These Times, "Snapshots of a World Coming Apart at the Seams," February 5, 1992.

Two-thirds
Source: Environmental Defense Fund

Bush's Polluter Protectionism Isn't Pro-Business

- By Michael Silverstein -

As a businessman, I find the attitude of George Bush toward environmental laws (and the environment generally) incomprehensible.

In recent months the Bush Administration has declared a moratorium on new environmental regulations and taken steps to ease enforcement of existing ones. Such an approach, from the business perspective, strikes me as absolutely cuckoo.

It seems based on a failure to understand the differences between environmental *spending* and environmental *potlatching*. Laws and regulations that force polluters to spend money on cleaning up the environment do not diminish the wealth of a nation. They transfer this wealth from polluters to polluter-cleaner-uppers and lay a foundation for greater future wealth.

The overall effects of this process over more than two decades have been extraordinarily positive. Not only have countless formerly high-polluting enterprises been forced to become less wasteful (i.e., less polluting) and thereby more competitive, not only has "the environment" become a force generating technological innovation on a scale as great as the defense or space program's, but also a vital new component of the US economy has emerged in the bargain—the environmental industry sector.

In 1991, this country's 65,000 to 70,000 environmental companies garnered an estimated $130 billion in sales. The 70 largest publicly traded firms in this group, with collective revenues of almost $30 billion, saw their revenues jump more than 18 percent last year. All told, some 2 million Americans now make their living doing some kind of environmental cleanup work.

The positive impact of this industry on the international trade standing of the US has also become striking. According to US Commerce Department data, the world market for pollution control products and environmental services ("green goods") reached $370 billion last year. It continues to grow rapidly. Of this total, some $50 billion is traded among nations, with the US winning a very respectable $6 billion of the business. And because of America's early involvement in this field, the US still has positive trade balances with virtually every other nation of the world in this category—including Japan.

Beneficial effects for the US economy of this trade go far beyond the well-known international involvement of a Waste Management Inc., whose foreign operations generated almost $1.1 billion in 1991. Increasingly, they also mean new business for such smaller firms as Isco (wastewater samplers), Safety-Kleen (solvent recycling) and Gundle Environmental (landfill liners).

The air, water and soil contamination problems becoming endemic around the world may be ecological disasters. But clearly, they represent enormous economic opportunities as well. Even countries like Taiwan and Mexico, which enjoyed-pollution-based prosperity for a few years, are now sharply boosting

Q: *What percentage of the world's population has no electricity at all?*

their spending on green goods.

Taiwan, for example, where industrial growth has soared since the mid-1970s, now discards an estimated 3 million metric tons of hazardous wastes into its national environment annually. In some parts of the country, only 1 percent of the wastewater and sewage is treated, and sulfur dioxide emissions from almost 12 million cars and motorcycles are staggeringly high. To combat these environmental ills, Taiwan's government and businesses plan to spend more than $20 billion by the end of this decade on pollution control equipment and environmental engineering and consulting.

US companies, of course, now stand to get a hefty share of this kind of spending. Reducing our own domestic demand for environmental services and pollution control equipment by scaling back on environmental regulations, however, is sure to make us a *less*-important player in such international green markets in times to come. After all, could a country be a successful car exporter without having a strong domestic car market or a developed road infrastructure? Why would anyone think a different set of rules applies to green goods?

The rationale for the Bush Administration's current anti-environmental policies is that they help keep us competitive, boost corporate profits and protect jobs. This is nonsense. What these policies *really* do is temporarily insulate inefficient producers from the need to innovate and invest in new equipment, while penalizing an industry that is arguably the most dynamic element of the entire US economy.

Bush Administration policies in this realm are no more "pro-business" than policies of Soviet-bloc regimes during the 1970s and 1980s were "pro-industry." Both simply were (and are) aimed at temporarily shielding ossified, entrenched interests from the dictates of a changing world economic order.

These are a commonplace set of observations among hundreds of college instructors and thousands of executives in scores of industries. They are routinely discussed not only by officials in the Bush Administration's own Environmental Protection Agency, but also by officials in its Commerce Department, which is today actively promoting green exports in a variety of ways. So why is it not obvious to people at the helm of the Bush Administration?

Regulation that promotes environmentally sound, efficiency-enhancing and innovation-producing activity is simply a jump start on the road toward a 21st-century production and transportation system that is far less wasteful. If you don't generate pollution in the first place, you don't have to worry about cleanup costs years later. Environmental regulation is a rocket assist until the *real* free market motors—consumer demand for greener products and international competition fostering manufacturing efficiency—kick in fully.

I can understand why Democratic activists, who see regulations as a punishment for corporate sin and an end in themselves, have not tumbled to the realities of a New Environmental Economics. But that people who claim to be good, solid, market-oriented Republicans are missing the boat here is astounding. And destructive. And just plain dumb.

Michael Silverstein is president of Environmental Economics, a Philadelphia consulting and research firm.

Reprinted with permission of The Wall Street Journal, © 1992 Dow Jones & Company, Inc. All rights reserved.

A: *33 percent, approximately 1.7 billion people*
Source: Union of Concerned Scientists

EARTH JOURNAL

An Urban Environmental Justice Perspective

- By Charles Lee -

The recent uprisings in South Central Los Angeles were massive tremors on the social seismograph of America, telltale indicators of the massive fault lines underneath America's inner cities ominously waiting to erupt. The violence in the wake of the Rodney King verdict came as no surprise to serious observers of urban affairs in the US. What was surprising was that it did not take place sooner and that it had not yet taken place elsewhere. In the aftermath of South Central Los Angeles, there appears to be an unquestioned drive toward economic development at all costs, i.e., enterprise zones free from any and all occupational and environmental safeguards.

An environmental justice perspective would be especially useful for better understanding of Los Angeles and America's urban crisis. Los Angeles is a car and tinsel town, famous for its never-ending freeways and its show-business glitter. Beginning in the 1930s, a consortium of auto, rubber and oil companies destroyed public transportation and made the city dependent on the automobile. Los Angeles has the worst air pollution in the nation, with the most grievous impact falling on communities of color. These communities are breeding grounds for unemployment, poverty, drugs, hopelessness and rage. South Central Los Angeles did not always have the urban decay it now suffers. Unemployment in South Central Los Angeles is no mystery when one realizes that not too long ago a General Motors plant there employed up to 4,000 workers. How ironic that General Motors saw fit to close shop in a city made dependent on the automobile and go overseas for cheaper labor and greater profits.

Halfway across the continent lies a city that is virtually 100 percent African-American. It has no obstetric services, no garbage collection, few jobs, and raw sewage regularly backs up into homes and schools. East St. Louis, Illinois, lies directly adjacent to Monsanto Chemical, Pfizer Chemical, Aluminum Ore, Big River Zinc and other industrial plants. Most of these plants have their own incorporated townships, where no one lives and which are no more than a legal fiction to provide tax shelters and immunity from the jurisdiction of East St. Louis. Raw sewage often floods the streets, parking lots and playgrounds of the city. Garbage is burnt in backyard lots. Lead is found in most playgrounds, sometimes at an astonishing 10,000 parts per million. Children play directly downstream from the chemical and metal-processing plants, leading to the highest rate of childhood asthma in the nation. Children also play in the aptly named Dead Creek, which received toxic discharges in the past and now smokes by day and glows on moonlit nights. It gained notoriety for instances of spontaneous combustion, created by friction when children ride their bicycles. The *St. Louis Post Dispatch* described East St. Louis as "America's Soweto."

Los Angeles, East St. Louis, New

 Q: *How long do chlorofluorocarbons (CFCs) remain in the Earth's atmosphere after being released?*

York, Chicago, Detroit and the other great urban centers of America were destinations for post-World War II streams of African Americans migrating from the deep South, seeking employment and economic opportunity. Racism and segregation have turned this odyssey into a bitter dead-end search. The inner-city communities into which they were forced have turned out like East St. Louis. Once a thriving transportation center that enticed African Americans for purposes of union breaking, East St. Louis is now described as "a repository for a nonwhite population now regarded as expendable."

People of color not only live in communities targeted for the disposal of environmental toxins and hazardous wastes, but also in fact live in fully *disposable communities*, to be thrown away when the population they hold has outlived its usefulness. No doubt, the pool of cheap labor today is fed by immigrants from Asia, Latin America and the Caribbean, whose value will soon deflate when they, too, have outlived their usefulness.

In October 1991, more than 600 persons from virtually every state in the US, plus Canada, Puerto Rico, Central America and the Marshall Islands, gathered in Washington, DC, at the First National People of Color Environmental Leadership Summit. It was a defining moment for a new environmental justice movement in the US. For people of color, the environment includes not only leisure space but also places where we live and work. Dana Alston of the Panos Institute proclaimed that, "The issues of the environment do not stand alone by themselves," but are "interwoven into an overall framework of social, racial and economic justice." According to sociologist Robert Bullard, "This country was built on three basic tenets: free land, free labor, free men. . . . There is a direct correlation between the exploitation of land and the exploitation of people. When we understand this link, we can begin to undo the damage heaped upon African Americans and other people of color."

Our inner cities are the products of the historical forces of greed, racism and genocide that Bullard talks about. They are the products of the accumulation of tons of industrial toxins and decades of environmental destruction. Entire communities like South Central Los Angeles, and entire cities like East St. Louis, are in effect Superfund sites, areas so threatening to human health and the environment that they need emergency cleanup.

Our task is to develop the visions and understandings that will effectively meet the challenges of Los Angeles and East St. Louis to revive, rebuild and renew our inner cities. We may have to recognize that there are some places such as East St. Louis that have been so destroyed they cannot be salvaged. In those cases, we have a special obligation to maintain and preserve community. Sustainable development is often used in an international context, but we must begin at home. This nation needs a comprehensive urban environmental agenda that includes jobs and justice, as well as equity and opportunity.

Charles Lee is the director of research for the United Church of Christ Commission for Racial Justice. He directed the commission's landmark 1987 study, Toxic Wastes and Race in the United States. *He also coordinated the First National People of Color Environmental Leadership Summit in 1991.*

A: *Up to 50 years*
Source: Protect Our Planet Calendar, Running Press Book Publishers

EARTH JOURNAL

How the People Saved the Earth Summit

- By Donella Meadows -

Was it all a waste of time, the enormous, contentious Earth Summit? One member of the German delegation summed it up: "Whether Rio was a disappointment depends on what one expected." Most of the participants I've talked to expected little. Therefore they were pleasantly surprised.

Knowing George Bush and the people to whom he listens, no one expected leadership from the US, and of course none was forthcoming. The unexpected result was that the Europeans and Japanese stopped capitulating to American stubbornness. "Europe did everything it could to keep from embarrassing Bush, but his administration's incompetence made that completely impossible," one delegate told me. So the rest of the industrialized world will go ahead of the US on the path toward sustainable development, at least until we elect a real environmental president.

Knowing governments in general, there was no reason to expect the Rio meeting to produce brilliant policy-making or heartwarming international cooperation, and it didn't. But what governments do is not all that happens. Especially in environment-development matters, the future is primarily in the hands not of governments, but of people who have babies, drive cars, turn lights on and off, buy stuff, generate garbage. On the level of people, the Earth Summit was a success.

Thousands of Brazilian schoolchildren trooped through the environmental exhibits. NBC found it necessary to explain on the nightly news what *biodiversity* means. BBC World Service, the one broadcast you can hear just about everywhere on the planet, did special environmental shows for weeks. One British reporter who normally covers business affairs told me that in preparation for Rio he finally got around to reading the environmental books piled up on his desk. He found himself shocked at the extent of the problems and excited by the notion of sustainable development.

If the Earth Summit did nothing more than advance the ecological literacy of the world's reporters, that would pay long-term dividends in the accuracy and frequency of environmental information. But it did more than that. In addition to the government summit there was a business summit, where corporations promoted new green technologies. There was a spiritual summit, attended by the Dalai Lama, Shirley MacLaine, rainforest shamans and earnest meditators who descended upon conference rooms two days in advance to fill them with vibrations of peace and compassion.

The most significant summit, I think, was the people's summit, the one that brought together the non-governmental organizations (NGOs), from the World Wildlife Fund to the Association of African Midwives. The preparations for this summit, like the others, went on for two years. Cynics looked down on these gatherings as "sandboxes for environmentalists." But don't underestimate the personal

: US citizens and the environment are currently exposed to how many synthetic chemicals?

connections and creativity that can arise in a sandbox.

At the NGO meetings, Africans with practical experience in village-level solar energy exchanged ideas with Danes demonstrating high-tech windmills. Women's groups found common ground. Indigenous people challenged the industrial world's condescension and pointed out that they know some things about coexisting with nature that everyone might need to understand. Coalitions formed. The NGO delegates took on with gusto all the topics the governments feared to touch—population, mindless consumption, economic justice, energy efficiency, stopping nuclear power, limiting carbon dioxide emissions. If you wanted to hear leading-edge ideas and watch them spread around the world, you had to be in the sandbox.

UN Secretary General Boutros Boutros-Ghali, who had to spend his time at the government summit, summarized it accurately: "The current level of commitment is not comparable to the size and gravity of the problems." Maurice Strong, the amazing diplomat who organized both the Rio meeting and the one in Stockholm 20 years ago, was exhausted by the last day at Rio and made one of the most negative comments of his upbeat career. "We don't have another 20 years now. I believe we are on the road to tragedy."

I agree with them both, but I also recall a remark that Czechoslovakia's former president, writer-philosopher Vaclav Havel, made when he addressed the US Congress a few years ago. He said, "Consciousness precedes being, and not the other way around as the Marxists claim." I wonder if the assembled legislators had any idea what he was talking about.

Consciousness precedes being. Treaties, institutions, governments, technologies, economies arise out of the human mind, in response to problems humans perceive, based upon human understanding of how the world works. The consciousness of the industrial age gives rise to steel girders and toxic wastes, desktop computers and ozone holes, mahogany furniture and burning rainforests. The new consciousness asks how to have steel, computers and furniture, how *everyone* can have them, and how that can happen in a way that does not degrade the material and the energy sources from which everything is made and from which all species live.

There are answers to tough questions like that. The answers won't be found until the questions are asked. The questions won't be asked out of a consciousness that is uninformed, or fearful, or resistant to new ways of thinking, or focused only on power and privilege. The Earth Summit raised consciousness. It directed the attention of the world's government, media, business, spiritual and grassroots leaders to the important questions. That was no small achievement.

Donella Meadows, an adjunct professor of environmental studies at Dartmouth College, is co-author of Beyond the Limits.

Reprinted by permission from New Age Journal, "How the People Saved the Earth Summit," September/October 1992.

A: *More than 65,000, with another 1,000 or so added each year*
Source: Greenpeace Action

EARTH JOURNAL

The Face of Gaia

- By Freeman Dyson

My story begins in the year 1978. I am lying on my back under some bushes on C Street, between the Department of the Interior and the statue of Simón Bolívar, in the city of Washington. It is two o'clock on a Saturday afternoon in June. All I can see is the brilliant green of sunlit leaves and the deep blue of the sky. Two friendly bandits have fractured my skull and jaw and relieved me of my wallet. I am expecting that one of them may shortly put a bullet into me to make sure I will not talk. And now, at this unlikely moment, my spirit is filled with peace. The green leaves and the blue sky are beautiful. Everything else fades into insignificance. This life is good and this death is good also. I am a leaf like the others. I am ready to float away on the blue wave of eternity.

This experience, which came to me as I walked, briefcase in hand, to a committee meeting of the National Academy of Sciences, ended happily for all concerned. My assailants escaped unharmed with $75 in cash and some photographs of my daughters. I made my entrance into the National Academy building, dramatically dripping blood upon the marble floor. Time healed my wounds, and the efficient Washington police retrieved my unbroken bifocal glasses from the bushes. Life returned quickly to its ordinary routines. But I have not forgotten that moment of illumination when the glory of earth and heaven was revealed to me.

What is one to make of such a revelation? It is not an uncommon experience among people who have come face-to-face with death. Tolstoy in *War and Peace* describes how Prince Andrei lies wounded on the battlefield of Austerlitz, one among thousands of soldiers, mostly dead, left behind after the battle is over. The prince, like me, gazes into the blue sky, unconcerned about his fate, conscious only of the beauty and greatness of that overarching sky. Tolstoy himself fought in the battles of the Crimean War. His description of Prince Andrei's state of mind was probably derived from experiences of men wounded beside him in the Crimea, if not from his own experience. The prince's peaceful contemplation of the sky is interrupted by the arrival of Napoleon, strutting over the scene of his famous victory. Napoleon, until that day, had been the prince's hero. But now, seeing his idol face-to-face, the prince is unimpressed. The prince sees only the littleness of the emperor under the greatness of the sky. When the emperor notices that he is not dead and addresses some friendly words to him, the prince does not bother to answer. The prince only wants Napoleon to move out of the way so that his view of the sky will be unobstructed.

My view of the sky was blocked not by Napoleon but by a passing motorist on C Street who kindly stopped and pulled me out of the bushes. I gladly accepted his offer of a ride to the National Academy three blocks away. I came back fast from the empyrean to the world of people and committees.

Quite apart from their possible religious significance, about which I am

: According to the Nuclear Regulatory Commission, what percentage of US nuclear reactors are 90 percent likely to be unable to contain a meltdown?

as skeptical as Tolstoy's prince, these revelations tell us something important about human nature. They tell us that we are better equipped for handling violence, mentally as well as physically, than we suppose. Nature designed human beings for living in a world of violence. We are designed to function well in good times and in bad. As Ecclesiastes said long ago, there is a time to be born and a time to die. When fear of death assails me, as it assails everyone from time to time, I take courage from that memory of green leaves and blue sky. Perhaps, when death comes, he will once again come as a friend.

The three most important things in my life have been, in this order, family, friends and work. When I think of family, I recall another happy moment when I am lying sprawled under a blue sky. This time I am not laid low by muggers but by my own incompetence. I came to grief skiing down an easy, wide-open snowfield in Austria. I went head over heels, scattering poles and skis in all directions. As I lie like a rag doll in the snow, I hear a raucous peal of laughter. My beloved youngest daughter, 7 years old, comes by me at top speed, her laughter ringing across the mountain. Into my head comes a favorite pair of lines from Shakespeare:

This is to be new born when thou art old,

To see thy blood warm when thou feel'st it cold.

With six healthy children, I have plenty of opportunities to see my blood warm. I am sorry for those of my contemporaries who grow old alone, without children to keep them young.

When I think of friendship, I am reminded of a third story. The worst indignity that can happen to a scientist has happened to me. I have published a paper in a leading scientific journal. After it is published, it turns out to be completely, irreparably wrong. I am overcome with misery. A scientist who publishes worthless papers must himself be worthless. I can expect nothing from my colleagues but ridicule and contempt. Then, while I am hiding my face in shame, my friend Res Jost, a Swiss physicist for whom I have a deep respect, comes to visit. "Cheer up," says Res. "There is nothing in the world more indestructible than a scientific reputation." And so I cheer up. I quickly discover that Res is right. My friends are still friends. Nobody cares about my wrong paper. I am still in good standing as a member of the scientific club.

Family is older than the human species, work is younger, friendship is about as old as we are. It is friendship that marks us as human. Human societies are glued together with conversation and friendship. Conversation is the natural and characteristic activity of human beings. Friendship is the milieu within which we function.

The central conflict in our nature is the conflict between the selfish individual and the group. Nature gave us greed, a robust desire to maximize our personal winnings. Without greed we would not have survived at the individual level. But Nature also gave us love in its many varieties, love of wife and husband and children to help us survive at the family level, love of friends to help us survive at the tribal level, love of conversation to help us survive at the cultural level, love of people in general to help us survive at the species level, love of nature to help us survive at the planetary level. Human beings cannot be human without a generous endowment of greed and love.

A: *22 percent*
Source: Nuclear Information and Resource Service

The central complexity of human nature lies in our emotions, not in our intelligence. Intellectual skills are a means to an end. Emotions determine what our ends shall be. Intelligence belongs to individual human beings. Emotions belong to the group, to the family, to the tribe, to the species or to Nature. Emotions have a longer history and deeper roots than intelligence. The limbic structures of our brain, where emotions are supposed to reside, are more ancient than the cerebral cortex that carries our intelligence. Emotions must be our pulley. Somehow or other, when we begin to improve artificially the physical and intellectual capacities of our children, we must learn to leave their emotional roots uncut.

To cut emotional roots is fatally easy. A drug addict is a person whose emotions have been deranged by a chemical acting on the limbic structures. Addiction is a vicious cycle because the deranged value system brings a continued craving for the drug. Any change in our natural value system carries the danger of running into such a vicious cycle. Any short circuit of the value system is a form of insanity. Any genetic modification of our inherited emotions brings danger that our whole society may become insane. To be sane means to possess a value system that allows us to survive on all time scales in harmony with Nature.

One hopeful sign of sanity in modern society is the popularity of the idea of Gaia, invented by James Lovelock to personify our living planet. Respect for Gaia is the beginning of wisdom. And the love of Gaia carries with it a love of trees. The town of Princeton where I live is full of trees. If you go up to the top of a tower in Princeton in summer, the town is almost invisible. You see then that the inhabitants of Princeton are actually living in a forest. All these bankers and stockbrokers, wealthy enough to choose where they want to live, chose to live in a forest. The love of trees is rooted deep in our value system, planted in us during the hundreds of thousands of years we spent hunting and gathering on a largely forested planet.

The climatic equilibrium of our planet is now threatened by the greenhouse effect of carbon dioxide accumulating in the atmosphere. The carbon dioxide comes partly from burning of coal and oil and partly from destruction of forests. Fortunately, there is a remedy. The quantity of carbon in the atmosphere is about equal to the quantity in living trees. This means that the problem of the greenhouse is essentially a problem of forest management. A large-scale international program of reforestation could hold the greenhouse in check, besides producing many other economic and environmental benefits. The cost of growing enough trees to nullify the greenhouse is not prohibitive. Only the will and the international consensus required to do the job are at present lacking. But we shall probably see the will and the consensus emerge, as soon as the climatic effects of the greenhouse become obvious and severe. When that happens, the whole world will begin planting trees and feeling an old familiar joy as they see the planet turning green. At that point we may say that Gaia is using us once again as her tools, using the love of trees that she implanted in us long ago as the means to keep herself alive.

As humanity moves into the future and takes control of its evolution, our first priority must be to preserve our emotional bond to Gaia. This bond

 According to a 1985 Nuclear Regulatory Commission study, what chance is there of a nuclear meltdown occurring in the US before the year 2005?

must be our pulley. If it stays intact, then our species will remain fundamentally sane. If Gaia survives, then human complexity will survive too. Perhaps, when I was lying under the bushes on C Street, the revelation that came to me was just Gaia showing her face.

Freeman Dyson has been a professor of physics at the Institute for Advanced Study in Princeton since 1953.

Reprinted by permission from From Eros to Gaia, by Freeman Dyson, Pantheon Books, New York, 1992.

Goodbye, Old Desert Rat

- By Charles Bowden -

He saw it first from the open door of a boxcar in 1944, when he was 17 years old. He saw it last from the floor of his shack in the very early morning hours of March 14, 1989, when he was 62 years old and blood gushed willy-nilly out the broken-down veins of his body and his flesh went cold. His name was Edward Abbey. He was my friend and this is no help in explaining either him or the ground that claimed him once and for all. The voice, it was very low and, oddly enough, soft. The words were precise, the manner at first low-key, but the face almost intense, a small smile screening a volcano of anger. He thought no one should be here, he thought he should not be here. He stops his words, looks at me in the noisy cafe and asks if I noticed that woman across the room eating a hamburger, the juice from the meat on her lips. Life is never a mess, but it is always messy, like a tidal pool.

He wrote about 20 books, but his reputation has remained largely regional, a legend locked into the small pockets of human beings in the largely empty ground we call the West, and more especially the Southwest. This fact did not sit well with him but he was helpless to remedy it because he was a captive of a place he said was empty of meaning and unfit to live in. He kept claiming he would change, that he would write no more about this ground, the rocks, the heat, the emptiness, that he had had his say. And I think he meant it, and truly he did write of other things. But he could not let it go.

He married often, divorced often, preached zero population growth and left five children in his wake. Sometimes he drank too much, sometimes he lost his temper and said and wrote hard things, ugly things. He would say that people who suffer from AIDS should be left in the desert to die. He would say that Mexican immigration must be stopped because Hispanic culture hates the Earth, hates nature and will bring its poison north and ruin us. At other times he played the recorder sitting on a rock in the middle of nowhere, and seemed as benign as a spring flower. Once he was late for lunch and apologized, and when we went out later to the parking lot the front of his truck was caved in. He explained he had had a head-on collision en route. Later he fastened a plastic flower to the busted prow of his old machine.

Mainly, he said whatever he had to say better than anyone else. He was as wily as a coyote, and as deadly as a

A: *45 percent*
Source: Union of Concerned Scientists

snake. The wit of the man was a pleasure, the venom alarming. Fortunately there were no big moments in our friendship to ruin its calm. But sometimes I worry that he may have managed to poison me with a fatal dose of, well, ethics.

I remember going one night to an environmental rally down at El Rio Neighborhood Center. The evening chugged along with the expected dose of environmental pep talks, sensitive poetry readings and we-ain't-going-to-take-this-anymore war cries. The audience was wall-to-wall waffle stompers and plaid flannel shirts, the women had long hair and no makeup. Then Abbey's turn came and he pulled some pages out of his pocket and started reading a long, shaggy-dog story about his earlier days in Albuquerque, about roaring down the road with a pal, tossing beer cans out the window and firing a pistol wildly into the countryside. I could feel the crowd get edgy. Abbey droned on seemingly oblivious, and his text somehow segued into the charms and joys of various sorority girls encountered in those college adventures. I sensed a sullen steam begin to rise up off the audience. Suddenly he was finished and the evening promptly returned to environmental proprieties.

I thought: Well, why pander?

Then there was the time he called me up and asked if I'd be interested in going to Mexico City. *Architectural Digest* had commissioned him to assess some fancy mansion there designed by a leading architect, and there were nice, crisp dollars to finance a reconnaissance. I thought, what the hell. A few weeks later, the trip was off. They'd sent him pictures of the house and he couldn't stand the way it looked. He told me he couldn't write about an obscene thing like that.

I thought: Well, don't just do it for the money.

A month or so before he died, he was showing me his toy, an old red Cadillac convertible. I said, "Christ, Ed, you've got no shame." And then he stopped living. So I'm left with: "You don't pander; you don't roll over for money; and you drive any damn thing you want."

And these words he left on paper.

The desert is also a-tonal, cruel, clear, inhuman, neither romantic nor classical, motionless and emotionless at one and the same time—another paradox—both agonized and deeply still. Like death? Perhaps.

He kept making this same point over and over: There is nothing really out there, it is just this blank we like to write our minds on. "But," he noted in explaining his attraction to the place with nothing, "there was nothing out there. Nothing at all. Nothing but desert. Nothing but the silent world. *That's why."*

But the silent world needs its friends. One day recently I came out of the Sierra Madre on foot down in Sonora and found a guy about 60 sitting by his VW camper having a cold beer. So I joined him. He said he'd run around the West all his life, and way back in the early 1950s he went to school up in Albuquerque with a guy named Ed Abbey, a writer guy. They started taking out billboards at night, first chopping them down, then getting chain saws, and finally, when the companies went to steel posts set in concrete, using dynamite. There are many such tales, and they accomplish two things: They've made Abbey a cult, and they scare the hell out of many other people. I suspect for Abbey they just happened. I do not understand why

 In the North Pacific alone, how many sharks die accidentally in driftnets each year?

the stories frighten anyone who looks and sees what is happening.

"I don't believe," he once wrote with the fine feeling of a son of Appalachia, "in doing work I don't want to do in order to live the way I don't want to live."

He had a gut hatred of modern industrialism, of a machine culture that creates machine people to tend to its needs. He could hardly write a page, comic or tragic, without this feeling lashing out: "Which was the worthier technological achievement, the moon landing or the invention of bread? The bake oven or the nuclear reactor? Only a fool would hesitate to answer."

He was careful of his denims and old shirts, careful of his hair and beard, more than careful of his writing. I am out at the house, his old black dog is dying of Valley fever, a local fungal plague of man and beast, and he jokes and complains of having to dose the dog with medicine and how much the medicine costs him and that he is a fool to persist in this indulgence. And then this fat novel comes out, *The Fool's Progress: An Honest Novel*, and the damn dog and the damn medicine and the damn complaint is there. Except for the joke. When he wrote the book he was dying, but forgot to tell anyone and the dog, well, the dog outlived him. A last cut of the cards, aces wild.

Abbey was a gambler and the desert was his last trick. He hates our cities, as do we all, and dreams of their ruin in book after book. He writes a novel called *The Monkey Wrench Gang*, which explores the destruction of Glen Canyon Dam. He writes a novel entitled *Good News*, which explores the American Southwest after the final collapse. And yet he lives in towns and cities, always publishing fictitious addresses. He takes hikes here and there and then carefully disguises the geography so we cannot find out exactly where he has been. He keeps telling us we do not belong here, that he does not belong here. And he won't leave.

He is our tongue, saying things we cannot bring ourselves to say. He is our sins written large and does not deny this fact. After all, the man drank too much, chased women, drove cars, threw trash out the windows, lived in a house not a wigwam, consumed things, made noise at times. But he felt in his bones what we feel. They predict now that world population will peak at 11 billion in the year 2150. They predict now that the last tropical tree will be felled in the year 2135. We have almost lost the will to even argue with these predictions. We feel their truths in our bones. So we turn to the casino, to the game, to the gambling we have always known and loved. We find it is easier that way.

The game is almost over and this is the last deal in one of the last places left. That is why it still persists—our ancestors couldn't figure out how to rape it. There they are cutting the cards, pick one up, put a house on it, build a hideaway, make it a park with nice paved roads, post little signs telling us things about nature we will never remember. Bury toxic wastes perhaps, test odd weapons—no one will notice here. Run strange machines, 4x4s, motorcycles, dune buggies, don't worry, for God's sake, there is nothing out there but cactus and snakes and scorpions. Build a gallery, hang large color photographs on the wall, sit there, right over there on that very long couch, sip a cold drink—we deserve that drink—and enjoy the sunset trapped inside the

A: *Approximately 1.8 million every year*
Source: Greenpeace Action

frame on the wall, the colors rich and intense thanks to that smear of chemicals. Write a book explaining a sensitive reaction to this dry ecosystem.

Ed, he just wouldn't listen. I talked to him one Wednesday night and he sounded fine. Thursday he went on the table. Friday his wife said he would be all right. Sunday—or was it Monday?—he looked up from his hospital bed, all those tubes connecting him to machines and jugs and juices that dripped into his veins, and he said to his wife, "I'm leaving. Are you coming with me?" He sat up all night under a tree waiting to die. The blood kept flowing out of him. But he failed, he failed at this simple task. He always was a kind of half-assed hillbilly at heart. They returned to the shack down by the wash where he wrote and listened to classical music and smoked bad cigars. He died on the floor just after dawn.

And then he pulled one more trick. They packed him in a sleeping bag, laid on some ice, and roared off into the white light. A long way off, they pulled over, dragged the body out there—way out there, no music now, no soft light, no air-conditioning, no more cards—and dug a hole and laid him in it. He lied to us, just as we all lie to us. He said there is nothing out there. A lie. He said we couldn't stay. And now he's checked into the place for what looks to be eternity.

But he understood the game. He saw through the phony house rules about progress, development, taking care of the environment, having our desert and factories too. He knew the only thing of value to be found here was the barest glimmer of an understanding about just what the word *place* might mean. Something that we do not conquer, something that does not care about us, something that might fill the emptiness that has driven us for so very long. Some place where we might belong even though we cannot stay except as a gesture. And the place has not suffered for this act. The silence is still there, the white glare, the songs of coyotes in the night, the track of the snake across the sand, the trail of ants carrying little bits of leaves.

We seem to want a world where there is a master plan and this plan states that our behavior will not be punished, our appetites will not be curbed, our present will not determine our future. We will be exempt from death, from hunger, from pain, from everything but love. And love will not be earned but freely given like that of a parent to a child. We say that such a world is our due, is our right, is part of the master plan. We say this to our politicians, to our friends, to our gods, to our dreams. But we do not go to the desert and say this, no, not there. Out on the hot ground we fall silent because there finally we know what we have always denied, though not a word is said to explain this fact to us. The desert will not sustain our lies but instead offers a taste of life. And out on the ground there is no master and there is no plan but there is death and hunger and pain. Because, as we always suspected, the desert does not care. And finally, love becomes a possibility.

We already know more than we will ever understand.

We already understand more than we will ever need in order to take action.

We are ready now.

We must do it.

Or . . . we will kill it.

: *How much water is needed to meet the American demand for meat?*

We have been at its throat so very, very long.

Press the pedal to the floor. The big brute motor will grumble like a lion, old, tired, hesitating, then catch fire and roar, eight-hearted in its block of iron, driving onward, westward always, into the sun.

Charles Bowden is author of six books of fiction and is a frequent contributor to Buzzworm.

Reprinted by permission from The Sonoran Desert, *photographs by Jack Dykinga, text by Charles Bowden, Harry N. Abrams, New York, 1992.*

Do We Really Need Zoos?

- By Colin Tudge -

If we define "zoo" narrowly, as a collection of animals in cages with bars, then the answer to that grand question is "no." But zoos do not have to be like that; should not be like that; and—if we are talking about good, modern zoos—are not like that. If we define "zoo" more broadly—as a place where animals live in a protected state, and are made accessible to human observation—then the zoo becomes self-justifying. If zoos did not exist, then any sensible conservation policy would lead inevitably to their creation.

Indeed, if we survey modern methods of conserving animals, we see an entire spectrum of approaches. At one extreme is the intensive breeding center. This does not have to be (and should not be) a row of cages; but if animals are kept in small spaces (and any artificial space is liable to be small compared with the wild) and yet are breeding, then this by definition is "an intensive breeding center." At the other end of the spectrum is the wilderness itself. In between is every kind of compromise: fenced reserves ("sanctuaries") for just one species; tightly managed reserves with a select list of species and natural vegetation; reserves with natural vegetation in contact with "wilderness," but with protection from predators; the national park, which resembles wilderness but must nonetheless be managed to maintain its diversity and prevent local extinctions.

Wilderness remains the "ideal," the dream, the representation of the pristine world. Wilderness must be allowed to prevail wherever possible (albeit visited and studied by human beings); and conservationists should endeavor to push as far as they can from the intensive end of the spectrum, toward the wilderness. Zoos in general should grow into national parks, and some national parks at least might eventually become so large that they can (almost) be left alone.

The spectrum exists; and each component of the spectrum can be justified in the present world, provided that each is good of its type. In short, there are advantages and disadvantages in each kind of sanctuary. The urban zoo, though its space is small, makes very good use of available cash; and cash is at a premium in conservation. Urban zoos, tiny though they are compared with continents, nowadays between them sustain populations that compare in size and in a few cases outstrip those of the wild.

A: *190 gallons per person per day*
Source: Worldwatch *magazine*

Single species sanctuaries (or zoos) can be good because they concentrate effort, and because conditions can be adjusted to a particular species' needs. Of course there is a case for sanctuaries for black rhinos, and for breeding cheetahs at De Wildt. But mixtures of species can be helpful too: Grazing animals in general make better use of herbage if complementary species graze together; woods have canopies for monkeys, but also leave floor space for deer and pigs; the scientists who study one species of cat can learn from looking at others—and through comparisons with totally different creatures; the vets who master the care of one exotic species use their time best if they extend their expertise to others.

Why, though, keep animals in Chester or Cincinnati that properly live in India or Brazil? If they have to be protected, why not keep them in their reserves in their own country? Of course there is a powerful case for conserving animals at home. Black rhinos in reserves in Kenya and South Africa; Sumatran rhinos and Bali starlings in Indonesia; European bisons in Germany; these and hundreds more are examples of legitimate in situ conservation. But the same problems that beset an animal in the wild can also threaten it in captivity, if it stays at home. Elephants in reserves in Vietnam, or scimitar-horned oryx in Chad, would not have escaped the ravages of war. The population of Puerto Rican parrots in a reserve in Puerto Rico was halved by Hurricane Hugo. Populations in situ are highly desirable—necessary, indeed—but it also makes sense to keep other populations elsewhere.

It would be perverse and gratuitously damaging to animals and to us not to take every opportunity to benefit from their existence. To put the matter crudely, conservation in poor countries (and conservation in general is most urgent in poor countries) must as far as possible pay its way. In rich countries, too, people must be persuaded to part with cash; for nothing succeeds in the present world without it.

Most of the ways in which we gain profit from animals are to some extent unpleasant for them. Some involve killing the animal, and some (as in circuses) involve its degradation. But if we are content simply to look at animals, and to be in their presence—which really ought to be reward enough—then it should be possible to gain from them without upsetting them. In practice, observation by more than a few specialists is not so simple to arrange, because animals do not go out of their way to be looked at. Yet with modern technology, a lot of ingenuity and a great deal of money, it is possible to reveal animals to large numbers of people without disturbing them. The techniques range from closed-circuit television in the eagle's nest at Cincinnati, to carefully conducted tours in national parks (as opposed to jamborees in land-rovers). Unless people care about animals, there is no hope for them; and they will care more for them if they are able to see them.

Colin Tudge is a zoologist and a scientific fellow of the Zoological Society of London.

Reprinted by permission from Last Animals at the Zoo, by Colin Tudge, Island Press, Washington, DC, 1992.

 Q: *In 1989, the US used over 12 billion pounds of plastic packaging. In the 1990s, what is expected to happen to that figure?*

EARTH DIGEST

The Real "Superwomen"

- By Jodi L. Jacobson -

The women of Sikandernagar, a village in the Indian state of Andhra Pradesh, work three shifts per day. Waking at 4 a.m., they light fires, milk buffaloes, sweep floors, fetch water and feed their families. From 8 a.m. until 5 p.m., they weed crops for a meager wage. In the evening they forage for branches, twigs and leaves to fuel their cooking fires, wild vegetables to nourish their children and grass to feed the buffalo. Finally, they return home to cook dinner and do evening chores. These women spend twice as many hours per week working to support their families as do the men in their village. But they do not own the land on which they labor, and every year, for all their effort, they find themselves poorer and less able to provide what their families need to survive.

As the 20th century draws to a close, some 3 billion people—more than half the Earth's population—live in the subsistence economies of the third world. The majority of these subsistence producers find themselves trapped in the same downward spiral as the women of Sikandernagar. Because they have cash incomes insufficient to meet their most basic needs, they must rely heavily on their own labor to secure whatever food, fuel and water they can from the surrounding environment.

In the not-so-distant past, subsistence farmers and forest dwellers were models of ecologically sustainable living, balancing their numbers against available resources. Today, however, the access of subsistence producers to the resources on which they depend for survival is eroding rapidly.

Because women in rural subsistence economies are the main providers of food, fuel and water, and the primary caretakers of their families, they depend heavily on community-owned croplands, grasslands and forests to fulfill their families' needs. The widespread depletion and degradation of these resources have led to equally widespread impoverishment of subsistence families throughout Africa, Asia and Latin America.

In rural areas, both men and women engage in agriculture, but women are the major producers of food for household consumption. In sub-Saharan Africa, women grow 80 percent of the food destined for their households. Women's labor produces 70 percent to 80 percent of food crops grown on the Indian subcontinent, and 50 percent of the food domestically consumed in Latin America and the Caribbean. In all regions, roughly half of all cash crops are cultivated by women farmers and agricultural laborers.

By custom, labor contributions are divided by gender. In sub-Saharan Africa, for example, males generally clear and till the land, while females are expected to carry out the bulk of the hoeing, weeding and harvesting of crops, the processing of food, and various other subsistence activities. Women, therefore, perform the majority of the work in African agriculture. Similar patterns in the division of labor are found throughout the subsistence economies of Asia and Latin America as well.

Using traditional methods, women farmers have been quite effective in

A: *It is expected to double*
Source: Greenpeace

conserving soil resources. Given access to appropriate resources, they employ "managed" fallowing (allowing land to rest between plantings), crop rotation, intercropping, mulching and a variety of other soil conservation and enrichment techniques. And they have played a leading role in maintaining crop diversity. In sub-Saharan Africa, for example, women cultivate as many as 120 different plants in the spaces alongside men's cash crops. And in the Andean regions of Bolivia, Colombia and Peru, women develop and maintain the seedbanks on which food production depends.

In subsistence households, therefore, it is largely up to women *either* to produce enough food to feed a family or to generate the income with which to purchase it. Faced with the endemic insecurity of their situation, they have evolved techniques to make efficient use of all available resources—cultivating a diverse array of crops, collecting wild fruits and vegetables, maintaining farm animals and earning whatever cash income they can to ensure a measure of food security. And while cultivating food is obviously difficult or impossible for women in families with little or no land, studies show land-poor women to be highly resourceful in devising ways to meet their families' needs. Solutions used include drawing more heavily on products gathered from commons, expanding their workloads and hiring themselves out as laborers in exchange for grain or cash.

Women in subsistence economies also are active managers of forest resources and traditionally play the leading role in their conservation. Forests provide a multitude of products to households. They are, for example, a major source of fuel for home consumption, without which none of the food grown and harvested could be cooked, nor many other essential tasks be carried out. In fact, lack of fuel with which to cook available food is itself a cause of malnutrition in some areas. "It's not what's in the pot, but what's under it, that worries you," say women in fuel-deficient areas of India.

The dependence of subsistence households on biomass—including wood, leaves and crop residues—as the traditional form of domestic energy remains widespread. Seventy-five percent of all household energy in Africa is derived from biomass, for example. And women use biomass fuel to support innumerable private enterprises, such as food processing and pottery, from which they also gain cash income.

Women depend heavily on the availability of non-wood forest products, too. They collect plant fibers, medicinal plants and herbs, seeds, oils, resins and a host of other materials used to produce goods or income for their families. The fruits, vegetables and nuts widely gathered as supplements to food crops are important sources of protein, fats, vitamins and minerals not found in some staple crops. In times of drought, flood or famine, these gathered foods have often made the difference between life and death.

These non-wood forest products make substantial contributions to local and national economies throughout the third world. Although unrecognized or unrecorded by national statistics, these activities often contribute more to national income than wood-based industries. A report by World Bank researchers Augusta Molnar and Gotz Schreiber estimates that in India, for example, non-timber products account for two-fifths of domestic forest

Q: *In Europe, how many acres of forest have died as a result of acid rain?*

revenues and three-fourths of net export earnings from forestry products.

In the true spirit of sustainability, female subsistence producers appear to be as careful in conserving forests as they are reliant on using them. Using traditional methods of extraction, for example, women in Africa and Asia obtain their fuel from branches and dead wood (often supplemented with crop residues, dried weeds or leaves), rather than live trees. Seventy-five percent of domestic fuel collected by women in northern India is in this form.

In surveys, women have consistently pointed to the value of ecosystem services provided by forests—such as their critical role in replenishing fresh water supplies—as reasons for their preservation. In fact, research on communal resource management systems—the "commons" on which women so heavily depend in subsistence economies—shows them to be more effective at protecting and regenerating the environment than management approaches taken by either the state or private landowners.

The reasons are obvious: Commons are as indispensable to land-poor women in subsistence economies as these women are to the maintenance of the commons. Commons lands are the one resource, apart from their children, to which women traditionally have had access relatively unfettered by the control of men. Unfortunately, women's access to these lands and the goods they yield is fast diminishing. The results are already evident in the declining food security among subsistence households.

Jodi L. Jacobson is a senior researcher at the Worldwatch Institute.

Reprinted by permission from "Gender Bias: Roadblock to Sustainable Development," by Jodi L. Jacobson, Worldwatch Paper No. 110, Worldwatch Institute, Washington, DC, 1992.

Act Locally

- By Wendell Berry -

We are living in an era of ecological crisis, and so it is understandable that much of our attention, anxiety and energy is focused on exceptional cases, the outrages and extreme abuses of the industrial economy: global warming, the global assault on the last remnants of wilderness, the extinction of species, oil spills, chemical spills, Love Canal, Bhopal, Chernobyl, the burning oil fields of Kuwait.

But a conservation effort that concentrates only on the extremes of industrial abuse tends to suggest to the suggestible that the only abuses are the extreme ones, when, in fact, the Earth is probably suffering more from many small abuses than from a few large ones. By treating the spectacular abuses as exceptional, the powers that be would like to keep us from seeing that the industrial system (capitalist or communist or socialist) is in itself, and by necessity of all of its assumptions, extremely dangerous and damaging, and that it exists to support an extremely dangerous and damaging way of life. The large abuses exist within, and because of, a pattern of smaller abuses.

Conservationists have won enough victories to give them heart and hope and a kind of accreditation, but they

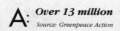
A: *Over 13 million*
Source: Greenpeace Action

know better than anybody how immense and how baffling their task has become. For all their efforts, our soils and waters, forests and grasslands, are being used up. Kinds of creatures, kinds of human life, good natural and human possibilities are being destroyed. Nothing now exists anywhere on Earth that is not under threat of human destruction. Poisons are everywhere. Junk is everywhere.

These dangers are large and public, and they inevitably cause us to think of changing public policy. This is good, as far as it goes. There should be no relenting in our efforts to influence politics and politicians. But in the name of honesty and sanity we must recognize the limits of politics. Think, for example, how much easier it is to improve a policy than it is to improve a community.

It is probable that some changes required by conservation cannot be politically made, and also that some necessary changes will have to be made by the governed without the help or approval of the government.

However destructive may be the policies of the government and the methods and products of the corporations, the root of the problem is always to be found in private life. We must learn to see that every problem that concerns us as conservationists always leads straight to the question of how we live. The world is being destroyed—no doubt about it—by the greed of the rich and powerful. It is also being destroyed by popular demand. There are not enough rich and powerful people to consume the whole world; for that, the rich and powerful need the help of countless ordinary people.

The world that environs us, that is around us, is also within us. We are made of it; we eat, drink and breathe it; it is bone of our bone and flesh of our flesh. It is also a Creation, a holy mystery, made for and to some extent by creatures, some but by no means all of whom are humans. This world, this Creation, belongs in a limited sense to us, for we may rightfully require certain things of it—the things necessary to keep us fully alive as the kind of creature we are; but we also belong to it, and it makes certain rightful claims upon us: that we care properly for it, that we leave it undiminished, not just to our children, but to all the creatures who will live in it after us. None of this intimacy and responsibility is conveyed by the word "environment."

The real names of the environment are the names of rivers and river valleys, creeks, ridges and mountains, towns and cities, lakes, woodlands, lanes, roads, creatures and people. The real name of our connection to this everywhere different and differently named Earth is "work." We are connected by work even to the places where we do not work, for all places are connected; it is clear by now that we cannot properly exempt one place from our ruin of another. The name of our *proper* connection to the Earth is "good work," for good work involves much giving of honor. It honors the source of its materials; it honors the place where it is done; it honors the art by which it is done; it honors the thing that it makes, and the user of the made thing. Good work is always modestly scaled, for it cannot ignore either the nature of individual places or the differences between places; and it always involves a sort of religious humility, for not everything is known. Good work can only be defined in particularity, for it must be defined a little differently for every

  What percentage of all car commuter trips made by Americans are single occupancy trips?

one of the places and every one of the workers on the Earth.

The name of our present society's connection to the Earth is "bad work"—work that is only generally and crudely defined, that enacts a dependence that is ill understood, that enacts no affection and gives no honor. Every one of us is to some extent guilty of this bad work. This guilt does not mean that we must indulge in a lot of breast-beating and confession; it means only that there is much good work to be done by every one of us, and that we must begin to do it. All of us are responsible for bad work, and not so much because we do it ourselves (though we all do it) as because we have it done for us by other people. And here we are bound to see our difficulty as almost overwhelming.

The only remedy for this that I can see is to draw in our economic boundaries and shorten our supply lines so as to permit us literally to know where we are economically. The closer we live to the ground that we live from, the more we will know about our economic life, the more able we will be to take responsibility for it. The way to bring discipline into one's personal or household or community economy is to limit one's economic geography.

If we define the task as beginning the reformation of our private or household economies, then the way is plain. What we must do is use well the considerable power we have as consumers: the power of choice. We can choose to buy or not to buy, and we can choose what to buy. The standard by which we choose must be the health of the community—and by that we must mean the *whole* community: ourselves, the place where we live and all the humans and other creatures who live there with us. In a healthy community, people would be richer in their neighbors, in neighborhood, in the health and pleasure of neighborhood, than in their bank accounts. And so it is better, even if the cost is greater, to buy near at hand than to buy at a distance. It is better to buy at a small, privately owned local store than from a chain store. It is better to buy a good product than a bad one. Do not buy anything you do not need. Do as much as you can for yourself. If you cannot do something for yourself, see if you have a neighbor who can do it for you. Do everything you can to see that your money stays as long as possible in the community. If you have money to invest, try to invest it locally, both to help the local community and to keep from helping the industrial economy that is destroying local communities. Begin to ask yourself how your money could be put at minimal interest into the hands of a young person who wants to start a farm, a store, a shop or a small business that the community needs. This agenda can be followed by individuals and single families. If it is followed by people in groups—churches, conservation organizations, neighborhood associations and the like—the possibilities multiply and the effects will be larger. Everything that is done by the standard of community health will make new possibilities for good work, the responsible use of the world.

Wendell Berry is a writer and poet based in Port Royal, Kentucky.

Reprinted by permission from Pantheon Books.

A: *80 percent*
Source: Union of Concerned Scientists

Index

A

Abbey, Edward, 413–417
Acid rain, 64, 71, 144
Afghanistan, 153
Africa, 6, 133–141
Agent Orange, 161
Agriculture, 116–118, 154, 167
AIDS, 6, 134
Albania, 147
Algeria, 140
Andorra, 144
Angola, 133
Animals, 148–149
  Rights, 65
  Testing, 67
Antarctica, 131–132
Antarctic Treaty and Convention, 130–132
Antigua and Barbuda, 167
Aral Sea, 152
Archaeology, 16–17
Arctic, 172–173
Arctic National Wildlife Refuge, 47, 172
Argentina, 164
Armenia, 150
Asia, 153–154, 156–160
Australia, 63, 162
Austria, 144
Automobiles, 64, 256–259
  Mileage, 259
  Tires, 256
Azerbaijan, 150

B

Bahamas, 167
Bahrain, 142
Bangladesh, 153
Barbados, 167
Belarus, 150
Belgium, 144
Belize, 167
Benin, 138
Bhopal, 15–16
Bhutan, 153
Bicycles, 257
Big Bang Theory, 11–12
Biodiversity, 44, 47, 60, 68–70, 88
Biodynamics, 253
Biosphere 2, 38–39
Black-Footed Ferrets, 86

Bolivia, 164
Books, 186–201
Bosnia-Herzegovina, 147
Botanical Medicines, 19, 22–23, 168
Botswana, 133
Brazil, 164
Brower, David, 66
Brunei, 159
Buffalo, 245–246
Bulgaria, 147
Burkina Faso, 140
Burundi, 133
Bush, George, 51

C

Caldicott, Dr. Helen, 178
California, 63–64
California Condor, 86
Cambodia, 159
Cameroon, 138
Canada, 64, 169–171
Cancer, 104-105
Cape Verde, 138
Capra, Fritjof, 180
Carbon Dioxide, 61, 63, 89–90
Carbon Monoxide, 63
Careers, 276–277
Catalogs, 282, 284–289
Center for Responsible Tourism, 309
Central African Republic, 133
Central America, 167–168
CERES Principles, 271–273
CFC, *see* Chlorofluorocarbons
Chad, 140
Chernobyl, 10–11, 150
Children's Defense Fund, 42
Children's Books, 197–201
Chile, 164
Chimpanzees, 19
China, 156
Chlorine, 5, 62, 104–106
Chlorofluorocarbons, 5, 62, 64, 104
Cholera, 18–19, 78
CIS, *see* Commonwealth of Independent States
CITES, *see* Convention on International Trade in Endangered Species of Wild Fauna and Flora
Cities, 260–263
Clean Air Act, 43, 64

Clean Water Act, 103
Cleansers, 237–238
Clothing, *see* Fashion
Clinton, Bill, 41, 51–52
Cockroaches, 239
Colombia, 69, 164
Columbus, 21–22
Commonwealth of Dominica, 167
Commonwealth of Independent States, 23–24, 150–152
Comoros, 136
Composting, 111
Computers, 209–213
Congo, 133
Congress, US, 322–324
Consumerism, 282–291
Contraception, 108–109
Convention Concerning the Protection of the World Cultural and Natural Heritage, 130
Convention on International Trade in Endangered Species of Wild Fauna and Flora, 130
Convention on the Conservation of Migratory Species of Wild Animals, 130
Convention on the Prevention of Marine Pollution by Dumping of Wastes and Other Matter, 130
Convention on Wetlands of International Importance Especially as Waterfowl Habitat, 130
Cosmetics, *see* Fashion
Costa Rica, 167, 301–306
Côte d'Ivoire, 138
Council on Competitiveness, 43–44
Croatia, 147
Cuba, 167
Cultural Survival, 309
Cyprus, 142
Czechoslovakia, 147

D

DDT, 78
Debt-for-Nature, 31, 149, 165
Defenders of Wildlife, 42
Deforestation, 60, 71–74, 153–154, 157, 159–161, 164–165, 168, 171
Denmark, 144
Deserts, 74, 140, 413–417
 Desertification, 74–76, *see also* Soil Degradation
Dioxin, 161
Discovery Channel, 205

Djibouti, 136
Dominican Republic, 167
Driftnets, 47, 156–157
Drought, 32–33, 53, 74–75, 133–134, 136, 162
Drugs, 19

E

E-Town, 202
Earth Day, 35
Earth Share, 267
Earth Summit, *see* United Nations Conference on Environment and Development
Earthquakes, 13–14
Earthwatch Radio, 203
Ecofeminism, 83–85
Economics, 51, 55, 232–233, 396–398, 421–423
Ecotourism, 168, 300–308
Ecotourism Society, 300–301, 306, 308
Ecuador, 164
Egypt, 140
El Niño, 7
El Salvador, 167
Electromagnetic Fields, 247–249
Elephants, 29–31, 134–135, 139
Endangered Species, 86–88, 138–139, 157–158
Endangered Species Act, 34, 43, 47, 52, 55, 68, 86–88
Energy, 51, 54, 60–61, 113–115, 145, 148
 Efficiency, 233, 268
 Policy, 98
 Renewable, 113–115
Environmental Defense Fund, 42
Environmental Disease, 77–79
Environmental Law, 52
Environmental Media Association, 178, 206
Environmental Movement, 60, 398–401
Environmental Protection Agency (EPA), 64
Equatorial Guinea, 138
Estonia, 150
Ethiopia, 136
Europe, 144–149
Everglades, 47
Exxon Valdez, 173

F

Famine, 136
Fashion, 219–223, 249–251
Federal Trade Commission (FTC), 286

INDEX

Fiji, 162, 301
Finland, 144
Food, 37, 241–246
  Additives, 241–243
Forests, 70–71, 419–421
  Old Growth, 71
  Rainforests, 22, 68, 164
Formaldehyde, 250–251
Fossil Fuels, 9, 23–24, 113, 115, 142
France, 144
French Guiana, 164
Friends of the Earth, 42
Friends of the Earth International, 266

G

Gabon, 138
Gambia, 138
Gaia, 62, 410–413
Gardening, 252–255
Georgia, 150
Germany, 144
Ghana, 138
Giardia, 79
Global Environmental Management Initiative, 270
Global Warming, 26, 55, 89–91, 131, 145–146, 162, 173
Gore, Al, 41, 47, 51–52, 178
Grand Canyon, 47
Greece, 144
Green Group, 42
Greenhouse Effect, 63
Grenada, 164
Guatemala, 167
Guinea, 138, 162
Guinea-Bissau, 138
Guyana, 164

H

Haiti, 167
Havens, Richie, 178
Hazardous Waste, 145
Health, 134, 247
Hepatitis, 79
Home, 232–240
  Cooling, 232
  Heating, 232
  Lighting, 233–236
Honduras, 167
Hong Kong, 156
House Plans, 234–236

Human Immunodeficiency Virus (HIV), *see* AIDS
Hungary, 147
Hydroelectric Power, 160–161

I

Iceland, 144
India, 153
Indigenous People, 23, 61, 92–94, 166, 300
  Aborigines, 163
  Inuit, 14
  Native Americans, 65
Indonesia, 159
International Whaling Commission, 37–38, 66
Investing, 278–281, 289
Iran, 142
Iraq, 142
Ireland, 144
Israel, 142
Italy, 144
Ivory, see Elephants
Izaak Walton League of America, 42

J

Jamaica, 167
Japan, 156
Jordan, 142

K

Kazakhstan, 150
Kenya, 136, 301
Kiribati, 162
Korea, 156
Kuwait, 142
Kyrgyzstan, 150

L

Landfills, 28–29
Laos, 159
Latvia, 150
Law of the Sea, 130
Lawns, 253–254
Lead, 237–239
Lebanon, 142
Lesotho, 133
Liberia, 138
Libya, 140
Liechtenstein, 144
Lithuania, 150

Logging, 47, *see also* Deforestation
Love Canal, 9
Luxembourg, 144

M

Maathai, Wangari, 179
Madagascar, 136
Magazines, 186–201
Malaria, 77–79
Malawi, 133
Malaysia, 159
Maldives, 153
Mali, 140
Malta, 144
Manuel Antonio National Park, 304
Maquiladoras, 170
Marketing, 267, 287–289
MARPOL, 130
Matthews, Jessica Tuchman, 180
Matthiessen, Peter, 178
Mauritania, 140
Mauritius, 133
Mexico, 25, 169–171, 269
Micronesia, 162
Middle East, 142–143
Migratory Species, 130
Military, 9
Mining, 132
Minorities, 95–97, 406–407
Moldova, 150
Monaco, 144
Mongolia, 156
Montreal Protocol, 5, 105–106
Morocco, 140
Movies, 202–209
Mozambique, 136
Music, 224–229
Myanmar (Burma), 159

N

Namibia, 133
National Audubon Society, 42
National Forests, 20–21
National Parks and Conservation Association, 42
National Toxics Campaign, 43
National Wildlife Federation, 43
Native American Rights Fund, 43
Natural Resources Defense Council, 43
Nauru, 162
Nepal, 153
Netherlands, 144

New Zealand, 162
Nicaragua, 167
Niger, 140
Nigeria, 138
Norberg-Hodge, Helena, 179
North America, 169–171
North American Free Trade Agreement (NAFTA), 169–171, 269
North Korea, 156
Norway, 144
Nuclear Issues
  Energy, 10, 54, 104, 148
  Waste, 9, 54, 98–100, 150–151
  Weapons, 48, 52, 54, 98–100, 155, 158
Nunavut, 14–15

O

Ocean Dumping, 130
Ocean Pollution, 101–103
Oman, 142
Organics, 116–118, 241
  Gardening, 252–253
  Cotton, 222–223
Overgrazing, 74, 76
Ozone
  Depletion, 104, 106
  Pollution, 5, 43, 62–63, 105, 131, 166

P

Pacific Yew Tree, 72
Packaging, 54, 269–270
Painting, 214–215
Pakistan, 153
Panama, 167
Papua New Guinea, 162
Paraguay, 164
People for the Ethical Treatment of Animals, 65
Perfume, 220
Permaculture, 252–253
Perot, Ross, 51
Peru, 18, 164
Pesticides, 255
Pests, 238–240, 254–255
Philippines, 159
Photography, 216–218
Pinatubo, Mt., 8, 26, 91
Planned Parenthood, 43
Poland, 31–32, 147
Pollution, 25–26, 62, 101–103, 119–121, 142, 144–147, 151–152, 156, 170–171, 274–277

INDEX

Air, 25–26, 62, 274–277
And Poverty, 53
Population Crisis Committee, 43
Population, 44, 51, 107–108, 137, 140, 153, 166
Portugal, 144
Poverty, 53, 141, 153, 155
Public Broadcasting System (PBS), 205
Public Lands, 52
Pulse of the Planet, 203

Q

Qatar, 142

R

Radio Expeditions, 203
Radio, 202–203
RAMSAR, 130
Recycling, 29, 54, 110–112, 233, 263
Romania, 147
Russia, 150
Rwanda, 133

S

San Marino, 144
Sao Tome and Principe, 138
Saudi Arabia, 142
Schmidheiny, Stephan, 180
Sculpture, 215–218
Senegal, 138
Sewage, 102–103
Seychelles, 136
Shopping, 290
Sierra Club, 43
Sierra Club Legal Defense Fund, 43
Sierra Leone, 138
Singapore, 159
Slovenia, 147
Soil Degradation, 74, 151, 162–163
Solomon Islands, 162
Somalia, 136
South Africa, 133, 135, 139
South America, 164–166
South Pacific, 162–163
Soviet Union, Former, *see* Commonwealth of Independent States
Spain, 144
Spirituality, 183, 185
Spotted Owl, Northern, 87
Sri Lanka, 153
St. Kitts and Nevis, Federation of, 167

St. Lucia, 167
St. Vincent and the Grenadines, 167
Strong, Maurice, 179
Sudan, 140
Suriname, 164
Swaziland, 133
Sweden, 144
Switzerland, 144
Syria, 142

T

Taiwan, 156
Tajikistan, 150
Tanzania, 136
Taxes, 266
Television, 204
Thailand, 159
Theater, 214–215
Tobacco, 64
Togo, 138
Tonga, 162
Tortuguero, 305
Tourism, 137, 153, 309
Toxics
  Pollution, 119–121
  Chemicals, 119
  Waste, 146–147, 166
  Waste Dumping, 138
Transportation, 36, 122–124, 257–258
  Mass, 257–258
Trinidad and Tobago, 164
Tunisia, 140–141, 311
Turkey, 142
Turkmenistan, 150
Turner Broadcasting System (TBS), 205
Tuvalu, 162

U

Uganda, 133
Ukraine, 150
Ultraviolet (UV) Radiation, 104
UNCED, *see* United Nations Conference on Environment and Development
Union Carbide, 15–16
Union of Concerned Scientists, 43
United Arab Emirates, 142
United Kingdom, 144
United Nations Conference on Environment and Development, 4, 43, 47, 53, 61, 70, 73, 83, 88–89, 91, 108, 145, 157, 179, 394–396, 408–409
United States, 169–171

Uruguay, 164
USA Outdoors, 204
Uzbekistan, 150

V

Vanuatu, 162
Vatican City, 144
Vegetarianism, 246
Venezuela, 164
Vietnam, 159

W

Walden Woods Project, 178
War, 136, 142, 149, 161, 167
Waste Disposal, 172
Water, 54, 80–82, 138, 140–141, 143, 166, 242–244
  Drinking, 80–82, 138, 242, 244
  Treatment Devices, 243
West Africa, 138
Western Europe, 144–146
Western Samoa, 162
Wetlands, 43
Whales, 16, 37–38
Wilderness, 125–127
Wilderness Society, 43
Wildlife, 42, 154–155, 163, 302
  Management, 302
  Trade, 138–139
Wise Use Movement, 44, 60
Wood, Megan Epler, 301
World Health Organization, 78
World Wildlife Fund, 43

X

Xeriscape, 253

Y

Yemen, 142
Youth, 181–182
Yugoslavia, Federal Republic of, 147

Z

Zaire, 34–35, 133
Zambia, 133
Zero Population Growth, 43
Zimbabwe, 133, 302
Zoos, 417–418

Keep up-to-date with the state of the Earth

If you liked this book, you'll love the *1994 Earth Journal*, another extraordinary volume of the environmental events of the year, completely new consumer information, updated directories and volunteer opportunities, a new collection of essays, and much more. Take another revealing look at the condition of the Earth with the editors of BUZZWORM magazine.

COMING DECEMBER 1, 1993!
Order Now. Call 1-800-333-8857
Or mail the coupon below.

1994 EARTH JOURNAL
ENVIRONMENTAL ALMANAC AND RESOURCE DIRECTORY

YES! Please send me ____ copies of the *1994 Earth Journal*, at $9.95 (*plus $1.50 shipping & handling*) each. Enclosed is my check/money order or please bill my credit card: ☐ Visa ☐ Mastercard

Card # _____ Exp. _____
Name _____
Address _____
City/State/Zip _____

Send to: BUZZWORM, 2305 Canyon Blvd., Suite 206, Boulder CO 80302